VOLUME FIVE HUNDRED AND NINETY FIVE

METHODS IN ENZYMOLOGY

Fe-S Cluster Enzymes Part A

METHODS IN ENZYMOLOGY

VOLUME FIVE HUNDRED AND NINETY FIVE

Methods in ENZYMOLOGY

Fe-S Cluster Enzymes Part A

Edited by

SHEILA S. DAVID

Department of Chemistry
University of California, Davis

ELSEVIER

ACADEMIC PRESS
An imprint of Elsevier

Academic Press is an imprint of Elsevier
50 Hampshire Street, 5th Floor, Cambridge, MA 02139, United States
525 B Street, Suite 1800, San Diego, CA 92101–4495, United States
The Boulevard, Langford Lane, Kidlington, Oxford OX5 1GB, United Kingdom
125 London Wall, London, EC2Y 5AS, United Kingdom

First edition 2017

Notices
Knowledge and best practice in this field are constantly changing. As new research and experience broaden our understanding, changes in research methods, professional practices, or medical treatment may become necessary.

Practitioners and researchers must always rely on their own experience and knowledge in evaluating and using any information, methods, compounds, or experiments described herein. In using such information or methods they should be mindful of their own safety and the safety of others, including parties for whom they have a professional responsibility.

To the fullest extent of the law, neither the Publisher nor the authors, contributors, or editors, assume any liability for any injury and/or damage to persons or property as a matter of products liability, negligence or otherwise, or from any use or operation of any methods, products, instructions, or ideas contained in the material herein.

ISBN: 978-0-12-811944-0
ISSN: 0076-6879

For information on all Academic Press publications
visit our website at https://www.elsevier.com/books-and-journals

Publisher: Zoe Kruze
Acquisition Editor: Zoe Kruze
Editorial Project Manager: Helene Kabes and Alina Cleju
Production Project Manager: Divya Krishna Kumar
Cover Designer: Greg Harris

Typeset by SPi Global, India

CONTENTS

CONTRIBUTORS

Jacob H. Artz
Institute of Biological Chemistry, Washington State University, Pullman, WA, United States

William Asma
School of Molecular Sciences, Arizona State University, Tempe, AZ, United States

Frédéric Barras
Laboratoire Chimie Bactérienne, Université Aix-Marseille, CNRS, Marseille, France

Anthony J. Blaszczyk
The Pennsylvania State University, University Park, PA, United States

Amie K. Boal
The Pennsylvania State University, University Park, PA, United States

Squire J. Booker
Howard Hughes Medical Institute, The Pennsylvania State University, University Park, PA, United States

Cécile Bouton
Institut de Chimie des Substances Naturelles (ICSN), CNRS UPR 2301, Univ. Paris-Sud, Université Paris-Saclay, Gif-sur-Yvette, France

Walter J. Chazin
Center for Structural Biology, Vanderbilt University, Nashville, TN, United States

James A. Cowan
The Ohio State Biochemistry Program, The Ohio State University, Columbus, OH, United States

Elizabeth A. Craig
University of Wisconsin–Madison, Madison, WI, United States

Katherine M. Davis
Princeton University, Princeton, NJ; The Pennsylvania State University, University Park, PA, United States

Zahra B. Dizicheh
School of Molecular Sciences, Arizona State University, Tempe, AZ, United States

Patricia C. Dos Santos
Wake Forest University, Winston-Salem, NC, United States

Rafal Dutkiewicz
Intercollegiate Faculty of Biotechnology, University of Gdansk and Medical University of Gdansk, Gdansk, Poland

Ioana Ferecatu
INSERM, UMR-S1139, Faculté des Sciences Pharmaceutiques et Biologiques, Université Paris Descartes, Sorbonne Paris Cité, Paris, France

Oleksandr Gakh
Mayo Clinic, Rochester, MN, United States

Belinda K. Galeano
Mayo Clinic Graduate School of Biomedical Sciences, Rochester, MN, United States

Giovanna Ghirlanda
School of Molecular Sciences, Arizona State University, Tempe, AZ, United States

Marie-Pierre Golinelli-Cohen
Institut de Chimie des Substances Naturelles (ICSN), CNRS UPR 2301, Univ. Paris-Sud, Université Paris-Saclay, Gif-sur-Yvette, France

Nicholas Halloran
School of Molecular Sciences, Arizona State University, Tempe, AZ, United States

Marilyn E. Holt
Center for Structural Biology, Vanderbilt University, Nashville, TN, United States

Yilin Hu
University of California, Irvine, Irvine, CA, United States

Grazia Isaya
Mayo Clinic; Mayo Clinic Graduate School of Biomedical Sciences; Mayo Clinic Children's Research Center, Rochester, MN, United States

Paul W. King
National Renewable Energy Laboratory, Biosciences Center, Golden, CO, United States

Ewen Lescop
Institut de Chimie des Substances Naturelles (ICSN), CNRS UPR 2301, Univ. Paris-Sud, Université Paris-Saclay, Gif-sur-Yvette, France

Jaroslaw Marszalek
Intercollegiate Faculty of Biotechnology, University of Gdansk and Medical University of Gdansk, Gdansk, Poland; University of Wisconsin–Madison, Madison, WI, United States

Cécile Mons
Institut de Chimie des Substances Naturelles (ICSN), CNRS UPR 2301, Univ. Paris-Sud, Université Paris-Saclay, Gif-sur-Yvette, France

David W. Mulder
National Renewable Energy Laboratory, Biosciences Center, Golden, CO, United States

Malgorzata Nowak
Intercollegiate Faculty of Biotechnology, University of Gdansk and Medical University of Gdansk, Gdansk, Poland

Sandrine Ollagnier de Choudens
Université Grenoble Alpes, Laboratoire de Chimie et Biologie des Métaux, BioCat; CNRS, Laboratoire de Chimie et Biologie des Métaux, BioCat, UMR; CEA-Grenoble, DRF/BIG/CBM, Grenoble, France

John W. Peters
Institute of Biological Chemistry, Washington State University, Pullman, WA, United States

Wasantha Ranatunga
Mayo Clinic, Rochester, MN, United States

Markus W. Ribbe
University of California, Irvine, Irvine, CA, United States

Sylvie Riquier
Institut de Chimie des Substances Naturelles (ICSN), CNRS UPR 2301, Univ. Paris-Sud, Université Paris-Saclay, Gif-sur-Yvette, France

Lauren E. Salay
Center for Structural Biology, Vanderbilt University, Nashville, TN, United States

Nathaniel S. Sickerman
University of California, Irvine, Irvine, CA, United States

Douglas S. Smith IV
Mayo Clinic, Rochester, MN, United States

James R. Thompson
Mayo Clinic, Rochester, MN, United States

Christine Wachnowsky
The Ohio State Biochemistry Program, The Ohio State University, Columbus, OH, United States

Roy X. Wang
The Pennsylvania State University, University Park, PA, United States

Oleg A. Zadvornyy
Institute of Biological Chemistry, Washington State University, Pullman, WA, United States

PREFACE

Iron–sulfur clusters in proteins are ancient cofactors with links to the very beginnings of life. Nature has adeptly used iron–sulfur cluster cofactors, like its heme cousins, for a wide-range of purposes. The most common Fe–S clusters contain Fe (II/III) and sulfide to make the common rhombic [2Fe–2S] or cuboidal [4Fe–4S] and [3Fe–4S] cores that are coordinated within the protein via primarily Cys ligands, but also His, Asp, or Ser, or with an open coordination site accessible to water. Many Fe–S proteins contain multiple Fe–S clusters, as well as more complex clusters of higher nuclearity and mixed metal clusters containing other transition metals such as Mo, V, and Ni. The ability of Fe–S cluster-containing proteins to access different oxidation states and tune the redox potential by the cluster structure and protein environment has made these clusters ideally suited to participate in electron-transfer reactions. Early in the history of Fe–S cofactors, it was thought that the functions were primarily related to their redox chemistry. However, the diversity of roles in biology for Fe–S clusters continues to grow, and nonredox roles for Fe–S clusters have been revealed. The rise of genomics has allowed for new Fe–S proteins to be identified and the membership of Fe–S families to grow. Major advances have also been made in deciphering the complexity of players involved in the vivo synthesis of Fe–S clusters and their insertion into their protein homes. Indeed, the critical functions of Fe–S cofactors are also underscored by the growing number of human diseases associated with genetic mutations in Fe–S cluster biogenesis genes. The scope of topics that interface with Fe–S clusters continues to expand.

In the two volumes of *Methods in Enzymology* (Volumes 595 and 599), leading investigators working on a diverse range of Fe–S proteins present important and insightful methods ranging from biochemistry, genetics, cell biology, and spectroscopy to study these fascinating systems. Though the topics are diverse, many common themes emerge such as special considerations for purification of Fe–S cofactor-containing proteins, and spectroscopic methods to probe the properties of Fe–S cluster cofactors in proteins. Due to the diversity of topics and techniques, the volumes are loosely organized into specific topics, but there are also topics that are covered from different angles in both volumes. As one might expect, some topics that could be covered in this volume have been or will be covered in other volumes organized by biological function or chemistry. In particular, there is an upcoming volume on radical *S*-adenosylmethionine (SAM) enzymes, and therefore, this important, diverse, and growing class of Fe–S cofactor-containing enzymes is only briefly covered in the two volumes dedicated to Fe–S enzymes.

Volume 595 (Fe–S Cluster Enzymes, Part A) starts with a review in Chapter 1 by Ollagnier-de Choudens and Barras on the genetic, biochemical, and biophysical methods commonly used to study Fe–S proteins with an emphasis on probing Fe–S assembly. Chapter 2 (Dizicheh et al.) turns toward designing Fe–S proteins de novo and characterizing their properties. The next five chapters discuss methods related to various aspects of Fe–S cluster biogenesis. Wachnowsky and Cowan in Chapter 3 describe a variety of in vitro methods for studying Fe–S proteins and monitoring kinetics and molecular mechanisms of Fe–S cluster transfer. The focus of Chapter 4 (Mons et al.) is focused on the unique mitochondrial membrane 2Fe–2S-containing protein mitoNEET. The chapter outlines methods for purification and monitoring cluster transfer from mitoNEET to ferredoxin as a model recipient protein. The fifth chapter by Gakh and colleagues focuses on the "big picture" in terms of characterizing the architecture of the human Fe–S cluster assembly machinery; the chapter outlines methods for reconstituting the very large multiprotein complex in vitro and then using various structural methods, such a electron microscopy, to visualize the entire assembly complex to define the key sites for protein–protein interactions and cluster assembly catalysis. The role of Hsp70 chaperones in facilitating the Fe–S cluster assembly dance is described in Chapter 6 by Dutkiewicz and coworkers. Chapter 7 is the last one in this volume to focus on Fe–S cluster assembly but stay-tuned for Volume 599 for more on this topic. This chapter by Patricia Dos Santos focuses on methods used to study functions of components of Fe–S cluster assembly in the Gram-positive bacteria *Bacillus subtilis*. The next two chapters shift focus to the complex Fe–S cofactors found in hydrogenases and nitrogenases. Chapter 8 by Artz and colleagues outlines strategies for crystallization of hydrogenases and nitrogenases under different conditions to trap distinct states for structural characterization. Sickerman, Hu, and Ribbe describe in detail procedures for purification of nitrogenase from *Azotobacter vinelandii*, and analysis of substrate reduction activity assays in Chapter 9. The subject of the next two chapters is the fascinating radical SAM enzymes. In Chapter 10, Blaszczyk, Wang, and Booker describe a method for purifying and characterizing the cobalamin-dependent radical SAM methylase TsrM. Davis and Boal describe in Chapter 11 mechanism-based strategies to capture structural snapshots of several different radical SAM enzymes. The volume concludes with the captivating subject of the role of Fe–S clusters in DNA polymerases in Chapter 12. Holt, Salay, and Chazin discuss methods for purification and structural characterization of DNA primase and its Fe–S cluster-containing domain, p58C, and the insights gained into the role of the Fe–S cluster in DNA replication.

Sheila David
University of California, Davis

CHAPTER ONE

Genetic, Biochemical, and Biophysical Methods for Studying Fe—S Proteins and Their Assembly

Sandrine Ollagnier de Choudens[*,†,‡,1], **Frédéric Barras**[§,1]
[*]Université Grenoble Alpes, Laboratoire de Chimie et Biologie des Métaux, BioCat, Grenoble, France
[†]CNRS, Laboratoire de Chimie et Biologie des Métaux, BioCat, UMR, Grenoble, France
[‡]CEA-Grenoble, DRF/BIG/CBM, Grenoble, France
[§]Laboratoire Chimie Bactérienne, Université Aix-Marseille, CNRS, Marseille, France
[1]Corresponding authors: e-mail address: sandrine.ollagnier@cea.fr; barras@imm.cnrs.fr

Contents

Methods in Enzymology, Volume 595
ISSN 0076-6879
http://dx.doi.org/10.1016/bs.mie.2017.07.015

Abstract

Fe—S clusters containing proteins are structurally and functionally diverse and present in most organisms. Our understanding of Fe—S cluster production and insertion into polypeptides has benefited from collaborative efforts between in vitro and in vivo studies. The former allows a detailed description of Fe—S-containing protein and a deep understanding of the molecular mechanisms catalyzing Fe—S cluster assembly. The second allows to include metabolic and environmental constraints within the analysis of Fe—S homeostasis. The interplay and the cross talk between the two approaches have been a key strategy to reach a multileveled integrated understanding of Fe—S cluster homeostasis. In this chapter, we describe the genetic and biochemical/biophysical strategies that were used in the field of Fe—S cluster biogenesis, with the aim of providing the reader with a critical view of both approaches. In addition to the description of classic tricks and a series of recommendations, we will also discuss models as well as spectroscopic techniques useful to characterize Fe—S clusters such as UV–visible, Mössbauer, electronic paramagnetic resonance, resonance Raman, circular dichroism, and nuclear magnetic resonance.

1. Fe—S CLUSTER BIOLOGY

Current understanding of Fe—S cluster biogenesis was made possible by decades of work in multiple laboratories focusing on different models and using diverse methods. Scheme 1 provides a summary of how Fe—S clusters get assembled and delivered to their apo-targets, and illustrates the different

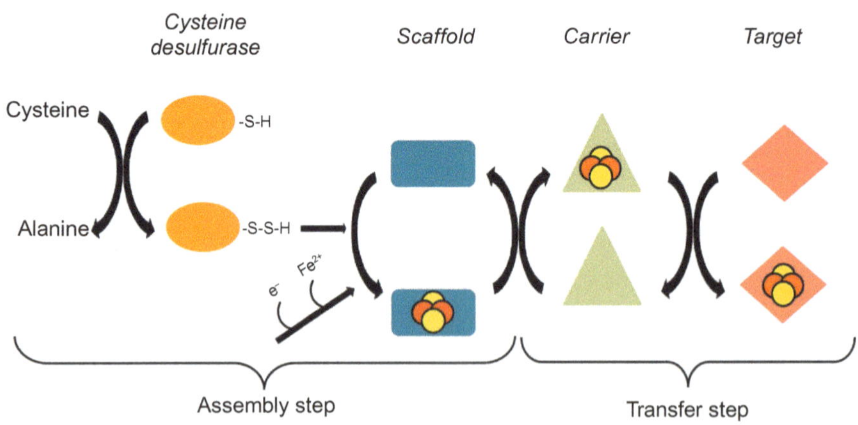

Scheme 1 General principles of Fe—S cluster biogenesis. The Fe—S cluster assembles on a scaffold protein (IscU, $SufBC_2D$, Isu), which receives sulfur from a cysteine desulfurase (IscS, SufSE, Nfs) and iron from a still nonidentified source. Then, the preformed Fe—S cluster is transferred to a carrier protein (IscA, Isa, SufA, NfuA, ErpA, Grx), which delivers it to the final apo-target.

actors and steps. Briefly, cluster biogenesis is catalyzed by two systems, ISC and SUF, which are well conserved in prokaryotes and eukaryotes. Two main steps can be differentiated. Assembling Fe—S involves cysteine desulfurase (IscS, Nfs, or SufSE), which produce sulfide from L-cysteine as a substrate, scaffold (IscU, SufBCD, Isu), which captures both sulfide and iron, whereas the iron donor remains unknown. Delivery of the cluster to the final target involves a series of carriers (IscA, Isa, SufA, ErpA, NfuA, Grx). A co-/chaperon duo (HscBA or Ssq/Jac) assists in dissociating the cluster from the scaffold.

2. GENETIC APPROACHES AND STRATEGIES

2.1 Genetic Selection and/or Screens Based on Fe—S Biogenesis Defects

Fe—S enzymes are found in several steps of respiratory metabolism. Succinate dehydrogenase, aconitase, and fumarate reductase as well as most electron-accepting terminal reductases are Fe—S enzymes (Magalon & Alberge, 2016). Therefore, *Escherichia coli* mutants altered in Fe—S biogenesis are unable to grow aerobically on acetate or anaerobically on nitrate or fumarate. These growth conditions could therefore constitute the basis for selection (or screening) protocols to isolate Fe—S biogenesis-deficient variants. Fe—S-dependent enzymes are present in pathways for biosynthesis of cofactors used by reductases, such as molybdenum or lipoate, and defects in these pathways might also impact respiratory processes.

Specific defects can also be used to design genetic protocols. D-Gluconate utilization relies on the Fe—S-containing phosphogluconate dehydratase Edd and a defect in Fe—S cluster biogenesis hampers gluconate utilization. This was used by Nachin et al. to assess the role of the SUF system under oxidative stress conditions (Nachin, Loiseau, Expert, & Barras, 2003). Other Fe—S-dependent phenotypes that could be exploited to design genetic screens or selections include (i) use of propionate as a carbon source, which is controlled by PrpD, a 2-methylcitrate dehydratase (Blank, Green, & Guest, 2002; Brock, Maerker, Schutz, Volker, & Buckel, 2002), (ii) phenyl acetic acid degradation, which is controlled by PaaE, a member of the ferredoxin–$NADP^+$ reductase family of proteins (Ferrandez et al., 1998), (iii) auxotrophy for isoleucine and valine, whose biosynthesis is controlled by IlvD, a Fe—S-bound dihydroxy-acid dehydratase (Flint, Emptage, Finnegan, Fu, & Johnson, 1993), and (iv) auxotrophy for leucine, the biosynthetic pathway of which includes LeuC, a Fe—S-dependent enzyme, isopropylmalate isomerase.

Fe—S-containing transcriptional regulators can also serve as tools. *E. coli* synthesizes IscR (Fe—S homeostasis sensor), NsrR (NO sensor), SoxR (NADH/NAD sensor), and Fnr (aerobic/anaerobic switch sensor). Level of expression of a target gene reflects the activity, hence the level of maturation of the regulator. For instance, the fusion *iscR::lacZ*, which is repressed by the Fe—S-bound IscR, reports on the cell's capacity to synthesize Fe—S clusters (Roche et al., 2015; Schwartz et al., 2001; Vinella, Brochier-Armanet, Loiseau, Talla, & Barras, 2009; Vinella, Loiseau, Ollagnier de Choudens, Fontecave, & Barras, 2013). Similarly, expression of the *hmp* gene reports on the maturation state of NsrR and that of the *yid* gene on Fnr maturation. Using these reporters permits quantitation of the expression of the target gene and evaluation of the contribution of Fe—S biogenesis factors to the maturation of the regulator used. Besides, the use of phenotypically tractable reporters, such as β-galactosidase, allows to identify positive and negative Fe—S biogenesis factors and/or study the effect of growth conditions.

Because Fe—S clusters play such a central cellular role, it is expected that global phenotypic screens will identify Fe—S biogenesis encoding genes and indeed cognate mutations altered antibiotic sensitivity, motility, phage infection resistance, or bacterial pathogenicity (Maynard et al., 2010; Nichols et al., 2011; Tamae et al., 2008).

2.2 The Use of Conditional Alleles

One way to overcome the barriers imposed by the essentiality of Fe—S cluster biogenesis is to use conditional alleles (Beckwith & Silhavy, 1992; Py & Barras, 2015). The Lill group was among the first to apply such an approach to the Fe—S field (Tamae et al., 2008). To study the role of the Yah1p protein for the activity of *Saccharomyces cerevisiae* mitochondrial Fe—S proteins, the YAH1 gene was put under the control of a galactose-inducible promoter. The strain containing Gal-YAH1 gene grew in the presence of galactose but could not use nonfermentable carbon source glycerol. Mitochondria isolated from yeast first grown in galactose and subsequently switched to galactose-free medium were isolated, activities of Fe—S enzymes measured and their level found to decrease after the switch. Quantitation of the amount of ^{55}Fe bound to a Fe—S protein is a way to assess maturation of a Fe—S protein. Cells are first grown for 1 h in the presence of (^{55}Fe) iron chloride, cell extracts prepared and submitted to immunoprecipitation with antibodies directed against a Fe—S protein model such as

Leu1p. Depletion of Yah1p led to a significant decrease in radiolabeled immunoprecipitated Leu1p, thereby demonstrating the importance of Yah1 for Leu1p maturation.

The association of conditional allele- and ^{55}Fe labeling was also elegantly used to assign a given step of the Fe—S cluster biogenesis process to the activity of a given cluster biogenesis factor (Muhlenhoff, Gerber, Richhardt, & Lill, 2003). Likewise, the amount of ^{55}Fe bound to overproduce Isu1p was estimated in different yeast mutants each containing conditionally expressed ISC gene. Decrease of ^{55}Fe labeling of Isu1p upon depleting ferredoxin Yah1 and the cysteine desulfurase Nfs1 indicated that Yah1 and Nfs1 are required for Fe—S building on Isu1p, hence that they act upstream Isu1p. In contrast, the chaperones Ssq1 were concluded to intervene downstream, i.e., to help dissociation of the cluster for delivery to apotargets (Muhlenhoff et al., 2003).

Genetic bypass (Beckwith & Silhavy, 1992) can also circumvent the essentiality of Fe—S biogenesis (Py & Barras, 2015). The two essential Fe—S proteins IspG and IspH catalyze the synthesis of isopentenyl phosphate (IPP), a precursor of secondary metabolites and component for cell envelope biosynthesis (Odom, 2011). The so-called 2-C-methylerythritol-4-phosphate pathway is present in most prokaryotes, whereas most eukaryotes synthesize IPP from mevalonate (MVA). The MVA-dependent pathway includes two kinases catalyzing formation of MVA diphosphate (5-MVPP) and a decarboxylase, none of them containing Fe—S cluster. The three cognate genes from humans and yeast were cloned as an operon and put under a bacterial arabinose-inducible promoter. The ectopic expression of this eukaryotic MVA-dependent pathway was able to suppress lethality due to a deficient IspG/H maturation in *E. coli* (Campos et al., 2001). This was subsequently exploited to assess the contribution of several Fe—S cluster biogenesis factors (Trotter et al., 2009; Vinella et al., 2009). Recently, a plasmidic version of the MVA pathway was made available (Tanaka et al., 2016).

2.3 Genetic Approaches to Investigate Functional Redundancy

Gene dosage has been exploited to study genetic interactions between alleles or genes with related functions. For instance, in their study on *Azotobacter vinelandii*, Dean and colleagues engineered an *nifUS*-conditional operon that allowed high level of NifU and NifS to be synthesized (using an arabinose-inducible promoter). They observed that an *iscU* mutant could be rescued by arabinose-induced *nifU* gene (Dos Santos, Johnson, Ragle, Unciuleac, &

Dean, 2007). Hence this study highlighted the potential redundancy between the two scaffolds IscU and NifU. Whether the high level reached by overproduction mimics levels reached under specific natural conditions is an open question. However, it is noticeable that overproduced NifS was unable to substitute for IscS, revealing differences in the in vivo specificities of two highly related enzymes (Dos Santos et al., 2007).

Combining mutations was also instrumental in delineating redundancy and specificity of Fe—S biogenesis factors (Vinella et al., 2013). Delivery of Fe—S clusters to apo-targets can be carried out by three so-called A-type carriers, IscA, SufA, and ErpA, which share multiple biochemical features. Yet, single mutation phenotypes diverge vastly: under aerobiosis *erpA* mutation is lethal whereas *sufA* or *iscA* mutations are neutral, but combining *iscA* and *sufA* mutations is lethal revealing redundancy between *iscA* and *sufA*. Unexpectedly, under anaerobiosis combining *iscA* and *erpA* is lethal, whereas single mutants are viable, indicating anaerobic redundancy between *iscA* and *erpA*. Thus, the cell, using biochemically related components, exploits genetic regulation to mount an appropriate response to fluctuating conditions.

2.4 Suppressor-Based Strategies

Suppressors are secondary mutations that compensate for the defect caused by the primary mutation. It has been a classic strategy to identify new proteins or to modify activity of already identified ones but somehow limited by the burden to localize the second-site mutation (Beckwith, 2007; Beckwith & Silhavy, 1992; Manson, 2000; Py & Barras, 2015). Genome sequencing has now reached such a low cost and high speed that hunt for suppressor-based strategy should be a privileged avenue for unearthing new aspects of Fe—S biogenesis.

Second-site suppressors allows discovery of new actors. The Fe—S cluster biogenesis system SUF was identified in *E. coli* by extragenic (second-site) suppressors (Takahashi & Tokumoto, 2002). From an *isc* mutant showing various auxotrophies (nicotinic acid, vitamin B1, and branched amino acids), Takahashi and Tokumoto selected revertants which grew on minimal medium. The suppressor mutations were all activating the expression of the *suf* operon.

Second-site suppressors can permit overpowering gene regulation-based barriers. An *erpA* and *iscA* double mutant is not viable. Suppressors were sought in an MVA-containing strain by selecting survivors on plates lacking both arabinose and MVA. Some of the suppressor mutations led to an enhanced level of SufA, which was apparently sufficient to compensate for the lack of the two other ATCs (Vinella et al., 2009). This illustrated that SufA has all features required to substitute for ErpA and IscA but that the cell prevents its use by downregulating its synthesis, presumably as an effort to reallocate resources, particularly iron.

Second-site suppressors can help structure/function relationship analysis of proteins. In the selection of suppressors from the *iscA erpA* mutant, two suppressing mutations changed a His into Asn at position 107 within the IscR sequence. It was concluded that this change shifted the equilibrium toward enhanced level of apo-IscR, which acts as an activator of *sufA* expression. The prediction then was that the His107 position was important for Fe—S binding, and this was eventually confirmed by structural characterization (Fleischhacker et al., 2012). Recently, starting from a nonviable *hscB suf* mutant, Tanaka et al. were able to select suppressing mutations that modify IscU (Tanaka et al., 2016). Because normally HscB and IscU interact, it was speculated that these bypass mutations modified IscU conformation such that it could now fulfill its function independently of the co−/chaperone duo HscBA.

2.5 *trans*-Dominant Mutations

Mutations that exert *trans*-dominant effects are useful to investigate protein/protein interactions (Beckwith & Silhavy, 1992). They can help identifying regulatory sites or interacting regions important for complex formation or functioning by altering sites of oligomerization, substrate binding, or catalysis. For instance, IscU variant with an Asp➔Ala mutation at position 39 is detrimental for growth when expressed in *trans*. The molecular explanation is that the [2Fe—2S] cluster formed on the variant cannot escape and the resulting IscU variant remains in complex with IscS, thereby sequestering it. Recently, a *trans*-dominant allele of Nfu1 (human NfuA) was obtained by changing a conserved Gly into a Cys residue at position 194. Biochemical characterization revealed that the G194C Nfu1 variant shows an enhanced interaction with Aco1, one of the Nfu1 substrates, thereby opening the possibility that the mutation identified an Nfu1 region that interacts with some of its targets (Melber et al., 2016).

2.6 *E. coli* Strains of Particular Interest for Fe—S Cluster Biogenesis Studies

2.6.1 The eSUF and eISC Strains

In *E. coli*, both the ISC and SUF systems are genetically connected. In an *isc* mutant, the IscR regulator is under its apo-form and it activates the expression of the *suf* operon; hence an *isc* mutant does not only cease making clusters with ISC but it makes them with SUF, even under conditions in which SUF is supposed not to be expressed. Two strains were constructed such as to interrupt this interdependence. The eSUF strain lacks the *iscSUA* gene and has the *suf* operon under the control of the arabinose-inducible pBAD promoter. The eISC strain lacks the *sufABCSDE* operon and the *iscR* gene has the *iscSUAhscBAfdx* operon under the pBAD promoter. Thus each strain relies on one single Fe—S biogenesis machinery. This pair of strains allowed to establish the apparent better capacity of ISC in maturating Nuo and Sdh complexes as compared with SUF, and as a consequence provided a molecular explanation as why an ISC-using strain shows enhanced sensitivity to gentamicin as compared with an SUF-using one (Ezraty et al., 2013). The former has a higher level of Nuo and Sdh activities and associated pmf, which allows Gm to be taken up by the cell.

2.6.2 The Hpx Strain

Aerobic metabolism continuously generates superoxide and hydrogen peroxide, which can be highly toxic to the cell. Organisms that live in aerobic habitats protect themselves from endogenous H_2O_2 by scavenging it with catalases and peroxidases. As a consequence, steady-state levels of H_2O_2 are kept as low as 20 n*M* (Seaver & Imlay, 2001). An *E. coli* strain which lacks catalase (*katE katG*) and peroxidase (*ahpCF*), referred to as Hpx, was constructed by the Imlay laboratory. Although Hpx mutants can grow in aerobic medium, Hpx *oxyR* mutants cannot, presumably because of Fenton-mediated damage to DNA and proteins (Park, You, & Imlay, 2005). The Hpx strain cannot catabolize TCA substrates, such as acetate, and they require supplementation with branched-chain (Ile, Leu, Val), aromatic (Phe, Trp, Tyr), and sulfur-containing (Met, Cys) amino acids. They also display a high rate of mutagenesis. When the Hpx strain is transferred from anaerobic to aerobic medium, endogenous H_2O_2 accumulates up to 1 m*M*. H_2O_2 oxidizes the solvent-exposed $[4Fe—4S]^{2+}$ clusters of a family of dehydratases (Flint et al., 1993; Jang & Imlay, 2007). The use of the Hpx strain allowed to investigate the sensitivity of various Fe—S enzymes to accumulate H_2O_2. It was also instrumental in establishing that *E. coli* uses the SUF system rather than the ISC when accumulating submicromolar H_2O_2.

2.6.3 The Iron-Minus Strain

A classic way to induce iron limitation is adding iron chelators such as 2,2-dipyridyl (DIP) in the growth medium. Although efficient, this method lacks specificity as DIP might strip iron from biologically important sites in addition to simply sequestering free iron. Therefore a series of strains were constructed, which contains a limited known iron level. These strains are lacking the Feo-, ferric-citrate-, ZupT-, and siderophore-dependent iron-import systems (Δ*tonB* Δ*feoABC* Δ*zupT*) (Martin, Waters, Storz, & Imlay, 2015). The resulting iron-minus strain has its Fur-regulon constitutively expressed. This iron-minus strain permits to investigate the role and/or the efficiency of a given Fe—S biogenesis factor under iron limitation.

2.7 Construction of Frataxin Models

Friedreich's ataxia (FA) is the most prevalent form of recessive ataxia in Caucasians causing severe neurological disabilities and reduced life expectancy. FA results from reduced level of the frataxin protein (FXN), which results in reduced Fe—S biogenesis. FXN acts in the early step of Fe—S biogenesis via the ISC pathway, interacting with cysteine desulfurase and activating Fe—S cluster formation. Most patients carry a homozygous GAA repeat expansion within the first intron of the FXN gene that leads to partial silencing of the gene through epigenetic mechanisms. During the past 20 years, a number of different FA cell and mouse models have been developed to investigate FA molecular disease mechanisms and therapy.

Conditional knockout (cKO) mice permitted to study the consequences of FXN deficiency (Puccio et al., 2001; Simon et al., 2004) and to develop therapeutic approaches (Perdomini et al., 2014; Seznec et al., 2005). They showed that FXN deficiency leads to primary Fe—S deficit, followed by mitochondrial dysfunction, disturbance in iron metabolism, cellular dysfunction, and cell death. These models contrast with GAA repeat expansion-based models, which do not present a clear robust phenotype, although they have been very useful to test epigenetic therapies targeted to restoring FXN expression (Miranda et al., 2002). Cellular models based on fibroblasts obtained from the cKO models exhibit key features of the disease (Fe—S cluster deficit, mitochondrial iron accumulation, increase sensitivity to oxidative stress) (Calmels et al., 2009; Schmucker et al., 2011).

3. BIOCHEMISTRY OF Fe—S CLUSTERS

3.1 Purification of Fe—S Cluster-Containing Proteins

Several [4Fe—4S] cluster-containing proteins are oxygen sensitive. Examples include radical SAM enzymes, nitrate reductase regulatory protein, aconitase, and quinolinate synthase (Beinert, Kennedy, & Stout, 1996; Cherrier et al., 2014; Gruner et al., 2011; Vey & Drennan, 2011). Oxygen degrades active [4Fe—4S] to a [3Fe—4S] form that can be further degraded into inactive [2Fe—2S] form and/or an apo-form.

In order to prevent oxygen degradation, many [4Fe—4S] proteins should be purified and characterized under anaerobic conditions using specific laboratory setups such as O_2-free glove boxes and vacuum gas manifold. Ultra-high purity gases with O_2 content lower than 3 ppm are required. The most popular inert gases for anaerobic biochemistry are nitrogen (N_2) and argon (Ar). O_2-free glove boxes, containing chromatographic columns and devices, are the most convenient equipment for purification and enzymatic assays of [4Fe—4S] proteins. A vacuum manifold is useful primarily for enzymatic assays and preparation of samples for various types of spectroscopy.

Redox potential of bacterial cytoplasm is sufficiently low to keep Fe—S clusters intact (−200 mV) even in aerobically grown cultures. In contrast, purification has to be done under anaerobiosis. Buffers must be made O_2 free by bubbling N_2 for 1 h using a manifold or by directly introducing them inside the glove box after switches of vacuum and flowing N_2 in the chamber associated to the glove box. Buffers can contain sodium dithionite to remove traces of oxygen. Soluble extracts are prepared by sonication (device inside the glove box) or by freeze (outside the box)–thawing (inside the box) cycles using hermetic tubes. After ultracentrifugation (outside the box), soluble extracts are loaded onto the adequate column equilibrated with a degassed buffer. After purification, the Fe—S protein is concentrated, aliquoted in sealed Eppendorf tubes inside the glove box, and frozen in liquid nitrogen. Such protocols allowed the purification of proteins in their Fe—S-bound native form, such as quinolinate synthase, carbon monoxide dehydrogenase, [Fe—Fe] hydrogenase, eukaryote ISCA (Beilschmidt et al., 2017; Dobbek, Svetlitchnyi, Gremer, Huber, & Meyer, 2001; Drennan, Heo, Sintchak, Schreiter, & Ludden, 2001; Ollagnier-de Choudens, Loiseau, Sanakis, Barras, & Fontecave, 2005; Peters, Lanzilotta, Lemon, & Seefeldt, 1998; Rousset, Fontecave, & Ollagnier de Choudens, 2008) (Scheme 2).

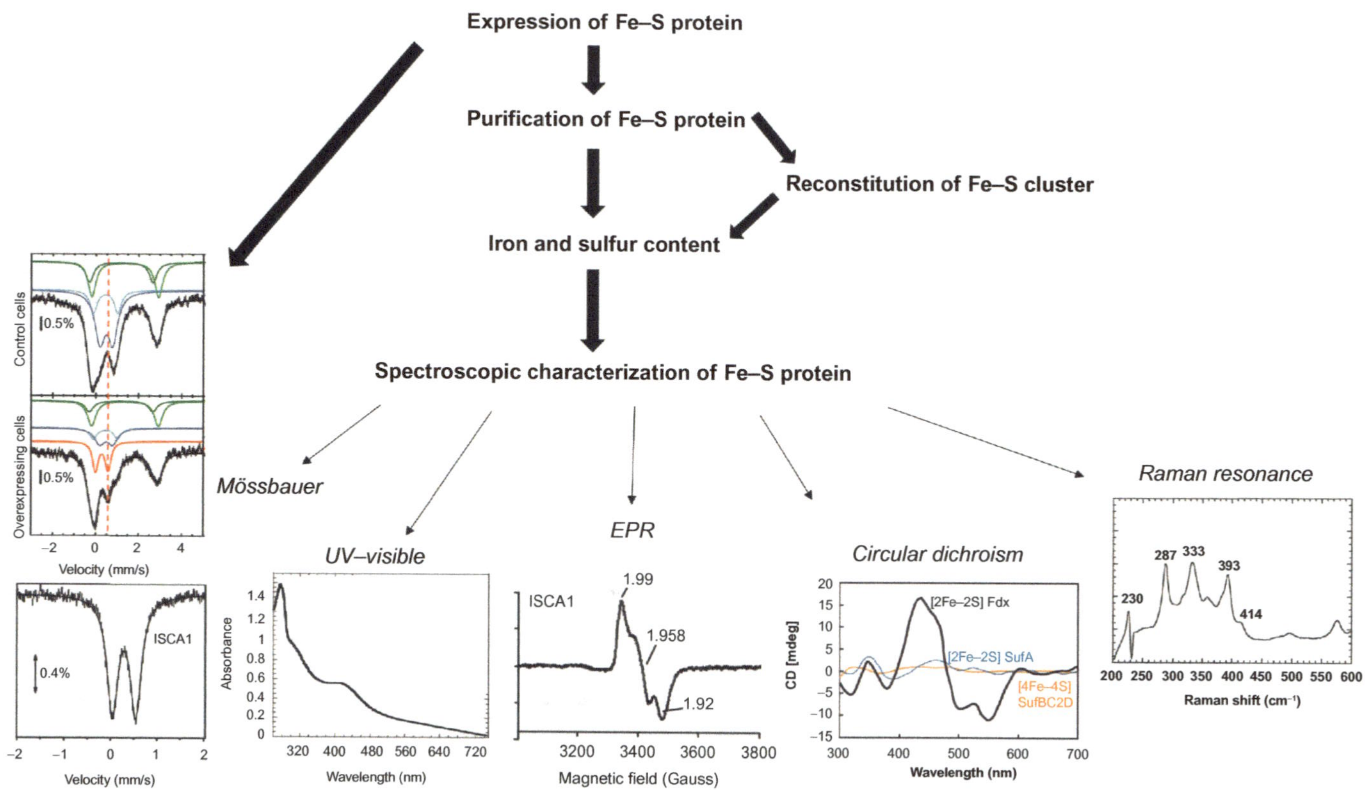

Scheme 2 In vitro characterization of Fe—S proteins. After expression, Fe—S protein is purified. Depending on its Fe and S content, Fe—S has to be reconstituted or not. Then, several spectroscopies (UV–visible (Py et al., 2012); EPR (Beilschmidt et al., 2017); Mössbauer (Beilschmidt et al., 2017); CD after (Chahal & Outten, 2012) modification and Raman (Ollagnier-de-Choudens, Mattioli, Takahashi, & Fontecave, 2001)) can be used to determine properties (type and quantity) of Fe—S cluster. Fe—S cluster can be directly characterized in cellulo by Mössbauer spectroscopy (Beilschmidt et al., 2017).

If proteins cannot be obtained under their Fe—S forms, clusters can be chemically reconstituted inside the glove box (Scheme 2) (Chandor et al., 2006; Perche-Letuvee et al., 2012; Pierrel, Bjork, Fontecave, & Atta, 2002; Rubach, Brazzolotto, Gaillard, & Fontecave, 2005). First, protein has to be prepared under an apo-form by treatment outside the glove box, using a strong reducing agent (dithionite) and an iron chelator (EDTA). The apo-protein is then transferred to anaerobiosis for at least 1 h. Fresh solutions of dithiothreitol (DTT) (500 m*M* in water), sodium sulfide (Na_2S), and ferrous iron (Mohr salt: $(NH_4)_2Fe^{II}(SO_4)_2$) (both 50 m*M* in water) are prepared. Sulfide can be provided to the apo-protein (10 mg/mL) either from Na_2S or using L-cysteine (50 m*M*) and catalytic amount of cysteine desulfurase. DTT (5 m*M* final) and subsequently iron and sulfide are added, in a slight excess with regard to the exact amount of Fe and S that the protein is supposed to accommodate (e.g., 5 *M* excess of Fe and S for proteins containing one [4Fe—4S]). In case of enzymatic reconstitution, cysteine desulfurase is added to initiate the reaction and the mixture incubated for several hours. Spectrophotometer and anaerobic cuvettes are used to monitor reconstitution. The reaction is pursued until absorbance at 420 nm reaches a plateau. Then the protein is desalted and aliquoted.

3.2 Characterization of Fe—S Proteins

A first biochemical analysis of the iron and sulfur content of the purified protein can provide indications about the nature of the cluster. Notice that in general [4Fe—4S] are yellow–brown in color, whereas [2Fe—2S] are generally pink–red.

For determination of the iron content,

- Prepare a 50 m*M* solution of Fe(II) using Mohr salt in water.
- Prepare a series of samples ranging from 0 to 12 nmol of Fe(II) (total volume 65 μL in water) for a calibration curve.
- Prepare appropriate dilutions (total volume 65 μL in water) of the sample(s) for analysis.
- Add perchloric acid (for denaturation of the protein) (45 μL at 1 *M*) and let the solution to incubate for 15 min.
- Centrifuge (5 min at 15,000 rpm), take out 90 μL of the supernatant, and add the ferrous iron chelator bathophenanthroline (72 μL at 1.7 mg/mL), the reductant sodium ascorbate (36 μL at 38 mg/mL), and a 1/3 saturated ammonium acetate solution (27 μL). A pink color appears attesting the formation of Fe(II)-bathophenanthroline complex.

- After 30-min incubation, absorbance of each sample is recorded at 535–680 nm. The calibration curve is constructed by plotting the iron in nmol (*x*-axis) vs $A_{535\text{–}680\text{nm}}$ (*y*-axis). The amount of iron present in the protein sample is determined by reference to this calibration curve. This protocol is based on Fish method (Fish, 1988).

For determination of the sulfide content,

- Weight out sodium sulfide nonhydrated and dissolve it in NaOH 0.03N in order to get a 200 μ*M* solution.
- Prepare sulfide standards in the range 0–20 nmol (total volume 100 μL in water) to enable the construction of a calibration curve.
- Prepare appropriate dilutions (total volume 100 μL in water) of the sample(s) for analysis.
- To each tube (standard and samples) add 300 μL of freshly prepared 1% (w/v) zinc acetate immediately followed by the addition of 15 μL of 3 *M* NaOH.
- Cap the tubes (to prevent H_2S departure) mix and incubate for 15 min.
- Add to each tube 75 μL of freshly prepared 0.1% (w/v) *N,N*-dimethyl-*p*-phenylenediamine dihydrochloride dissolved in 5N HCl. Immediately add 16 μL of freshly prepared 23 m*M* $FeCl_3$ prepared in HCl 1.2 *M*.
- After 30-min incubation in centrifuge tubes, record the absorbance of each standard sample at 670 nm. The calibration curve is constructed by plotting the sulfide in nmol (*x*-axis) vs $A_{670\text{ nm}}$ (*y*-axis). The amount of sulfide present in the protein sample is determined by reference to this calibration curve. This protocol was described first by Beinert (1983).

4. SPECTROSCOPY OF Fe—S PROTEINS

Spectroscopy of Fe—S cluster exploits electronic properties of the iron atom. Iron is a transition metal in the d block elements of the periodic table whose atomic number (*Z*) is 26. Its electronic configuration is $4s^2\ 3d^6$. It can lose from two to six electrons, the most common oxidation states being +2 ($S=2$) and +3 ($S=5/2$) corresponding to the loss of 2 and 3 electrons, respectively. In Fe—S clusters, iron atom exists under +2 and/or +3 oxidation state. Taking into account that sulfur atom is under sulfide form (S^{2-}) the corresponding electronic structures for [2Fe—2S] and [4Fe—4S] are given in Table 1.

Table 1 Formal Oxidation States and Electronic Spin of [4Fe—4S] and [2Fe—2S] Clusters (Iron in *Red* and Sulfide in *Yellow*)

Cluster Type	Formal Fe Oxidation State	*n* or Fe—S Oxidation State	$Spin_{tot}$ (*S*)
$[2Fe—2S]^n$	$Fe^{2+}—Fe^{2+}$	0	0
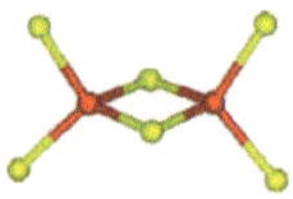	$Fe^{2+}—Fe^{3+}$	+1	1/2
	$Fe^{3+}—Fe^{3+}$	+2	0
$[4Fe—4S]^n$	$Fe^{2+}—Fe^{2+}—Fe^{2+}—Fe^{2+}$	0	4
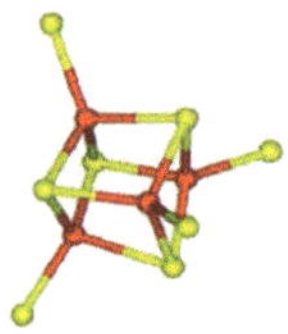	$Fe^{2+}—Fe^{2+}—Fe^{2+}—Fe^{3+}$	+1	1/2
	$Fe^{2+}—Fe^{2+}—Fe^{3+}—Fe^{3+}$	+2	0
	$Fe^{2+}—Fe^{3+}—Fe^{3+}—Fe^{3+}$	+3	1/2
	$Fe^{3+}—Fe^{3+}—Fe^{3+}—Fe^{3+}$	+4	0

For oxidation states those found in nature are indicated in *green*.

4.1 UV-Visible Absorption Spectroscopy

The absorption of a photon allows the transition from a fundamental electron state S0 to an excited electron state S1 by passing an electron from one orbital to another. Photon energy absorption during a transition between two levels of energy is given by relation $E = hC/\lambda$ (λ the wavelength, h the Planck constant, and C the light celerity in the vacuum). Different absorbed energies allow different transitions, and in case of proteins, some of these transitions are permitted with the absorption of photons close to UV and visible light. Two important parameters of the absorption band, position and intensity, are obtained from electronic spectra. From the intensity and the concentration of the sample, an extinction coefficient, ε, for each band is calculated according to the Beer–Lambert Law: $A = \varepsilon lC$ (ε, extinction coefficient ($M^{-1}\,cm^{-1}$); l, path length (cm); C, molar concentration (M)) (Que, 2000). Concerning Fe—S proteins, the absorption spectrum is characterized by a band at 280 nm due to the conjugated double bonds of the aromatic residues (tryptophan, tyrosine, phenylalanine) and by bands characteristics of the Fe—S cluster that correspond to charge transfer bands (passage of an electron from one atom to another during the excitation) from the ligand → metal (Fe^{3+}) and referred as LMCT (ligand metal charge transfer). LMCT causes absorption between 300 and 600 nm, depending on the type of cluster. Both the position of their maxima and the general shape of the spectrum provide information about the Fe—S center type. $[4Fe—4S]^{+2}$ clusters display a main maximal absorption around 410 nm (ε_{410}: 9–16 mM^{-1}-cm^{-1}) (Agar et al., 2000; Antonkine et al., 2009; Crack et al., 2008, 2015; Mapolelo, Zhang, Naik, Huynh, & Johnson, 2012; Shen et al., 2007; Tedeschi et al., 1997; Zhang et al., 2013), whereas $[2Fe—2S]^{2+}$ present several bands at 320, 410, 450–460 nm and a shoulder around 550 nm (ε_{452}: 7–10 mM^{-1}cm^{-1} for complete cysteinyl coordination (Orme-Johnson & Orme-Johnson, 1982), lower values (5.3–6.3 mM^{-1}cm^{-1}) for incomplete cysteine coordination (Agar et al., 2000; Nechushtai et al., 2012). Absorption around 410 nm is lost for $[4Fe—4S]^{1+}$. For reduced $[2Fe—2S]^{1+}$, one absorption band might remain at 550 nm (Ollagnier-de-Choudens et al., 2001). For $[4Fe—4S]^{2+}$ clusters, the absorption ratio, Abs410/280, is sometimes used to estimate the Fe—S cluster incorporation: for proteins containing one [4Fe—4S], ratios vary from 0.17 to 0.36 depending the protein sequence (Martinez-Gomez & Downs, 2008; Morra, Arizzi, Valetti, & Gilardi, 2016).

UV–visible absorption spectroscopy is carried out in a cell of known path length (usually 1 cm). Such spectroscopy can be done both under aerobic and anaerobic conditions with small volumes (100 μL) and at low protein concentration (10–50 μM).

4.2 Mössbauer Spectroscopy

The underlying process is the absorption by a nucleus of a gamma radiation emitted by another decaying nucleus of the same isotope. ^{57}Fe is the best-suited isotope for Mössbauer spectroscopy and the γ radiation is produced by decaying ^{57}Co nuclei (half-life of 271.8 days) (Que, 2000). It is a very sensitive technique. It is possible to study with great precision the hyperfine structure of the energy levels of the atomic nucleus and their perturbations under the effect of the chemical and magnetic environment of the atom, its oxidation state, and ligands. Environments of the source and of the nuclei are usually different, and the hyperfine interactions between the nucleus and its environment change the nuclear transition energy. To detect this change, the source is accelerated through a range of velocities changing slightly the energy of the γ-ray (Doppler effect). This explains why the energy scale of a Mössbauer spectrum is reported in terms of velocity. At velocities corresponding to the resonant energy levels of the sample, γ rays are absorbed, with a corresponding peak in the spectrum (Fig. 1). Number,

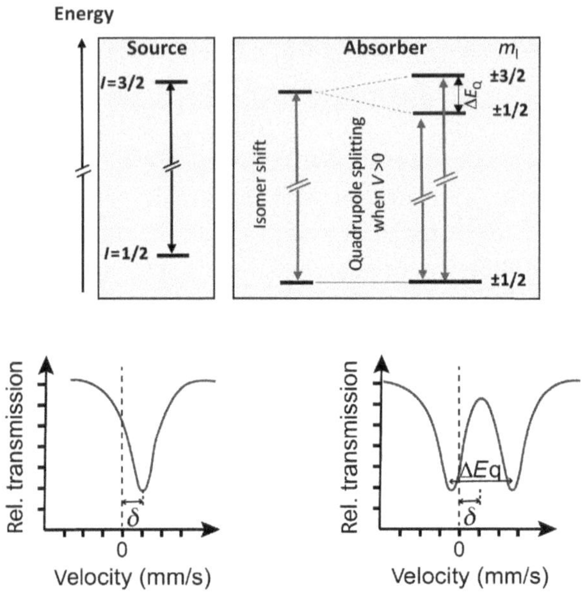

Fig. 1 Chemical shift and quadrupole splitting of the nuclear energy levels and corresponding Mössbauer spectra.

positions, and intensities of the peaks indicate the type of Fe species present. At 4.2 K it is estimated that the cumulated absorption is strictly proportional to the number of ^{57}Fe nuclei that account for that signal. Thus, Mössbauer spectroscopy is both a qualitative and quantitative technique in terms of abundance of individual iron species. In the absence of magnetic interaction, a Mössbauer spectrum is characterized by two parameters: called isomer shift, noted δ (in mm/s), and "quadrupole splitting," noted ΔE_Q (in mm/s). Isomer shift (δ) is useful for determining oxidation state, valence states, and ligand nature. The quadrupole splitting (ΔE_Q) is used for determining symmetry and charge distribution (Fig. 1). δ and ΔE_Q depend mostly on the oxidation state of each iron site: Fe^{2+}, Fe^{3+}, or delocalized $Fe^{+2.5}$ (mixed valence pair) (Table 2). In the presence of magnetic interaction, a third parameter called "hyperfine splitting" is calculated. It results from the interaction between the nucleus and any surrounding magnetic field. Altogether, these parameters can be used to identify a particular species. For an in-depth theoretical introduction in Mössbauer spectroscopy and details on Fe—S proteins, the reader can refer to some nice reviews (Clémancey, Latour, & Garcia-Serres, 2014; Gütlich, Bill, & Trautwein, 2011; Pandelia, Lanz, Booker, & Krebs, 2015).

Table 2 Mössbauer and Electronic Parameters of [2Fe—2S] and [4Fe—4S] Clusters at 4.2 K

Type	Valences	S_{total}	δ (mm/s)	ΔE_Q (mm/s)	Reference
$[2Fe—2S]^{2+}_{(cys)4}$	Fe^{3+} Fe^{3+}	0	0.27	0.32, 0.91	Munck, Debrunner, Tsibris, and Gunsalus (1972)
$[2Fe—2S]^{2+}_{Rieske}$	Fe^{3+} Fe^{3+}	0	0.24, 0.32	0.6	Fee et al. (1984)
$[2Fe—2S]^{+}_{(cys)4}$	Fe^{3+} Fe^{2+}	½	0.35, 0.6	0.65, 2.70	Munck et al. (1972)
$[2Fe—2S]^{+}_{Rieske}$	Fe^{3+} Fe^{2+}	½	0.31, 0.74	0.63, 3.05	Fee et al. (1984)
$[2Fe—2S]^{0}$	Fe^{2+} Fe^{2+}	0	0.7, 0.81	2.76, 2.32	Leggate, Bill, Essigke, Ullmann, and Hirst (2004)
$[4Fe—4S]^{3+}$	$(Fe^{3+}—Fe^{3+})$ $(Fe^{2.5+}—Fe^{2.5+})$	½	0.29, 0.4	0.88, 1.03	Middleton, Dickson, Johnson, and Rush (1980)
$[4Fe—4S]^{2+}$	2× $(Fe^{2.5+}—Fe^{2.5+})$	0	0.42	1.12	Middleton et al. (1980)
$[4Fe—4S]^{1+}$	$(Fe^{2+}—Fe^{2+})$ $(Fe^{2.5+}—Fe^{2.5+})$	½	0.58, 0.49	1.89, 1.32	Middleton et al. (1980)
$[4Fe—4S]^{0}$	2× $(Fe^{2+}—Fe^{2+})$	4	0.65	1.51, 2.19	Hans, Buckel, and Bill (2008)

(Cys)4, four cysteine as ligands; *Rieske*, two cysteine and two histidine as ligands.

[2Fe—2S] clusters. A cluster in the +2 oxidation states consists of two antiferromagnetically coupled Fe^{3+} ($S=0$). The resulting spectrum is composed of two equal-intensity quadrupole doublets whose parameters depend upon ligands (Table 2) (Fee et al., 1984; Tinberg et al., 2010). A cluster in the +1 oxidation state comprises a Fe^{2+} antiferromagnetically coupled to a Fe^{3+} (Table 2).

[4Fe—4S] clusters. A [4Fe—4S] cluster exists in four oxidation states (0/+1/+2/+3), the +1 and +2 being the most common ones. Cluster in the 0 oxidation state consists of 4 × Fe^{2+} ($S=4$). The resulting spectrum is a broad quadrupole doublet. In the +1 oxidation state, cluster consists of two Fe^{2+} coupled to a delocalized $2Fe^{2.5+}$ pair ($S=1/2$ or $3/2$ that can be differentiated with a magnetic field). The Mossbauer spectrum is a superposition of two six-line spectra, one for the ferrous pair and one of the delocalized one. Cluster in the +2 oxidation state consists of two delocalized $2Fe^{2.5+}$ pairs antiferromagnetically coupled ($S=0$). The

corresponding spectrum appears as a single (sometimes asymmetric) doublet. Cluster in the +3 oxidation state consists of two Fe^{3+} coupled to a delocalized $2Fe^{2.5+}$ pair ($S=1/2$) with a spectrum similar to that of $[4Fe—4S]^{1+}$ but shifted to lower velocities (Table 2).

For Mössbauer study, purified Fe—S proteins have to be enriched with ^{57}Fe since natural iron contains only 2.2% of ^{57}Fe. Enrichment can be achieved either by in vitro anaerobic reconstitution using $^{57}FeCl_3$ and sulfide salts (see above), or by anaerobic purification from a culture in minimal medium (M9) supplemented with 35 μ*M* $^{57}FeCl_3$. Typical requirements for biological samples of Fe—S proteins are volumes of 400 μL and ^{57}Fe concentrations of 0.5–1 *M*, which represents a significant amount of protein. Mössbauer sample is prepared either aerobically (for O_2-insensitive Fe—S cluster) or anaerobically inside a glove box and frozen using liquid isopentane cooled with liquid nitrogen for anaerobic conditions. Numerous Fe—S proteins involved in Fe—S biogenesis were characterized by Mössbauer spectroscopy, either as purified or after chemical reconstitution, such as mammalian recombinant ISCA1 and ISCA2 (Beilschmidt et al., 2017), *E. coli* NfuA (Angelini et al., 2008), ErpA (Loiseau et al., 2007), SufA (Gupta et al., 2009), IscU (Agar et al., 2000), IscR regulator (Fleischhacker et al., 2012), SufBC2D (Wollers et al., 2010), *Arabidopsis thaliana* Nfu2 (Gao et al., 2013).

Some Fe—S cluster forms easily interconvert or degrade during purification. Therefore, Mössbauer spectroscopy analysis of whole cells may provide additional important information. Because Mössbauer spectroscopy is sensitive to all ^{57}Fe-labeled centers, identification of authentic signals associated to the Fe—S protein of interest requires comparison of spectra of whole-cell samples before and after overexpression of the protein of interest. Protocol consists in a bacterial culture in MOPS medium supplemented with 20 μ*M* $^{57}FeCl_3$, at 37°C with agitation (200 rpm). Cells harvested by centrifugation (4000 × *g*, 20 min, 4°C) are washed twice with MOPS buffer containing 50 m*M* NaCl and 22 m*M* glucose and centrifuged for 10 min at 5000 × *g*, 4°C. Samples are in the form of a solid paste that is packed in a Mössbauer cup (0.6–0.7 g of cells per Mössbauer cup) and frozen in liquid nitrogen. The Fnr transcription regulator system (Popescu, Bates, Beinert, Munck, & Kiley, 1998) contains a Fe—S cluster that interconvert between [4Fe—4S] and [2Fe—2S] forms. By employing Mössbauer spectroscopy on whole cells, it was demonstrated that the $[4Fe—4S]^{2+}$ to $[2Fe—2S]^{2+}$ interconversion upon exposure to O_2 takes place in vivo and is relevant to Fnr function. Other systems such as biotin synthase (BioB) and the pyruvate formate lyase-activating enzyme were also studied by whole-cell Mössbauer spectroscopy (Benda et al., 2002; Yang et al., 2009). Recently, mammalian recombinant ISCA1 and ISCA2, for which Fe—S type is under debate, were shown to be produced as [2Fe—2S] proteins (Beilschmidt et al., 2017).

4.3 Electronic Paramagnetic Resonance Spectroscopy

Electronic paramagnetic resonance (EPR) spectroscopy can be applied to any paramagnetic species as it studies the interaction of unpaired electrons from the sample with a magnetic field that can split the energy levels of the ground state by a quantity given by the following equation: $\Delta E = h\nu = gB0\beta$ (Zeeman effect) (Que, 2000). It is possible to calculate the g factor considered as a fingerprint of the molecule, containing the information on the chemical nature of the bond between the electron and the molecule. The intensity of the EPR signal is closely linked to the temperature of the sample and the power of microwaves entering the microwave cavity, and these two parameters are optimized together to maximize signal intensity. EPR spectroscopy is particularly suitable for the study of proteins in solution that have a nonnull electronic spin, such as $[4Fe—4S]^{1+}$ ($S = ½$) and $[4Fe—4S]^{3+}$ ($S = ½$), and reduced $[2Fe—2S]^{1+}$ ($S = ½$) (Table 3). In general, clusters are observable between 4 and 10 K and using a low microwave power. Higher temperature increases the relaxation rate and expands signals to the point where they are no longer observable. Some $[2Fe—2S]^{1+}$ clusters remain however observable around 50–60 K in contrast to $[4Fe—4S]^{1+}$. For long time it was hard to observe Fe—S cluster of scaffold and carrier proteins involved in Fe—S biogenesis likely due to their function to assemble/accept transiently a Fe—S cluster for a fast transfer to apo-targets. This is made possible by the nature of ligands and geometry of the cluster but inevitably leads to some lability of the cluster that may be enhanced after reduction for EPR spectroscopy. Using short-time reduction or softer reducing agents such as DTT, Fe—S clusters of ISCA, NfuA

Table 3 Oxidation States of Fe—S Clusters and Their Properties in EPR Spectroscopy

Type of Cluster	EPR Silent	EPR Active	g_m Value (4 K)
[2Fe—2S]	+2 (oxidized)	+1 (reduced)	1.96
$[2Fe—2S]_{Rieske}$	+2 (oxidized)	+1 (reduced)	1.91
[4Fe—4S]	+2 (oxidized)	+1 (reduced)	≍2
$[4Fe—4S]_{HiPiPs}$	+2 (reduced)	+3 (oxidized)	>2

carriers (Beilschmidt et al., 2017; Brancaccio et al., 2014; Qian, Zheng, & Liu, 2013; Zimbler et al., 2012), and glutaredoxins (Banci et al., 2015; Bandyopadhyay, Gama, et al., 2008) could be recently observed and characterized by EPR spectroscopy.

A limitation of the EPR-based study is that it is difficult to determine cluster coordination with noncysteinyl ligands such as oxygen-based (aspartate, tyrosinate, glutamate) and nitrogen-based (histidine, arginine) residues.

4.4 Hyperfine Sublevel Correlation Spectroscopy

Hyperfine sublevel correlation (HYSCORE) is a two-dimensional pulsed EPR technique that monitors nuclei interacting with paramagnetic Fe—S clusters. In particular, it allows the detection of hyperfine coupling between the Fe—S cluster and species possessing a nonzero nuclear spin, such as ^{1}H, ^{2}N, ^{13}C, ^{14}N, ^{15}N, or ^{77}Se (Cutsail, Telser, & Hoffman, 2015). This technique is based on the analysis of the return to equilibrium of the electron spin of the studied paramagnetic species after a perturbation. This species is first placed in a defined magnetic field (conventionally, for a species with $S=½$ the field for which the maximum response is observed by RPE in continuous wave (cw)). The system is then submitted to a series of pulses of an electromagnetic wave at high power that will disrupt the orientation of the electronic spin of the analyzed species. The return to equilibrium allows to visualize the hyperfine couplings between the electronic spin and the possible species possessing a nonzero nuclear spin in the vicinity. The main advantage of HYSCORE is that it sorts three types of nuclei: the strongly, the weakly coupled, and the "distant" nuclei. In the "distant" case, the signal lies on the diagonal of the (+, +) quadrant, whereas the strongly and weakly coupled nuclei appear off the diagonal in the (−, +) and (+, +) quadrants, respectively (Fig. 2). This technique is therefore useful in the study of natural nitrogen or exchangeable ligands. It was especially used to characterize HydF, a [4Fe—4S] protein involved in the maturation of [Fe—Fe] Hydrogenase (Berto et al., 2012; Brazzolotto et al., 2006), and very helpful to investigate ligand environment of biotin synthase clusters during catalysis (Fugate et al., 2012). This spectroscopy allowed to support for the MiaB [4Fe—4S] enzyme a sulfation mechanism in which an exogenous sulfur cosubstrate is activated at an exchangeable coordination site on its cluster II (Forouhar et al., 2013). More recently, this spectroscopy was used in the mechanism characterization of quinolinate synthase (NadA), a [4Fe—4S] enzyme involved in the biogenesis of the key cofactor NAD in biology.

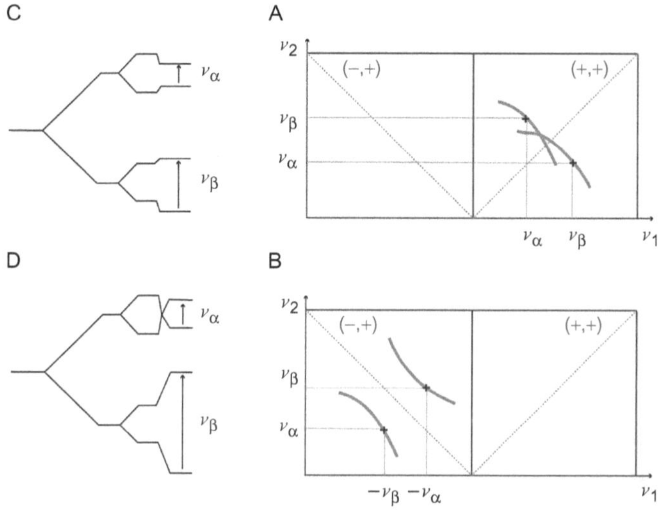

Fig. 2 HYSCORE spectroscopy (system $I = ½$ and $S = ½$). A weak coupling (A) is observed in the dial (+, +) with its corresponding energy level diagram (C). A strong coupling (B) is observed in the dial (−, +) with its corresponding energy level diagram (D).

Hyscore allowed to unambiguously demonstrate that the product of the reaction catalyzed by NadA, quinolinate, coordinates the catalytic iron site of the cluster through its N1 and C7 carboxylate group (Esakova et al., 2016).

4.5 Raman Resonance Spectroscopy

Raman resonance (RR) is a vibrational spectroscopy, which reports transitions between different vibrational energy levels in a molecule (Que, 2000; Spiro & Czernuszewicz, 1995). Information is derived from light scattering rather than a direct absorption process as in infrared spectroscopy (IR). The main advantage of RR spectroscopy is that the wavelength of the exciting laser line (monochromatic light corresponding to an allowed absorption band region) is adjusted such that it coincides with that of an allowed electronic transition in the sample. Vibrational frequencies are sensitive to bond strength, number of atoms, geometry, and coordination environment.

A limitation of RR experiments is that illumination of the sample within an electronic absorption band can induce local heating of the sample. Freezing of the sample in the focused laser beam might be used. Also, in particular for Fe—S proteins the RR device might be put under an argon flux to prevent any traces of O_2. The most common and reliable laser sources are cw noble gas lasers pumped by an electrical discharge. Ar + and Kr + ion lasers (>1 W) emit at a series of wavelengths in the visible and near-UV region, and the He—Ne (543.5 and 632.8 nm) and He—Cd (441.6 and 325.0 nm) lasers have useful wavelengths, albeit at lower power levels. For Fe—S proteins, the excitation wavelength is generally obtained from the 441.6 nm line of a He—Cd laser.

For canonical $[2Fe—2S]_{(Cys)4}$ and $[4Fe—4S]_{(Cys)4}$, the absorption bands arise from S→Fe charge transfer transitions, and RR spectra show enhancement of modes in the 200–450 cm^{-1} region, associated with the stretching of the Fe—S bonds. However, [2Fe—2S] and [4Fe—4S] RR spectra are very different (Fig. 3). RR spectrum of $[2Fe—2S]_{(Cys)4}$ proteins comprises terminal modes (from 290 to 340 cm^{-1}), whereas the bridging modes are somewhat higher at 315–425 cm^{-1}. There is also the prominence of a band near 290 cm^{-1} (286–295 cm^{-1}), a diagnostic of $[2Fe—2S]^{2+}$ clusters (not seen for $[4Fe—4S]^{2+}$ clusters). The RR spectrum of $[4Fe—4S]_{(Cys)4}$ has a dominant band (breathing mode) at around 336 cm^{-1}, and other modes that range up to 400 cm^{-1} and down to 240 cm^{-1}. Sometimes the symmetric breathing mode for [4Fe—4S] cluster is out of the normal range (333–339 cm^{-1}) with an intense band at higher frequency (344–353 cm^{-1}). This can reflect clusters with oxygenic ligand (OH^-, serine, or aspartate) at a unique Fe site (340–343 cm^{-1}) (Vey & Drennan, 2011), or cluster distortion associated with a subunit-bridging $[4Fe—4S]^{2+}$ cluster ligated symmetrically by cysteines, as for the cluster in *Av*NfuA (353 cm^{-1}) (Bandyopadhyay, Naik, et al., 2008), *A. thaliana* Nfu2 (344 cm^{-1}) (Gao et al., 2013) involved in Fe—S biogenesis or MgATP-bound nitrogenase Fe-protein (intense RR band at 350 cm^{-1}) (Sen et al., 2004). Aconitase, in which one of the Fe atoms is catalytic and coordinated to a hydroxide, has an RR spectrum resembling those of canonical [4Fe—4S] with a shift in one of the cluster modes, reflecting the altered interaction with a hydroxide instead of a cysteine ligand. Also, a change in the pattern of terminal Fe—S modes is observed on substrate or inhibitor binding, attributed to anomalous cysteinyl Fe—Sγ—Cβ—Cα dihedral angles (Fu, Morgan, Mortenson, & Johnson, 1991). From a practical point of view, RR spectroscopy requires very small amounts (20 μL) of samples but at very high concentration (100 m*M*–1 *M*).

4.6 Circular Dichroism

Study of Fe—S proteins by circular dichroism (CD) is quite recent. CD depends on differential interaction between sample and right vs left

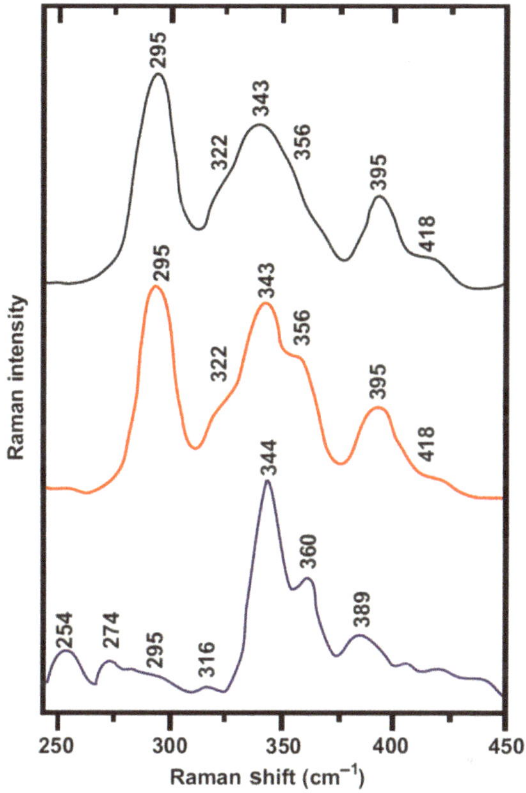

Fig. 3 Low-temperature resonance Raman spectra of as-isolated [2Fe—2S] AtNfu2 (*black line*), reconstituted [2Fe—2S] AtNfu2 (*red line*), and reconstituted [4Fe—4S] AtNfu2 (*blue line*) after (Gao et al., 2013) modification. Spectra were recorded at 17 K using 457.9-nm laser excitation. Each spectrum is the sum of 80–120 scans.

circularly polarized light. The extrinsic or induced CD observed in the visible and near-UV spectra of the proteins provides information about the conformation and/or environment of the bound chromophore. Fe—S transfer between *E. coli* scaffold IscU and the ferredoxin Fdx was studied by CD. The near UV–visible region CD of the IscU [2Fe—2S] is qualitatively similar to that of Fdx [2Fe—2S], consistent with similar cluster coordination, but the CD intensity of IscU [2Fe—2S] is weaker due to different cluster environments in the two proteins. Thus, Fe—S transfer velocity could be measured (Bonomi, Iametti, Ta, & Vickery, 2005). Similar work was done with IscA, which has a CD spectrum much different from Fdx and IscU. Thanks to CD, HscB was shown recently to slow down a CD-active cluster on apo-IscU (Iametti, Barbiroli, & Bonomi, 2015), and used to monitor *A. thaliana* Nfu2 Fe—S conversion, due to exceptional difference in intensity between its [4Fe—4S] and its [2Fe—2S] (Gao et al., 2013). Finally, CD spectroscopy allowed to monitor Fe—S biogenesis within the ISC assembly complex, formed by cysteine desulfurase, frataxin, and the scaffold (Fox, Chakrabarti, McCormick, Lindahl, & Barondeau, 2015). Like for [2Fe—2S], CD spectrum of [4Fe—4S] clusters can vary among proteins (Stephens et al., 1979), for $SufBC_2D$ it is of low intensity and with broad features around 400–500 nm (Chahal & Outten, 2012).

The instrumentation of CD is very well established, nondestructive, and not very demanding in terms of sample workup and consumption, thus enabling an application on a routine basis. Two beams of monochromatic left and right circularly polarized light are provided by phase modulation of monochromatic, linearly polarized light in the modulator. These beams are traversing the sample that, following Lambert–Beer's law, absorbs light from both components to a different extent. Although instruments directly measure the difference in absorption ΔA, for historical reasons CD is usually expressed in degrees of ellipticity $[\theta]$. Buffers and salts must be selected with care. 10 mM phosphate buffer is widely used, but Tris buffers are tolerated. CD measurements are recorded at 25°C in 1-cm path anaerobic cuvettes. For CD spectroscopy principles, see reviews Stephens et al. (1978) and Woody (1995).

4.7 Nuclear Magnetic Resonance Spectroscopy

Nuclear magnetic resonance (NMR) spectroscopy exploits the magnetic properties of certain atomic nuclei. It relies on the phenomenon of NMR and provides information at an atomic level about the structure, dynamics, and chemical environment of molecules. The general strategy is to use multidimensional (2D, 3D, and 4D) multinuclear magnetic resonance techniques to detect and assign resonances from atoms of biological interest (^{1}H, ^{13}C, ^{15}N, and ^{31}P). With these assignments in hand, one can interpret the spectral information present in coupling constants, relaxation rates, cross-relaxation rates, and chemical shifts. Proton–proton cross-relaxation rates and coupling constants are used to derive 3D structures. Relaxation rates, lineshapes, and nuclear Overhauser effect measurements provide information about molecular motions and conformational changes. Fe—S proteins have attracted the interest of NMR investigators from the very beginning of NMR studies of paramagnetic systems (Phillips, Poe,

Weiher, McDonald, & Lovenberg, 1970; Poe, Phillips, McDonald, & Lovenberg, 1970) and since then many biochemists have used NMR in an attempt to characterize them (see some excellent reviews on NMR of Fe—S proteins: Bertini, Turano, & Vila, 1993; Cheng & Markley, 1995). Recently, NMR proved to be very useful in Fe—S biogenesis area, especially to investigate the Fe—S cluster transfer mechanism. Important data were obtained with IscU and the chaperones HscA/B (Kim, Tonelli, Frederick, Chow, & Markley, 2012) and more recently on the Fe—S transfer from glutaredoxin to ISCA and anamorsin (Banci et al., 2014, 2015; Brancaccio et al., 2014, 2017) demonstrating the involvement of a specific protein–protein recognition process. Such investigations are possible since NMR is highly predictable for small proteins (20–30 kDa) that constitute the major part of Fe—S biogenesis proteins (scaffolds, carriers, etc.). A disadvantage of NMR is that a relatively large amount (2–50 mg) of a purified protein is required and the timescale is relatively long. Also, among atoms constituting proteins only hydrogen atom has a nuclear spin ½ and therefore proteins have to be isotopically labeled for NMR study.

ACKNOWLEDGMENTS

Thanks are due to all members, past and present, of the Barras and Ollagnier groups who contributed to studies on Fe—S clusters, and especially to Béatrice Py (Marseille) who pointed out several of the genetic screens described. This chapter was written during a stay of F.B. in Prof. B. Colonna lab at La Sapienza; thanks are due for her support and space. Thanks to Yann Rolland for his help in figure drafting.

REFERENCES

Agar, J. N., Krebs, C., Frazzon, J., Huynh, B. H., Dean, D. R., & Johnson, M. K. (2000). IscU as a scaffold for iron-sulfur cluster biosynthesis: Sequential assembly of [2Fe-2S] and [4Fe-4S] clusters in IscU. *Biochemistry*, *39*, 7856–7862.

Angelini, S., Gerez, C., Ollagnier-de Choudens, S., Sanakis, Y., Fontecave, M., Barras, F., et al. (2008). NfuA, a new factor required for maturing Fe/S proteins in Escherichia coli under oxidative stress and iron starvation conditions. *The Journal of Biological Chemistry*, *283*, 14084–14091.

Antonkine, M. L., Koay, M. S., Epel, B., Breitenstein, C., Gopta, O., Gartner, W., et al. (2009). Synthesis and characterization of de novo designed peptides modelling the binding sites of [4Fe-4S] clusters in photosystem I. *Biochimica et Biophysica Acta*, *1787*, 995–1008.

Banci, L., Brancaccio, D., Ciofi-Baffoni, S., Del Conte, R., Gadepalli, R., Mikolajczyk, M., et al. (2014). [2Fe-2S] cluster transfer in iron-sulfur protein biogenesis. *Proceedings of the National Academy of Sciences of the United States of America*, *111*, 6203–6208.

Banci, L., Ciofi-Baffoni, S., Gajda, K., Muzzioli, R., Peruzzini, R., & Winkelmann, J. (2015). N-terminal domains mediate [2Fe-2S] cluster transfer from glutaredoxin-3 to anamorsin. *Nature Chemical Biology*, *11*, 772–778.

Bandyopadhyay, S., Gama, F., Molina-Navarro, M. M., Gualberto, J. M., Claxton, R., Naik, S. G., et al. (2008). Chloroplast monothiol glutaredoxins as scaffold proteins for the assembly and delivery of [2Fe-2S] clusters. *The EMBO Journal, 27*, 1122–1133.

Bandyopadhyay, S., Naik, S. G., O'Carroll, I. P., Huynh, B. H., Dean, D. R., Johnson, M. K., et al. (2008). A proposed role for the Azotobacter vinelandii NfuA protein as an intermediate iron-sulfur cluster carrier. *The Journal of Biological Chemistry, 283*, 14092–14099.

Beckwith, J. (2007). What lies beyond uranus? Preconceptions, ignorance, serendipity and suppressors in the search for biology's secrets. *Genetics, 176*, 733–740.

Beckwith, J. R., & Silhavy, T. J. (Eds.), (1992). *The power of bacterial genetics. A litterature based-course*: New York, USA: Cold Spring Harbor Laboratory.

Beilschmidt, L. K., Ollagnier de Choudens, S., Fournier, M., Sanakis, I., Hograindleur, M. A., Clemancey, M., et al. (2017). ISCA1 is essential for mitochondrial Fe4S4 biogenesis in vivo. *Nature Communications, 8*, 15124.

Beinert, H. (1983). Semi-micro methods for analysis of labile sulfide and of labile sulfide plus sulfane sulfur in unusually stable iron-sulfur proteins. *Analytical Biochemistry, 131*, 373–378.

Beinert, H., Kennedy, M. C., & Stout, C. D. (1996). Aconitase as iron sulfur protein: Enzyme and iron-regulatory protein. *Chemical Reviews, 96*, 2335–2374.

Benda, R., Tse Sum Bui, B., Schunemann, V., Florentin, D., Marquet, A., & Trautwein, A. X. (2002). Iron-sulfur clusters of biotin synthase in vivo: A Mossbauer study. *Biochemistry, 41*, 15000–15006.

Bertini, I., Turano, P., & Vila, A. J. (1993). Nuclear-magnetic-resonance of paramagnetic metalloproteins. *Chemical Reviews, 93*, 2833–2932.

Berto, P., Di Valentin, M., Cendron, L., Vallese, F., Albertini, M., Salvadori, E., et al. (2012). The [4Fe-4S]-cluster coordination of [FeFe]-hydrogenase maturation protein HydF as revealed by EPR and HYSCORE spectroscopies. *Biochimica et Biophysica Acta, 1817*, 2149–2157.

Blank, L., Green, J., & Guest, J. R. (2002). AcnC of Escherichia coli is a 2-methylcitrate dehydratase (PrpD) that can use citrate and isocitrate as substrates. *Microbiology, 148*, 133–146.

Bonomi, F., Iametti, S., Ta, D., & Vickery, L. E. (2005). Multiple turnover transfer of [2Fe2S] clusters by the iron-sulfur cluster assembly scaffold proteins IscU and IscA. *The Journal of Biological Chemistry, 280*, 29513–29518.

Brancaccio, D., Gallo, A., Mikolajczyk, M., Zovo, K., Palumaa, P., Novellino, E., et al. (2014). Formation of [4Fe-4S] clusters in the mitochondrial iron-sulfur cluster assembly machinery. *Journal of the American Chemical Society, 136*, 16240–16250.

Brancaccio, D., Gallo, A., Piccioli, M., Novellino, E., Ciofi-Baffoni, S., & Banci, L. (2017). [4Fe-4S] cluster assembly in mitochondria and its impairment by copper. *Journal of the American Chemical Society, 139*, 719–730.

Brazzolotto, X., Rubach, J. K., Gaillard, J., Gambarelli, S., Atta, M., & Fontecave, M. (2006). The [Fe-Fe]-hydrogenase maturation protein HydF from Thermotoga maritima is a GTPase with an iron-sulfur cluster. *The Journal of Biological Chemistry, 281*, 769–774.

Brock, M., Maerker, C., Schutz, A., Volker, U., & Buckel, W. (2002). Oxidation of propionate to pyruvate in Escherichia coli. Involvement of methylcitrate dehydratase and aconitase. *European Journal of Biochemistry, 269*, 6184–6194.

Calmels, N., Schmucker, S., Wattenhofer-Donze, M., Martelli, A., Vaucamps, N., Reutenauer, L., et al. (2009). The first cellular models based on frataxin missense mutations that reproduce spontaneously the defects associated with Friedreich ataxia. *PLoS One, 4*, e6379.

Campos, N., Rodriguez-Concepcion, M., Sauret-Gueto, S., Gallego, F., Lois, L. M., & Boronat, A. (2001). Escherichia coli engineered to synthesize isopentenyl diphosphate and dimethylallyl diphosphate from mevalonate: A novel system for the genetic analysis of the 2-C-methyl-d-erythritol 4-phosphate pathway for isoprenoid biosynthesis. *The Biochemical Journal*, *353*, 59–67.

Chahal, H. K., & Outten, F. W. (2012). Separate FeS scaffold and carrier functions for SufB(2)C(2) and SufA during in vitro maturation of [2Fe2S] Fdx. *Journal of Inorganic Biochemistry*, *116*, 126–134.

Chandor, A., Berteau, O., Douki, T., Gasparutto, D., Sanakis, Y., Ollagnier-de-Choudens, S., et al. (2006). Dinucleotide spore photoproduct, a minimal substrate of the DNA repair spore photoproduct lyase enzyme from Bacillus subtilis. *The Journal of Biological Chemistry*, *281*, 26922–26931.

Cheng, H., & Markley, J. L. (1995). NMR spectroscopic studies of paramagnetic proteins—Iron-sulfur proteins. *Annual Review of Biophysics and Biomolecular Structure*, *24*, 209–237.

Cherrier, M. V., Chan, A., Darnault, C., Reichmann, D., Amara, P., Ollagnier de Choudens, S., et al. (2014). The crystal structure of Fe(4)S(4) quinolinate synthase unravels an enzymatic dehydration mechanism that uses tyrosine and a hydrolase-type triad. *Journal of the American Chemical Society*, *136*, 5253–5256.

Clémancey, M. B. G., Latour, J. M., & Garcia-Serres, R. (2014). Mössbauer spectroscopy. In Y. Nicolet & J. C. Fontecilla-Camps (Eds.), *Metalloproteins: Methods and protocols* (pp. 153–170). Totowa, NJ: Humana Press.

Crack, J. C., Gaskell, A. A., Green, J., Cheesman, M. R., Le Brun, N. E., & Thomson, A. J. (2008). Influence of the environment on the [4Fe-4S]2+ to [2Fe-2S]2+ cluster switch in the transcriptional regulator FNR. *Journal of the American Chemical Society*, *130*, 1749–1758.

Crack, J. C., Munnoch, J., Dodd, E. L., Knowles, F., Al Bassam, M. M., Kamali, S., et al. (2015). NsrR from Streptomyces coelicolor is a nitric oxide-sensing [4Fe-4S] cluster protein with a specialized regulatory function. *The Journal of Biological Chemistry*, *290*, 12689–12704.

Cutsail, G. E., 3rd, Telser, J., & Hoffman, B. M. (2015). Advanced paramagnetic resonance spectroscopies of iron-sulfur proteins: Electron nuclear double resonance (ENDOR) and electron spin echo envelope modulation (ESEEM). *Biochimica et Biophysica Acta*, *1853*, 1370–1394.

Dobbek, H., Svetlitchnyi, V., Gremer, L., Huber, R., & Meyer, O. (2001). Crystal structure of a carbon monoxide dehydrogenase reveals a [Ni-4Fe-5S] cluster. *Science*, *293*, 1281–1285.

Dos Santos, P. C., Johnson, D. C., Ragle, B. E., Unciuleac, M. C., & Dean, D. R. (2007). Controlled expression of nif and isc iron-sulfur protein maturation components reveals target specificity and limited functional replacement between the two systems. *Journal of Bacteriology*, *189*, 2854–2862.

Drennan, C. L., Heo, J., Sintchak, M. D., Schreiter, E., & Ludden, P. W. (2001). Life on carbon monoxide: X-ray structure of Rhodospirillum rubrum Ni-Fe-S carbon monoxide dehydrogenase. *Proceedings of the National Academy of Sciences of the United States of America*, *98*, 11973–11978.

Esakova, O. A., Silakov, A., Grove, T. L., Saunders, A. H., McLaughlin, M. I., Yennawar, N. H., et al. (2016). Structure of quinolinate synthase from Pyrococcus horikoshii in the presence of its product, quinolinic acid. *Journal of the American Chemical Society*, *138*, 7224–7227.

Ezraty, B., Vergnes, A., Banzhaf, M., Duverger, Y., Huguenot, A., Brochado, A. R., et al. (2013). Fe-S cluster biosynthesis controls uptake of aminoglycosides in a ROS-less death pathway. *Science*, *340*, 1583–1587.

Fee, J. A., Findling, K. L., Yoshida, T., Hille, R., Tarr, G. E., Hearshen, D. O., et al. (1984). Purification and characterization of the Rieske iron-sulfur protein from Thermus thermophilus. Evidence for a [2Fe-2S] cluster having non-cysteine ligands. *The Journal of Biological Chemistry*, *259*, 124–133.

Ferrandez, A., Minambres, B., Garcia, B., Olivera, E. R., Luengo, J. M., Garcia, J. L., et al. (1998). Catabolism of phenylacetic acid in Escherichia coli. Characterization of a new aerobic hybrid pathway. *The Journal of Biological Chemistry, 273*, 25974–25986.

Fish, W. W. (1988). Rapid colorimetric micromethod for the quantitation of complexed iron in biological samples. *Methods in Enzymology, 158*, 357–364.

Fleischhacker, A. S., Stubna, A., Hsueh, K. L., Guo, Y., Teter, S. J., Rose, J. C., et al. (2012). Characterization of the [2Fe-2S] cluster of Escherichia coli transcription factor IscR. *Biochemistry, 51*, 4453–4462.

Flint, D. H., Emptage, M. H., Finnegan, M. G., Fu, W., & Johnson, M. K. (1993). The role and properties of the iron-sulfur cluster in Escherichia coli dihydroxy-acid dehydratase. *The Journal of Biological Chemistry, 268*, 14732–14742.

Forouhar, F., Arragain, S., Atta, M., Gambarelli, S., Mouesca, J. M., Hussain, M., et al. (2013). Two Fe-S clusters catalyze sulfur insertion by radical-SAM methylthiotransferases. *Nature Chemical Biology, 9*, 333–338.

Fox, N. G., Chakrabarti, M., McCormick, S. P., Lindahl, P. A., & Barondeau, D. P. (2015). The human iron-sulfur assembly complex catalyzes the synthesis of [2Fe-2S] clusters on ISCU2 that can be transferred to acceptor molecules. *Biochemistry, 54*, 3871–3879.

Fu, W. G., Morgan, T. V., Mortenson, L. E., & Johnson, M. K. (1991). Resonance Raman studies of the [4Fe-4S] to [2Fe-2S] cluster conversion in the iron protein of nitrogenase. *FEBS Letters, 284*, 165–168.

Fugate, C. J., Stich, T. A., Kim, E. G., Myers, W. K., Britt, R. D., & Jarrett, J. T. (2012). 9-Mercaptodethiobiotin is generated as a ligand to the [2Fe-2S]+ cluster during the reaction catalyzed by biotin synthase from Escherichia coli. *Journal of the American Chemical Society, 134*, 9042–9045.

Gao, H., Subramanian, S., Couturier, J., Naik, S. G., Kim, S. K., Leustek, T., et al. (2013). Arabidopsis thaliana Nfu2 accommodates [2Fe-2S] or [4Fe-4S] clusters and is competent for in vitro maturation of chloroplast [2Fe-2S] and [4Fe-4S] cluster-containing proteins. *Biochemistry, 52*, 6633–6645.

Gruner, I., Fradrich, C., Bottger, L. H., Trautwein, A. X., Jahn, D., & Hartig, E. (2011). Aspartate 141 is the fourth ligand of the oxygen-sensing [4Fe-4S]2+ cluster of Bacillus subtilis transcriptional regulator Fnr. *The Journal of Biological Chemistry, 286*, 2017–2021.

Gupta, V., Sendra, M., Naik, S. G., Chahal, H. K., Huynh, B. H., Outten, F. W., et al. (2009). Native Escherichia coli SufA, coexpressed with SufBCDSE, purifies as a [2Fe-2S] protein and acts as an Fe-S transporter to Fe-S target enzymes. *Journal of the American Chemical Society, 131*, 6149–6153.

Gütlich, P., Bill, E., & Trautwein, A. X. (2011). *Mössbauer spectroscopy and transition metal chemistry. Vol. 15*(p. 569). Berlin, Heidelberg: Springer-Verlag.

Hans, M., Buckel, W., & Bill, E. (2008). Spectroscopic evidence for an all-ferrous [4Fe-4S] 0 cluster in the superreduced activator of 2-hydroxyglutaryl-CoA dehydratase from Acidaminococcus fermentans. *Journal of Biological Inorganic Chemistry, 13*, 563–574.

Iametti, S., Barbiroli, A., & Bonomi, F. (2015). Functional implications of the interaction between HscB and IscU in the biosynthesis of FeS clusters. *Journal of Biological Inorganic Chemistry, 20*, 1039–1048.

Jang, S., & Imlay, J. A. (2007). Micromolar intracellular hydrogen peroxide disrupts metabolism by damaging iron-sulfur enzymes. *The Journal of Biological Chemistry, 282*, 929–937.

Kim, J. H., Tonelli, M., Frederick, R. O., Chow, D. C., & Markley, J. L. (2012). Specialized Hsp70 chaperone (HscA) binds preferentially to the disordered form, whereas J-protein (HscB) binds preferentially to the structured form of the iron-sulfur cluster scaffold protein (IscU). *The Journal of Biological Chemistry, 287*, 31406–31413.

Leggate, E. J., Bill, E., Essigke, T., Ullmann, G. M., & Hirst, J. (2004). Formation and characterization of an all-ferrous Rieske cluster and stabilization of the [2Fe-2S]0 core by protonation. *Proceedings of the National Academy of Sciences of the United States of America, 101*, 10913–10918.

Loiseau, L., Gerez, C., Bekker, M., Ollagnier-de Choudens, S., Py, B., Sanakis, Y., et al. (2007). ErpA, an iron sulfur (Fe S) protein of the A-type essential for respiratory metabolism in Escherichia coli. *Proceedings of the National Academy of Sciences of the United States of America, 104*, 13626–13631.

Magalon, A., & Alberge, F. (2016). Distribution and dynamics of OXPHOS complexes in the bacterial cytoplasmic membrane. *Biochimica et Biophysica Acta, 1857*, 198–213.

Manson, M. D. (2000). Allele-specific suppression as a tool to study protein-protein interactions in bacteria. *Methods, 20*, 18–34.

Mapolelo, D. T., Zhang, B., Naik, S. G., Huynh, B. H., & Johnson, M. K. (2012). Spectroscopic and functional characterization of iron-bound forms of Azotobacter vinelandii (Nif)IscA. *Biochemistry, 51*, 8056–8070.

Martin, J. E., Waters, L. S., Storz, G., & Imlay, J. A. (2015). The Escherichia coli small protein MntS and exporter MntP optimize the intracellular concentration of manganese. *PLoS Genetics, 11*, e1004977.

Martinez-Gomez, N. C., & Downs, D. M. (2008). ThiC is an [Fe-S] cluster protein that requires AdoMet to generate the 4-amino-5-hydroxymethyl-2-methylpyrimidine moiety in thiamin synthesis. *Biochemistry, 47*, 9054–9056.

Maynard, N. D., Birch, E. W., Sanghvi, J. C., Chen, L., Gutschow, M. V., & Covert, M. W. (2010). A forward-genetic screen and dynamic analysis of lambda phage host-dependencies reveals an extensive interaction network and a new anti-viral strategy. *PLoS Genetics, 6*, e1001017.

Melber, A., Na, U., Vashisht, A., Weiler, B. D., Lill, R., Wohlschlegel, J. A., et al. (2016). Role of Nfu1 and Bol3 in iron-sulfur cluster transfer to mitochondrial clients. *eLife, 5*.

Middleton, P., Dickson, D. P., Johnson, C. E., & Rush, J. D. (1980). Interpretation of the Mossbauer spectra of the high-potential iron protein from Chromatium. *European Journal of Biochemistry, 104*, 289–296.

Miranda, C. J., Santos, M. M., Ohshima, K., Smith, J., Li, L., Bunting, M., et al. (2002). Frataxin knockin mouse. *FEBS Letters, 512*, 291–297.

Morra, S., Arizzi, M., Valetti, F., & Gilardi, G. (2016). Oxygen stability in the new [FeFe]-hydrogenase from Clostridium beijerinckii SM10 (CbA5H). *Biochemistry, 55*, 5897–5900.

Muhlenhoff, U., Gerber, J., Richhardt, N., & Lill, R. (2003). Components involved in assembly and dislocation of iron-sulfur clusters on the scaffold protein Isu1p. *The EMBO Journal, 22*, 4815–4825.

Munck, E., Debrunner, P. G., Tsibris, J. C., & Gunsalus, I. C. (1972). Mossbauer parameters of putidaredoxin and its selenium analog. *Biochemistry, 11*, 855–863.

Nachin, L., Loiseau, L., Expert, D., & Barras, F. (2003). SufC: An unorthodox cytoplasmic ABC/ATPase required for [Fe-S] biogenesis under oxidative stress. *The EMBO Journal, 22*, 427–437.

Nechushtai, R., Conlan, A. R., Harir, Y., Song, L., Yogev, O., Eisenberg-Domovich, Y., et al. (2012). Characterization of Arabidopsis NEET reveals an ancient role for NEET proteins in iron metabolism. *Plant Cell, 24*, 2139–2154.

Nichols, R. J., Sen, S., Choo, Y. J., Beltrao, P., Zietek, M., Chaba, R., et al. (2011). Phenotypic landscape of a bacterial cell. *Cell, 144*, 143–156.

Odom, A. R. (2011). Five questions about non-mevalonate isoprenoid biosynthesis. *PLoS Pathogens, 7*, e1002323.

Ollagnier-de Choudens, S., Loiseau, L., Sanakis, Y., Barras, F., & Fontecave, M. (2005). Quinolinate synthetase, an iron-sulfur enzyme in NAD biosynthesis. *FEBS Letters, 579*, 3737–3743.

Ollagnier-de-Choudens, S., Mattioli, T., Takahashi, Y., & Fontecave, M. (2001). Iron-sulfur cluster assembly: Characterization of IscA and evidence for a specific and functional complex with ferredoxin. *The Journal of Biological Chemistry, 276*, 22604–22607.

Orme-Johnson, W. H., & Orme-Johnson, N. R. (1982). Iron-sulfur proteins. In T. G. Spiro (Ed.), *Metal ion in biology series: Vol. 4*. New York: John Wiley & Sons.

Pandelia, M. E., Lanz, N. D., Booker, S. J., & Krebs, C. (2015). Mossbauer spectroscopy of Fe/S proteins. *Biochimica et Biophysica Acta, 1853*, 1395–1405.

Park, S., You, X., & Imlay, J. A. (2005). Substantial DNA damage from submicromolar intracellular hydrogen peroxide detected in Hpx- mutants of Escherichia coli. *Proceedings of the National Academy of Sciences of the United States of America, 102*, 9317–9322.

Perche-Letuvee, P., Kathirvelu, V., Berggren, G., Clemancey, M., Latour, J. M., Maurel, V., et al. (2012). 4-Demethylwyosine synthase from Pyrococcus abyssi is a radical-S-adenosyl-L-methionine enzyme with an additional [4Fe-4S](+2) cluster that interacts with the pyruvate co-substrate. *The Journal of Biological Chemistry, 287*, 41174–41185.

Perdomini, M., Belbellaa, B., Monassier, L., Reutenauer, L., Messaddeq, N., Cartier, N., et al. (2014). Prevention and reversal of severe mitochondrial cardiomyopathy by gene therapy in a mouse model of Friedreich's ataxia. *Nature Medicine, 20*, 542–547.

Peters, J. W., Lanzilotta, W. N., Lemon, B. J., & Seefeldt, L. C. (1998). X-ray crystal structure of the Fe-only hydrogenase (CpI) from clostridium pasteurianum to 1.8 angstrom resolution. *Science, 282*, 1853–1858.

Phillips, W. D., Poe, M., Weiher, J. F., McDonald, C. C., & Lovenberg, W. (1970). Proton magnetic resonance, magnetic susceptibility and Mossbauer studies of clostridium pasteurianum rubredoxin. *Nature, 227*, 574–577.

Pierrel, F., Bjork, G. R., Fontecave, M., & Atta, M. (2002). Enzymatic modification of tRNAs: MiaB is an iron-sulfur protein. *The Journal of Biological Chemistry, 277*, 13367–13370.

Poe, M., Phillips, W. D., McDonald, C. C., & Lovenberg, W. (1970). Proton magnetic resonance study of ferredoxin from clostridium pasteurianum. *Proceedings of the National Academy of Sciences of the United States of America, 65*, 797–804.

Popescu, C. V., Bates, D. M., Beinert, H., Munck, E., & Kiley, P. J. (1998). Mossbauer spectroscopy as a tool for the study of activation/inactivation of the transcription regulator FNR in whole cells of Escherichia coli. *Proceedings of the National Academy of Sciences of the United States of America, 95*, 13431–13435.

Puccio, H., Simon, D., Cossee, M., Criqui-Filipe, P., Tiziano, F., Melki, J., et al. (2001). Mouse models for Friedreich ataxia exhibit cardiomyopathy, sensory nerve defect and Fe-S enzyme deficiency followed by intramitochondrial iron deposits. *Nature Genetics, 27*, 181–186.

Py, B., & Barras, F. (2015). Genetic approaches of the Fe-S cluster biogenesis process in bacteria: Historical account, methodological aspects and future challenges. *Biochimica et Biophysica Acta, 1853*, 1429–1435.

Py, B., Gerez, C., Angelini, S., Planel, R., Vinella, D., Loiseau, L., et al. (2012). Molecular organization, biochemical function, cellular role and evolution of NfuA, an atypical Fe-S carrier. *Molecular Microbiology, 86*, 155–171.

Qian, L., Zheng, C. L., & Liu, J. S. (2013). Characterization of iron-sulfur cluster assembly protein isca from Acidithiobacillus ferrooxidans. *Biochemistry (Moscow), 78*, 244–251.

Que, L. J. (2000). *Physical methods in bioinorganic chemistry*. University Science Books.

Roche, B., Agrebi, R., Huguenot, A., Ollagnier de Choudens, S., Barras, F., & Py, B. (2015). Turning Escherichia coli into a Frataxin-dependent organism. *PLoS Genetics, 11*, e1005134.

Rousset, C., Fontecave, M., & Ollagnier de Choudens, S. (2008). The [4Fe-4S] cluster of quinolinate synthase from Escherichia coli: Investigation of cluster ligands. *FEBS Letters, 582*, 2937–2944.

Rubach, J. K., Brazzolotto, X., Gaillard, J., & Fontecave, M. (2005). Biochemical characterization of the HydE and HydG iron-only hydrogenase maturation enzymes from Thermatoga maritima. *FEBS Letters*, *579*, 5055–5060.

Schmucker, S., Martelli, A., Colin, F., Page, A., Wattenhofer-Donze, M., Reutenauer, L., et al. (2011). Mammalian frataxin: An essential function for cellular viability through an interaction with a preformed ISCU/NFS1/ISD11 iron-sulfur assembly complex. *PLoS One*, *6*, e16199.

Schwartz, C. J., Giel, J. L., Patschkowski, T., Luther, C., Ruzicka, F. J., Beinert, H., et al. (2001). IscR, an Fe-S cluster-containing transcription factor, represses expression of Escherichia coli genes encoding Fe-S cluster assembly proteins. *Proceedings of the National Academy of Sciences of the United States of America*, *98*, 14895–14900.

Seaver, L. C., & Imlay, J. A. (2001). Hydrogen peroxide fluxes and compartmentalization inside growing Escherichia coli. *Journal of Bacteriology*, *183*, 7182–7189.

Sen, S., Igarashi, R., Smith, A., Johnson, M. K., Seefeldt, L. C., & Peters, J. W. (2004). A conformational mimic of the MgATP-bound "on state" of the nitrogenase iron protein. *Biochemistry*, *43*, 1787–1797.

Seznec, H., Simon, D., Bouton, C., Reutenauer, L., Hertzog, A., Golik, P., et al. (2005). Friedreich ataxia: The oxidative stress paradox. *Human Molecular Genetics*, *14*, 463–474.

Shen, G., Balasubramanian, R., Wang, T., Wu, Y., Hoffart, L. M., Krebs, C., et al. (2007). SufR coordinates two [4Fe-4S]2+, 1+ clusters and functions as a transcriptional repressor of the sufBCDS operon and an autoregulator of sufR in cyanobacteria. *The Journal of Biological Chemistry*, *282*, 31909–31919.

Simon, D., Seznec, H., Gansmuller, A., Carelle, N., Weber, P., Metzger, D., et al. (2004). Friedreich ataxia mouse models with progressive cerebellar and sensory ataxia reveal autophagic neurodegeneration in dorsal root ganglia. *The Journal of Neuroscience*, *24*, 1987–1995.

Spiro, T. G., & Czernuszewicz, R. S. (1995). Resonance Raman spectroscopy of metalloproteins. *Methods in Enzymology*, *246*, 416–460.

Stephens, P. J., McKenna, C. E., Smith, B. E., Nguyen, H. T., McKenna, M. C., Thomson, A. J., et al. (1979). Circular dichroism and magnetic circular dichroism of nitrogenase proteins. *Proceedings of the National Academy of Sciences of the United States of America*, *76*, 2585–2589.

Stephens, P. J., Thomson, A. J., Dunn, J. B., Keiderling, T. A., Rawlings, J., Rao, K. K., et al. (1978). Circular dichroism and magnetic circular dichroism of iron-sulfur proteins. *Biochemistry*, *17*, 4770–4778.

Takahashi, Y., & Tokumoto, U. (2002). A third bacterial system for the assembly of iron-sulfur clusters with homologs in archaea and plastids. *The Journal of Biological Chemistry*, *277*, 28380–28383.

Tamae, C., Liu, A., Kim, K., Sitz, D., Hong, J., Becket, E., et al. (2008). Determination of antibiotic hypersensitivity among 4,000 single-gene-knockout mutants of Escherichia coli. *Journal of Bacteriology*, *190*, 5981–5988.

Tanaka, N., Kanazawa, M., Tonosaki, K., Yokoyama, N., Kuzuyama, T., & Takahashi, Y. (2016). Novel features of the ISC machinery revealed by characterization of Escherichia Coli mutants that survive without iron-sulfur clusters. *Molecular Microbiology*, *99*, 835–848.

Tedeschi, G., Zetta, L., Negri, A., Mortarino, M., Ceciliani, F., & Ronchi, S. (1997). Redox potentials and quinone reductase activity of L-aspartate oxidase from Escherichia coli. *Biochemistry*, *36*, 16221–16230.

Tinberg, C. E., Tonzetich, Z. J., Wang, H., Do, L. H., Yoda, Y., Cramer, S. P., et al. (2010). Characterization of iron dinitrosyl species formed in the reaction of nitric oxide with a biological Rieske center. *Journal of the American Chemical Society*, *132*, 18168–18176.

Trotter, V., Vinella, D., Loiseau, L., Ollagnier de Choudens, S., Fontecave, M., & Barras, F. (2009). The CsdA cysteine desulphurase promotes Fe/S biogenesis by recruiting Suf components and participates to a new sulphur transfer pathway by recruiting CsdL (ex-YgdL), a ubiquitin-modifying-like protein. *Molecular Microbiology*, *74*, 1527–1542.

Vey, J. L., & Drennan, C. L. (2011). Structural insights into radical generation by the radical SAM superfamily. *Chemical Reviews*, *111*, 2487–2506.

Vinella, D., Brochier-Armanet, C., Loiseau, L., Talla, E., & Barras, F. (2009). Iron-sulfur (Fe/S) protein biogenesis: Phylogenomic and genetic studies of A-type carriers. *PLoS Genetics*, *5*, e1000497.

Vinella, D., Loiseau, L., Ollagnier de Choudens, S., Fontecave, M., & Barras, F. (2013). In vivo [Fe-S] cluster acquisition by IscR and NsrR, two stress regulators in Escherichia coli. *Molecular Microbiology*, *87*, 493–508.

Wollers, S., Layer, G., Garcia-Serres, R., Signor, L., Clemancey, M., Latour, J. M., et al. (2010). Iron-sulfur (Fe-S) cluster assembly: The SufBCD complex is a new type of Fe-S scaffold with a flavin redox cofactor. *The Journal of Biological Chemistry*, *285*, 23331–23341.

Woody, R. W. (1995). Circular dichroism. *Methods in Enzymology*, *246*, 34–71.

Yang, J., Naik, S. G., Ortillo, D. O., Garcia-Serres, R., Li, M., Broderick, W. E., et al. (2009). The iron-sulfur cluster of pyruvate formate-lyase activating enzyme in whole cells: Cluster interconversion and a valence-localized [4Fe-4S]2+ state. *Biochemistry*, *48*, 9234–9241.

Zhang, B., Bandyopadhyay, S., Shakamuri, P., Naik, S. G., Huynh, B. H., Couturier, J., et al. (2013). Monothiol glutaredoxins can bind linear [Fe3S4]+ and [Fe4S4]2+ clusters in addition to [Fe2S2]2+ clusters: Spectroscopic characterization and functional implications. *Journal of the American Chemical Society*, *135*, 15153–15164.

Zimbler, D. L., Park, T. M., Arivett, B. A., Penwell, W. F., Greer, S. M., Woodruff, T. M., et al. (2012). Stress response and virulence functions of the Acinetobacter baumannii NfuA Fe-S scaffold protein. *Journal of Bacteriology*, *194*, 2884–2893.

CHAPTER TWO

De Novo Design of Iron–Sulfur Proteins

Zahra B. Dizicheh, Nicholas Halloran, William Asma, Giovanna Ghirlanda[1]
School of Molecular Sciences, Arizona State University, Tempe, AZ, United States
[1]Corresponding author: e-mail address: giovanna.ghirlanda@asu.edu

Contents

Abstract

Iron–sulfur proteins are one of the most abundant and functionally pliable redox proteins found in all living organisms. Because of their crucial role in mediating electron transfer processes, minimalist model systems have been developed as a proxy to study natural Fe–S redox proteins and to dissect rules to enable tuning of their redox and electron transfer activities. This goal has been pursued through computational design, mutagenesis in the first and second coordination sphere, metal substitution, cofactor replacement, and the use of unnatural amino acids to stabilize a given cluster. In this chapter, we discuss the most recent design strategies to introduce various Fe–S clusters into natural and artificial protein scaffolds. Practical approaches for the cluster reconstitution, hydrogen production, and electrochemical characterization are mentioned.

Methods in Enzymology, Volume 595
ISSN 0076-6879
http://dx.doi.org/10.1016/bs.mie.2017.07.014

1. INTRODUCTION

Iron–sulfur proteins are one of the most abundant and versatile classes of proteins. They are implicated in fundamental biological processes, with an array of functions ranging from electron transfer to catalysis, iron level regulation, and storage, as well as transport of ligands within the cell. Remarkably, they are often used in multielectron redox processes, most prominently proton reduction, sulfite reduction, and elemental nitrogen reduction (Beinert, 1997; Lill, 2009). The chemical properties of iron–sulfur proteins, such as generally low redox potential and high oxygen sensitivity, their prevalence, structural flexibility, and functional versatility, suggest that they emerged early in evolution. They are thought to have evolved in a prephotosynthetic world by incorporation of inorganic precursors found in the highly reducing, iron and sulfur-rich environment into primitive peptide sequences (Eck & Dayhoff, 1966; Meyer, 2008). Iron–sulfur clusters are found in a wide range of forms in proteins, including [2Fe–2S], [3Fe–4S], [4Fe–4S], and higher order clusters. The clusters are generally embedded into protein scaffolds via ligation of iron through cysteine side chains, and contain bridging sulfur atoms.

Iron–sulfur clusters often mediate the delivery of reducing equivalents from metabolic pathways to the active site of redox proteins. In soluble small proteins such as the ferredoxins, this is achieved by docking to the surface of their metabolic partners, while in other complex redox proteins such as the hydrogenases iron–sulfur clusters are embedded within the protein and organized as chains positioned at distances favorable for fast electron transfer, and work as conduit from the protein surface to the active site. The redox potentials of the clusters often control the reactivity of the whole enzyme: the oxygen tolerance of membrane associated [Ni–Fe] hydrogenase has been ascribed to the peculiar coordination sphere of one of its iron–sulfur clusters, which is surrounded by six cysteines rather than by the customary four.

These and other observations suggest that the de novo design of enzymes catalyzing redox reactions should include the engineering of means for the delivery of electrons to and from the active site in order to optimize catalytic activity (Yu et al., 2015). In this context, the ability to design artificial iron–sulfur proteins in which multiple redox-active centers are arranged in a controlled manner, and their properties tailored to the intended application, is becoming increasingly crucial.

The rational design of metal-binding sites is well established in the literature (Calhoun et al., 2005; Chakraborty, Iranzo, Zuiderweg, & Pecoraro, 2012; Coldren, Hellinga, & Caradonna, 1997; Der et al., 2012; Ghirlanda et al., 2004; Grzyb et al., 2010; Salgado, Radford, & Tezcan, 2010; Shiga et al., 2012; Zastrow, Peacock, Stuckey, & Pecoraro, 2011). Our design approach adapts lesson learned through those efforts to the specific issues peculiar of iron–sulfur clusters. We start with the coordinates of a metal center and its first coordination sphere, and identify a suitable inert protein scaffold. We then replace the hydrophobic core of the scaffold protein with a sterically compatible metal-binding center. In our work, we chose as scaffolds de novo designed coiled coils, because they are ideally suited to accommodate metal-binding sites: the interhelix distances in the core are compatible with the typical bond length in metal–ligand complexes, and the regular structure repeat of the coiled coil lends itself to parametrization and thus to computational design (Tebo & Pecoraro, 2015). Further, the rules governing oligomerization, assembly, and stability are well understood, thus allowing their use in designed multimeric proteins (Boyken et al., 2016; Crick, 1952; Lupas & Bassler, 2017; Moutevelis & Woolfson, 2009; Woolfson, 2017). Here, we present examples from our lab and others of designed iron–sulfur proteins, with an eye toward applications in complex redox processes.

2. [4Fe–4S]

The cubane-type [4Fe–4S] cluster is the most abundant and versatile type of Fe–S clusters. Clusters are found either isolated, or in pairs, or arranged in multicluster chains within protein structure ranging from small and water soluble, such as the ferredoxins, to large multiprotein complexes such as hydrogenases, nitrogenases, and photosystem II (Hosseinzadeh & Lu, 2016; Nanda et al., 2016). The wide range of redox potentials exhibited by proteins containing [4Fe–4S] clusters points to the role of second shell and long range interactions in modulating their intrinsic reactivity: iron–sulfur proteins exhibit redox potentials ranging from −700 to 450 mV. These properties have spurred intensive investigations of natural cluster-binding proteins, leading to elucidating several distinctive features. For example, this range of potential is achieved in part by accessing three different redox states: $[4Fe–4S]^{+3/+2/+1}$. In bacterial ferredoxins the $[4Fe–4S]^{1+}/[4Fe–4S]^{2+}$ transition can occur near −600 mV, while in high potential iron–sulfur proteins (HiPIPs) redox potentials near +500 mV have been measured for the $[4Fe–4S]^{2+}/[4Fe–4S]^{3+}$ couple. The similarities in structure and geometry of the clusters embedded in ferredoxins and HiPIPs suggest that such large differences in redox potential are due to the surrounding protein environment (Dey et al., 2007).

Building on the information gleaned on natural proteins, our group and others have designed model proteins, with the dual purpose of further unmasking relationships between sequence and structural motifs on one side, and function on the other, as well as generating proteins tailored to specific applications.

The [4Fe–4S] clusters are composed of four iron atoms, each coordinated by three inorganic sulfurs and one cysteine residue; positioning the cluster within the hydrophobic core of the protein is achieved by anchoring through four cysteines that can be part of a number of secondary structure motives (Beinert, 1997). Protein sequence analysis in combination with the elucidation of a number of X-ray and NMR structures of natural [4Fe–4S] proteins have highlighted common arrangements of the four coordinating residues. One of the first sequence motives identified from sequence analysis is the conserved CXXCXXC···C, which is found in ferredoxins among other proteins. The motif was extensively investigated using ferredoxin model peptides spanning 16 amino acids and designed to coordinate a single cluster. Nonligand positions were systematically mutated to determine the influence of nonbinding amino acids on the cluster's properties. One of the early discoveries was that a bulky hydrophobic amino acid in the second position of the motif plays an important part in facilitating cluster incorporation, most likely by stabilizing the protein-bound form. Sequences including CIACGAC showed a fivefold increase in yield of cluster incorporation compared to the CGGCGGC and CAACAAC variants, proving that structural and steric effects contribute to stabilization of [4Fe–4S] within de novo proteins (Mulholland, Gibney, Rabanal, & Dutton, 1999). Continuing this approach, Antonkine et al. incorporated [4Fe–4S] clusters into two 16 amino acid model peptides derived from the conserved cluster-binding motif of photosystem I. The redox potentials of these two peptides were -0.44 and -0.47 V, which are within the range of native ferredoxin potentials, supporting the feasibility of using designed proteins as canvas for manipulating the properties of the bound clusters (Antonkine et al., 2009).

A different approach to the de novo design of [4Fe–4S] proteins uses both rational and computational strategies: rather than replicating natural sequence motifs, the examples described later start from the geometric requirements of a disembodied cluster—that is, a cluster and the four cysteines that form the first coordination sphere—and search for protein scaffolds that are compatible with the geometry. The protein structure is then carved out to accommodate the cluster. Coldren et al. incorporated a [4Fe–4S] cluster in *Escherichia coli* HiPIP thioredoxin by selecting and mutating four residues in the protein core to cysteine so that cluster coordination is sterically possible. Importantly, thioredoxin does not contain a [4Fe–4S] cluster, or other cofactors. Using the DEZYMER algorithm, the appropriate residues were selected and mapped out to the X-ray structure. In addition to the four cysteine residues, several stabilizing mutations were made to allow [4Fe–4S] incorporation while maintaining the native protein fold (Coldren et al., 1997).

Grzyb et al. further extended this approach by designing a four-helix bundle around a disembodied [4Fe–4S] cluster, rather than relying on the PDB as a source of protein scaffolds. Starting with an analysis of bound clusters within protein structures, they focused on the binding motif of tryptophanyl-tRNA synthase (PDB 2G46) and discovered that the cluster, including the cysteine ligands and [4Fe–4S] atoms, had inherent tetrahedral symmetry. The first coordination sphere could be described as two CXXC motives within a helical structure and related by C_2 symmetry. The CXXC motifs were extended into two parallel helices, and the structure of the four-helix bundle was finalized using protCAD to build the backbone of two additional helices and the loops connecting them. Finally, the best sequence for the backbone structure was selected using protCAD for core residues, and Rosetta for surface exposed amino acids. The resulting protein was experimentally characterized. UV–CD analysis upon [4Fe–4S] cluster incorporation showed an increase in alpha-helical structure compared to the apoprotein, indicating higher stability (Grzyb et al., 2010). The metal-first approach used here introduced an alternative method for [4Fe–4S] protein design, described in detail in a recent review (Nanda et al., 2016). A direct comparison of proteins built on this principle and designed either rationally or computationally highlights benefits and drawbacks of the two methods: the redox-chain maquette (RCM) was designed empirically, while the coiled-coil iron–sulfur protein (CCIS) was designed computationally. Both were able to incorporate [4Fe–4S] clusters; however, RCM was not able to fully account for the spatial constraints and was prone to aggregation, while the orientation of the cysteine ligands in CCIS diverged from the initially specified rotamers (Grzyb et al., 2012).

To facilitate electron transfer to and from catalytic sites embedded within proteins, or across cellular membranes, iron–sulfur clusters are arranged in chains of two or more. The design of proteins containing multiple clusters, thus, is an important goal in protein design. Coiled-coil proteins are ideally suited to accommodate isostructural cofactors because they are defined by a repetitive helical structure encoded in the "heptad repeat" (Crick, 1952). Tweaks in the sequence of the heptad repeat, in turns, help determining whether a single helix will assembly in dimers, trimers, tetramers, or higher order oligomers (Woolfson et al., 2015).

Our group has exploited the coiled-coil structure to design a series of three-helix bundles, derived from a de novo designed structure (Ogihara et al., 2001), which can house two iron–sulfur clusters within their hydrophobic core (Roy, Sarrou, Vaughn, Astashkin, & Ghirlanda, 2013; Roy et al., 2014). The scaffold chosen as a starting point is a homodimeric three-helix bundle (domain-swapped dimer, DSD, PDB 1G6U), formed by helix–loop–helix peptides and containing a hydrophobic core formed exclusively by leucine residues (Ogihara et al., 2001). One of the helices is approximately twice the length of the other, and the two peptides pair up by these long helices; the second helices then dock against this helical dimer, forming a three-helix bundle with internal twofold symmetry. To design a [4Fe–4S] cluster-binding site, we grafted a cluster and its coordinating cysteine ligand into the leucine-rich core. We used the cluster contained in *Thermotoga maritima* tryptophanyl tRNA synthase (PDB entry 2G36), in which the four cysteine ligands to the [4Fe–4S] cofactor are part of a C···C···CXXC sequence motif (Han et al., 2010); analysis of the structure using Pymol shows that the CXXC motif is in helical conformation, as assessed by backbone dihedral angles, while the other two ligands are part of loops and are in a nonhelical conformation (Grzyb et al., 2010). However, the distances between the Cα of the ligands are compatible with the interhelix distance in DSD. The model of our cluster-binding protein was built in Pymol by overlaying the cluster and its ligands onto the structure of DSD. First, the CXXC motif of 2G36 was overlaid with LXXL motifs of DSD: because of compatible distances, the Cα atoms of the two remaining cysteines overlapped with the Cα atoms of leucines on neighboring helices. The process was manually repeated for LXXL motives located in the long helix as well as the short helix of DSD, resulting in placement of the cluster in either of two layers within the core of the coiled coil. Both alignments—one resulting in intercluster distances of 30 Å, and the second in intercluster distances of 12 Å—are permissible, and were explored in two separate designs (Fig. 1). Once the Cα of C and L were overlayed, using the Pairwise alignment function in Pymol, we mutated the L residues of DSD to C and

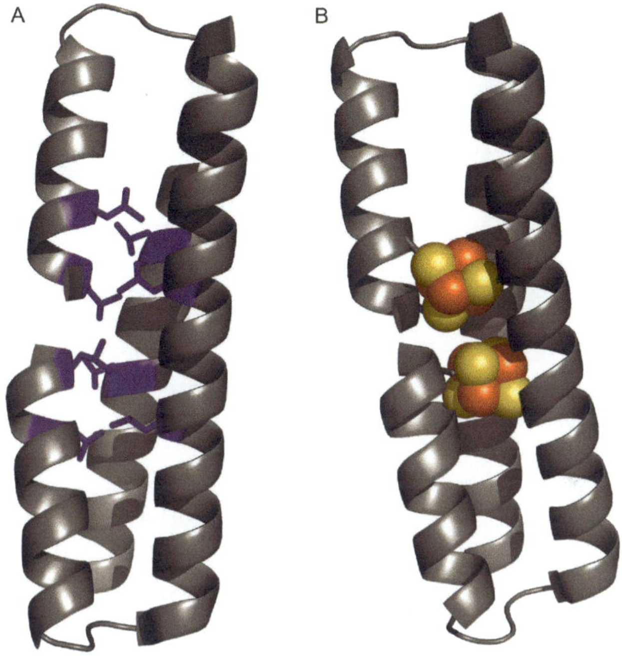

Fig. 1 (A) Model of the DSD family of iron–sulfur cluster-binding site. (B) Cluster ligation to the cluster-binding site of DDS.

selected the appropriate rotamers to coordinate the iron atoms in the cluster. Bonds between the sulfurs of 2G36 and the irons on the cluster were swapped to the sulfurs on the newly inserted cysteines, and the resulting model was subjected to minimization.

This procedure is compatible with the pseudotetrahedral geometry preference of the iron–sulfur cluster. Notably, the symmetry of the bundle results in a sequence motif reminiscent of the arrangement found in ferredoxins: three cysteines are located on one monomer, and the fourth on the other monomer. The symmetry of the bundle also results in duplication of the binding site on the other end of the bundle.

The mutations inserted in DSD resulted in destabilization of the apo structure compared to the parent peptide; however, cluster incorporation increased the stability of the designed peptide by substituting metal coordination for hydrophobic interactions. Analyzing the models with Q-SiteFinder29 revealed the presence of a 58 $Å^3$ cavity in the core of DSD-4Cys corresponding to the cluster-binding site completely occupied by the [4Fe–4S] cluster in the model of DSD-bis[4Fe–4S].

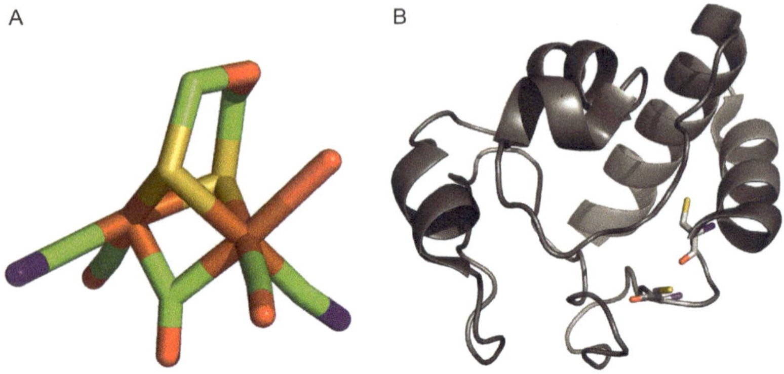

Fig. 2 The structure of (A) H-cluster of hydrogenase obtained from PDB 3C8Y. (B) Apo cytochrome *c* obtained from PDB 1HRC, showing two cysteine residues for incorporation of a [Fe–Fe]-hydrogenase mimic.

DSD-bis[4Fe–4S] conserved the oligomeric state and overall structure of the parent peptide. Pulse electron–electron double-resonance spectroscopy was used to assess the distance between the clusters by measuring magnetic dipole interactions. The cluster–cluster distance is 29–34 Å, in agreement with the estimated distance of ~30 Å in the model, and the clusters are weakly coupled (Roy et al., 2013).

The heptad repeat pattern of coiled coils also allows to move the cluster-binding sites along the longitudinal axis of the DSD scaffold, resulting in an intercluster distance shortened from 30 Å to within 12 Å of one another, closer to the distance of 15 Å or less observed in natural proteins (Fig. 2B); using these principles we designed a second peptide, DSD-Fdm (Roy et al., 2014). The cluster-binding motif was moved one heptad closer to the center of the bundle by DSD-Fdm is able to participate in electron transfer processes, and show promising features for applications in technology. DSD-Fdm is able to transfer energy from photosensitized dyes, a process reminiscent of the electron-accepting ability of ferredoxins from photosystem I. The reduced form of DSD-Fdm acts as a reducing agent for cytochrome c_{550}. Both clusters transfer one electron each as shown by the stoichiometry of the reaction.

3. [3Fe–4S] CLUSTERS

Proteins containing [3Fe–4S] clusters can be obtained from the sequence of [4Fe–4S]-binding proteins by substituting one of the ligand cysteines with serine or leucine, resulting in destabilization of the cubane-type cluster. However, incorporation of the less stable [3Fe–4S] cluster is in competition with the more stable [4Fe–4S], and usually is incomplete. This idea was demonstrated in 33 amino acid model peptide containing the cluster binding site of dimethyl sulfoxide (DMSO) reductase subunit B (DmsB) as scaffold. Mutations in the binding motif of the peptide have indicated which parameters affect the cluster type and its coordination. For example, C27S lead to the incorporation of [3Fe–4S] cluster in a relatively large yield. However, longer incubation times resulted in the formation of [4Fe–4S] clusters as evidenced by EPR, suggesting that the [3Fe–4S] cluster is an intermediate species in the formation pathway of the more stable [4Fe–4S] cluster (Hoppe, Pandelia, Gärtner, & Lubitz, 2011). Our group utilized the DSD scaffold to see whether mutations to polar (Ser) or hydrophobic (Leu) residues could stabilize [3Fe–4S] clusters to different extent. The two variants, DSD-3CSer and DSD-3CLeu, exhibit similar stability to chemical denaturation; in both cases, incorporation of iron–sulfur clusters results in increase in protein stability. In contrast with previous work on peptides derived from DmsB, which showed incorporation only when a coordinating Cys is exchanged with polar residues, cluster incorporation proceeds readily in both cases, although to a limited extent (Sommer, Roy, Astashkin, & Ghirlanda, 2015)

Protein-bound [4Fe–4S] clusters can be converted to heterometallic Fe–S clusters: in wild-type *Pyrococcus furious* ferredoxin, as well as two other mutants D14C and D14H, [$AgFe_3S_4$] cluster was formed by reducing the [3Fe–4S] cluster and incubating with $AgNO_3$. A crystal structure of the construct showed that silver (I) is coordinated to the thiolate group of Cys14 and is part of the cluster. The oxidized form of the [$AgFe_3S_4$] D14C is shown to be intact in presence of dioxygen, whereas the wild-type variant converted to [Fe_3S_4] when exposed to air (Martic, Jakab-Simon, Haahr, Hagen, & Christensen, 2013).

4. [2Fe–2S] CLUSTERS

Iron–sulfur proteins containing [2Fe–2S] clusters generally serve as electron carriers, and present significant variations in their first coordination shell: in ferredoxins, each iron atom is coordinated by one inorganic sulfur and two cysteine thiolates in a tetrahedral geometry, whereas in Rieske proteins one iron is coordinated by two histidine residues (Liu et al., 2014). The cluster is typically surrounded by hydrophobic residues and is located near the surface of the protein to facilitate electron transfer with redox partners, as well as to stabilize the cluster. Similarly, to the cubane type, the [2Fe–2S] clusters also span a broad range of redox potential.

The simplest bioinspired model of [2Fe–2S] cluster was obtained by simply combining ferric ions and sulfide with glutathione, resulting in a 4 glutathione:1 iron–sulfur cluster complex stable in aqueous environments. The stability of the binding pocket arises from intermolecular salt bridge and hydrogen bond formation that leads to a tetrameric species soluble in aqueous solutions (Qi et al., 2013).

The hydrophobicity of the amino acids surrounding a cluster, as well as the ability to form hydrogen bonds, modulates both the cluster's redox potential and its identity. Elegant work on *TmHydC*, a [2Fe–2S] containing subunit of the trimeric [Fe–Fe]-hydrogenases from *T. maritima*, showed that bulky hydrophobic amino acids close to the bridging sulfur in the cluster as well as eliminating hydrogen bonding decrease the redox potential significantly (Birrell, Laurich, Reijerse, Ogata, & Lubitz, 2016). Small alterations in amino acids adjacent to [2Fe–2S] clusters are shown to lead to cluster's functional changes. SoxR, a [2Fe–2S] transcriptional factor in *E.coli*, acts as a sensor of oxidative stress (Fujikawa, Kobayashi, Tsutsui, Tanaka, & Kozawa, 2017).

Rieske-type [2Fe–2S] clusters are ligated by two histidine-imidazole on one iron and by two cysteine thiolates on the second. Histidine ligation results in pH-dependent redox potentials because the protonation state of the imidazole leads to proton-coupled electron transfer (PCET; Saouma, Kaminsky, & Mayer, 2012). The Rieske-type redox potential spans from −150 to 490 mV depending on the protein environment of the cluster (Iwasaki, Fukazawa, & Miyajima-Nakano, 2012).

A novel category of [2Fe–2S] clusters was discovered recently in MitoNEET, a protein involved in regulating metabolism and found in the outer mitochondrial membrane. This protein presents an unusual coordination shell containing three cysteines and one histidine, and displays PCET features (Bak, Zuris, Paddock, Jennings, & Elliott, 2009). The role of the coordinating histidine in imparting PCET has been ascertained through small molecule mimics (Bergner et al., 2017).

A unique [2Fe–2S] center forms the catalytic site of the [Fe–Fe] hydrogenases. These enzymes catalyze the reversible reduction of protons with high efficiency, displaying turnover numbers as high as $10^4\,s^{-1}$, a feat unparalleled by small-molecule catalysts for this reaction. The catalytic site, known as the H-cluster, is composed by a [4Fe–4S] cluster and by a [Fe–Fe] site coordinated by a nonproteinogenic azadithiolate bridging ligand, known as the [2Fe–2S] subcluster. The [2Fe–2S] subcluster is coordinated by carbon monoxide (CO) and cyanide (CN) ligands, with the [Fe–Fe] site anchored to the protein through a cysteine (Cys) that bridges the iron–sulfur cluster and one iron, designated as proximal; depending on the redox states sampled during the catalytic cycle, one of the CO ligands bridges the two metals. During the catalytic cycle, the distal iron is thought to undergo a conformational change to assume an inverted square pyramidal conformation, or "rotated" state, which is poised for transfer of electrons through the proximal iron to a coordinated proton (Esselborn et al., 2013; Singleton, Reibenspies, & Darensbourg, 2010). The role of the protein matrix is to stabilize the catalytically active "rotated" conformation of the distal iron and to facilitate the reaction by rapidly transferring electrons and protons to the active site (Mulder et al., 2011). The azadithiolane-bridged [2Fe–2S] subcluster obtained by chemical synthesis is inactive in isolation, but spontaneous assembly within the protein scaffold results in an intact, functional H-cluster as confirmed by spectroscopic evidence and activity assays (Esselborn et al., 2013; Singleton et al., 2010).

The protein scaffold is crucial in stabilizing the catalytically active conformation of the [2Fe–2S] scaffold. The design of protein-based artificial hydrogenases, however, presents unusual challenges: the diiron site is coordinated by nonproteinogenic ligands, and is linked to the protein mainly through noncovalent interactions, with only a single cysteine shared with the cubane cluster component of the H-cluster providing a covalent link to the protein. To circumvent this problem, several groups have sought to utilize naturally occurring cysteines rather than the azathiolane-bridging ligand to coordinate and anchor the diiron hexacarbonyl site within a protein scaffold. Boc-protected L-cysteine can reconstitute the diiron site starting from an inorganic precursor, $[Fe_3(CO)_{12}]$, obtaining two stereoisomers; both were inactive (He et al., 2004). Preorganization of the cysteine ligands to restrict the conformational space available was obtained by utilizing a CXXC motif within a model helix, resulting in the formation of a structurally intact albeit inactive diiron hexacarbonyl center (Jones, Lichtenstein, Dutta, Gordon, & Dutton, 2007). The observation that CXXC motifs occur in *c*-type cytochromes led the Hayashi group to incorporate the diiron hexacarbonyl site into apo cytochrome *c*, resulting in a hybrid protein that catalyzes proton reduction in the presence of $Ru(bpy)_3$ as photosensitizer, and ascorbate as sacrificial electron donor (Fig. 2); turnover numbers of approximately 80 are observed over 2 h in water at pH 4.7 (Sano et al., 2011).

Our group has taken a different approach to anchoring the active site to the peptide scaffold by using unnatural dithiol amino acids, which can be incorporated directly into peptide sequences by solid-phase peptide synthesis. Using this strategy we designed a hydrogenase mimic, HDt1, by incorporating an artificial amino acid containing a 1,3-dithiol moiety into an alpha helical model peptide scaffold, in which the dithiol amino acid coordinates a functional (μ-S–$(CH_2)_3$–S)[$Fe_2(CO)_6$] cluster (Fig. 3). The [2Fe–2S] cluster was incorporated into purified HDt1 by direct reaction with $Fe_3(CO)_{12}$ under strictly anaerobic conditions, and spectroscopic characterization verified the integrity of the HDt1-[$Fe_2(CO)_6$] complex. HDt1 catalyzed photo-induced hydrogen production with a TON of 84 in 2.30 h (Roy, Madden, & Ghirlanda, 2012).

We used a related approach to generate an artificial hydrogenase within an existing protein scaffold: StrepHyd1 is obtained by tethering an organometallic diiron hexacarbonyl center to biotin, and reconstituting it into streptavidin. First, a hydrogenase mimic is modified by chemical biotinylation, then the complex is incorporated readily into streptavidin; spectroscopic studies show features typical of intact diiron hexacarbonyl centers. StrepHyd1 catalyzes hydrogen production under photocatalytic conditions with enhanced efficiency compared to the isolated diiron hexacarbonyl center (TON of 350

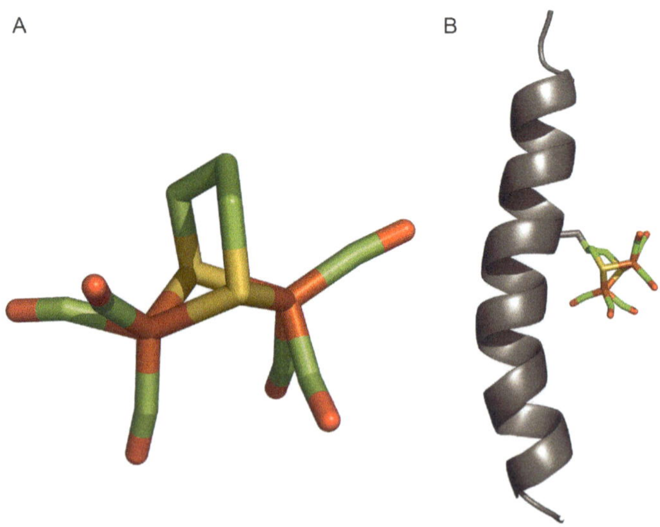

Fig. 3 Structure of the (A) hydrogenase [2Fe–2S] subcluster in the active site (B) model of peptide 1-[$Fe_2(CO)_6$].

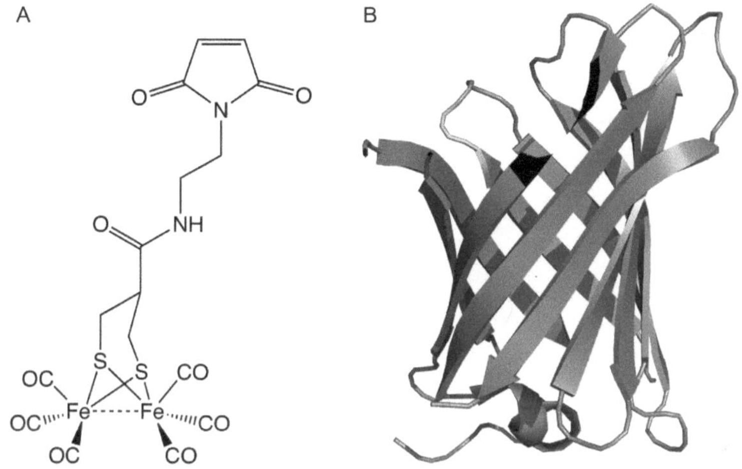

Fig. 4 (A) Maleimide-modified synthetic diiron catalyst. (B) Protein scaffold of nitrobindin. The structure obtained from PDB 3WJB.

vs 60). The crystal structure of StrepHyd1 provides blueprints for the design of improved versions by selective mutation of residues in contact with the active site (Roy, 2014).

Covalent attachment was used to anchor $(\mu\text{-S})_2Fe_2(CO)_6$ to a cysteine residue of nitrobindin (NB) (Fig. 4). The embedded diiron complex catalyzes photo-induced H_2 production with TON up to 130 for 6 h irradiation. NB has a rigid cavity in its β-barrel which provides an appropriate host for the diiron cluster (Onoda, Kihara, Fukumoto, Sano, & Hayashi, 2014).

5. [Fe_8S_7] CLUSTERS

Nitrogenase is a multicomponent enzyme system that fixes atmospheric dinitrogen into biologically available ammonia at ambient temperatures and pressures, using an unusual Fe–S cluster at the active site, either containing Mo and Fe, or V and Fe, or Fe only (Djurdjevic, Einsle, & Decamps, 2017; Hu & Ribbe, 2016a, 2016b). The [MoFe] nitrogenase system includes a homodimeric reductase component (Fe protein) and a tetrameric catalytic component (MoFe protein) (Hu & Ribbe, 2016a). Upon association of the two components, an electron is transferred from the Fe protein's [4Fe–4S] cluster to the MoFe protein's M-cluster via the intermediate MoFe protein's P-cluster (Fig. 5) (Katz, Owens, & Tezcan, 2016). The

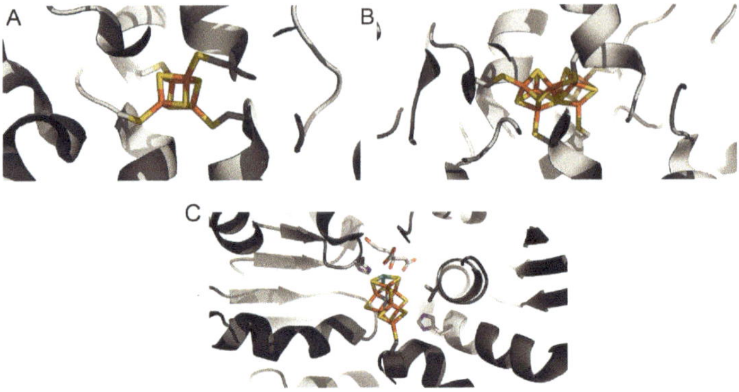

Fig. 5 (A) Organometallic centers of wild-type nitrogenase include a cubane cluster [Fe_4S_4] coordinated by four cysteines, (B) a P-cluster [Fe_8S_7] coordinated by six cysteines, and (C) an M-cluster [homocitrate:$MoFe_7S_9C$] coordinated by cysteine and histidine.

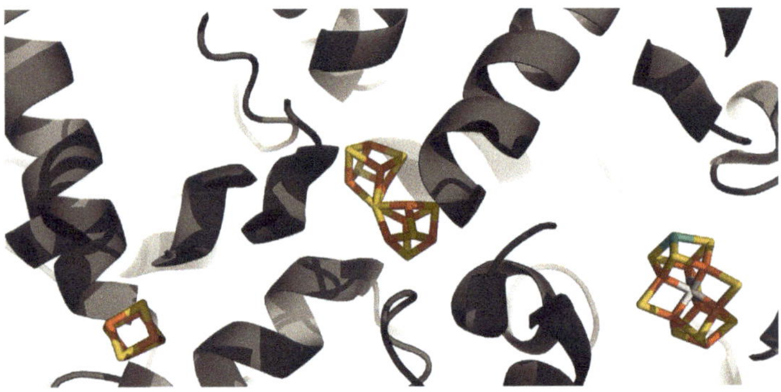

Fig. 6 Association of the reductase domain (Fe protein) with the catalytic domain (MoFe protein) initiates electron transfer to the M-cluster via the intermediate P-cluster.

cofactors which facilitate this electron transfer include the cubane-type [Fe–S] cluster, P-cluster [Fe_8S_7], and M-cluster [homocitrate:$MoFe_7S_9C$] (Fig. 6). Within the catalytic MoFe protein, the M-cluster is ligated by an organic homocitrate and two residues: C275 and H442. Structural homologs to the M-cluster include the native V-cluster [VFe_7S_9C] (substitution of molybdenum with vanadium) and symmetric L-cluster [Fe_8S_9C].

In addition to reducing dinitrogen, nitrogenase has demonstrated the ability to reduce single-carbon substrates into hydrocarbons of various lengths. Wild-type M-nitrogenase was demonstrated to reduce carbon dioxide (CO_2) to formate (HCO_2^-), CO, and methane (CH_4) (Khadka et al., 2016). Moreover, wild-type V-nitrogenase was demonstrated to facilitate carbon–carbon bond formation by reducing CO_2 into hydrocarbon chains (C_1–C_4) (Rebelein, Hu, & Ribbe, 2015). Other proven single-carbon substrates include cyanide (CN^-) and even CO, which has been demonstrated as a reversible, noncompetitive inhibitor of the MoFe protein (Spatzal, Perez, Einsle, Howard, & Rees, 2014). To demonstrate enzyme catalytic functionality, the reduction of acetylene (C_2H_2) to ethylene (C_2H_4) can be monitored (Tanifuji et al., 2015).

Remarkably, the nitrogenase cofactors can be extracted into an organic solvent and catalyze the reduction of monocarbon compounds to C_1–C_4 hydrocarbons in the presence of a reductant and a proton source. This discovery suggests the possibility to develop nitrogenase-based catalysts by inserting the cofactors into a protein scaffold, thus allowing reactivity in water (Tanifuji et al., 2015).

6. PROTOCOLS

6.1 Diiron Hexacarbonyl Cluster Incorporation Into Peptide (Roy et al., 2012)

1. Dissolve 1.2 μmol of peptide in 500 μL of degassed water in a round-bottom flask.
2. Dissolve 8.8 μmol of $Fe_3(CO)_{12}$ in 1.7 mL of methanol and add to round-bottom flask.
3. Reflux under argon for 2 h and observe a green to red color change.
4. Evaporate solvent under reduced pressure to yield a brown solid, and then redissolve in 10 mL degassed water.
5. Centrifuge solution and remove supernatant containing apo and holo peptide.
6. HPLC purify supernatant to obtain pure peptide-bound diiron cluster.
7. Verify incorporation using ESI-MS.

6.2 [4Fe–4S] Cluster Incorporation Into Peptide

The procedure for cluster incorporation has been well established and summarized later. In anaerobic conditions, combination of β-mercaptoethanol, ferric chloride, sodium sulfide, and peptide allows for the spontaneous formation of the desired cluster. Initially, β-mercaptoethanol ligates the cluster; however, coordination to the designed cysteines is more entropically favorable. Therefore due to ligand exchange, the cluster is incorporated into the protein scaffold and can be verified by a number of spectroscopic methods

(Roy et al., 2013). This protocol can also be used to incorporate [3Fe–4S] clusters, albeit with a lower yield than the [4Fe–4S] clusters (Sommer et al., 2015).

1. In an anaerobic chamber (5% H_2/95% N_2), dissolve peptide in 100 m*M* Tris, pH 8.5, to a final concentration of 150 μ*M*.
2. Degas and add the following reagents sequentially, pausing for 20 min after each addition to a final volume of 2.5 mL.
   a. 0.8% (v/v) β-mercaptoethanol.
   b. 4 m*M* ferric chloride ($FeCl_3$) added dropwise.
   c. 4 m*M* sodium sulfide (Na_2S)
3. Incubate the dark brown/black solution overnight at 4°C.
4. In anaerobic conditions, desalt the solution using a GE Healthcare PD10 G25 column equilibrated with 100 m*M* Tris, pH 7.5, obtaining a final volume of 3.5 mL.

6.3 EPR [4Fe–4S] Sample Preparation

The incorporated clusters can be identified using EPR spectroscopy. A number of protocols are available for sample preparation, and we have had success using the following procedure. Due to antiferromagnetic coupling the $[4Fe–4S]^{2+}$ cluster is not EPR active, therefore, reduction to $[4Fe–4S]^{1+}$ is needed to obtain a ½ spin state.

1. Following desalting and maintaining anaerobic conditions, check the UV–Vis spectrum of the incorporated cluster
   a. A broad absorption at around 410 nm is indicative of the $[4Fe–4S]^{2+}$ cluster.
   b. Reduction by sodium dithionite results in elimination of 410 nm absorbance and shows a redox-active cluster (Nakamaru-Ogiso, Yano, Ohnishi, & Yagi, 2002).
2. EPR samples should be concentrated to a final volume of 150 μL containing 10 absorbance units at 410 nm.
3. Reduced samples are prepared by adding the following reagents:
   a. Glycerol (20%) as a cryoprotectant
   b. 300 m*M* glycine
   c. 30 m*M* sodium dithionite ($Na_2S_2O_4$)
4. Solid sodium dithionite can be dissolved directly into glycine buffer
5. Oxidized samples can be prepared simultaneously by dilution to the same protein concentration.

6. Concentrate samples to 150 μL using a spin concentrator.
7. Transport samples to quartz EPR tubes and flash freeze in liquid nitrogen.

6.4 Hydrogen Production With Diiron Hexacarbonyl

The ability and efficiency of hydrogen gas generation by a specific catalyst can be quantified using gas chromatography. A sealed cuvette contains the reagents required for catalysis, and the reaction occurs in the presence of light. The cuvette contents in a citrate buffer include a known amount of catalyst, a photosensitizer $Ru(bpy)_3$, and a sacrificial electron donor (sodium ascorbate) to replenish the photosensitizer. Later is a protocol using a GC (SRI Instruments Model 310C), with a 5 Å molecular sieve column and a thermal conductivity detector. In order to calculate the amount of H_2 produced, a standard curve must be obtained using known amounts of hydrogen gas.

1. In an anaerobic chamber (5% H_2/95% N_2), add 10 μ*M* peptide-$[Fe_2(CO)_6]$, 150 μ*M* $Ru(bpy)_3$, and 50 m*M* sodium ascorbate to a 50 m*M* citrate buffer pH 4.5 and final volume of 400 μL in a 1-mm path length airtight cuvette with rubber septum.
2. Degas cuvette extensively with argon for at least 30 min to remove any residual gasses.
3. Illuminate the cuvette with a 450 W xenon lamp with 400 nm cutoff filter at a distance providing 1100 W/m^2 for the duration of the experiment.
4. Prepare H_2 standard curve for GC(SRI Instruments Model 310C, 5 Å molecular sieve column with thermal conductivity detector and argon carrier gas) by injecting varying amounts of 1% H_2/99% N_2 gas mix.
5. During the course of the experiment to measure H_2 generation, degas a gas-tight syringe with argon and inject 100 uL of argon through the septum into the cuvette.
6. Without removing the syringe, mix headspace gas of cuvette and remove 100 μL.
7. Inject 100 μL of headspace gas into GC and calculate H_2 concentration using standard curve.
8. Continue measurements at regular intervals (1 h) until H_2 production ceases.
9. Turnover number can be calculated after normalizing for diluted headspace due to argon injections.

6.5 Electrochemical Measurements

Direct electrochemistry of ferredoxin type proteins is challenging and tricky, due to the weak interaction between the electrode surface and protein. In order to enhance the interaction of protein and electrode surface and improve the quasi-reversible signal, we add 3–5 m*M* neomycin to the electrolyte solution. Later is a protocol for doing electrochemical measurements with ferredoxin-type proteins. This protocol can be easily adopted for various Fe–S containing proteins and peptides:

1. In an anaerobic chamber (5% H_2/95% N_2), 100 m*M* Tris, 100 m*M* NaCl into an electrochemical cell, and adjust the pH to the desired value
2. Degas buffers extensively with argon following by incubation for 2 days in the anaerobic chamber prior to use
3. Polish the working electrode with 1 um alumina for 5 min following by sonication for 10 min
4. Add 5–10 μ*M* peptide to the buffer
5. Set up the three electrode cell by placing the reference, counter, and working electrode in the cell
6. Potential range depends on the experiment, typical scan rate is 100 mV s^{-1}

Note: 5 m*M* neomycin can be added to the buffer to stabilize and enhance the interaction of the peptide with electrode surface (Roy et al., 2014).

REFERENCES

Antonkine, M. L., Koay, M. S., Epel, B., Breitenstein, C., Gopta, O., Gärtner, W., et al. (2009). Synthesis and characterization of de novo designed peptides modelling the binding sites of [4Fe-4S] clusters in photosystem I. *Biochimica et Biophysica Acta—Bioenergetics*, *1787*(8), 995–1008.

Bak, D. W., Zuris, J. A., Paddock, M. L., Jennings, P. A., & Elliott, S. J. (2009). Redox characterization of the FeS protein MitoNEET and impact of thiazolidinedione drug binding. *Biochemistry*, *48*(43), 10193–10195.

Beinert, H. (1997). Iron-sulfur clusters: Nature's modular, multipurpose structures. *Science*, *277*(5326), 653–659.

Bergner, M., Dechert, S., Demeshko, S., Kupper, C., Mayer, J. M., & Meyer, F. (2017). Model of the MitoNEET [2Fe – 2S] cluster shows proton coupled electron transfer. *Journal of the American Chemical Society*, *139*(2), 701–707.

Birrell, J. A., Laurich, C., Reijerse, E. J., Ogata, H., & Lubitz, W. (2016). Importance of hydrogen bonding in fine tuning the [2Fe-2S] cluster redox potential of HydC from Thermotoga maritima. *Biochemistry*, *55*(31), 4344–4355.

Boyken, S. E., Chen, Z., Groves, B., Langan, R. A., Oberdorfer, G., Ford, A., et al. (2016). De novo design of protein homo-oligomers with modular hydrogen-bond network-mediated specificity. *Science*, *352*(6286), 680–687.

Calhoun, J. R., Nastri, F., Maglio, O., Pavone, V., Lombardi, A., & DeGrado, W. F. (2005). Artificial diiron proteins: From structure to function. *Biopolymers—Peptide Science Section*, *80*(2–3), 264–278.

Chakraborty, S., Iranzo, O., Zuiderweg, E. R. P., & Pecoraro, V. L. (2012). Experimental and theoretical evaluation of multisite cadmium(II) exchange in designed three-stranded coiled-coil peptides. *Journal of the American Chemical Society*, *134*(14), 6191–6203.

Coldren, C. D., Hellinga, H. W., & Caradonna, J. P. (1997). The rational design and construction of a cuboidal iron-sulfur protein. *Proceedings of the National Academy of Sciences*, *94*(13), 6635–6640.

Crick, F. H. C. (1952). The packing of a-helices: Simple coiled-coils. *Acta Cryst*, *5*(6), 581.

Der, B. S., MacHius, M., Miley, M. J., Mills, J. L., Szyperski, T., & Kuhlman, B. (2012). Metal-mediated affinity and orientation specificity in a computationally designed protein homodimer. *Journal of the American Chemical Society*, *134*(1), 375–385.

Dey, A., Jenney, F. E., Adams, A. W. W., Bbini, E., Takahashi, Y., & Fekuyama, K. (2007). Solvent tuning of electrochemical potentials in the active sites of HiPIP versus ferredoxin. *Science*, *318*(November), 1464–1469.

Djurdjevic, I., Einsle, O., & Decamps, L. (2017). *Nitrogenase cofactor: Inspiration for model chemistry*. [Chemistry—An Asian Journal].

Eck, R. V., & Dayhoff, M. O. (1966). Evolution of the structure of ferredoxin based on living relics of primitive amino acid sequences. *Science*, *152*(3720), 363–366.

Esselborn, J., Lambertz, C., Adamska-Venkatesh, A., Simmons, T., Berggren, G., Noth, J., et al. (2013). Spontaneous activation of [FeFe]-hydrogenases by an inorganic [2Fe] active site mimic. *Nature Chemical Biology*, *9*(10), 607–609.

Fujikawa, M., Kobayashi, K., Tsutsui, Y., Tanaka, T., & Kozawa, T. (2017). Rational tuning of superoxide sensitivity in SoxR, the [2Fe-2S] transcription factor: Implications of species-specific lysine residues. *Biochemistry*, *56*(2), 403–410.

Ghirlanda, G., Osyczka, A., Liu, W., Antolovich, M., Smith, K. M., Dutton, P. L., et al. (2004). De novo design of a D2-symmetrical protein that reproduces the diheme four-helix bundle in cytochrome bc1. *Journal of the American Chemical Society*, *126*(26), 8141–8147.

Grzyb, J., Xu, F., Nanda, V., Łuczkowska, R., Reijerse, E., Lubitz, W., et al. (2012). Empirical and computational design of iron-sulfur cluster proteins. *Biochimica et Biophysica Acta—Bioenergetics*, *1817*(8), 1256–1262.

Grzyb, J., Xu, F., Weiner, L., Reijerse, E. J., Lubitz, W., Nanda, V., et al. (2010). De novo design of a non-natural fold for an iron-sulfur protein: Alpha-helical coiled-coil with a four-iron four-sulfur cluster binding site in its central core. *Biochimica et Biophysica Acta—Bioenergetics*, *1797*(3), 406–413.

Han, G. W., Yang, X. L., McMullan, D., Chong, Y. E., Krishna, S. S., Rife, C. L., et al. (2010). Structure of a tryptophanyl-tRNA synthetase containing an iron-sulfur cluster. *Acta Crystallographica Section F: Structural Biology and Crystallization Communications*, *66*(10), 1326–1334.

He, C., Wang, M., Zhang, X., Wang, Z., Chen, C., Liu, J., et al. (2004). An unusual cyclization in a bis(cysteinyl-S) diiron complex related to the active site of Fe-only hydrogenases. *Angewandte Chemie—International Edition*, *43*(27), 3571–3574.

Hoppe, A., Pandelia, M. E., Gärtner, W., & Lubitz, W. (2011). [Fe_4S_4] and [Fe_3S_4]-cluster formation in synthetic peptides. *Biochimica et Biophysica Acta—Bioenergetics*, *1807*(11), 1414–1422.

Hosseinzadeh, P., & Lu, Y. (2016). Design and fine-tuning redox potentials of metalloproteins involved in electron transfer in bioenergetics. *Biochimica et Biophysica Acta—Bioenergetics*, *1857*(5), 557–581.

Hu, Y., & Ribbe, M. W. (2016a). Biosynthesis of the Metalloclusters of Nitrogenases. *Annual Review of Biochemistry*, *85*(1), 455–483.

Hu, Y., & Ribbe, M. W. (2016b). Nitrogenases—A tale of carbon atom(s). *Angewandte Chemie—International Edition*, *55*(29), 8216–8226.

Iwasaki, T., Fukazawa, R., & Miyajima-Nakano, Y. (2012). Dissection of hydrogen bond interaction network around an iron-sulfur cluster. *Journal of the American Chemical Society, 134*(48), 19731–19738.

Jones, A. K., Lichtenstein, B. R., Dutta, A., Gordon, G., & Dutton, P. L. (2007). Synthetic hydrogenases: Incorporation of an iron carbonyl thiolate into a designed peptide. *Journal of the American Chemical Society, 129*(48), 14844–14845.

Katz, F. E. H., Owens, C. P., & Tezcan, F. A. (2016). Electron transfer reactions in biological nitrogen fixation. *Israel Journal of Chemistry, 56*(9–10), 682–692.

Khadka, N., Dean, D. R., Smith, D., Hoffman, B. M., Raugei, S., & Seefeldt, L. C. (2016). CO_2 reduction catalyzed by nitrogenase: pathways to formate, carbon monoxide, and methane. *Inorganic Chemistry, 55*(17), 8321–8330.

Lill, R. (2009). Function and biogenesis of iron–sulphur proteins. *Nature, 460*(7257), 831–838.

Liu, J., Chakraborty, S., Hosseinzadeh, P., Yu, Y., Tian, S., Petrik, I., et al. (2014). Metalloproteins containing cytochrome, iron–sulfur, or copper redox centers. *Chemical Reviews, 114*(8), 4366–4469.

Lupas, A. N., & Bassler, J. (2017, February). Coiled coils—A model system for the 21st century. *Trends in Biochemical Sciences, 42*(2), 130–140.

Martic, M., Jakab-Simon, I. N., Haahr, L. T., Hagen, W. R., & Christensen, H. E. M. (2013). Heterometallic [$AgFe_3S_4$] ferredoxin variants: Synthesis, characterization, and the first crystal structure of an engineered heterometallic iron-sulfur protein. *Journal of Biological Inorganic Chemistry, 18*(2), 261–276.

Meyer, J. (2008). Iron-sulfur protein folds, iron-sulfur chemistry, and evolution. *Journal of Biological Inorganic Chemistry, 13*(2), 157–170.

Moutevelis, E., & Woolfson, D. N. (2009). A periodic table of coiled-coil protein structures. *Journal of Molecular Biology, 385*(3), 726–732.

Mulder, D. W., Shepard, E. M., Meuser, J. E., Joshi, N., King, P. W., Posewitz, M. C., et al. (2011). Insights into [FeFe]-hydrogenase structure, mechanism, and maturation. *Structure, 19*, 1038–1052.

Mulholland, S. E., Gibney, B. R., Rabanal, F., & Dutton, P. L. (1999). Determination of nonligand amino acids critical to [4Fe-4S]2+/+ assembly in ferredoxin maquettes. *Biochemistry, 38*, 10442–10448.

Nakamaru-Ogiso, E., Yano, T., Ohnishi, T., & Yagi, T. (2002). Characterization of the iron-sulfur cluster coordinated by a cysteine cluster motif(CXXCXXXCX27C) in the Nqo3 subunit in the proton-translocating NADH-quinone oxidoreductase (NDH-1) of Thermus thermophilus HB-8. *Journal of Biological Chemistry, 277*(3), 1680–1688.

Nanda, V., Senn, S., Pike, D. H., Rodriguez-Granillo, A., Hansen, W. A., Khare, S. D., et al. (2016). Structural principles for computational and de novo design of 4Fe-4S metalloproteins. *Biochimica et Biophysica Acta—Bioenergetics, 1857*(5), 531–538.

Ogihara, N. L., Ghirlanda, G., Bryson, J. W., Gingery, M., DeGrado, W. F., & Eisenberg, D. (2001). Design of three-dimensional domain-swapped dimers and fibrous oligomers. *Proceedings of the National Academy of Sciences of the United States of America, 98*(4), 1404–1409.

Onoda, A., Kihara, Y., Fukumoto, K., Sano, Y., & Hayashi, T. (2014). Photoinduced hydrogen evolution catalyzed by a synthetic diiron dithiolate complex embedded within a protein matrix. *ACS Catalysis, 4*(8), 2645–2648.

Qi, W., Li, J., Chain, C. Y., Pasquevich, G. A., Pasquevich, A. F., & Cowan, J. A. (2013). Glutathione-complexed iron–sulfur clusters. Reaction intermediates and evidence for a template effect promoting assembly and stability. *Chemical Communications, 49*(56), 6313.

Rebelein, J. G., Hu, Y., & Ribbe, M. W. (2015). Widening the product profile of carbon dioxide reduction by vanadium nitrogenase. *Chembiochem, 16*(14), 1993–1996.

Roy, A. (2014). *Engineering Oxidoreductases: Towards artificial hydrogenases*. Tempe, Arizina: Arizona State University.

Roy, A., Madden, C., & Ghirlanda, G. (2012). Photo-induced hydrogen production in a helical peptide incorporating a [FeFe] hydrogenase active site mimic. *Chemical Communications*, *48*(79), 9816–9818.

Roy, A., Sarrou, I., Vaughn, M. D., Astashkin, A. V., & Ghirlanda, G. (2013). De novo design of an artificial Bis[4Fe-4S] binding protein. *Biochemistry*, *52*, 7586–7594.

Roy, A., Sommer, D. J., Schmitz, R. A., Brown, C. L., Gust, D., Astashkin, A., et al. (2014). A de novo designed 2[4Fe-4S] ferredoxin mimic mediates electron transfer. *Journal of the American Chemical Society*, *136*(49), 17343–17349.

Salgado, E. N., Radford, R. J., & Tezcan, F. A. (2010). Metal-directed protein self-assembly. *Accounts of Chemical Research*, *43*(5), 661–672.

Sano, Y., Onoda, A., Hayashi, T., Ikeda, N., Harada, A., Huang, C.-Y., et al. (2011). A hydrogenase model system based on the sequence of cytochrome c: Photochemical hydrogen evolution in aqueous media. *Chemical Communications*, *47*(29), 8229.

Saouma, C. T., Kaminsky, W., & Mayer, J. M. (2012). Protonation and concerted proton-electron transfer reactivity of a bis-benzimidazolate ligated [2Fe-2S] model for rieske clusters. *Journal of the American Chemical Society*, *134*(17), 7293–7296.

Shiga, D., Funahashi, Y., Masuda, H., Kikuchi, A., Noda, M., Uchiyama, S., et al. (2012). Creation of a binuclear purple copper site within a de novo coiled-coil protein. *Biochemistry*, *51*(40), 7901–7907.

Singleton, M. L., Reibenspies, J. H., & Darensbourg, M. Y. (2010). A Cyclodextrin host/guest approach to a Hydrogenase active site biomimetic cavity. *Journal of the American Chemical Society*, *132*(26), 8870–8871.

Sommer, D. J. o., Roy, A., Astashkin, A., & Ghirlanda, G. (2015). Modulation of cluster incorporation specificity in a de novo iron-sulfur cluster binding peptide. *Biopolymers*, *104*(4), 412–418.

Spatzal, T., Perez, K. A., Einsle, O., Howard, J. B., & Rees, D. C. (2014). Ligand binding to the FeMo-cofactor: Structures of CO-bound and reactivated nitrogenase. *Science*, *345*(6204), 1620–1623.

Tanifuji, K., Lee, C. C., Ohki, Y., Tatsumi, K., Hu, Y., & Ribbe, M. W. (2015). Combining a nitrogenase scaffold and a synthetic compound into an artificial enzyme. *Angewandte Chemie—International Edition*, *54*(47), 14022–14025.

Tebo, A. G., & Pecoraro, V. L. (2015). Artificial metalloenzymes derived from three-helix bundles. *Current Opinion in Chemical Biology*, *25*, 65–70.

Woolfson, D. N. (2017). Coiled-coil design: Updated and upgraded. *Sub-Cellular Biochemistry*, *82*, 35–61.

Woolfson, D. N., Bartlett, G. J., Burton, A. J., Heal, J. W., Niitsu, A., Thomson, A. R., et al. (2015). De novo protein design: How do we expand into the universe of possible protein structures? *Current Opinion in Structural Biology*, *33*, 16–26.

Yu, Y., Cui, C., Liu, X., Petrik, I. D., Wang, J., & Lu, Y. (2015). A designed metalloenzyme achieving the catalytic rate of a native enzyme. *Journal of the American Chemical Society*, *137*(36), 11570–11573.

Zastrow, M. L., Peacock, A. F. A., Stuckey, J. A., & Pecoraro, V. L. (2011). Hydrolytic catalysis and structural stabilization in a designed metalloprotein. *Nature Chemistry*, *4*(2), 118–123.

CHAPTER THREE

In Vitro Studies of Cellular Iron–Sulfur Cluster Biosynthesis, Trafficking, and Transport

Christine Wachnowsky*,†, James A. Cowan*,†,1
*The Ohio State University, Columbus, OH, United States
†The Ohio State Biochemistry Program, The Ohio State University, Columbus, OH, United States
[1]Corresponding author: e-mail address: cowan@chemistry.ohio-state.edu

Contents

Abstract

Iron–sulfur clusters are metal cofactors that comprise the largest class of metalloproteins and are utilized for a wide variety of functions ranging from electron transport to DNA repair. These clusters and their respective cluster-binding proteins are highly conserved and are produced in the mitochondria via an evolutionarily conserved process for export to the cytosol and delivery to other organelles, including the nucleus. Disruption of the biosynthetic pathway results in a number of disease conditions that reflect the essential requirements of cluster function and trafficking within the cell. In vivo studies are limited in their ability to examine the detailed molecular mechanisms of protein–protein interactions, since they often focus on the downstream effects of protein depletion or mutation. As such, in vitro analyses are essential for defining the roles of specific Fe–S proteins in trafficking events and supporting in vivo analyses of disease conditions arising from aberrant Fe–S assembly and trafficking. In this chapter, we describe a variety of methods for the analysis of structure–function relationships in holo Fe–S cluster proteins, as well as monitoring the kinetics and molecular mechanisms of Fe–S cluster transfer.

Methods in Enzymology, Volume 595
ISSN 0076-6879
http://dx.doi.org/10.1016/bs.mie.2017.06.045

1. INTRODUCTION

Many proteins require cofactors for proper function, such as metals or metal complexes, with well-known examples including myoglobin, azurin, and flavodoxin (Cowan, 1997). Iron–sulfur (Fe–S) cluster proteins, in particular, comprise the largest class of metalloproteins (Napier, 2005) and demonstrate unique functional roles that include electron transport, substrate binding and catalysis, structural stabilization, and DNA-mediated charge transfer (Fuss, Tsai, Ishida, & Tainer, 2015; Johnson, Dean, Smith, & Johnson, 2005; Lill & Muehlenhoff, 2008; Qi & Cowan, 2011b). These functions underlie essential biological processes, such as transcriptional and translational regulation, DNA replication, genome maintenance, and metabolism (Fuss et al., 2015; Maio & Rouault, 2015). Fe–S clusters are typically found in the form of [2Fe–2S], [3Fe–4S], and [4Fe–4S] centers (Johnson et al., 2005; Liu et al., 2005; Fig. 1), with the iron ions linked via inorganic sulfide and typically coordinated to proteins via cysteine residues. Alternative ligands have been observed with water (Huang & Cowan, 2009; Robbins & Stout, 1989), histidine (Tan et al., 2016), serine (Harmer et al., 2014), other amino acids (Bak & Elliott, 2014; Johnson et al., 2005; Qi & Cowan, 2011b), and even small molecules (Harmer et al., 2014), where these changes can greatly modulate protein function.

Although tremendous variety exists among the functional properties of Fe–S cluster proteins, Fe–S clusters, and their family of cluster-binding proteins are found ubiquitously in nature and exhibit high conservation in terms of cluster biosynthesis and trafficking (Lill & Muehlenhoff, 2008). Because free iron and sulfide are toxic, the cellular pathway for cluster biogenesis and trafficking is inherently complex (Johnson et al., 2005; Stehling,

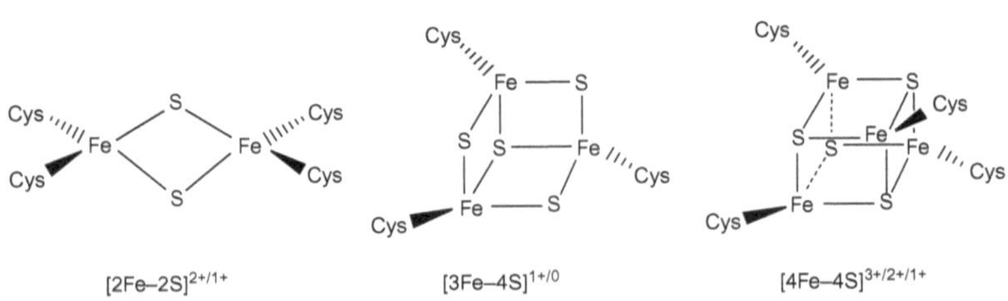

Fig. 1 Typical forms of Fe–S clusters and their oxidation states. Protein ligands are shown for simplicity as cysteine residues and connect to the iron atoms via the indicated bonds. Other amino acids, small molecules, or even water may coordinate the cluster in place of the cysteines.

Wilbrecht, & Lill, 2014). Three different assembly systems have been identified for Fe–S cluster biosynthesis, but the ISC (iron–sulfur cluster) system is primarily utilized in eukaryotes (Johnson et al., 2005; Qi & Cowan, 2011b). From the scaffold protein, a de novo Fe–S cluster is formed and transferred through a series of chaperone/delivery proteins to final target proteins that employ the clusters for function (Maio & Rouault, 2015; Stehling et al., 2014). Many of the proteins involved in cluster biogenesis tend to bind more labile Fe–S clusters and/or clusters that bridge protein–protein interfaces, which presents additional challenges for cluster characterization and mapping of the trafficking pathway. Furthermore, iron–sulfur clusters, even when ligated to proteins, are susceptible to oxidative damage (Baltscheffsky, 1996; Maio & Rouault, 2015), presenting a unique challenge for the synthesis and transport of these small molecules in the cell, and mechanistic analysis of this process in vitro. Fe–S clusters are often too unstable to survive aerobic in vitro protein purification and isolation methods, and so special care must be taken when attempting to characterize such proteins in vitro. Likewise, these precautions must also be considered when attempting to monitor cluster trafficking between proteins to examine cellular cluster pathways.

In order to conduct a mechanistic analysis of Fe–S cluster trafficking, our laboratory has modified a series of biochemical characterization techniques to develop a standard suite of experiments to: (1) analyze the apo protein stability and direct-binding interactions with partner proteins; (2) examine the ability to reconstitute an Fe–S cluster on the apo protein, quantify the amount of iron, and relate it to the Fe–S cluster content and oligomeric state of the holo protein; and (3) refine methods to monitor and quantitate cluster trafficking and determine the kinetic parameters of the process (Fig. 2), which is the focus of this chapter.

2. PREPARATION OF Fe–S CLUSTERS AND KINETIC DETERMINATION OF Fe–S CLUSTER TRANSFER REACTIONS

Fe–S clusters in proteins perform a variety of cellular functions due to their unique structural and electronic properties that allow them to carry out their essential roles. The majority of final target proteins typically contain clusters that are encapsulated inside the protein and are not exposed to solvent (Berkovitch, Nicolet, Wan, Jarrett, & Drennan, 2004; Harmer et al., 2014; Qi, Li, & Cowan, 2013; Takeda & Miki, 2017). However, proteins involved in cluster biogenesis must be able to bind and transfer Fe–S clusters

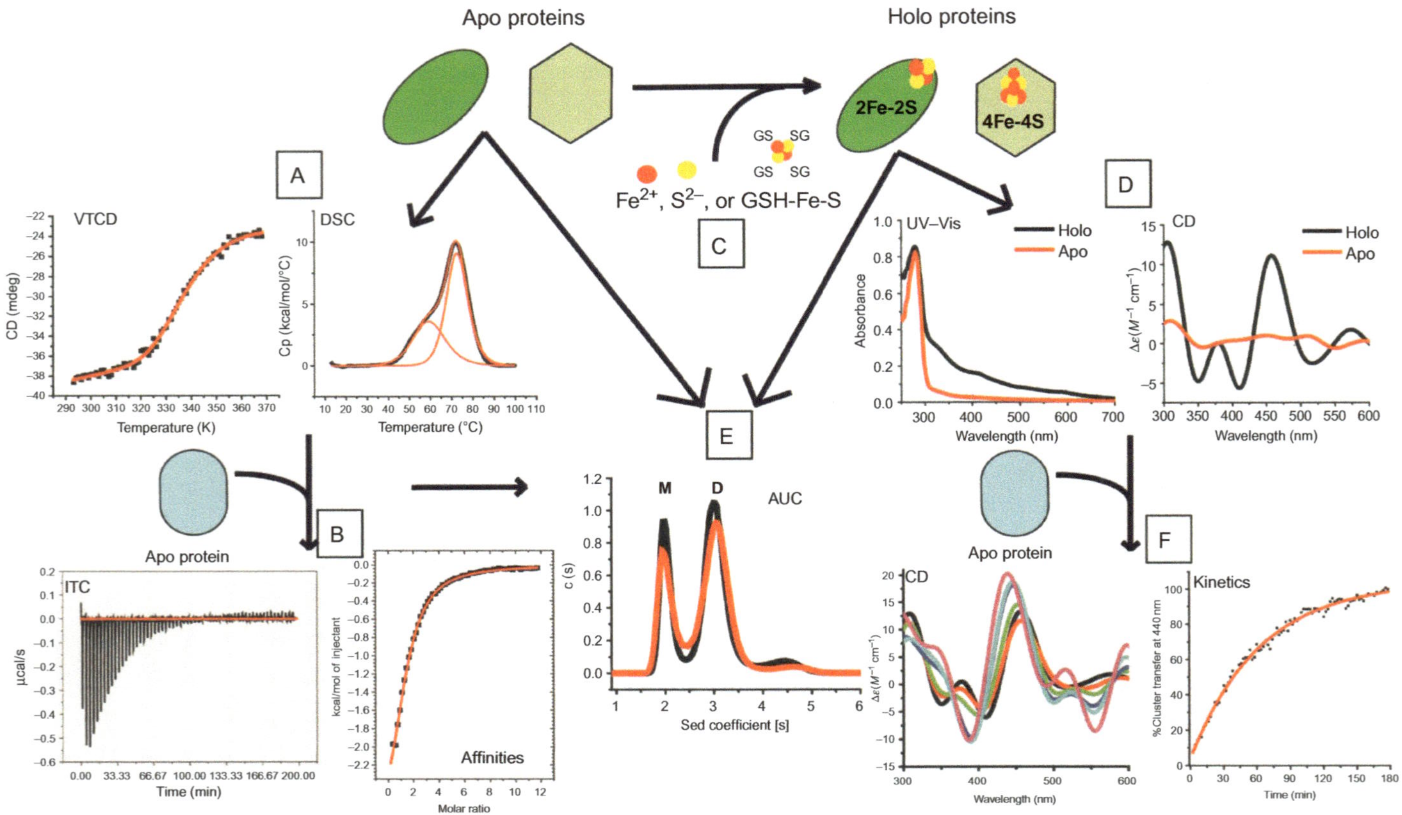

Fig. 2 See legend on opposite page.

for proper cluster trafficking, and so they typically present their clusters at surface exposed sites (Brancaccio et al., 2014; Cai et al., 2016; Johansson et al., 2011; Kim, Bothe, Alderson, & Markley, 2015). To understand structure–function relationships in the various classes of Fe–S cluster proteins, they first need to be examined in the apo state to assess conformational states and flexibility, as a prelude to understanding their mechanisms and roles in cluster trafficking by use of an array of biophysical techniques, such as variable temperature circular dichroism (VTCD) (Li, Ding, & Cowan, 2013), differential scanning calorimetry (DSC) (Wachnowsky, Wesley, Fidai, & Cowan, 2017), and isothermal titration calorimetry (ITC) (Qi & Cowan, 2011a). These techniques can also be used for comparison across mutant constructs for understanding stability determinants and disease conditions (Huang & Cowan, 2009; Kim, Tonelli, Kim, & Markley, 2012; Wachnowsky et al., 2017) and have been thoroughly described elsewhere (Greenfield, 2007; Ibarra-Molero, Naganathan, Sanchez-Ruiz, & Muñoz, 2016; Quinn, Carpenter, Croteau, & Wilcox, 2016).

Under aerobic conditions, a number of Fe–S cluster-binding proteins purify in the apo state, unless they bind the cluster in a particularly stable conformation that protects it from solvent and air, as commonly found for simple ferredoxin proteins (Qi, Li, & Cowan, 2013). Alternatively, purification under anaerobic conditions can result in isolation of holo protein; however, the yield will vary according to the intrinsic stability of the holo

Fig. 2 Workflow for Fe–S cluster protein characterization. (A) Initial analysis of apo proteins consists of thermodynamic experiments by use of variable temperature circular dichroism (VTCD) and differential scanning calorimetry (DSC), which provide information regarding secondary structure and tertiary structure, respectively. (B) Two putative apo protein partners are examined using isothermal titration calorimetry (ITC) to determine interaction thermodynamics and binding affinities. ITC can also be utilized to measure ligand binding, such as glutathione, to apo proteins. (C) Apo proteins are converted to their holo forms in reconstitution reactions using either inorganic sources of iron and sulfide or the physiologically relevant $[Fe_2S_2](GS)_4$ cluster complex. (D) Analysis of holo protein forms by use of UV–visible spectrophotometry or circular dichroism (CD) spectroscopy, utilizing the unique spectra observed from the charge-transfer bands present when Fe–S clusters are bound. In both the UV and CD spectra, the holo protein is shown in *black*, while the apo protein is in *red*. (E) Apo proteins, apo protein complexes, or holo proteins can be monitored using analytical ultracentrifugation (AUC) for determination of oligomeric state based on the molecular weight distribution present as the complexes sediment. In this example, the M peak represents a monomer and the D peak illustrates a dimer. (F) Kinetics of Fe–S cluster uptake and transfer between putative-binding partners can be determined by use of CD spectroscopy for mapping of cellular pathways.

form and often a mixture of states may be isolated. To achieve a homogeneous state it is often more expedient to remove the cluster and form apo and then follow a reconstitution protocol to reform the holo state. Therefore, procedures have been developed that allow for the chemical reconstitution of Fe–S clusters for the generation of the holo protein forms for characterization and analysis (Krebs et al., 2001; Rabinowitz, 1972; Webert et al., 2014; Wu, Wu, Surerus, & Cowan, 2002). Once reconstituted, the holo protein concentration can be determined either from a known extinction coefficient at one of the Fe–S cluster charge-transfer UV wavelengths, which are specific for each protein, or from iron quantitation (Beinert, 1978; Berger & McKay, 1974; Brumby & Massey, 1967; Fish, 1988; Moulis & Meyer, 1982; Stookey, 1970). Finally, the mode of cluster binding must be examined, since protein organization around the Fe–S cluster can occur via a variety of oligomeric states that correlate with protein roles in cluster assembly, trafficking, and use of the Fe–S cluster for function, with literature precedent for monomeric Fe–S cluster proteins (Harmer et al., 2014; Kim et al., 2012; Qi, Li, & Cowan, 2013), dimers (Brancaccio et al., 2014; Johansson et al., 2011), and higher order complexes (Galeano et al., 2017; Gnandt, Dörner, Strampraad, de Vries, & Friedrich, 2016; Netz et al., 2012). The oligomeric state for cluster binding can be confirmed from either native PAGE or AUC for insight into cluster transfer processes (Cai et al., 2016; Johansson et al., 2011; Muthurajan et al., 2016; Wachnowsky, Fidai, & Cowan, 2016b; Wachnowsky et al., 2017). This preliminary characterization of the apo and holo states is essential for mechanistic determination of cluster trafficking.

Traditional Fe–S cluster reconstitution methods, using iron salts and either inorganic sulfide or enzymatically generated sulfide, have previously been described (Krebs et al., 2001; Rabinowitz, 1972; Webert et al., 2014; Wu et al., 2002). An additional mechanism for chemical reconstitution of Fe–S clusters is via a synthetic glutathione-complexed iron–sulfur cluster (Fig. 3; Qi et al., 2012; Qi, Li, Chain, et al., 2013). Such a cluster has been shown to be capable of reconstituting a variety of Fe–S cluster proteins, but with unique exchange rates for each protein, while also demonstrating a high degree of selectivity (Fidai, Wachnowsky, & Cowan, 2016a; Qi et al., 2012; Wachnowsky et al., 2016b). Furthermore, the $[Fe_2S_2](GS)_4$ cluster demonstrates physiological relevance as a potential component of the labile iron pool, reflecting its potential biosynthesis via glutathione extraction from the IscU scaffold protein and export by the ABCB7 mitochondrial transporter (Li & Cowan, 2015; Qi, Li, & Cowan, 2014).

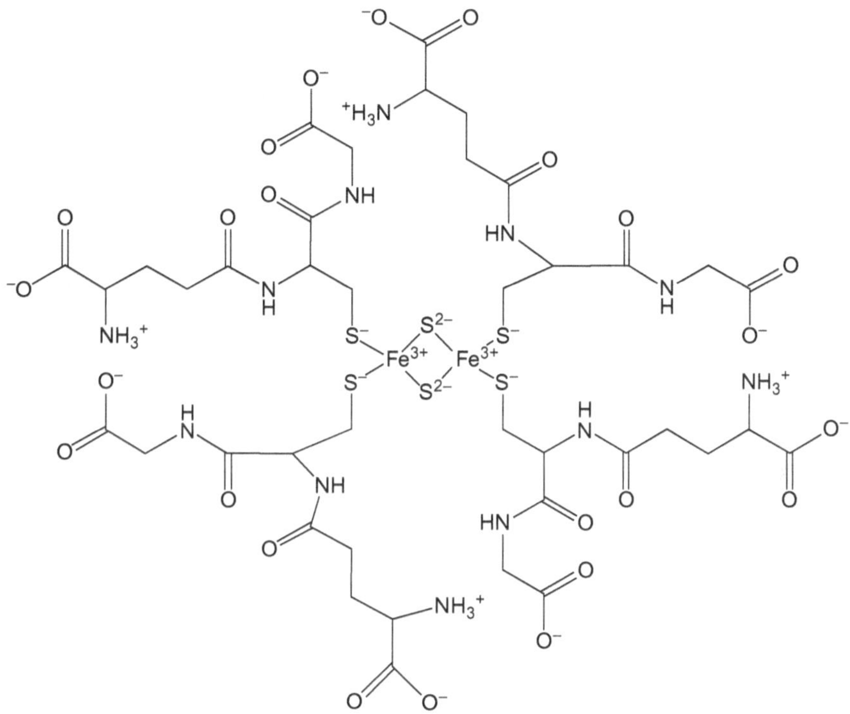

Fig. 3 Schematic illustration of a two-dimensional representation of the proposed structure of the $[Fe_2S_2](GS)_4$ cluster complex.

2.1 $Na_6[Fe_2S_2](GS)_4$ Synthesis

The $Na_6([Fe_2S_2](GS)_4)$ cluster (hereafter abbreviated as $[Fe_2S_2](GS)_4$ is readily synthesized and soluble in aqueous solution to permit examination of its role in the process of Fe–S cluster biosynthesis and trafficking.

2.1.1 Equipment

- Microcentrifuge
- pH meter
- Vortex
- SpeedVac vacuum concentrator

2.1.2 Buffers and Reagents

- Ethanol
- $FeCl_3$
- Na_2S
- Glutathione

2.1.3 Procedure

1. Prepare 10 mL of a 10 m*M* GSH solution and adjust the pH to 8.6.
2. Add $FeCl_3$ and Na_2S to final concentrations of 20 m*M*, vortexing between each addition. The solution should be black after both components have been added.
3. Add 40 mL of ethanol and vortex.
4. Centrifuge at 13,000 rpm for 10 min, and pour off the supernatant.
5. Wash with ethanol and centrifuge again. Repeat.
6. Remove the supernatant and dry the precipitate overnight in a SpeedVac.

2.1.4 Notes

1. All of the steps can be performed aerobically. The dried $[Fe_2S_2](GS)_4$ cluster can be stored long-term in a desiccator. The color of the solid cluster can be an indication of the stability. It should remain black in color; a gray color is usually indicative of degradation.
2. Must prepare GSH, $FeCl_3$, and Na_2S fresh for each synthesis reaction.
3. After the first ethanol wash, a portion of the cluster can be checked for successful synthesis by dissolving it in water, where it must readily go in to solution and be black in color. Otherwise, there has been a problem at one step in the synthesis and the cluster will be of low quality.
4. Confirm the success of the cluster synthesis after from the characteristic UV–visible spectrum, which should have Fe–S cluster charge-transfer bands at 330 and 420 nm ($\varepsilon_{330} = 7600\,M^{-1}\,cm^{-1}$, $\varepsilon_{420} = 4900\,M^{-1}\,cm^{-1}$) (Fidai et al., 2016a; Qi et al., 2012).

2.2 Cluster Uptake and Protein–Protein Transfer Reactions

As a possible component of the labile iron pool, the $[Fe_2S_2](GS)_4$ cluster complex can freely and selectively exchange with select Fe–S cluster trafficking proteins (Fidai et al., 2016a; Qi et al., 2012; Wachnowsky et al., 2016b). This cluster complex also can react with cytosolic Fe–S cluster proteins (Fidai et al., 2016a; Wachnowsky, Fidai, & Cowan, 2016a), a cellular compartment that has not been examined in depth for $[Fe_2S_2](GS)_4$ cluster reactivity. These cluster exchange processes can be monitored by use of circular dichroism (CD) spectroscopy. Although CD is most often used for secondary structure analysis of macromolecules by monitoring signals in the far-UV region (190–250 nm) (Greenfield, 2007), Fe–S cluster proteins display unique signals in the near-UV region (300–600 nm). Each Fe–S cluster protein can be characterized by signature spectral features (Fig. 4), since they all

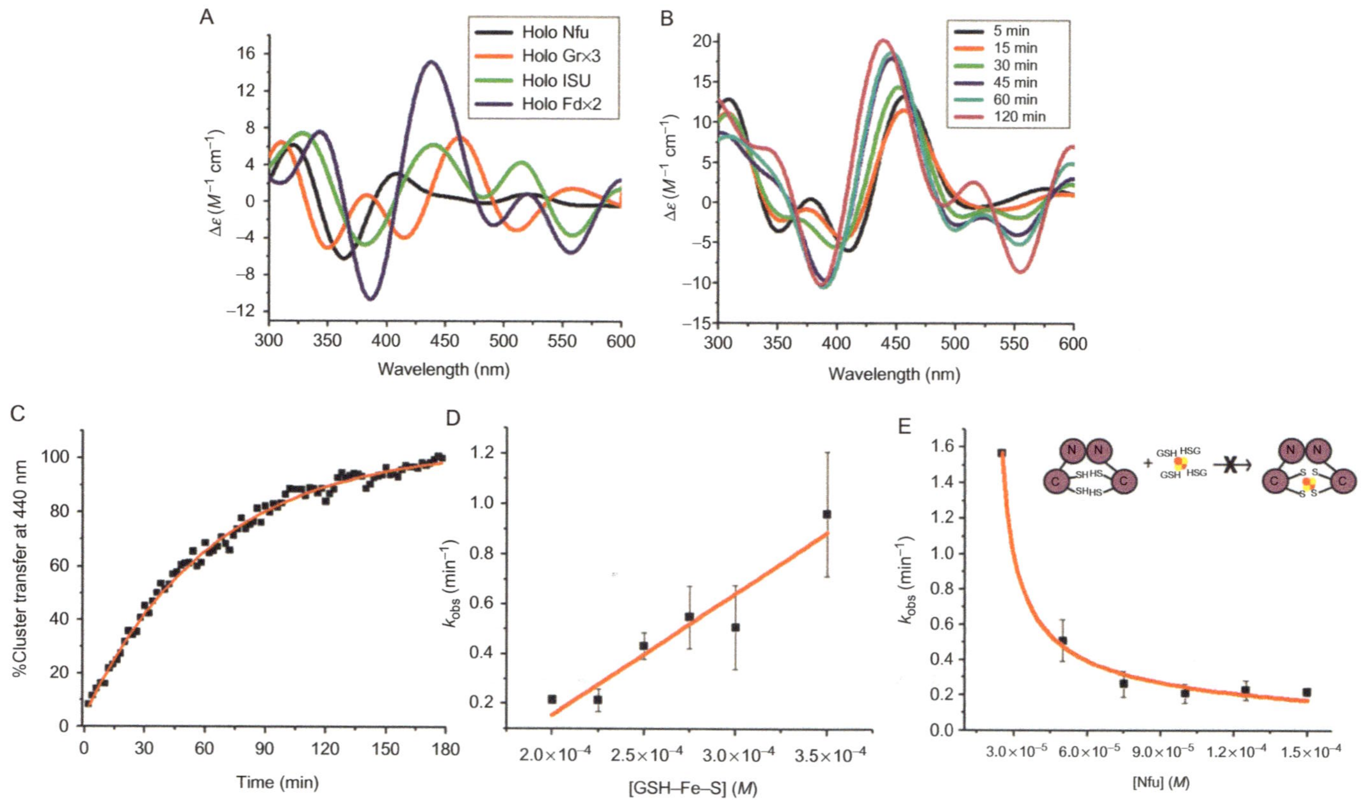

Fig. 4 See legend on next page.

demonstrate distinct protein folding motifs around the bound Fe–S cluster to create a defined chiral environment for detection by use of polarized light (Stephens et al., 1978). These signatures can be monitored for their appearance or disappearance in chemical reactions to provide kinetic information regarding cluster trafficking among Fe–S cluster proteins.

CD provides a unique advantage for monitoring $[Fe_2S_2](GS)_4$ cluster transfer because it is symmetric and therefore CD silent (Qi et al., 2012). As such, $[Fe_2S_2](GS)_4$ cluster reactivity and exchange with Fe–S cluster proteins can be easily and unambiguously monitored using CD to determine accurate second-order rate constants. Either the concentration of the cluster complex, or the apo protein, can be maintained at a constant excess while the concentration of the other species is varied (and vice versa) to establish pseudo-first order conditions (Fidai et al., 2016a; Wachnowsky et al., 2016b). Observed rates from simple cluster uptake by a monomeric protein plotted against the concentration of cluster or protein can be fit linearly to obtain the second-order rate constant for the uptake reaction (Fig. 4D). In more complex scenarios, such as cluster uptake to form a dimeric holo protein, two different mechanisms of uptake must be considered: (1) a

Fig. 4 (A) Overlay of holo CD spectra for common [2Fe–2S] cluster trafficking proteins. All of the proteins were prepared in 50 m*M* HEPES, 100 m*M* NaCl, pH 7.5, with the $\Delta\varepsilon$ values based on the concentration of the [2Fe–2S] cluster for each particular protein. (B) Cluster transfer reaction from [2Fe–2S] holo *Saccharomyces cerevisiae* glutaredoxin 3 to apo human ferredoxin 2 monitored by CD over the course of 1 h. (C) Spectra were converted to percent cluster transfer at the peak characteristic of ferredoxin 2 (440 nm) based on the initial [2Fe–2S] cluster concentration of Grx3 for calculation of the second-order rate constant using DynaFit (Kuzmic, 1996). (D) $[Fe_2S_2](GS)_4$ cluster concentration was plotted against k_{obs} from each concentration and fit linearly to determine the overall second-order rate constant from the reaction. (E) Protein concentration was plotted against k_{obs} from each concentration. Two different mechanisms were considered: (1) cluster uptake proceeds via a monomeric intermediate with cluster bound to one Nfu monomer and ligated by two GSH molecules from the $[Fe_2S_2](GS)_4$ complex followed by addition of a second Nfu monomer and (2) two Nfu monomers form a dimer that receives the $[Fe_2S_2](GS)_4$ complex directly (*inset*). Since the k_{obs} actually decrease as the protein concentration increases, the formation of dimers at higher concentration inhibits cluster uptake, eliminating mechanism 2. The data were fit with a substrate inhibition model, and the second-order rate constant was determined by fitting the initial slope of the plot. *Panel D: Adapted from Wachnowsky, C., Fidai, I., & Cowan, J. A. (2016a). Cytosolic iron-sulfur cluster transfer: A proposed kinetic pathway for the reconstitution of glutaredoxin 3.* FEBS Letters, 590*(24), 4531–4540. Panel E: Adapted from Wachnowsky, C., Fidai, I., & Cowan, J. A. (2016b). Iron-sulfur cluster exchange reactions mediated by the human Nfu protein.* Journal of Biological Inorganic Chemistry, 21, *825–836.*

monomeric intermediate (Fig. 4E) and (2) a dimeric intermediate (Fidai et al., 2016a; Wachnowsky et al., 2016b). Fitting the data with rate laws that represent these two different intermediate states can support the mechanism of cluster uptake by the particular protein under investigation (Wachnowsky et al., 2016b). Similar experiments can be conducted to analyze ligand effects, when small molecules are required for cluster binding, such as for the role of GSH in mediating cluster uptake by glutaredoxin proteins, by varying the species in question (Fidai et al., 2016a).

Similar to the cluster uptake experiments, cluster exchange between Fe–S proteins can be monitored using CD as a preferred spectroscopic method, relative to UV experiments, since each protein demonstrates a unique signature in the holo state (Fig. 4), for easy tracking of movement of the cluster from one protein to another. One challenge with the use of CD in kinetic cluster transfer experiments is the concentration limitations of the instrument, which prevents the use of ideal pseudo-first order conditions. As such, we have built from the work of others (Gao et al., 2013; Mapolelo et al., 2013; Shakamuri, Zhang, & Johnson, 2012) to develop a method for second-order rate constant determination while working within these limits. This method utilizes a holo protein donor in a 1:1 ratio with an apo protein acceptor and a kinetics simulation software, DynaFit (Kuzmic, 1996), to fit the raw kinetic data using a series of rate-laws corresponding to a second-order reaction to yield the second-order rate constant. The values obtained in this manner agree within error with rate constants obtained in test cases, where the second-order rate constant was determined using proper pseudo-first order kinetic conditions (Fidai et al., 2016a; Wachnowsky et al., 2016b). These second-order rate constants can be utilized for mapping of cellular kinetic pathways (Fidai, Wachnowsky, & Cowan, 2016b; Wachnowsky et al., 2016a).

2.2.1 Equipment

- Circular Dichroism Spectrometer
- CD data analysis software
- 1 cm anaerobic cuvette
- Gas-tight syringe
- Glove Box/Coy Chamber/Schlenk line
- Anaerobic vials
- Septa
- DynaFit or other chemical kinetics simulation software

2.2.2 Buffers and Reagents

- $[Fe_2S_2](GS)_4$ cluster
- Holo protein
- Apo protein
- DTT
- GSH
- 50 m*M* HEPES, 100 m*M* NaCl, pH 7.5

2.2.3 Procedure

1. Degas a mixture of 100 μ*M* apo protein in 50 m*M* HEPES, 100 m*M* NaCl, pH 7.5 and 5 m*M* DTT, and transfer using a gas-tight syringe to an argon-purged anaerobic cuvette. For cluster transfer, use 40 μ*M* apo protein with the same buffer conditions.
2. Set the CD to the following parameters: scan from 600 to 300 nm at a rate of 200 nm/min at 25°C with 2 min intervals between scans.
3. Baseline the CD using the apo mixture.
4. For cluster uptake, resuspend solid $[Fe_2S_2](GS)_4$ in degassed 50 m*M* HEPES, 100 m*M* NaCl, pH 7.5 to a concentration of 1 m*M* and transfer it to the anaerobic cuvette to have a final concentration of 100 μ*M* and begin data accumulation.
5. For cluster transfer from a holo protein, degas holo protein in 50 m*M* HEPES, 100 m*M* NaCl, pH 7.5 and transfer it to the anaerobic cuvette to a final concentration of 40 μ*M*, based on the [2Fe–2S] cluster concentration, and begin data accumulation.
6. Process the data using the CD software provided with the instrument by deconvoluting the spectra to distinguish overlapping peaks in the spectra.
7. Export the CD data and convert CD signal in millidegrees (also known as ellipticity, θ) to change in extinction ($\Delta\varepsilon$) (Chakraborty & Lentz, 2012). To convert the raw CD data, first convert from millidegrees to degrees. From there, the molar ellipticity ([θ]) can be calculated by use of Eq. (1), where C is the concentration of the Fe–S cluster in molarity and l is the path length in centimeters. [θ] can directly be converted to $\Delta\varepsilon$ by dividing by 3298.2.

$$[\theta] = \frac{100^*\theta}{C^*l} \tag{1}$$

8. Average a 10 nm window around the peak maximum, and convert to percent cluster transfer for fitting by DynaFit software (Kuzmic, 1996) to a second-order equation to obtain a second-order rate constant. A sample script is included below. Each of the places below, with a

question mark, indicate that a guess is required to run the script. The concentrations, the output file, and the raw data can be modified for each individual CD run.

```
[task]
  data = progress
  task = fit
[mechanism]
  A + H ----> P    :   k2
[constants]     ; units: M, min
  k2 =  ?
[concentrations]  ; units: M
  H = 0.000025
  A = 0.000025
[responses]     ; percent product/M substrate
  P = ?
[data]
  set  percentProduct
  offset auto ?
[output]
  directory ./manual/task/data/progress/output/01
[settings]
{Output}
  XAxisLabel = time, min
  YAxisLabel = Product, %
[set:percentProduct]
t,min   P,%
0  0
4  0.94036
6  8.18004
8  12.96791
10  31.54376
12  32.09425
14  42.93384
[end]
```

2.2.4 Notes

1. For conversion of raw CD data to percent cluster transfer, the CD spectra for the reconstituted holo protein must be recorded at a variety of concentrations to have a scale for measurement and comparison at the end of a CD transfer reaction.

2. Alternative concentration ratios can be explored, as long as the rate law is accurately represented in the simulation software, since the equilibrium for the putative partner pair and their relative cellular concentrations are unknown.
3. Since proteins such as glutaredoxins require GSH as a ligand (Johansson et al., 2011; Li & Outten, 2012; Wang et al., 2012), GSH should be included in the reaction. This would apply to other systems that use small molecule ligands.
4. For rapid cluster uptake reactions, a smaller range of wavelengths can be monitored. Once the prominent cluster-bound peaks have been determined from the holo protein CD spectra, a 10 nm window around the peak maximum can be the scan range instead of the full 300–600 nm and intensity monitored at 10 s intervals.
5. The cluster uptake or transfer reaction can only occur when the cluster ligating residues, usually cysteine, are reduced. Therefore, we have primarily utilized DTT in the reactions to facilitate that process. In cluster transfer reactions, 5 m*M* DTT has been used as a standard concentration. For cluster uptake, the DTT concentration may require optimization over a range of 200 μ*M* to 5 m*M* depending on the particular protein (Fidai et al., 2016a); however, maintaining some excess is the key. Alternatively, other reducing agents, such as TCEP, can be utilized, but the appropriate concentrations will need to be optimized for each protein, or the samples can be prereduced with DTT. Both of these alternatives are effective (Wachnowsky et al., 2016b); however, DTT is the more effective reducing agent and inclusion of DTT in the reaction often provides the most consistent results (Fidai et al., 2016a, 2016b; Wachnowsky et al., 2016a, 2016b). However, anaerobic conditions must be maintained to prevent oxidation of the cysteine residues, which would inhibit cluster uptake.
6. If required, UV–visible spectrophotometry can be used in place of CD spectroscopy (Fidai et al., 2016a; Qi et al., 2012); however, careful consideration of spectra may be required if there is extensive spectral overlap or similarity in extinction coefficients, although a change in absorbance at some wavelength is typically expected.

2.3 Glutathione Extraction

Glutathione is also capable of extracting Fe–S clusters from holo proteins to complete the exchange reaction and form the $[Fe_2S_2](GS)_4$ cluster complex (Fidai et al., 2016a; Qi et al., 2012). These reactions are typically slower than those found for cluster uptake or protein–protein cluster exchange rate constants, suggesting that the former reactions are kinetically preferred; most likely reflecting an internal chelate effect from the protein ligands. These extraction reactions are preferentially conducted on UV as opposed to CD, since the $[Fe_2S_2](GS)_4$ cluster complex is CD silent and formation cannot be directly monitored. However by use of UV–Vis, the $[Fe_2S_2](GS)_4$ cluster complex signals can be observed and, based on extinction coefficients, the yield of the cluster products can be determined (Fidai et al., 2016a; Wachnowsky et al., 2016b).

2.3.1 Equipment

- UV–Visible spectrophotometer
- 1 cm anaerobic cuvette
- Gas-tight syringe
- Glove Box/Coy Chamber/Schlenk line
- Anaerobic vials
- Septa
- DynaFit or other chemical kinetics simulation software

2.3.2 Buffers and Reagents

- Holo protein
- GSH
- 50 m*M* HEPES, 100 m*M* NaCl, pH 7.5

2.3.3 Procedure

1. Degas a mixture of GSH (concentration can range from 0.2 to 2 m*M*) in 50 m*M* HEPES, 100 m*M* NaCl, pH 7.5 and transfer using a gas-tight syringe to an argon-purged anaerobic cuvette.
2. Set the UV to the following parameters: scan from 800 to 200 nm at a rate of 600 nm/min at 25°C with 2 min intervals between scans.
3. Degas holo protein in 50 m*M* HEPES, 100 m*M* NaCl, pH 7.5 and transfer to the anaerobic cuvette to have a final concentration of 10 μ*M*, based on the [2Fe–2S] cluster concentration, and begin data accumulation.
4. Monitor the decrease in absorbance at 420 nm and convert to change in absorbance.
5. Plot the change in absorbance against time and fit to an exponential decay to determine the k_{obs}.
6. Determine k_{obs} for a variety of GSH concentrations to fit k_{obs} against GSH concentration either linearly or to saturation kinetics (Fig. 5; Fidai et al., 2016a).

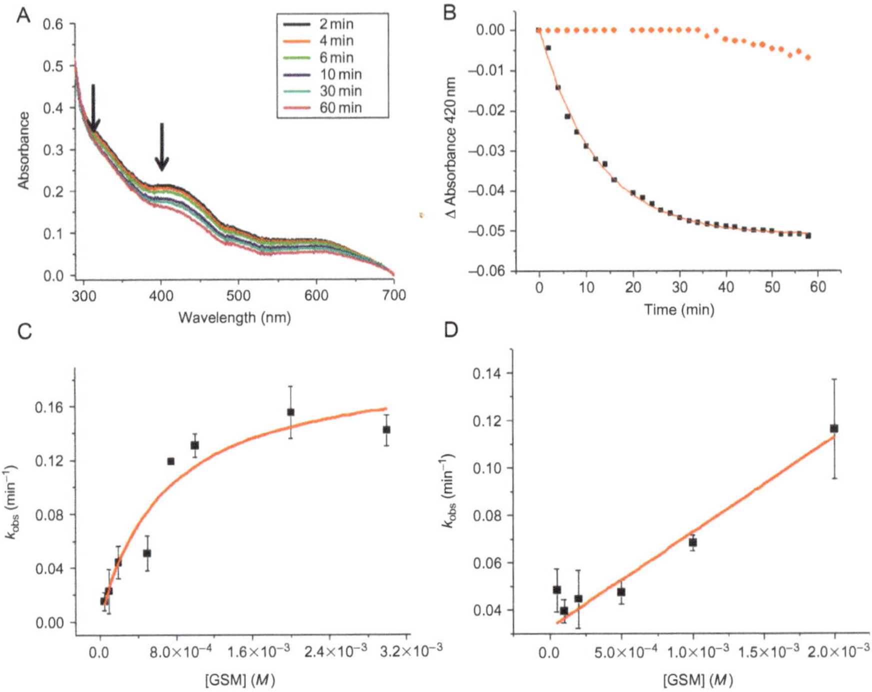

Fig. 5 Glutathione extraction of a [2Fe–2S] cluster from holo Grx3 to form the $[Fe_2S_2](GS)_4$ cluster. (A) UV–Vis spectrum of 20 μ*M* holo Grx3 incubated with 1 m*M* GSH in 50 m*M* Hepes, 100 m*M* NaCl, pH 7.5. *Black arrows* indicate the charge-transfer bands at 330 nm and 420 nm that decrease over time as the cluster moves from holo Grx3 to the $[Fe_2S_2](GS)_4$ cluster complex with the change in extinction coefficients. (B) The change in absorbance at 420 nm was plotted against time for each of the different concentrations of GSH and fit to an exponential to obtain the k_{obs} for each concentration. Sample traces are shown for 1 m*M* GSH (*black*) and no GSH (*red*). k_{obs} data were plotted against GSH concentration and fit to either (C) an exponential for calculation of an apparent K_D and a first-order rate constant (Fidai et al., 2016a) or (D) *a line*. In both cases, the overall second-order rate constant was determined from the initial slope. *Adapted from Fidai, I., Wachnowsky, C., & Cowan, J.A. (2016a). Glutathione-complexed [2Fe-2S] clusters function in Fe-S cluster storage and trafficking.* Journal of Biological Inorganic Chemistry, 21, *887–901.*

2.3.4 Notes

1. The GSH extraction experiment also requires a control conducted in the absence of GSH to monitor the inherent instability of the holo protein.
2. Formation of the $[Fe_2S_2](GS)_4$ cluster complex should be confirmed using mass spectrometry (Fidai et al., 2016a; Qi et al., 2012; Qi, Li, Chain, et al., 2013; Wachnowsky et al., 2016b).

3. Fe–S CLUSTER STIMULATION AND TRANSPORT IN LIPOSOME ASSAYS

Fe–S cluster trafficking not only occurs between putative partner proteins in separate cellular compartments, but clusters must move between the compartments as well, since they are primarily made in the mitochondria and transported to other areas (Lill et al., 2015; Maio & Rouault, 2015; Stehling et al., 2014). Fe–S cluster trafficking out of the mitochondria is believed to occur through an ABC transporter; ABCB7 in humans and Atm1p in yeast (Csere, Lill, & Kispal, 1998). Yeast cells deficient in Atm1p have a general deficiency of heme and iron–sulfur cluster containing proteins, are sensitive to oxidizing agents, have an increased level of glutathione, have a drastic increase in mitochondrial iron levels, and have decreased levels of cytosolic Fe–S proteins (Cavadini et al., 2007), indicating a crucial role for Atm1p and ABCB7 in iron homeostasis (Kispal, Csere, Guiard, & Lill, 1997; Miao et al., 2009). To date the substrate for this transporter has yet to be identified; however, it is believed that the substrate most likely contains cysteine, with much of the evidence suggesting a glutathione species (Lee, Yang, Zhitnitsky, Lewinson, & Rees, 2014; Lill et al., 2015; Srinivasan, Pierik, & Lill, 2014). Our laboratory has previously demonstrated the possibility of glutathione-complexed $[Fe_2S_2](GS)_4$ cluster as a substrate for the mitochondrial exporter, thereby providing a source of intact Fe–S clusters for the cytosolic Fe–S cluster machinery.

3.1 ATPase Stimulation Assays

As an ABC transporter, Atm1p (ABCB7) is part of the ATP "binding cassette" (ABC) transporter family. These are integral membrane proteins that function as homodimers and exhibit a conserved core structure with two transmembrane (TM) domains that confer substrate binding, and two highly conserved nucleotide binding domains (NBD) that utilize the energy of ATP to initiate conformational changes to propel substrates across the membrane (Fig. 6; Jones & George, 2004). Once the substrate is bound, hydrolysis of ATP promotes the conformational change required to move the substrate to the cytosol. Our previous work with Atm1p has shown that both GSH and the $[Fe_2S_2](GS)_4$ cluster complex are capable of stimulating this ATP hydrolysis in an ATPase assay (Li & Cowan, 2015; Qi et al., 2014).

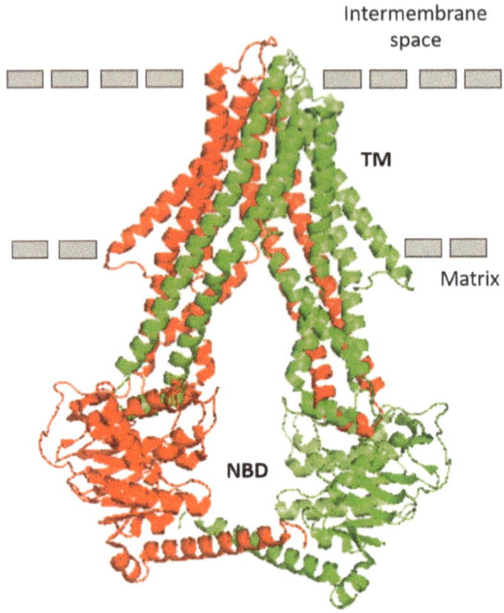

Fig. 6 Depiction of homodimeric *S. cerevisiae* Atm1p based on the crystal structure (PDB ID 4MYC) determined in the open conformation (Srinivasan et al., 2014), with one monomer shown in *red* and the other in *green*. The transmembrane domains are indicated by the TM, and the nucleotide-binding domains are indicated by NBD.

3.1.1 Equipment

- UV–Visible Spectrophotometer

3.1.2 Buffers and Reagents

- ENZChek Phosphate Assay Kit
- 100 μ*M* MESG
- 100 m*M* ATP in 1 × buffer (from the ENZChek kit)
- 100 m*M* $MgCl_2$ in 1 × buffer (from the ENZChek kit)
- 500 μ*M* purine nucleoside phosphorylase
- GSH
- $[Fe_2S_2](GS)_4$ clusters
- 2.5 μ*M* Atm1p or ABCB7

3.1.3 Procedure

1. Dissolve the $[Fe_2S_2](GS)_4$ cluster or GSH in degassed water to desired stock concentrations.
2. Prepare Mg-ATP by mixing $MgCl_2$ and ATP in a 1:1 ratio.

3. Mix 7.2 μL of 2.5 μ*M* protein, 3 μL of 20 × buffer from the kit, 3 μL of 100 μ*M* MESG, 1.2 μL of Mg-ATP, 0.6 μL of 500 μ*M* purine nucleoside phosphorylase, varying amounts of GSH or $[Fe_2S_2](GS)_4$ cluster complex, and water to the final volume of 60 μL. Vary GSH from 0 to 2.5 m*M* and the $[Fe_2S_2](GS)_4$ cluster from 0 to 100 μ*M*.
4. Monitor the absorbance at 360 nm for 10 min.
5. Calculate the slope from 5 to 10 min to determine the ATPase activity.
6. Determine the stimulation and affinity by plotting the activity against the concentration of either GSH or $[Fe_2S_2](GS)_4$ cluster complex and fitting the data to Eq. (2), where c is the initial ATPase activity in the absence of either GSH or cluster complex and $d(x)$ is a term to account for any cluster breakdown.

$$y = \left(\frac{V_{\max}(x)}{K_d + x}\right) + c - d(x) \tag{2}$$

3.1.4 Notes

1. To fully explore the stimulation kinetic parameters, ATP, GSH, and $[Fe_2S_2](GS)_4$ cluster can all be varied.
2. As an enzymatic assay, the concentrations of protein, GSH and cluster may need to be optimized to cover a suitable signal range for the concentrations of interest.

3.2 Liposome Synthesis and Protein Incorporation

Although ATPase activity can be examined in isolation with detergents present to enhance stability, membrane proteins, such as Atm1p, are more stable in their native form in a lipid bilayer. Therefore, a variety of methods exist to replicate lipid systems in vitro to reconstitute membrane proteins for experimentation that would reflect the cellular system (Poolman et al., 2005). Our previous work with Atm1p has demonstrated the successful reconstitution of Atm1p into proteoliposomes (Fig. 7A), which can be used for in-depth protein functional assays (Li & Cowan, 2015; Qi et al., 2014).

3.2.1 Equipment

- Vortex
- Argon
- Desiccator
- 400 nm membranes

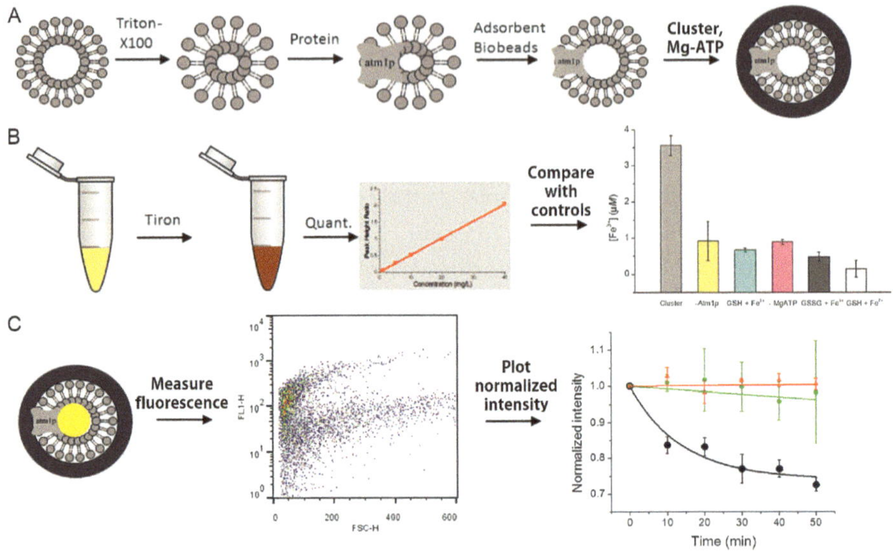

Fig. 7 Illustration depicting the proteoliposome assays conducted with Atm1p. (A) Preparation of the proteoliposomes begins by the addition of 10% Triton-X to the lipid solution, followed by the addition of Atm1p and final incubation with BioBeads. The following day, reactions can be monitored after the addition of the $[Fe_2S_2](GS)_4$ cluster complex and Mg-ATP either by (B) Tiron assay or (C) flow cytometry. In (B), the amount of iron transported into the liposome via Atm1p in the form of the $[Fe_2S_2](GS)_4$ cluster complex can be measured by the addition of the chelating agent Tiron, which results in a colored solution. Quantitative values are then obtained from absorbance with a standard curve of various Fe concentrations bound to Tiron for comparison with control reactions. (C) Proteoliposomes with fluorescein on the inside were monitored for fluorescence intensity every 10 min over the course of 1 h by use of flow cytometry. The area of fluorescence was normalized against the background fluorescence from control reactions using FlowJo software and plotted against time to fit the initial velocity. *Adapted from Li, J. & Cowan, J. A. (2015). Glutathione-coordinated [2Fe–2S] cluster: A viable physiological substrate for mitochondrial ABCB7 transport.* Chemical Communications, 51, *2253–2255.*

- Avanti membrane filter supports
- Two gas-tight syringes
- Microcentrifuge
- UV–Visible Spectrophotometer
- Ultracentrifuge

3.2.2 *Buffers and Reagents*

- 15 m*M* DOPG (dioleoylphosphatidyl-glycerol)
- 15 m*M* DOPC (dioleoylphosphatidyl-choline)

- 15 m*M* DOPE (dioleoylphosphatidyl-ethanolamine)
- 50 m*M* HEPES, 100 m*M* NaCl, pH 7.5
- 10% Triton-X
- 2.5 μ*M* Atm1p
- BioBeads™ SM-2 Adsorbent Media, Bio-Rad
- Methanol
- Ethanol

3.2.3 Procedure

1. Mix DOPG, DOPC, and DOPE at equimolar ratios, vortex, and dry over Argon gas.
2. Store the mixture in a desiccator for 1 day to allow the lipids to dry completely.
3. Dissolve the lipid mixture in 1 mL of 50 m*M* HEPES, 100 m*M* NaCl, pH 7.5 and extrude 21 times through a 400 nm membrane (Geertsma, Nik Mahmood, Schuurman-Wolters, & Poolman, 2008).
4. Dilute the liposomes two-fold in the HEPES buffer.
5. Titrate a 10% solution of Triton-X into the liposome mixture, kept on ice, at 4 μL increments with gentle mixing between each addition until a maximum OD_{550} is reached.
6. Titrate with 2 μL increments of 10% Triton-X until the OD_{550} is between 0 and −0.5.
7. Add 80 μL of 2.5 μ*M* Atm1p per 1 mL of liposome and incubate the mixture for 15 min at 4°C.
8. Prepare the BioBeads by washing them with at least 10 mL of methanol, followed by at least 10 mL ethanol and ending with two washes of a minimum of 10 mL of water.
9. Add 75 mg of the washed BioBeads to the proteoliposomes and incubate at 4°C.
10. Add another 75 mg of BioBeads after 30 min, 1 h, and the following day. After the last addition, incubate at 4°C for 2 h.
11. Remove the BioBeads by centrifugation at 3000 rpm for 1 min.
12. Isolate the proteoliposomes by ultracentrifugation at 80,000 rpm for 20 min at 4°C.
13. Redissolve the pelleted proteoliposomes in 1 mL of fresh 50 m*M* HEPES, 100 m*M* NaCl, pH 7.5, and store on ice. The reconstituted proteoliposomes must be used within 24 h (see Section 3.3).

3.2.4 Notes

1. The lipid mixture can only be stored in the desiccator for 2 weeks.
2. All vortex or mixture steps must be conducted at low speeds to prevent destruction of the lipid mixture.
3. While preparing the liposomes, equilibrate the extrusion apparatus with buffer and check for any leaks in the membrane or supports.
4. In preparation of the proteoliposomes, prepare a negative control, in which only buffer is added to the liposome that can be used in the transfer assays (Section 3.3).
5. After protein incorporation into the liposome, check the continuity of the mixture using dynamic light scattering to estimate the size of the liposomes (Li & Cowan, 2015).

3.3 Cluster Transfer Monitored by Tiron Assay and Flow Cytometry

After preparation of the proteoliposomes, the experimental assays can be conducted. Thus far, we have utilized two assays that in tandem can measure the amount of cluster transported via iron quantitation in the Tiron assay (Fig. 7B), and the rate of cluster transfer through flow cytometry, based on the quenching of fluorescein signal by the $[Fe_2S_2](GS)_4$ cluster (Fig. 7C; Li & Cowan, 2015). These assays can be adapted for monitoring the transport of other ligands or by alternative fluorescence methods.

3.3.1 Equipment

- Microcentrifuge
- Water bath at 95°C
- Ultracentrifuge
- UV–Visible Spectrophotometer
- FACS Calibur Flow Cytometer (BD Biosciences) (Lasers: 488 nm/633 nm)
- Computer with Cell Quest Pro (BD Biosciences) and FlowJo software

3.3.2 Buffers and Reagents

- Proteoliposome solution from Section 3.2
- GSH
- 2 *M* HCl
- Tiron™ (4,5-dihydroxy-1,3-benzenedisulfonic acid, disodium salt monohydrate), ACROS Organics
- 5 *M* NaOH

- 10 m*M* Mg-ATP
- 50 m*M* HEPES, 100 m*M* NaCl, pH 7.5
- 10 m*M* MES, pH 6.5
- 1 m*M* fluorescein
- $[Fe_2S_2](GS)_4$ cluster complex

3.3.3 Procedure

For Tiron assay and direct quantitation of iron transferred, see step 6. For flow cytometry, and kinetics of transfer skip to step 11.

1. To the liposome solution from Section 3.2, add Mg-ATP and the $[Fe_2S_2](GS)_4$ cluster complex, resuspended in 10 m*M* GSH solution, pH 8.6, to final concentrations of 10 m*M* each.
2. Vortex and incubate the reaction for 1 h at 25°C in the dark, especially in cases where fluorescein is included.
3. Centrifuge the reaction for 5 min at 6000 rpm to remove precipitate from cluster hydrolysis.
4. Centrifuge the supernatant further at 80,000 rpm for 20 min at 4°C to isolate the proteoliposomes.
5. Resuspend the pellet in half of the original liposome volume (see step 7, Section 3.2) in 50 m*M* HEPES, 100 m*M* NaCl, pH 7.5.
6. Denature the proteoliposomes by the addition of 150 μL of 2 *M* HCl per 500 μL liposome resuspension and boil at 95°C for 15 min for iron release.
7. Centrifuge the denatured proteoliposome solution at 13,000 rpm for 10 min.
8. Neutralize 100 μL of the reaction supernatant with 9.2 μL of 5 *M* NaOH and 800 μL of 10 m*M* MES, pH 6.5.
9. Add 100 μL of 100 m*M* Tiron, also prepared in the MES buffer, to the solution for chelation of the released ferric ions.
10. Incubate the solution in the dark for 10 min at 25°C and measure the absorbance at 550 nm (Fig. 7B).
11. For the fluorescein proteoliposomes, initiate cluster transfer by the addition of Mg-ATP and $[Fe_2S_2](GS)_4$ cluster complex to final concentrations of 10 m*M* each to 1 mL of fluorescein proteoliposome suspension.
12. Monitor the kinetic profile for cluster transfer by collecting data for 10,000 events every 10 min for a total of 50 min, keeping the proteoliposome solutions in the dark on ice. Instrument parameters are described in depth elsewhere (Li & Cowan, 2015).

13. Plot the normalized intensities, obtained via FlowJo flow cytometry software and accounting for control reactions (Li & Cowan, 2015), against time and fit the initial slope to obtain velocity and rate of transport (Fig. 7C).

3.3.4 Notes

1. For flow cytometry experiments, fluorescein must be added when dissolving the lipid residue (Section 3.2), such that it can be incorporated on the inside of the membrane.
2. Iron quantitation via the Tiron assay is extremely sensitive to pH. If no color change is observed, check the pH of all of the solutions.
3. As mentioned in Section 3.2, both assays require a negative control of only buffer added to the liposome.

4. SUMMARY AND CONCLUSIONS

Iron–sulfur cluster biogenesis is a highly conserved process that is essential in trafficking clusters to their final targets for proper cellular function. Because of their roles in cluster biogenesis, these proteins demonstrate structural flexibility and unique conformational states that can be examined even in their apo form. Challenges arise when trying to examine these structural aspects in holo proteins, due to the potential for oxidative damage to the cluster and the existence of a mixture of apo and holo protein forms. Many of the relevant characteristics can be examined by use of thermodynamic techniques that probe secondary and tertiary structural properties, as well as the stability of these motifs. Likewise, end-point product apo–holo interactions can be examined similar to the apo–apo protein forms, although thermodynamic analysis of starting holo–apo mixtures would typically be precluded as these represent active transfer complexes.

In contrast, the active transfer complex is better examined by use of kinetics techniques. A holo protein sample must first be prepared and the amount of bound cluster quantified prior to taking kinetic measurements. Evaluating reactivity toward a series of putative partner proteins allows the determination of a kinetic map that reflects and models cellular cluster trafficking pathways (Fidai et al., 2016a; Wachnowsky et al., 2016b). It also highlights the intrinsic redundancy in many cellular pathways and identifies partners of interest for in-depth study of the mechanism of transfer, as well as highlighting cluster trafficking between cellular compartments via membrane transporters (Li & Cowan, 2015; Qi et al., 2014).

Furthermore, mapping these transfer partners and pathways can subsequently be utilized to analyze both disease-causing mutants, and mutants known to perturb the conformational state of individual Fe–S proteins. In turn, such insights promote our understanding of Fe–S cluster biogenesis at the molecular level, and further validates and rationalizes trends observed at the cellular level.

REFERENCES

Bak, D. W., & Elliott, S. J. (2014). Alternative FeS cluster ligands: Tuning redox potentials and chemistry. *Current Opinion in Chemical Biology*, *19*, 50–58.

Baltscheffsky, H. (1996). *Origin and evolution of biological energy conversion*. New York: Wiley-VCH.

Beinert, H. (1978). Micro methods for the quantitative determination of iron and copper in biological material. *Methods in Enzymology*, *54*, 435–445.

Berger, S. A., & McKay, J. B. (1974). The photometric titration of Fe(III) with EDTA using tiron indicator. *Mikrochimica Acta*, *62*(4), 665–670.

Berkovitch, F., Nicolet, Y., Wan, J. T., Jarrett, J. T., & Drennan, C. L. (2004). Crystal structure of biotin synthase, an S-adenosylmethionine-dependent radical enzyme. *Science*, *303*, 76–79.

Brancaccio, D., Gallo, A., Mikolajczyk, M., Zovo, K., Palumaa, P., Novellino, E., et al. (2014). Formation of [4Fe-4S] clusters in the mitochondrial iron–sulfur cluster assembly machinery. *Journal of the American Chemical Society*, *136*(46), 16240–16250.

Brumby, P. E., & Massey, V. (1967). Determination of nonheme iron, total iron, and copper. *Methods in Enzymology*, *10*, 463–474.

Cai, K., Liu, G., Frederick, R. O., Xiao, R., Montelione, G. T., & Markley, J. L. (2016). Structural/functional properties of human NFU1, an intermediate [4Fe-4S] carrier in human mitochondrial iron-sulfur cluster biogenesis. *Structure*, *24*, 2080–2091.

Cavadini, P., Biasiotto, G., Poli, M., Levi, S., Verardi, R., Zanella, I., et al. (2007). RNA silencing of the mitochondrial ABCB7 transporter in HeLa cells causes an iron-deficient phenotype with mitochondrial iron overload. *Blood*, *109*, 3552–3559.

Chakraborty, H., & Lentz, B. R. (2012). A simple method for correction of circular dichroism spectra obtained from membrane-containing samples. *Biochemistry*, *51*(5), 1005–1008.

Cowan, J. A. (1997). *Inorganic biochemistry: An introduction* (2nd ed.). New York: Wiley-VCH, Inc.

Csere, P., Lill, R., & Kispal, G. (1998). Identification of a human mitochondrial ABC transporter, the functional orthologue of yeast Atm1p. *FEBS Letters*, *441*, 266–270.

Fidai, I., Wachnowsky, C., & Cowan, J. A. (2016a). Glutathione-complexed [2Fe-2S] clusters function in Fe-S cluster storage and trafficking. *Journal of Biological Inorganic Chemistry*, *21*, 887–901.

Fidai, I., Wachnowsky, C., & Cowan, J. A. (2016b). Mapping cellular Fe-S cluster uptake and exchange reactions—Divergent pathways for iron-sulfur cluster delivery to human ferredoxins. *Metallomics*, *8*(12), 1283–1293.

Fish, W. W. (1988). Rapid colorimetric micromethod for the quantitation of complexed iron in biological samples. *Methods in Enzymology*, *158*, 357–364.

Fuss, J. O., Tsai, C.-L., Ishida, J. P., & Tainer, J. A. (2015). Emerging critical roles of Fe-S clusters in DNA replication and repair. *Biochimica et Biophysica Acta*, *1853*(6), 1253–1271.

Galeano, B. K., Ranatunga, W., Gakh, O., Smith, D. Y., Thompson, J. R., & Isaya, G. (2017). Zinc and the iron donor frataxin regulate oligomerization of the scaffold protein to form new Fe-S cluster assembly centers. *Metallomics*, *9*, 773–801.

Gao, H., Subramanian, S., Couturier, J., Naik, S. G., Kim, S.-K., Leustek, T., et al. (2013). Arabidopsis thaliana Nfu2 accommodates [2Fe-2S] or [4Fe-4S] clusters and is competent for in vitro maturation of chloroplast [2Fe-2S] and [4Fe-4S] cluster-containing proteins. *Biochemistry*, *52*(38), 6633–6645.

Geertsma, E. R., Nik Mahmood, N. A., Schuurman-Wolters, G. K., & Poolman, B. (2008). Membrane reconstitution of ABC transporters and assays of translocator function. *Nature Protocols*, *3*, 256–266.

Gnandt, E., Dörner, K., Strampraad, M. F. J., de Vries, S., & Friedrich, T. (2016). The multitude of iron-sulfur clusters in respiratory complex I. *Biochimica et Biophysica Acta*, *1857*(8), 1068–1072.

Greenfield, N. J. (2007). Using circular dichroism collected as a function of temperature to determine the thermodynamics of protein unfolding and binding interactions. *Nature Protocols*, *1*(6), 2527–2535.

Harmer, J. E., Hiscox, M. J., Dinis, P. C., Fox, S. J., Iliopoulos, A., Hussey, J. E., et al. (2014). Structures of lipoyl synthase reveal a compact active site for controlling sequential sulfur insertion reactions. *The Biochemical Journal*, *464*(1), 123–133.

Huang, J., & Cowan, J. A. (2009). Iron-sulfur cluster biosynthesis: Role of a semi-conserved histidine. *Chemical Communications*, *21*, 3071–3073.

Ibarra-Molero, B., Naganathan, A. N., Sanchez-Ruiz, J. M., & Muñoz, V. (2016). Modern analysis of protein folding by differential scanning calorimetry. *Methods in Enzymology*, *567*, 281–318.

Johansson, C., Roos, A. K., Montano, S. J., Sengupta, R., Filippakopoulos, P., Guo, K., et al. (2011). The crystal structure of human GLRX5: Iron-sulfur cluster co-ordination, tetrameric assembly and monomer activity. *Biochemical Journal*, *433*(2), 303–311.

Johnson, D. C., Dean, D. R., Smith, A. D., & Johnson, M. K. (2005). Structure, function, and formation of biological iron-sulfur clusters. *Annual Review of Biochemistry*, *74*(1), 247–281.

Jones, P. M., & George, A. M. (2004). The ABC transporter structure and mechanism: Perspectives on recent research. *Cellular and Molecular Life Sciences: CMLS*, *61*(6), 682–699.

Kim, J. H., Bothe, J. R., Alderson, T. R., & Markley, J. L. (2015). Tangled web of interactions among proteins involved in iron–sulfur cluster assembly as unraveled by NMR, SAXS, chemical crosslinking, and functional studies. *Biochimica et Biophysica Acta*, *1853*(6), 1416–1428.

Kim, J. H., Tonelli, M., Kim, T., & Markley, J. L. (2012). Three-dimensional structure and determinants of stability of the iron-sulfur cluster scaffold protein IscU from Escherichia coli. *Biochemistry*, *51*(28), 5557–5563.

Kispal, G., Csere, P., Guiard, B., & Lill, R. (1997). The ABC transporter Atm1p is required for mitochondrial iron homeostasis. *FEBS Letters*, *418*(3), 346–350.

Krebs, C., Agar, J. N., Smith, A. D., Frazzon, J., Dean, D. R., Huynh, B. H., et al. (2001). IscA, an alternate scaffold for Fe-S cluster biosynthesis. *Biochemistry*, *40*, 14069–14080.

Kuzmic, P. (1996). Program DYNAFIT for the analysis of enzyme kinetic data: Application to HIV proteinase. *Analytical Biochemistry*, *237*, 260–273.

Lee, J. Y., Yang, J. G., Zhitnitsky, D., Lewinson, O., & Rees, D. C. (2014). Structural basis for heavy metal detoxification by an Atm1-type ABC exporter. *Science*, *343*, 1133–1136.

Li, J., & Cowan, J. A. (2015). Glutathione-coordinated [2Fe–2S] cluster: A viable physiological substrate for mitochondrial ABCB7 transport. *Chemical Communications*, *51*, 2253–2255.

Li, J., Ding, S., & Cowan, J. A. (2013). Thermodynamic and structural analysis of human NFU conformational chemistry. *Biochemistry*, *52*(29), 4904–4913.

Li, H., & Outten, C. E. (2012). Monothiol CGFS glutaredoxins and BolA-like proteins: [2Fe-2S] binding partners in iron homeostasis. *Biochemistry*, *51*(22), 4377–4389.
Lill, R., Dutkiewicz, R., Freibert, S. A., Heidenreich, T., Mascarenhas, J., Netz, D. J., et al. (2015). The role of mitochondria and the CIA machinery in the maturation of cytosolic and nuclear iron–sulfur proteins. *European Journal of Cell Biology*, *94*, 280–291.
Lill, R., & Muehlenhoff, U. (2008). Maturation of iron-sulfur proteins in eukaryotes: Mechanisms, connected processes, and diseases. *Annual Review of Biochemistry*, *77*, 669–700.
Liu, J., Oganesyan, N., Shin, D.-H., Jancarik, J., Yokota, H., Kim, R., et al. (2005). Structural characterization of an iron-sulfur cluster assembly protein IscU in a zinc-bound form. *Proteins*, *59*(4), 875–881.
Maio, N., & Rouault, T. A. (2015). Iron–sulfur cluster biogenesis in mammalian cells: New insights into the molecular mechanisms of cluster delivery. *Biochimica et Biophysica Acta*, *1853*(6), 1493–1512.
Mapolelo, D. T., Zhang, B., Randeniya, S., Albetel, A. N., Li, H., Couturier, J., et al. (2013). Monothiol glutaredoxins and A-type proteins: Partners in Fe-S cluster trafficking. *Dalton Transactions*, *42*(9), 3107–3115.
Miao, R., Kim, H., Koppolu, U. M. K., Ellis, E. A., Scott, R. A., & Lindahl, P. A. (2009). Biophysical characterization of the iron in mitochondria from Atm1p-depleted Saccharomyces cerevisiae. *Biochemistry*, *48*(40), 9556–9568.
Moulis, J.-M., & Meyer, J. (1982). Characterization of the selenium-substituted 2[4Fe-4Se] ferredoxin from *Clostridium pasteurianum*. *Biochemistry*, *21*, 4762–4771.
Muthurajan, U., Mattiroli, F., Bergeron, S., Zhou, K., Gu, Y., Chakravarthy, S., et al. (2016). In vitro chromatin assembly: Strategies and quality control. *Methods in Enzymology*, *573*, 3–41.
Napier, I. (2005). Iron trafficking in the mitochondrion: Novel pathways revealed by disease. *Blood*, *105*(5), 1867–1874.
Netz, D. J. A., Pierik, A. J., Stumpfig, M., Bill, E., Sharma, A. K., Pallesen, L. J., et al. (2012). A bridging [4Fe-4S] cluster and nucleotide binding are essential for function of the Cfd1-Nbp35 complex as a scaffold in iron-sulfur protein maturation. *The Journal of Biological Chemistry*, *287*(15), 12365–12378.
Poolman, B., Doeven, M. K., Geertsma, E. R., Biemans-Oldehinkel, E., Konings, W. N., & Rees, D. C. (2005). Functional analysis of detergent-solubilized and membrane-reconstituted ATP-binding cassette transporters. *Methods in Enzymology*, *400*, 429–459.
Qi, W., & Cowan, J. A. (2011a). Mechanism of glutaredoxin-ISU [2Fe-2S] cluster exchange. *Chemical Communications*, *47*(17), 4989–4991.
Qi, W., & Cowan, J. A. (2011b). Structural, mechanistic and coordination chemistry of relevance to the biosynthesis of iron-sulfur and related iron cofactors. *Coordination Chemistry Reviews*, *255*(7–8), 688–699.
Qi, W., Li, J., Chain, C. Y., Pasquevich, G. A., Pasquevich, A. F., & Cowan, J. A. (2012). Glutathione complexed Fe-S centers. *Journal of the American Chemical Society*, *134*, 10745–10748.
Qi, W., Li, J., Chain, C. Y., Pasquevich, G. A., Pasquevich, A. F., & Cowan, J. A. (2013). Glutathione-complexed iron-sulfur clusters. Reaction intermediates and evidence for a template effect promoting assembly and stability. *Chemical Communications*, *49*(56), 6313.
Qi, W., Li, J., & Cowan, J. A. (2013). Human ferredoxin-2 displays a unique conformational change. *Dalton Transactions*, *42*(9), 3088–3091.
Qi, W., Li, J., & Cowan, J. A. (2014). A structural model for glutathione-complexed iron–sulfur cluster as a substrate for ABCB7-type transporters. *Chemical Communications*, *50*(29), 3795.
Quinn, C. F., Carpenter, M. C., Croteau, M. L., & Wilcox, D. E. (2016). Isothermal titration calorimetry measurements of metal ions binding to proteins. *Methods in Enzymology*, *567*, 3–21.

Rabinowitz, J. (1972). Preparation and properties of clostridial ferredoxins. *Methods in Enzymology*, *24*, 431–436.

Robbins, A. H., & Stout, C. D. (1989). Structure of activated aconitase: Formation of the [4Fe-4S] cluster in the crystal. *Proceedings of the National Academy of Sciences of the United States of America*, *86*, 3639–3643.

Shakamuri, P., Zhang, B., & Johnson, M. K. (2012). Monothiol glutaredoxins function in storing and transporting [2Fe-2S] clusters assembled on IscU scaffold proteins. *Journal of the American Chemical Society*, *134*(37), 15213–15216.

Srinivasan, V., Pierik, A. J., & Lill, R. (2014). Crystal structures of nucleotide-free and glutathione-bound mitochondrial ABC transporter Atm1. *Science*, *343*, 1137–1140.

Stehling, O., Wilbrecht, C., & Lill, R. (2014). Mitochondrial iron-sulfur protein biogenesis and human disease. *Biochimie*, *100*, 61–77.

Stephens, P. J., Thomson, A. J., Dunn, J. B. R., Keiderling, T. A., Rawlings, J., Rao, K. K., et al. (1978). Circular dichroism and magnetic circular dichroism of iron-sulfur proteins. *Biochemistry*, *22*, 4770–4778.

Stookey, L. L. (1970). Ferrozine—A new spectrophotometric reagent for iron. *Analytical Chemistry*, *42*, 779–781.

Takeda, K., & Miki, K. (2017). Ultra-high-resolution structure and charge-density analysis of high-potential iron-sulfur protein. *The FEBS Journal*, *284*, 2163–2166,

Tan, G., Liu, D., Pan, F., Zhao, J., Li, T., Ma, Y., et al. (2016). His-87 ligand in mitoNEET is crucial for the transfer of iron sulfur clusters from mitochondria to cytosolic aconitase. *Biochemical and Biophysical Research Communications*, *470*, 226–232.

Wachnowsky, C., Fidai, I., & Cowan, J. A. (2016a). Cytosolic iron-sulfur cluster transfer: A proposed kinetic pathway for the reconstitution of glutaredoxin 3. *FEBS Letters*, *590*(24), 4531–4540.

Wachnowsky, C., Fidai, I., & Cowan, J. A. (2016b). Iron-sulfur cluster exchange reactions mediated by the human Nfu protein. *Journal of Biological Inorganic Chemistry*, *21*, 825–836.

Wachnowsky, C., Wesley, N. A., Fidai, I., & Cowan, J. A. (2017). Understanding the molecular basis for multiple mitochondrial dysfunctions syndrome 1 (MMDS1)—Impact of a disease-causing Gly208Cys substitution on structure and activity of NFU1 in the Fe/S cluster biosynthetic pathway. *Journal of Molecular Biology*, *429*, 790–807.

Wang, L., Ouyang, B., Li, Y., Feng, Y., Jacquot, J. P., Rouhier, N., et al. (2012). Glutathione regulates the transfer of iron-sulfur cluster from monothiol and dithiol glutaredoxins to apo ferredoxin. *Protein & Cell*, *3*(9), 714–721.

Webert, H., Freibert, S.-A., Gallo, A., Heidenreich, T., Linne, U., Amlacher, S., et al. (2014). Functional reconstitution of mitochondrial Fe/S cluster synthesis on Isu1 reveals the involvement of ferredoxin. *Nature Communications*, *5*, 5013.

Wu, S.-p., Wu, G., Surerus, K. K., & Cowan, J. A. (2002). Iron-sulfur cluster biosynthesis. Kinetic analysis of [2Fe-2S] cluster transfer from holo ISU to apo Fd: Role of redox chemistry and a conserved aspartate. *Biochemistry*, *41*, 8876–8885.

CHAPTER FOUR

Combined Biochemical, Biophysical, and Cellular Methods to Study Fe–S Cluster Transfer and Cytosolic Aconitase Repair by MitoNEET

Cécile Mons*, Ioana Ferecatu†, Sylvie Riquier*, Ewen Lescop*, Cécile Bouton*, Marie-Pierre Golinelli-Cohen*,[1]

*Institut de Chimie des Substances Naturelles (ICSN), CNRS UPR 2301, Univ. Paris-Sud, Université Paris-Saclay, Gif-sur-Yvette, France
†INSERM, UMR-S1139, Faculté des Sciences Pharmaceutiques et Biologiques, Université Paris Descartes, Sorbonne Paris Cité, Paris, France
[1]Corresponding author: e-mail address: marie-pierre.golinelli@cnrs.fr

Contents

Abstract

MitoNEET is the first identified Fe–S protein anchored to mammalian outer mitochondrial membranes with the vast majority of the protein polypeptide located in the cytosol, including its [2Fe–2S] cluster-binding domain. The coordination of the cluster is unusual and involves three cysteines and one histidine. MitoNEET is capable of transferring its redox-active Fe–S cluster to a bacterial apo-ferredoxin in vitro even under aerobic conditions, unlike other Fe–S transfer proteins such as ISCU. This specificity suggests its possible involvement in Fe–S repair after oxidative and/or nitrosative stress. Recently, we identified cytosolic aconitase/iron regulatory protein 1 (IRP1) as the first physiological protein acceptor of the mitoNEET Fe–S cluster in an Fe–S repair process.

Methods in Enzymology, Volume 595
ISSN 0076-6879
http://dx.doi.org/10.1016/bs.mie.2017.07.010

This chapter describes methods to study in vitro mitoNEET Fe–S cluster transfer/repair to a bacterial ferredoxin used as a model aporeceptor and in a more comprehensive manner to cytosolic aconitase/IRP1 after a nitrosative stress using in vitro, in cellulo, and in vivo methods.

1. INTRODUCTION

For the first time in 1978, the presence of a [2Fe–2S] protein in outer mitochondrial membranes (OMMs) from beef heart was detected by electronic paramagnetic resonance spectroscopy (Heidrich, Albracht, & Backstrom, 1978). It was only in 2004 that this Fe–S protein, called mitoNEET (mNT), was isolated from rat liver mitochondria and shown to be a potential target of the type II diabetes drug pioglitazone (Colca et al., 2004). Biophysical and biochemical studies characterized mNT as a small homodimeric protein (13 kDa for each monomer), which assembles a [2Fe–2S] cluster per monomer. In mammalian cells, mNT is anchored to the OMM by its N-terminus part (32 amino acids), with the vast majority of the protein located in the cytosol, including its Fe–S cluster-binding domain (Wiley, Murphy, Ross, van der Geer, & Dixon, 2007). This was the first identified Fe–S protein of the OMM.

MNT belongs to the NEET protein family. Its members are characterized by the presence of at least one CDGSH iron–sulfur domain (CISD) motif ([**C**-X-**C**-X2-(S/T)-X3-P-X-**C**-D-G-(S/A/T)-**H**]) involved in the coordination of a [2Fe–2S] cluster by three cysteines (Cys-72, Cys-74, and Cys-83 for *Homo sapiens* mNT) and one histidine (His-87 for *H. sapiens* mNT) (Paddock et al., 2007; Wiley et al., 2007). Present in all kingdoms of life, this family is composed of only three members in mammals: mNT (CISD1), Miner1 (CISD2, ERIS, Noxp70 or NAF-1), and Miner2 (CISD3) (Inupakutika et al., 2017).

MNT is a cluster donor protein whose cellular function is linked to iron and ROS homeostasis (Tamir et al., 2015). Holo-mNT (i.e., the form with the cluster) is capable of transferring in vitro its redox-active Fe–S cluster to a bacterial apo-ferredoxin (FDX), an FDX that has lost its cluster (Zuris et al., 2011). In contrast to the strict anoxia required for other Fe–S transfer proteins such as ISCU, mNT is able to perform cluster transfer under aerobic conditions, suggesting its possible involvement in Fe–S repair after oxidative stress. Recently, we identified the cytosolic iron regulator, named cytosolic aconitase/iron regulatory protein 1 (IRP1) as the first physiological protein acceptor of the mNT Fe–S cluster in an Fe–S cluster repair pathway (Ferecatu et al., 2014).

The first part of this chapter describes methods to study Fe–S cluster transfer/repair in vitro using *Escherichia coli* FDX as model recipient protein. The UV–visible absorption spectroscopy allows following the evolution of holo-protein populations (holo-mNT vs holo-FDX) as well as the redox state of the cluster. With native PAGE and nuclear magnetic resonance (NMR) spectroscopy, the populations of protein species (holo- and apo-mNT, holo- and apo-FDX) are precisely quantified. In the second part, we will describe in cellulo methods to study the involvement of mNT in the repair of the Fe–S cluster of cytosolic aconitase after stress challenge. Finally, we will present protocols to check the expression of the two partners in animal tissues.

2. IN VITRO Fe–S CLUSTER TRANSFER FROM mNT TO A MODEL RECIPIENT

The properties of the mNT cluster transfer reaction can be properly characterized in vitro using a model recipient apo-protein. This protein must (1) be easy to purify in large amounts, (2) have a UV–visible absorption spectrum and a native PAGE migration different from those of holo-mNT, (3) have native PAGE migrations that differ between the apo- and holo-forms, (4) have a holo-form that is stable over time. We used the [2Fe–2S]-containing *E. coli* FDX, which fulfills all these criteria. This protein is stable under aerobic conditions up to roughly 25°C–30°C. A hyperthermophilic FDX will be preferred for studies at higher temperatures. In Section 2.1, we describe the purification of the two proteins (holo-mNT and apo-FDX) and three methods to study the cluster transfer reaction (Fig. 1).

2.1 Purifications of mNT and FDX

The proteins are purified after overexpression in *E. coli* BL21 Star (DE3). For FDX, we have chosen to purify the holo-form, and then to remove the Fe–S cluster to obtain a homogeneous apo-FDX.

Equipment and chemical products

- Protease inhibitor SIGMA FAST EDTA free (Sigma, St. Louis, MO)
- Cell disruptor: Constant Cell Disruption Systems (Constant Systems Ltd., Daventry, UK)
- Äkta fast protein liquid chromatography (FPLC) system (GE Healthcare, Uppsala, Sweden)

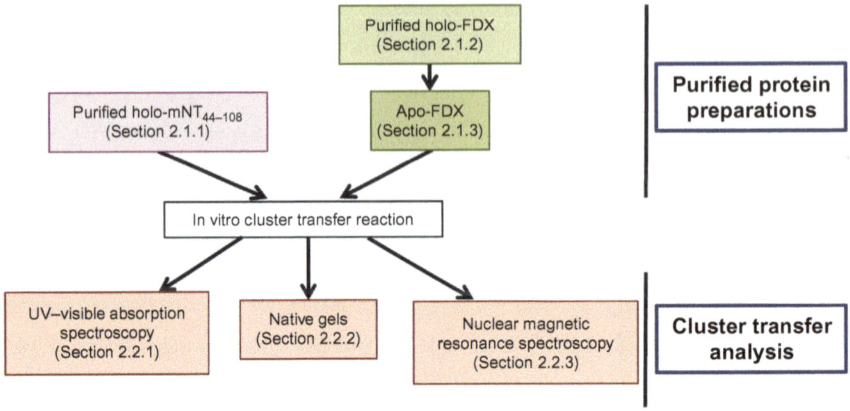

Fig. 1 Scheme summarizing the different steps and methods for studying in vitro Fe–S cluster transfer from holo-mNT to apo-FDX.

General considerations

- All buffers for FPLC have to be 0.45 μm filtered and flushed with nitrogen
- Cell lysates have to be 0.45 μm filtered before loading on columns
- Purification of both proteins can be performed at room temperature. If a multiwavelength detector is used with FPLC, set it to 280 nm (polypeptide) and 460 nm (cluster)
- Proteins were concentrated by centrifugation using Amicon Ultra 15 mL 3 K (Sigma) at 4°C
- Protein concentrations were measured using the Bradford method and BSA as standard

2.1.1 Purification of H. sapiens *Holo-mNT*

We overexpressed a truncated form of *H. sapiens* (*Hs*) mNT (mNT_{44-108}) without its N-terminal mitochondrial anchor, which was therefore restricted to the well-folded part of the protein (Paddock et al., 2007). Codon usage was optimized for its expression in *E. coli* (GenScript). The open reading frame coding for the fragment M44-T108 was cloned between *Nde*I and *Xho*I of pET22b(+) vector. Using this construct, a His-tag (LEHHHHHH) was added to its C-terminal end without effect on the biochemical properties of the protein.

Columns, buffers, and reagents

- HiTrap Chelating HP 5 mL and Superdex 75 HiLoad 16/60 (GE Healthcare)

- Lysis Buffer: 50 m*M* $NaPO_4$ pH 8.0, 500 m*M* NaCl, 15% glycerol
- Buffer A: 50 m*M* $NaPO_4$ pH 8.0, 500 m*M* NaCl, 10 m*M* imidazole
- Buffer B: 50 m*M* $NaPO_4$ pH 8.0, 500 m*M* NaCl, 500 m*M* imidazole
- Buffer C: 50 m*M* $NaPO_4$ pH 8.0, 50 m*M* NaCl

Procedures

β-Mercaptoethanol cannot be used because it increases cluster degradation. The purification is performed in phosphate buffer for increased protein stability.

(1) Grow 3 L of LB culture of *E. coli* transformed by the plasmid with 100 mg/L ampicillin at 37°C with an agitation of 150 rpm. 3 h later, add 500 μ*M* $FeCl_3$ and decrease the temperature to 20°C. When the optical density at 600 nm reaches 0.7, add 0.5 m*M* IPTG. 20 h after, centrifuge the cells at 5000 rpm for 15 min. For NMR studies, mNT_{44-108} is expressed in M9 minimal medium enriched with 0.001% thiamine-HCl and 10 μ*M* $FeCl_3$ and supplemented with 1 g/L of $^{15}NH_4Cl$ and 4 g/L of [^{13}C] D-glucose as the sole nitrogen and carbon sources, respectively.

(2) Resuspend the cells on ice in at least 100 mL Lysis Buffer containing one tablet of protease inhibitors and disrupt them using the cell disruptor. Pass the sample through the cell disruptor until it is no longer viscous (at least three times).

(3) Centrifuge the cell lysate for 1 h at 4°C and 40,000 rpm. Keep the slightly red supernatant.

(4) Equilibrate the HiTrap Chelating HP column with Buffer A, which contains 10 m*M* imidazole and 500 m*M* NaCl to limit nonspecific interactions. Load the supernatant on the column then wash it with Buffer A until stable low absorbance is reached. Apply a 30-mL linear gradient between Buffer A and Buffer B at 1 mL/min.

(5) Pool the red fractions containing mNT (around 40% Buffer B) and concentrate them down to 2 mL.

(6) Equilibrate the Superdex 75 HiLoad 16/60 column with Buffer C. Load the proteins and run the column with Buffer C. Apo-mNT will be eluted before the holo-form (50 and 70 mL, respectively).

(7) Check the purity of the colored fractions on 16% SDS-PAGE gels. Then, analyze the fractions of interest on a 12% native PAGE gel to check the absence of apo-mNT. Pool pure holo-mNT fractions and concentrate them down to no more than 2 m*M* to avoid aggregation. Freeze in dry ice and store at −80°C. We obtain roughly 12 mg of pure holo-mNT per liter of LB culture.

2.1.2 Purification of E. coli Holo-FDX

E. coli FDX is overexpressed without a tag using the pET21a-FDX construct (Nakamura, Saeki, & Takahashi, 1999).

Columns and buffers

- 5 mL HiTrap DEAE Sepharose FF, Superdex 75 HiLoad 16/60, and 1 mL MonoQ columns (GE Healthcare)
- Buffer D: 50 m*M* Tris–HCl pH 7.2, 30 m*M* NaCl, 0.2 m*M* EDTA
- Buffer E: 50 m*M* Tris–HCl pH 7.2, 1 *M* NaCl, 0.2 m*M* EDTA
- Buffer F: 50 m*M* Tris–HCl pH 7.2, 0.1 *M* NaCl, 0.2 m*M* EDTA
- Buffer G: 50 m*M* Tris–HCl pH 7.2, 25 m*M* NaCl, 0.1 m*M* EDTA
- Buffer H: 50 m*M* Tris–HCl pH 7.2, 1 *M* NaCl, 0.1 m*M* EDTA

Procedures

Add 10 m*M* β-mercaptoethanol to all buffers just before use.

(1) Grow 6 L of LB culture of *E. coli* transformed with pET21a-FDX supplemented with 100 mg/L ampicillin. When the optical density at 600 nm reaches 0.7, lower the temperature to 18°C and add 500 μ*M* $FeCl_3$. 40 min later, induce protein overexpression by adding 0.5 m*M* IPTG. After 20 h, centrifuge the culture.

(2) Cells are resuspended on ice in at least 200 mL of Buffer D with antiprotease, disrupted using the cell disruptor and then lysed and centrifuged as described for mNT.

(3) Equilibrate the HiTrap DEAE Sepharose FF column with Buffer D. Load the cell lysate on the column and wash the column with Buffer D until stable low absorbance is reached. Apply an 80-mL linear gradient between 0% and 40% Buffer E. Pool and concentrate the red fractions down to 2 mL.

(4) Equilibrate the Superdex 75 HiLoad 16/60 column with Buffer F. Load the concentrated eluate and run the column with Buffer F. The apo-FDX will be eluted before the holo-form. Analyze the colored fractions on 16% SDS-PAGE, then pool and concentrate them.

(5) Equilibrate the MonoQ column with 93% Buffer G/7% Buffer H. Dilute the protein sample at least fourfold with Buffer G without NaCl, load it on the column, then wash it with 10 mL of 93% Buffer G/7% Buffer H. Apply a 60-mL linear gradient from 7% to 50% Buffer H.

(6) Analyze colored fractions on 16% SDS-PAGE and on a 16% native PAGE gel. Pool the holo-FDX fractions and concentrate them. Freeze in dry ice and store at −80°C. We obtain 6–7 mg of pure holo-FDX per liter of culture.

2.1.3 Preparation of Apo-FDX

Equipment and buffers

- NAP-5 column (GE Healthcare)
- Centricon 5 kDa (Millipore, Billerica, MA)
- Exchange Buffer: 50 m*M* Tris–HCl pH 7.0, 100 m*M* NaCl

Procedure

(1) Incubate 450 μL of holo-FDX (at roughly 1 m*M*) with 10 m*M* dithiothreitol (DTT), 10 m*M* EDTA at 80°C until the protein becomes colorless, which is an evidence of cluster disassembly (at least 5 min).

(2) Equilibrate the NAP-5 column with 10 mL of Exchange Buffer. Load the protein and elute it with Exchange Buffer. Collect the eluate in 0.5 mL fractions and check for the presence of protein using the Bradford assay. Pool the fractions containing apo-FDX and concentrate them using a Centricon 5 kDa down to roughly 450 μL. Aliquot the protein (10 μL fractions), freeze in dry ice, and store at −80°C.

2.2 In Vitro Cluster Transfer Reaction Between Holo-mNT and Apo-FDX

For this reaction, holo-mNT is used as Fe–S cluster donor and DTT-reduced apo-FDX (reduction of the cysteines) as recipient. Recently, we showed that only the oxidized cluster can be transferred and that oxygen does not affect the transfer reaction (Golinelli-Cohen et al., 2016). To prevent cluster disassembly in holo-mNT or holo-FDX, we perform the transfer reaction at a slightly acidic pH (between 6.0 and 6.7; the mNT cluster is too labile at lower pH under aerobic conditions) and 25°C (holo-FDX is not stable enough at higher temperature) to have complete cluster transfer in a few hours. This timescale allows study of the effect of compounds on the transfer with limited change in the compound during the course of the reaction. Finally, we recommend performing the reaction under anaerobic conditions when studying the effect of oxygen-sensitive molecules on the reaction.

2.2.1 UV–Visible Absorption Spectroscopy

This method uses the different spectral properties in the visible region (350–800 nm) of: (1) holo-mNT (maximum absorption at 458 nm for the oxidized form) and holo-FDX (415 and 458 nm for the oxidized form) (Fig. 2A), (2) the different redox states of the proteins (Fig. 2A and B), and (3) the apo- (no absorbance in the visible) and holo-proteins. Cluster transfer is monitored by recording the UV–visible absorption spectra of

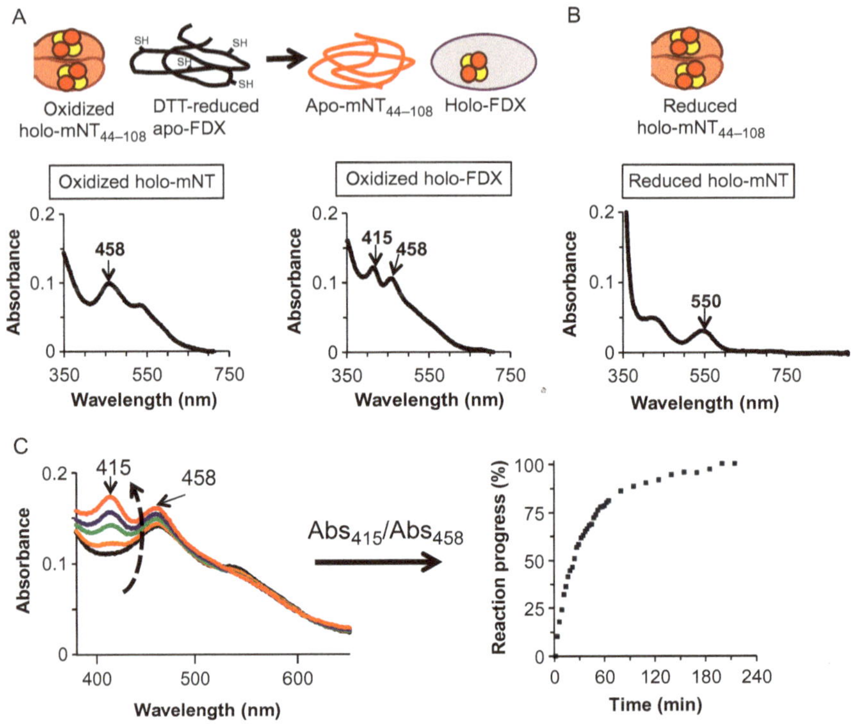

Fig. 2 UV–visible absorption spectroscopy used to monitor in vitro Fe–S cluster transfer between holo-mNT and apo-FDX. (A) Reaction of cluster transfer between oxidized holo-mNT and DTT-reduced apo-FDX (20 μ*M* each) in 50 m*M* Bis–Tris pH 6.2, 100 m*M* NaCl at 25°C under aerobic conditions. At the beginning of the reaction, only oxidized holo-mNT absorbs in the visible part of the spectrum (350–750 nm), with a maximum at 458 nm. When holo-mNT has completely transferred its cluster, only holo-FDX contributes to the absorption in the visible part of the spectrum, with two maxima at 415 and 458 nm. (B) Visible part of the absorption spectrum of 20 μ*M* dithionite-reduced holo-mNT. (C) Kinetics of UV–visible absorption spectra of the transfer reaction (only the 380–680 nm part is shown) (*Left panel*). The *curved arrow* indicates the change in absorbance at 415 nm over time. (*Right panel*) Reaction progress as a function of time. *This research was originally published in Golinelli-Cohen, M.-P., Lescop, E., Mons, C., Goncalves, S., Clemancey, M., Santolini, J., et al. (2016). Redox control of the human iron-sulfur repair protein MitoNEET activity via its iron-sulfur cluster.* The Journal of Biological Chemistry, 291*(14), 7583–7593. © the American Society for Biochemistry and Molecular Biology.*

the reaction over time (Fig. 2C). Initially, only holo-mNT contributes to the visible part of the spectrum, whereas only holo-FDX contributes when all mNT clusters have been transferred (Fig. 2A). Thus, at intermediate times, the spectrum is a combination of the contribution of both holo-proteins. While this spectroscopic method can easily follow cluster transfer, it is worth mentioning that one drawback is the difficulty of detecting species at low concentrations.

Equipment

- UV–visible absorption spectrophotometer Cary 100 (Agilent, Santa Clara, CA) with a temperature control apparatus. A multicell holder allows multiple reactions to be followed in parallel
- High-precision cell (Hellma Analytics, Plainview, NY,reference 105-201-QS). This cuvette is adapted to a small reaction volume (roughly 160 μL) with a reasonable background absorbance
- NAP-5 (GE Healthcare), Micro Bio-Spin 6 columns (Bio-Rad, Hercules, CA)

Procedure

(1) If necessary change the holo-mNT buffer using a NAP-5 column equilibrated with the desired buffer, and proceed as described for the preparation of apo-FDX. If highly concentrated proteins are used, the contribution of protein storage buffers to the reaction is negligible.

(2) Reduce apo-FDX with 5 m*M* DTT for 30 min at room temperature. If necessary, remove DTT using a Micro Bio-Spin 6 columns.

(3) Mix holo-mNT and reduced apo-FDX together (typically 20 μ*M* of each protein) in the desired transfer reaction buffer. Transfer the reaction mixture into the cell and place it in the thermostated spectrophotometer.

(4) Record full spectra between 240 and 900 nm (every 5 min during 2 h and then every 15 min, for example) (Fig. 2C).

(5) At the end of the experiment, correct the spectra for baseline variations (for each wavelength, remove the average of the absorbances between 850 and 900 nm at the measured absorbance).

(6) Calculate and plot the transfer reaction progress *R* over time (Fig. 2C):

$$R = \left(R_{(t)} - R_{\text{initial}}\right) / \left(R_{\text{final}} - R_{\text{initial}}\right)$$

- $R_{(t)}$, ratio between the absorbances at 415 and 458 nm at time t.
- R_{initial} and R_{final}, R at the beginning and at the end (when the transfer is complete) of the reaction, respectively.

2.2.2 Native PAGE

This method allows separation of the apo- and holo-forms of mNT and FDX and easy quantification of each species even if they are present in low amounts, but it does not give information on the redox state of the clusters (Fig. 3).

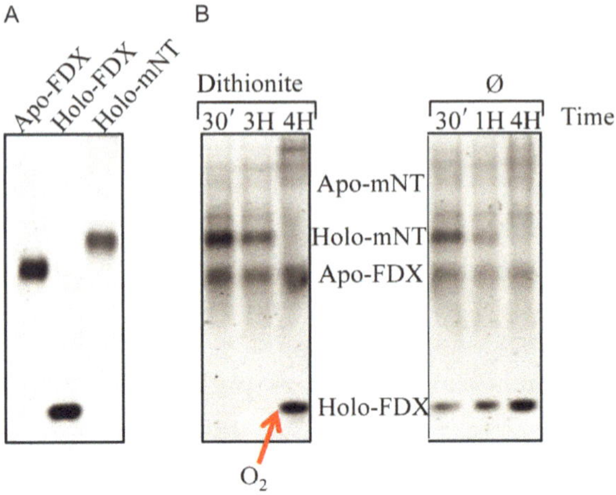

Fig. 3 Native polyacrylamide gels used to follow in vitro Fe–S cluster transfer between holo-mNT and apo-FDX. (A) Differences in migration between apo-, holo-FDX, and holo-mNT on a 16% reducing native gel. (B) Reaction of cluster transfer between holo-mNT and DTT-reduced apo-FDX (20 μ*M* each) in 50 m*M* Bis–Tris pH 6.2, 100 m*M* NaCl at 25°C. (*Left panel*) mNT was reduced with 2 m*M* dithionite before starting the reaction performed under anaerobic conditions up to 3 h, then under aerobic conditions. (*Right panel*) The same reaction is performed under aerobic conditions without reduction of mNT. *Panel A: This research was originally published in Golinelli-Cohen, M.-P., Lescop, E., Mons, C., Goncalves, S., Clemancey, M., Santolini, J., et al. (2016). Redox control of the human iron-sulfur repair protein MitoNEET activity via its iron-sulfur cluster.* The Journal of Biological Chemistry, 291*(14), 7583–7593. © the American Society for Biochemistry and Molecular Biology.*

Equipment

- 16% native PAGE gel, loading and electrophoresis buffers (Arndt, Koristka, Bartsch, & Bachmann, 2012)
- Odyssey Infrared Imaging System (LI-COR, Lincoln, NE)

Procedure

(1) The reaction mixture is prepared as described previously (see Section 2.2.1), except that the reaction takes place in a 1.5-mL centrifuge tube in a thermostated plate.

(2) Aliquots (8 μL for reactions performed with 20 μ*M* proteins) are removed from the reaction mixture at regular times and immediately frozen in dry ice.

(3) When the reaction is complete, add 5 m*M* DTT to the samples (to obtain thinner bands on gels) and load them on a 16% native PAGE gel. Run it at 70 V on ice until elution of the bromophenol blue present in the loading buffer.

(4) Stain the gel using the highly sensitive Blue Silver staining (Candiano et al., 2004).

(5) After destaining, scan the gel using the Odyssey Infrared Imaging System. Quantify each band (Fig. 3) using LI-COR Odyssey software, calculate the ratio [holo-FDX]/[Total FDX] and plot it over time.

2.2.3 NMR Spectroscopy

1D and multidimensional NMR techniques allow the direct and real-time observation and quantification of the various states of mNT and FDX in a single experiment. Here, we use ^{15}N-labeled holo-mNT$_{(44–108)}$ and apo-FDX at natural abundance. Each mNT state gives rise to a distinct ^{15}N HSQC spectrum, and their relative proportions can be assessed from cross-peak intensity. The ^{15}N HSQC spectra of holo-mNT in its oxidized or reduced states are very similar and well dispersed, indicating a well-folded protein. Nevertheless, the ^{1}H/^{15}N chemical shifts of residues V57, H58, and I102 are distinct in the reduced and oxidized states and are convenient probes of the mNT redox state (see Fig. 4A). Upon cluster loss, the ^{15}N HSQC spectrum shows substantial reduction in ^{1}H spectral dispersion, due to its disordered nature (Ferecatu et al., 2014; Golinelli-Cohen et al., 2016), as shown in Fig. 4B.

Equipment

- A high-field NMR spectrometer (at least 600 MHz) equipped with a triple resonance, preferentially cryogenically cooled, probe.
- For aerobic conditions, standard 3-mm or 5-mm NMR tubes can be used. For anaerobic conditions, we use a 5-mm NMR tube equipped with a controlled atmosphere valve (for example, the NE-CAV5 series from New Era, Vineland, NJ) and we prepare the sample in a glove box.

Procedure

(1) Prepare a solution (200 μL for a 3-mm tube, 500 μL for a 5-mm tube) of 50–250 μ*M* ^{15}N-labeled holo-mNT in the required reaction buffer with 95%/5% H_2O/D_2O.

(2) Check the initial redox state of mNT by transferring the solution to the NMR tube and collect reference NMR spectra including the water-suppressed ^{15}N-decoupled ^{1}H spectrum, water-suppressed ^{15}N-filtered ^{1}H spectrum, and 2D ^{15}N SOFAST-HMQC (Schanda & Brutscher, 2005; Schanda, Kupce, & Brutscher, 2005).

(3) Prepare a stock solution of 1–2 m*M* apo-FDX in 50 m*M* Tris–HCl pH 8.0, 100 m*M* NaCl 95%/5% H_2O/D_2O. Treat apo-FDX with 5 m*M*

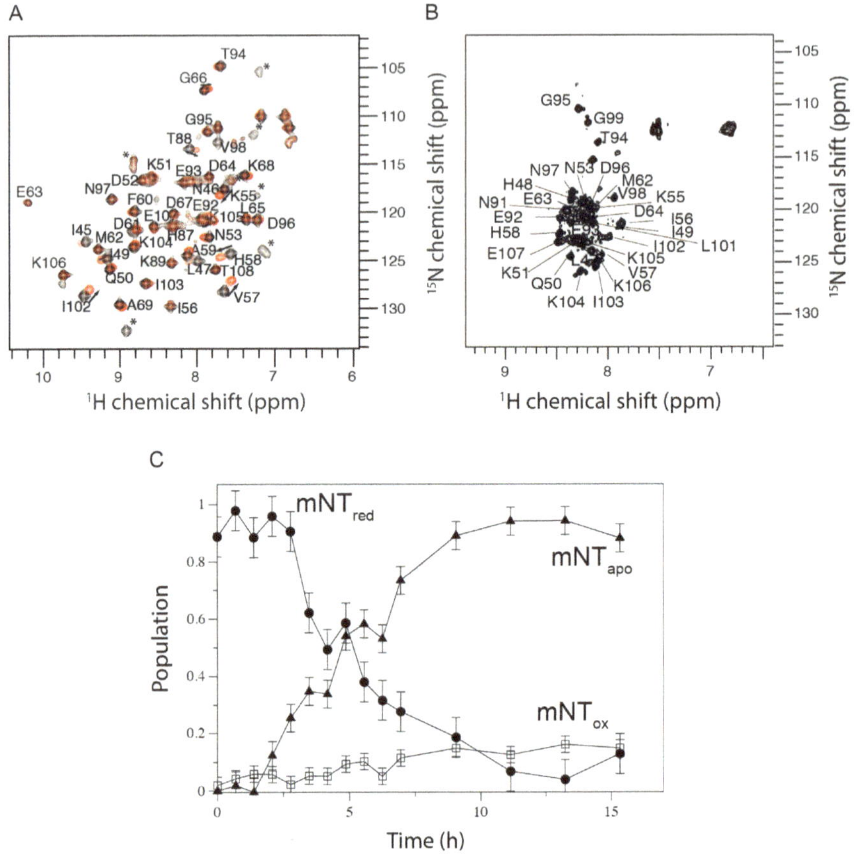

Fig. 4 NMR fingerprints of mNT. (A) Superimposition of $^{1}H/^{15}N$ correlation spectra of oxidized (*black*) and reduced (*gray* or *red*) holo-mNT in 50 m*M* Tris–HCl pH 8.0, 100 m*M* NaCl at 298 K. (B) $^{1}H/^{15}N$ correlation spectrum of apo-mNT in 50 m*M* Bis–Tris pH 6.2, 100 m*M* NaCl at 298 K. The $^{1}H/^{15}N$ cross-peaks in panels A and B are labeled according to their amino acid number. (C) Example of the real-time monitoring of Fe–S cluster transfer from holo-mNT to apo-FDX. In this experiment, reduced holo-mNT was mixed at 1:1 *M* ratio with apo-FDX under anaerobiosis. At $t=0$, the NMR tube was opened and the relative populations of the reduced (mNT$_{red}$), and oxidized (mNT$_{ox}$) mNT, and apo-mNT (mNT$_{apo}$) were estimated from the cross-peak intensity in $^{1}H/^{15}N$ correlation spectra, highlighting the cluster transfer from mNT to FDX upon mNT reoxidation. *Results presented in panels A and B were originally published in Golinelli-Cohen, M.-P., Lescop, E., Mons, C., Goncalves, S., Clemancey, M., Santolini, J., et al. (2016). Redox control of the human iron-sulfur repair protein MitoNEET activity via its iron-sulfur cluster.* The Journal of Biological Chemistry, 291*(14), 7583–7593. © The American Society for Biochemistry and Molecular Biology. Panel C is the quantification of the experiment originally published as fig. 4 in the same publication.*

DTT for 30 min to ensure free thiols. If required, excess DTT can be removed using Micro Bio-Spin 6.

(4) Add the required volume of apo-FDX to the mNT sample to achieve the expected mNT/apo-FDX molar ratio and transfer the reaction mixture into the NMR tube. This operation represents time = 0 h of the reaction.

(5) Collect NMR spectra including the water-suppressed ^{15}N-decoupled ^{1}H spectrum and 2D ^{15}N SOFAST-HMQC. The reaction rate being rather slow, a series of NMR experiments can last up to 15 min and be repeated as long as the reaction is not complete.

(6) The progress of the reaction can be followed by monitoring peak intensity variations in the 2D ^{15}N SOFAST-HMQC for signals specific to the three mNT states. For the holo-mNT states, peak intensities can be normalized to the initial intensities, assuming that reduced and oxidized states have similar relaxation properties. The apo-mNT is highly dynamic and normalization can be done using intensities at the end of the reaction when holo-mNT has been completely converted into apo-mNT (Fig. 4C). The water-suppressed ^{15}N-decoupled ^{1}H spectra contain contributions from the side chain of the single tryptophan residue (W77) in the apo- (at 10.08 ppm) and holo-FDX (at 10.02 ppm) forms that can be used to estimate the population of FDX forms.

3. IDENTIFICATION OF A PHYSIOLOGICAL RECIPIENT OF THE mNT CLUSTER

A physiological recipient of the mNT cluster has to have the following characteristics: (1) it has to be cytosolic (localization of mNT cluster), (2) its apo-form should not be degraded in living cells, (3) its cluster has to be partially or totally disassembled by stress. Considering all these conditions, the first candidate recipient we have identified is cytosolic aconitase/IRP1 (Ferecatu et al., 2014). In mammals, this bifunctional protein readily commutes from a [4Fe–4S]-containing form (aconitase activity) to an apo-form (RNA-binding activity) upon exposure to NO or H_2O_2 (Pantopoulos & Hentze, 1995; Soum & Drapier, 2003). Interestingly, IRP1 can recycle into cytosolic aconitase after nitrosative stress involving energized mitochondria (Bouton, Chauveau, Lazereg, & Drapier, 2002). This part will be dedicated to the methods we used to identify and characterize IRP1 as a physiological partner of mNT after nitrosative damage of its cluster.

3.1 In Vitro Studies

The first step is to check if holo-mNT can transfer in vitro its cluster to IRP1, which has been obtained from cytosolic aconitase exposed to nitrosative stress leading to complete Fe–S cluster disassembly (Fig. 5).

General considerations

- The disassembly of the cytosolic aconitase cluster was performed by treatment of the protein using a diazeniumdiolate NO donor, diethylamine

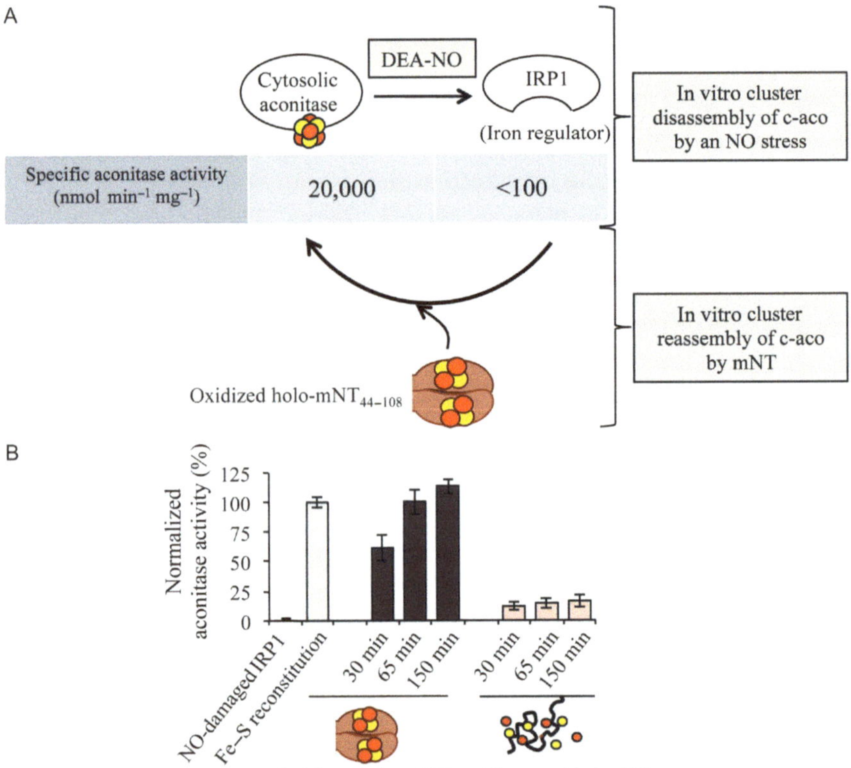

Fig. 5 (A) Scheme of the protocol of in vitro reactivation of the activity of cytosolic aconitase by transfer of the mNT cluster. First, purified cytosolic aconitase is treated with DEA-NO under anaerobic conditions (IRP1). Then, IRP1 is incubated with oxidized mNT at 25°C under anaerobic conditions at pH 6.2. (B) Kinetics of aconitase activity after incubation of IRP1 with either Na_2S and Mohr's salt, oxidized holo-mNT, or disassembled mNT. *This research was originally published in Ferecatu, I., Gonçalves, S., Golinelli-Cohen, M.-P., Clémancey, M., Martelli, A., Riquier, S., et al. (2014). The drug diabetes MitoNEET governs a novel trafficking pathway to rebuild an Fe-S cluster into cytosolic aconitase/iron regulatory protein 1. The Journal of Biological Chemistry, 289(41), 28070–28086. © The American Society for Biochemistry and Molecular Biology.*

(DEA) NONOate. This donor has a half-life of roughly 2 min at room temperature and neutral pH (Soum & Drapier, 2003).

- The insertion of a functional [4Fe–4S] cluster into IRP1 is followed by measurement of the aconitase activity of the reaction. UV–visible absorption spectroscopy (poor absorption of the [4Fe–4S] cluster compared to the [2Fe–2S] cluster of mNT) and native gels (no difference in migration between the cytosolic aconitase and IRP1) cannot be used to monitor the reaction.
- All the steps until the measurement of aconitase activity require a glove box due to the high oxygen sensitivity of cytosolic aconitase.

3.1.1 *In Vitro* Cluster Transfer Reaction Between Holo-mNT and IRP1

Proteins and buffers

- Human cytosolic aconitase was expressed and purified as described by Brazzolotto, Gaillard, Pantopoulos, Hentze, and Moulis (1999). Its specific activity was around 20,000 nmol/min/mg
- Purified holo-mNT_{44-108}: see Section 2.1.1
- Disassembled mNT: Incubate holo-mNT in Transfer Buffer at 65°C for 10 min
- DEA-NONOate (DEA-NO) (Cayman Chemical, Ann Arbor, MI, Cat. #82100). Prepare the stock solution in a basic buffer (1 *M* Bis–Tris propane pH 9.0, for example)
- Buffer A: 20 m*M* Hepes pH 7.3, 0.2 m*M* citrate, 30 m*M* ammonium acetate
- Buffer B: 10 m*M* Hepes pH 7.6, 40 m*M* KCl, 3 m*M* $MgCl_2$
- Transfer Buffer: 100 m*M* Bis–Tris pH 6.2, 100 m*M* NaCl, 5 m*M* DTT

Procedure

(1) Treat 290 μ*M* human holo-IRP1 with 7 m*M* DEA-NO in Buffer A for 1 h at 29°C in a glove box and purify the reaction on a Micro Bio-Spin 6 equilibrated with Buffer B. No residual aconitase activity should be detected.
(2) Incubate IRP1 with 5 m*M* DTT for 30 min at room temperature.
(3) Incubate 100 μ*M* holo-mNT at 25°C with 25 μ*M* IRP1 in Transfer Buffer in a total volume of 20 μL.
(4) After 30 min, 1 h, and 2.5 h, remove 2.5 μL aliquots of the reaction mixture to measure aconitase activity (see Section 3.1.2).
(5) The negative control reaction is performed in parallel using disassembled mNT.

(6) Check the ability of IRP1 to receive a functional cluster by chemical reconstitution using 250 μ*M* Na2S and Mohr's salt for 1 h at 25°C in a glove box. One reaction has to be performed in Buffer B (highest reactivation expected) and one in the Transfer Buffer reaction (positive control of the experiment).

3.1.2 Measurement of Aconitase Activity

Several aconitase assays based on the conversion of *cis*-aconitate to isocitrate could be routinely used. One is indirect and uses excess isocitrate dehydrogenase, which catalyzes the oxidation of isocitrate into α-ketoglutarate, coupled to the reduction of $NADP^+$, which is monitored at 340 nm (Fansler & Lowenstein, 1969). Another method uses in-gel detection of aconitase activity after migration of total cell extracts on native gels (differential migration of mitochondrial and cytosolic aconitases) and incubation of the gels in a buffer containing *cis*-aconitic acid and isocitrate dehydrogenase (Tong & Rouault, 2006). In our case, we directly monitor the disappearance of *cis*-aconitate at 240 nm (Drapier & Hibbs, 1996).

Buffers and chemicals

- Aconitase Buffer: 100 m*M* Tris–HCl pH 7.4 preheated at 37°C. Flush it with an inert gas as nitrogen before use
- *cis*-aconitate solution: 20 m*M* *cis*-aconitic acid pH 7.4

Equipment

- SAFAS Monaco UV–visible spectrophotometer MC 98000 (Monaco) with a temperature controller set at 37°C.
- High-precision cell (Hellma Analytics, reference 104-QS).

Procedure

(1) Add (685–V_e) μL of Aconitase Buffer to the cell.
(2) Add V_e of the protein to be tested.
(3) Add 15 μL of the *cis*-aconitate solution. Use a 200-μL pipetman or invert the cell at least three times to mix the reaction. Be careful not to form bubbles.
(4) Start the spectrophotometer and follow the absorbance at 240 nm for 20 min.
(5) Make a control cell without protein using 685 μL of Aconitase Buffer and 15 μL of *cis*-aconitate solution. The absorbance of the control reaction should be stable over time.
(6) Calculate the specific activity of aconitase (SA) using the slope of the linear part of the kinetics (ΔA, the initial rate). 1 unit corresponds to 1 nmol of substrate consumed per minute.

$SA\left(\frac{U}{mg}\right) = \frac{\Delta A * 700}{\varepsilon * [c] * Ve}$; $\varepsilon = 3.6\,mM^{-1}\,cm^{-1}$; V_e: volume of the protein to be tested; [c]: protein concentration.

3.2 In Cellulo Studies

For in cellulo studies, we use HeLa cells, a human cell line from cervical cancer easily transfectable by plasmids or siRNA. Cells were exposed to NO challenge, which disrupts the Fe–S cluster of cytosolic aconitase. Then, we followed the kinetics of recovery of cytosolic aconitase activity (Section 3.2.1). To look at the involvement of mNT in the cluster repair process, we did this experiment with cells depleted or not in mNT using RNAi technology (Section 3.2.2) and we compared the kinetics of activity recovery in the absence and presence of mNT.

3.2.1 Effect of NO Challenge on Cytosolic Aconitase Activity

In these experiments, cells are treated with an NO donor (Section 3.2.1.1) (Phase 1, Fig. 6A), then the NO stress is removed and we monitor the kinetics of cytosolic aconitase activity recovery (Section 3.2.1.2) (Phase 2, Fig. 6A).

3.2.1.1 NO Challenge Assay

Reagents

- Diethylenetriamine NONOate (DETA-NO) (Cayman Chemical, Cat. #82120). Prepare a stock solution in basic buffer (1 *M* Bis–Tris propane pH 9.0, for example) or directly in the appropriate cell medium (physiological pH around 7.4).

Procedure

Day 1. Seed HeLa cells at 3.5×10^5 cells/cm^2 in DMEM containing 4.5 g/L glucose and 1 m*M* stable L-glutamine (Sigma, Cat. #D0819), supplemented with 1% penicillin–streptomycin (Sigma, Cat. #P0781) and 10% fetal bovine serum (Lonza, Gampel, Switzerland, Cat. #DE14-801F). Culture them overnight under 5% CO_2 and humidified atmosphere at 37°C.

Day 2. Add 250 μ*M* DETA-NO in a fresh medium and check the DETA-NO concentration by measuring the absorbance at 252 nm ($\varepsilon = 7640\,M^{-1}\,cm^{-1}$) using 100 μL of cell medium in 900 μL of distilled water. The absorbance should be around 0.4 (0.2 in the absence of DETA-NO). Add the medium to the cells and incubate them for 16 h.

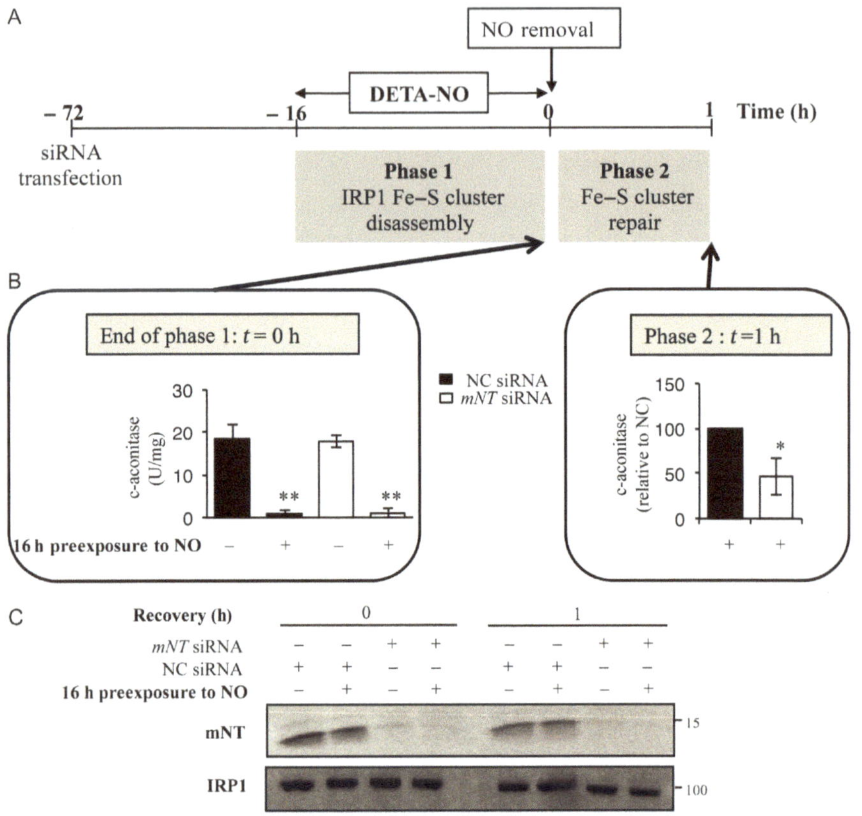

Fig. 6 mNT is required to restore cytosolic aconitase activity after NO challenge in cellulo. (A) Scheme of the protocol. Cells transfected for 72 h with mNT or control siRNA were exposed to DETA-NO for 16 h. NO was removed ($t=0$) and the cells were incubated further in fresh culture medium for an additional hour. (B) (*Left graph*) Cytosolic aconitase activity at time 0 h (phase 1) represented as means ± SD, $n=4$, $^{**}P<0.001$. (*Right graph*) Relative activities of cytosolic aconitase at 1 h after NO removal (phase 2). Results are presented as the ratio of cytosolic aconitase recovery for *mNT*-depleted cells over NC cells ± SD $n=4$, $^{*}P<0.05$. (C) Immunoblotting of mNT (mitochondrial fractions) and IRP1 (cytosolic fractions). *This research was originally published in Ferecatu, I., Gonçalves, S., Golinelli-Cohen, M.-P., Clémancey, M., Martelli, A., Riquier, S., et al. (2014). The drug diabetes MitoNEET governs a novel trafficking pathway to rebuild an Fe-S cluster into cytosolic aconitase/iron regulatory protein 1.* The Journal of Biological Chemistry, 289(41), 28070–28086. *© The American Society for Biochemistry and Molecular Biology.*

Day 3. Remove the media, wash the cells twice with medium (the absorbance at 252 nm should be at 0.2), add fresh cell culture medium ($t=0$ of the kinetics of aconitase recovery), and then incubate them further (for 1, 2, 4, 6, and 8 h, for example) at 37°C.

3.2.1.2 Preparation of Mitochondria-Enriched Fractions and Cytosolic Extracts

General considerations

- Protein concentrations are measured using the BCA assay (Smith et al., 1985).

Equipment

- Beckman Coulter LE80K Ultracentrifuge with 70Ti rotor
- 10-mL thick wall polycarbonate tubes (Beckman, Fullerton, CA, Cat. #355640)

Buffers and reagents

- Buffer A: 0.25 *M* sucrose, 100 m*M* HEPES pH 7.4, 1/200 (v/v) protease inhibitors (Sigma, Cat. #P8340)
- Digitonin (Merck Millipore, Cat. #300410) 1% solution in DMSO stored in dark at room temperature
- Trypan blue 0.4% (Sigma, Cat. #93595)
- Buffer B: 100 m*M* Tris–HCl pH 7.5, 0.5% Triton X-100, 1/200 (v/v) protease inhibitors

Procedure

(1) At the desired time, remove the media and wash the cells twice with cold PBS buffer. Keep cells on ice throughout.

(2) Harvest cells directly in cold Buffer A. Add digitonin (cell permeabilization) at a final percentage of 0.007%, mix gently several times for 15 min on ice.

(3) Check the permeabilization of cell membranes by staining an aliquot with trypan blue (v/v) and estimate the percentage of permeabilized (blue) cells under a photonic microscope. If less than 90% of the cells are permeabilized, continue the incubation.

(4) Centrifuge the cell lysate at 10,000 × *g* for 10 min at 4°C then separate the mitochondria-enriched fractions (pellet, see step 6) from the cytosolic fractions (supernatant, see step 5).

(5) Ultracentrifuge the resulting supernatant at 100,000 × *g* for 1 h at 4°C in order to remove cell particles. Recover the supernatant (pure cytosol), determine the protein concentration, measure the cytosolic aconitase activity using 50 μg of protein (see Section 3.1.2), calculate the specific activities and plot them over time.

(6) Wash the mitochondria-enriched pellets with cold Buffer A twice and lyse the pellets with degassed Buffer B (usually with 80–100 μL, depending on the size of the pellet) and incubate for 5 min on ice.

Centrifuge at 10,000 × *g* for 5 min at 4°C, recover the supernatant and measure immediately the mitochondrial aconitase activity using 15 μL of the mitochondria-enriched fraction (see Section 3.1.2) and degassed buffer. The mitochondrial aconitase is particularly oxygen sensitive; therefore, mitochondria lysis and enzymatic activity measurements have to be performed under an anaerobic atmosphere using buffers flushed with nitrogen gas.

(7) Measure protein concentration and calculate the specific activities and plot them over time.

Notes

- Keep one dish untreated to measure the specific activity of cytosolic aconitase in control cytosolic extract (roughly 20 U/mg with HeLa cells). Just after NO stress relief, the activity should be very low (1–2 U/mg).
- NO challenge should not affect the cytosolic aconitase protein level (Fig. 6C). We recommend checking it by immunoblotting using standard procedures. Load 40 μg of protein extracts on an 8% reducing SDS-PAGE gel, transfer the proteins onto a membrane (Immobilon PVDF 0.45 μm, Millipore) and use specific antibodies against IRP1 (from Agro-Bio, La Ferté Saint-Aubin, France, for example) and mitochondrial aconitase.

3.2.2 Involvement of mNT in the Fe–S Repair of Cytosolic Aconitase After Nitrosative Stress

To check the involvement of mNT in a repair pathway of the cytosolic aconitase cluster after nitrosative stress, we compare the aconitase activity recovery kinetics after NO challenge in HeLa cells with and without siRNA depletion of mNT. In the absence of mNT, slower recovery of aconitase activity indicates involvement of mNT in the repair pathway (Fig. 6B).

General considerations

- In order to optimize endogenous gene silencing, a range of siRNA duplex concentrations (1–20 n*M*) and different volumes of the transfection reagent (30–50 μL for a 100-mm diameter dish) should be tested beforehand. In the following protocol, we specify the optimized concentrations we use to deplete mNT in HeLa cells.
- At the end of the reaction, check by immunoblot the level of mNT depletion in the mitochondrial extracts (Fig. 6C).

Reagents

- siRNA duplexes (Life Technologies, Carlsbad, CA): CISD1 (mNT-0, Cat. #s31650; mNT-1, Cat. #s31651), negative control (NC, Cat. #4390843).
- INTERFERin™ Transfection Reagent (Polyplus Transfection, Illkirch, France, Ozyme Cat. #409-10)

Procedure

Day 1. Seed cells in a 100 mm Petri dish to reach approximately 30% confluence at the time of transfection (3.5×10^5 cells/cm^2 for HeLa cells) and incubate under 5% CO_2 and humidified atmosphere at 37°C.

Day 2. Transfect cells with 10 n*M* siRNA duplexes and 40 μL INTERFERin™ Transfection Reagent (the presence of serum and penicillin–streptomycin is not an impediment). Incubate the plate at 37°C. After 5 h, change the medium to ensure lower toxicity.

Day 3. Proceed to an NO challenge (see Section 3.2.1.1).

3.3 Partner Protein Abundance in Mammalian Tissues

In the previous part, the involvement of mNT in the repair of the cytosolic aconitase was proved using cultured cell lines. To definitively demonstrate that both proteins are partners in mammals, it is necessary to check that they can be both found in at least one common tissue. Thus, we looked at the profile of expression of mNT and IRP1 in different mouse tissues by immunoblotting. Neither is ubiquitous, but they are both well expressed in several tissues, including kidney, liver, heart, and brain (Fig. 7).

Equipment

- Sonicator ultrasonic processor XL (Misonix Incorporated, Farmingdale, NY). Use it at position 7.

Reagents

- Transfer Buffer: 12.5 m*M* Tris–HCl, 96 m*M* glycine, variable percentage of ethanol
- Buffer A: 50 m*M* NaCl, 25 m*M* Tris–HCl pH 7.5, 2 m*M* EDTA, 0.5% Triton X-100
- Primary antibodies against IRP1 (Medimabs, Montréal, Canada, #MM-0074, used at a 1/1000 dilution) and mNT (Eurogentec, Liège, Belgium, antibody raised against the following peptide "CKAVYCRCWRSKKFPF," used at a 1/2000 dilution)

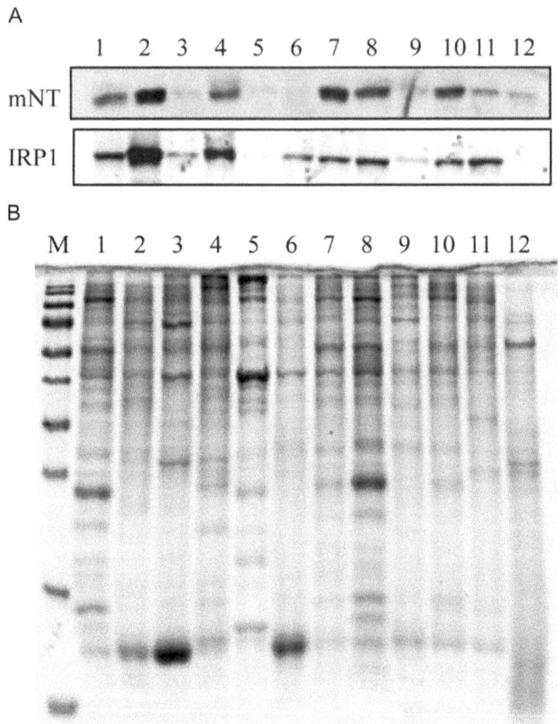

Fig. 7 Level of expression of mNT and IRP1 in different mouse tissues. (A) Immunoblotting against mNT and IRP1. (B) SDS-PAGE gel stained with Simply Blue. (1) brain stem, (2) kidney, (3) lung, (4) liver, (5) muscle, (6) spleen, (7) brain, (8) spinal cord, (9) heart, (10) cerebellum, (11) testis, and (12) pancreas.

- Fluorescent secondary antibodies coupled to IRDye 800CW (LI-COR Biosciences)
- Simply Blue kit (Invitrogen, Life Technologies)

Procedure

(1) Sonicate 30–100 mg of each tissue in 1 mL of Buffer A with cocktail protease inhibitor (Sigma, Cat. #P8340) (1/200) on ice for 15 s and then centrifuge at 14,000 × *g* at 4°C for 15 min.

(2) Determine the protein content of supernatant using the BCA protein assay.

(3) Separate 10 μg of protein extracts on two reducing polyacrylamide SDS-PAGE gels (one 16% for detection of mNT and one 8% for IRP1).

(4) Transfer proteins onto Immobilon-FL (Millipore) membranes for 90 min at 100 mA/gel using Transblot Cell (Life Technology) in Transfer Buffer containing 10% and 20% ethanol for IRP1 and mNT detection, respectively.

(5) Perform immunoblotting using standard procedures. Incubate with primary antibodies overnight at 4°C. After washing in TBS-T, fluorescent secondary antibodies coupled to LI-COR IRDye 800CW are used at a 1/20,000 dilution (Fig. 7).

(6) No protein is expressed at exactly the same level in all tissues and can be used as loading control. Thus, as loading control, perform another SDS-PAGE under the same conditions, but stain the gel using the Simply Blue kit.

4. CONCLUSIONS

In this chapter, we described methods to study in vitro Fe–S cluster transfer from holo-mNT to a recipient model, *E. coli* apo-FDX, and to the first identified physiological recipient, cytosolic aconitase/IRP1 in cellulo and in vivo. More generally, methods described in the first part can be easily adapted to study any protein involved in Fe–S cluster transfer/repair. The second part gives a general scheme to assess physiological Fe–S cluster recipients.

REFERENCES

Arndt, C., Koristka, S., Bartsch, H., & Bachmann, M. (2012). Native polyacrylamide gels. In B. T. Kurien & R. H. Scofield (Eds.), *Protein electrophoresis: Methods and protocols* (pp. 49–53). Totowa, NJ: Humana Press.

Bouton, C., Chauveau, M. J., Lazereg, S., & Drapier, J. C. (2002). Recycling of RNA binding iron regulatory protein 1 into an aconitase after nitric oxide removal depends on mitochondrial ATP. *The Journal of Biological Chemistry*, *277*(34), 31220–31227.

Brazzolotto, X., Gaillard, J., Pantopoulos, K., Hentze, M. W., & Moulis, J. M. (1999). Human cytoplasmic aconitase (iron regulatory protein 1) is converted into its [3Fe-4S] form by hydrogen peroxide in vitro but is not activated for iron-responsive element binding. *The Journal of Biological Chemistry*, *274*(31), 21625–21630.

Candiano, G., Bruschi, M., Musante, L., Santucci, L., Ghiggeri, G. M., Carnemolla, B., et al. (2004). Blue silver: A very sensitive colloidal Coomassie G-250 staining for proteome analysis. *Electrophoresis*, *25*(9), 1327–1333.

Colca, J. R., McDonald, W. G., Waldon, D. J., Leone, J. W., Lull, J. M., Bannow, C. A., et al. (2004). Identification of a novel mitochondrial protein ("mitoNEET") cross-linked specifically by a thiazolidinedione photoprobe. American Journal of Physiology. Endocrinology and Metabolism, 286(2), E252–E260.

Drapier, J. C., & Hibbs, J. B., Jr. (1996). Aconitases: A class of metalloproteins highly sensitive to nitric oxide synthesis. *Methods in Enzymology*, *269*, 26–36.

Fansler, B., & Lowenstein, J. (1969). Aconitase from pig heart. *Methods in Enzymology, 13*, 26–30.

Ferecatu, I., Gonçalves, S., Golinelli-Cohen, M.-P., Clémancey, M., Martelli, A., Riquier, S., et al. (2014). The drug diabetes MitoNEET governs a novel trafficking pathway to rebuild an Fe-S cluster into cytosolic aconitase/iron regulatory protein 1. *The Journal of Biological Chemistry, 289*(41), 28070–28086.

Golinelli-Cohen, M.-P., Lescop, E., Mons, C., Goncalves, S., Clemancey, M., Santolini, J., et al. (2016). Redox control of the human iron-sulfur repair protein MitoNEET activity via its iron-sulfur cluster. *The Journal of Biological Chemistry, 291*(14), 7583–7593.

Heidrich, H. G., Albracht, S. P., & Backstrom, D. (1978). Two iron-sulfur centers in mitochondrial outer membranes from beef heart as prepared by free-flow electrophoresis. *FEBS Letters, 95*(2), 314–318.

Inupakutika, M. A., Sengupta, S., Nechushtai, R., Jennings, P. A., Onuchic, J. N., Azad, R. K., et al. (2017). Phylogenetic analysis of eukaryotic NEET proteins uncovers a link between a key gene duplication event and the evolution of vertebrates. *Scientific Reports, 7*, 42571.

Nakamura, M., Saeki, K., & Takahashi, Y. (1999). Hyperproduction of recombinant ferredoxins in Escherichia coli by coexpression of the ORF1-ORF2-iscS-iscU-iscA-hscB-hs cA-fdx-ORF3 gene cluster. *Journal of Biochemistry, 126*(1), 10–18.

Paddock, M. L., Wiley, S. E., Axelrod, H. L., Cohen, A. E., Roy, M., Abresch, E. C., et al. (2007). MitoNEET is a uniquely folded 2Fe 2S outer mitochondrial membrane protein stabilized by pioglitazone. *Proceedings of the National Academy of Sciences of the United States of America, 104*(36), 14342–14347.

Pantopoulos, K., & Hentze, M. W. (1995). Rapid responses to oxidative stress mediated by iron regulatory protein. *The EMBO Journal, 14*(12), 2917–2924.

Schanda, P., & Brutscher, B. (2005). Very fast two-dimensional NMR spectroscopy for real-time investigation of dynamic events in proteins on the time scale of seconds. *Journal of the American Chemical Society, 127*(22), 8014–8015.

Schanda, P., Kupce, E., & Brutscher, B. (2005). SOFAST-HMQC experiments for recording two-dimensional heteronuclear correlation spectra of proteins within a few seconds. *Journal of Biomolecular NMR, 33*(4), 199–211.

Smith, P. K., Krohn, R. I., Hermanson, G. T., Mallia, A. K., Gartner, F. H., Provenzano, M. D., et al. (1985). Measurement of protein using bicinchoninic acid. *Analytical Biochemistry, 150*(1), 76–85.

Soum, E., & Drapier, J. C. (2003). Nitric oxide and peroxynitrite promote complete disruption of the [4Fe-4S] cluster of recombinant human iron regulatory protein 1. *Journal of Biological Inorganic Chemistry, 8*(1–2), 226–232.

Tamir, S., Paddock, M. L., Darash-Yahana-Baram, M., Holt, S. H., Sohn, Y. S., Agranat, L., et al. (2015). Structure-function analysis of NEET proteins uncovers their role as key regulators of iron and ROS homeostasis in health and disease. *Biochimica et Biophysica Acta, 1853*, 1294–1315.

Tong, W. H., & Rouault, T. A. (2006). Functions of mitochondrial ISCU and cytosolic ISCU in mammalian iron-sulfur cluster biogenesis and iron homeostasis. *Cell Metabolism, 3*(3), 199–210.

Wiley, S. E., Murphy, A. N., Ross, S. A., van der Geer, P., & Dixon, J. E. (2007). MitoNEET is an iron-containing outer mitochondrial membrane protein that regulates oxidative capacity. *Proceedings of the National Academy of Sciences of the United States of America, 104*(13), 5318–5323.

Zuris, J. A., Harir, Y., Conlan, A. R., Shvartsman, M., Michaeli, D., Tamir, S., et al. (2011). Facile transfer of [2Fe-2S] clusters from the diabetes drug target mitoNEET to an apo-acceptor protein. *Proceedings of the National Academy of Sciences of the United States of America, 108*(32), 13047–13052.

CHAPTER FIVE

Defining the Architecture of the Core Machinery for the Assembly of Fe–S Clusters in Human Mitochondria

Oleksandr Gakh*[,1], Wasantha Ranatunga*, Belinda K. Galeano[†], Douglas S. Smith IV*, James R. Thompson*, Grazia Isaya*[,†,‡,1]

*Mayo Clinic, Rochester, MN, United States
[†]Mayo Clinic Graduate School of Biomedical Sciences, Rochester, MN, United States
[‡]Mayo Clinic Children's Research Center, Rochester, MN, United States
[1]Corresponding authors: e-mail address: gakh.oleksandr@mayo.edu; isaya@mayo.edu

Contents

Abstract

Although Fe–S clusters may assemble spontaneously from elemental iron and sulfur in protein-free systems, the potential toxicity of free Fe^{2+}, Fe^{3+}, and S^{2-} ions in aerobic

Methods in Enzymology, Volume 595
ISSN 0076-6879
http://dx.doi.org/10.1016/bs.mie.2017.07.003

environments underscores the requirement for specialized proteins to oversee the safe assembly of Fe–S clusters in living cells. Prokaryotes first developed multiprotein systems for Fe–S cluster assembly, from which mitochondria later derived their own system and became the main Fe–S cluster suppliers for eukaryotic cells. Early studies in yeast and human mitochondria indicated that Fe–S cluster assembly in eukaryotes is centered around highly conserved Fe–S proteins (human ISCU) that serve as scaffolds upon which new Fe–S clusters are assembled from (i) elemental sulfur, provided by a pyridoxal phosphate-dependent cysteine desulfurase (human NFS1) and its stabilizing-binding partner (human ISD11), and (ii) elemental iron, provided by an iron-binding protein of the frataxin family (human FXN). Further studies revealed that all of these proteins could form stable complexes that could reach molecular masses of megadaltons. However, the protein–protein interaction surfaces, catalytic mechanisms, and overall architecture of these macromolecular machines remained undefined for quite some time. The delay was due to difficulties inherent in reconstituting these very large multiprotein complexes in vitro or isolating them from cells in sufficient quantities to enable biochemical and structural studies. Here, we describe approaches we developed to reconstitute the human Fe–S cluster assembly machinery in *Escherichia coli* and to define its remarkable architecture.

1. INTRODUCTION

In human mitochondria, two different isoforms of frataxin, FXN^{42-210} and FXN^{81-210}, are involved in Fe–S cluster assembly, serving as elemental iron donors as well as regulators of the cysteine desulfurase activity of NFS1 (Gakh et al., 2010; Li, Gakh, Smith, Ranatunga, & Isaya, 2013). Interestingly, they are produced via two sequential cleavages of the FXN protein precursor (FXN^{1-210}) that are carried out by the mitochondrial processing peptidase in the mitochondrial matrix (Cavadini, Adamec, Taroni, Gakh, & Isaya, 2000; Schmucker, Argentini, Carelle-Calmels, Martelli, & Puccio, 2008). The presence (or absence) of only 39 extra amino terminal residues has profound effects on the physical properties of the two isoforms, namely, the ability of FXN^{42-210} to oligomerize in a metal-independent manner and the inability of FXN^{81-210} to do so (Gakh et al., 2010; Tsai & Barondeau, 2010). Thus, in native conditions, FXN^{42-210} is present as a monomer–oligomer equilibrium, while FXN^{81-210} is present as a stable monomer (Gakh et al., 2010); and the same is true for the respective recombinant proteins when overexpressed in *Escherichia coli* (Gakh et al., 2010; Tsai & Barondeau, 2010). In turn, the ability or inability to oligomerize has distinct effects on the biochemical properties of the two isoforms, in particular their modes of interacting with the

[NFS1]•[ISD11] complex and the ISCU scaffold. In vitro, oligomeric FXN^{42-210} was found to bind to its protein partners with high affinity (Gakh et al., 2010), while a large excess of FXN^{81-210} monomer was required to form a [FXN^{81-210}]•[NFS1]•[ISD11]•[ISCU] complex (Colin et al., 2013). Accordingly, only complexes containing native oligomeric FXN^{42-210} together with NFS1 and ISCU could be isolated from normal human cells by use of size-exclusion chromatography and coimmunoprecipitation, while no complexes containing FXN^{81-210} together with NFS1 and ISCU could be detected under the same experimental conditions (Gakh et al., 2010; Li et al., 2013).

In spite of these differences, both FXN^{42-210} and FXN^{81-210} stimulate the cysteine desulfurase activity of the [NFS1]•[ISD11] complex in the presence of ISCU and L-cysteine in vitro (Li et al., 2013; Tsai & Barondeau, 2010). In addition, both isoforms stimulate Fe–S cluster synthesis in the presence of [NFS1]•[ISD11], ISCU, L-cysteine, and ferrous iron under solution conditions that inhibit iron oxidation (Gakh et al., 2010). However, FXN^{81-210} monomer is inactive in solution conditions that promote oxidation of Fe^{2+} to Fe^{3+}, in which the protein only binds ≤ 1 iron equivalents (Colin et al., 2013; Gakh et al., 2010) and as such it is unable to stimulate Fe–S cluster synthesis (Gakh et al., 2010). In contrast, under the same solution conditions, oligomeric FXN^{42-210} can bind up to $10\,M$ equivalents of iron, and this ferric iron-loaded oligomer is able to support robust Fe–S cluster synthesis (Gakh et al., 2010). Based on the striking functional differences summarized earlier, we proposed that inside mitochondria the two FXN isoforms might fulfill complementary roles. On one hand, FXN^{81-210} would normally bind low levels of iron and support minimal levels of Fe–S cluster assembly via transient interactions with [NFS1]•[ISD11]•[ISCU] complexes. On the other hand, oligomeric FXN^{42-210}, by storing substantial amounts of iron and forming stable contacts with [NFS1]•[ISD11]•[ISCU] complexes, would provide higher rates of Fe–S cluster assembly as needed during cell growth or mitochondrial biogenesis (Vaubel & Isaya, 2013) (Fig. 1).

Although it has long been accepted that the iron donor must work together with the cysteine desulfurase to ensure the concerted delivery of elemental iron and sulfur to the scaffold, the architecture of the complexes and subcomplexes formed by these proteins has remained poorly defined, not only for the human system but also the yeast and bacterial systems. Consequently, an integrated mechanism for Fe–S cluster assembly has been lacking. The crystal structures of two bacterial complexes suggested that the cysteine desulfurase (IscS) and the scaffold (IscU) may form elongated

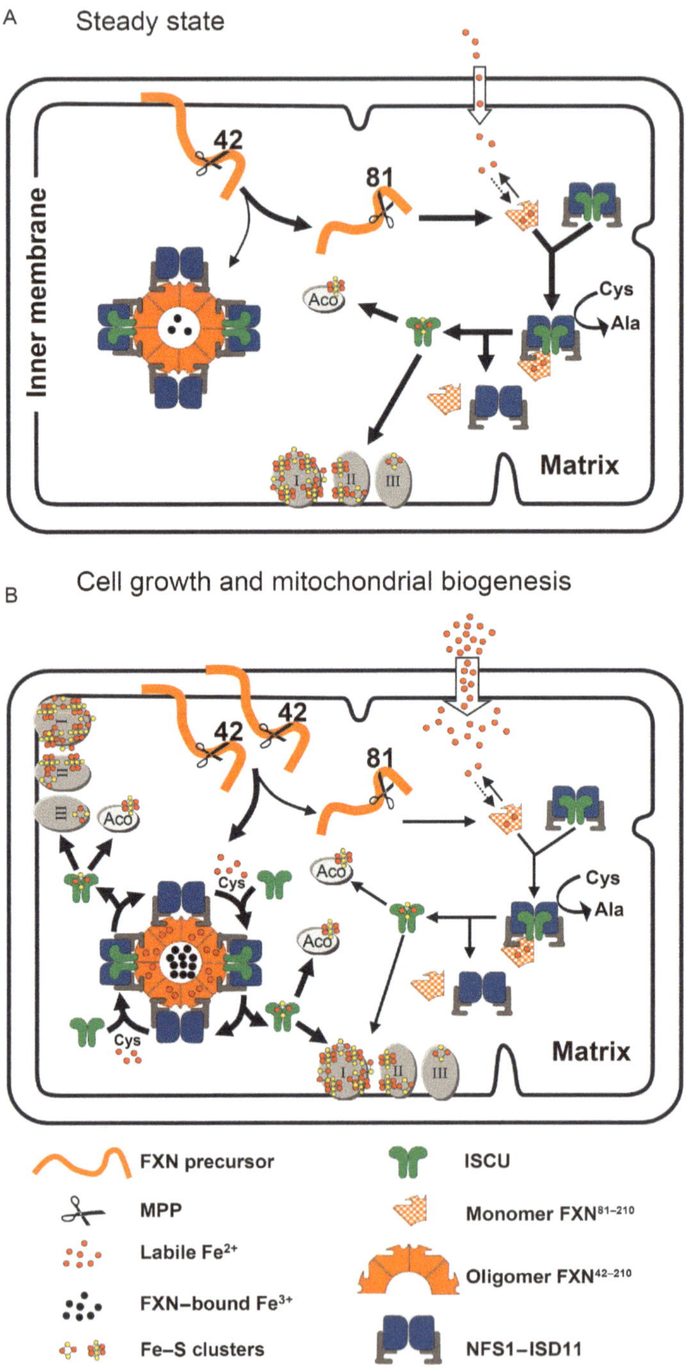

Fig. 1 See legend on opposite page.

heterodimers consisting of one IscS and one IscU subunit (PDB: 3LVL) (Shi et al., 2010) as well as elongated heterotetramers consisting of two IscS and two IscU subunits (PDB: 4EB5) (Marinoni et al., 2012). Another study further suggested that one monomer of bacterial frataxin, CyaY, could bind in a pocket between IscS and IscU (Prischi et al., 2010). Based on these studies, a cartoon model of a $[FXN^{81-210}]_2$•$[NFS1]_2$•$[ISD11]_4$•$[ISCU]_2$ complex was proposed consisting of two antiparallel [NFS1]•$[ISD11]_2$ subcomplexes each binding to one ISCU subunit and one FXN^{81-210} subunit (Colin et al., 2013). This configuration would result in the presence of one Fe–S cluster assembly site at each end of the $[FXN^{81-210}]_2$•$[NFS1]_2$•$[ISD11]_4$•$[ISCU]_2$ complex (Colin et al., 2013) and would be compatible with a proposed mechanism for the delivery of elemental sulfur from the cysteine desulfurase to the scaffold (Amela, Delicado, Gomez, Querol, & Cedano, 2013). The mechanism involves a flexible loop of the cysteine desulfurase containing a highly conserved catalytic cysteine residue (Cys-381 in NFS1), which becomes persulfurated through the cysteine desulfurase reaction (Marinoni et al., 2012; Pandey et al., 2013; Shi et al., 2010; Tsai & Barondeau, 2010). It is believed that binding of FXN^{81-210} to a preformed $[NFS1]_2$•$[ISD11]_4$•$[ISCU]_2$ complex would accelerate persulfide formation on NFS1 and would also induce a conformational change in ISCU that would enhance the transfer of sulfur from NFS1 to ISCU via persulfuration of Cys-138 (Cys-104 if the

Fig. 1 Proposed functions of human frataxin isoforms FXN^{42-210} and FXN^{81-210} in mitochondrial Fe–S cluster assembly. The frataxin precursor protein, FXN^{1-210}, is encoded in the nucleus and translated in the cytoplasm. Upon import to the mitochondrial matrix it is processed by MPP to FXN^{42-210}, which can undergo further processing by MPP yielding FXN^{81-210} or can oligomerize. (A) Under basal conditions, most FXN^{42-210} is cleaved to FXN^{81-210}, which may be sufficient to control the labile Fe^{2+} pool and to support low levels of [2Fe–2S] cluster assembly through dynamic interactions with [NFS1]•[ISD11]•[ISCU] complexes. (B) Under conditions that require higher rates of Fe–S cluster synthesis (e.g., rapidly dividing cells and/or increased mitochondrial biogenesis), a larger proportion of FXN^{42-210} is converted to oligomeric FXN^{42-210} via product inhibition of MPP. Oligomeric FXN^{42-210} binds the iron that exceeds the small iron-binding capacity of FXN^{81-210}, and stable $[FXN^{42-210}]$•[NFS1]•[ISD11]•[ISCU] complexes become a major site of Fe–S cluster assembly. This augments the rate of Fe–S cluster synthesis above the low rate allowed by the small iron-binding capacity of monomeric FXN^{81-210}. *I, II,* and *III,* respiratory chain complexes I, II, and III; *Aco,* mitochondrial aconitase. Components involved in the transfer of [2Fe–2S] clusters from ISCU to Apo enzymes are not shown. *This research was originally published in The Journal of Biological Chemistry. Gakh, O., et al. (2010). Normal and Friedreich ataxia cells express different isoforms of frataxin with complementary roles in iron-sulfur cluster assembly.* The Journal of Biological Chemistry, *285(49), 38486–38501. © The American Society for Biochemistry and Molecular Biology.*

ISCU amino acid sequence is numbered starting from the precursor rather than the mature protein N terminus) (Bridwell-Rabb, Fox, Tsai, Winn, & Barondeau, 2014; Fox, Das, Chakrabarti, Lindahl, & Barondeau, 2015; Parent et al., 2015). Additional studies suggest that sulfur donation to the scaffold would also require conformational changes in the cysteine desulfurase itself to bring its flexible loop close to its substrate-binding site and enable persulfuration of the catalytic cysteine residue. The flexible loop would then have to swing back to get closer to the Fe–S cluster coordination site on ISCU and enable sulfur transfer from the desulfurase to the scaffold (Amela et al., 2013; Pandey et al., 2013).

All of the studies described earlier did not provide any conclusive clues about the mechanism for the delivery of elemental iron from the iron donor to the scaffold. In the specific case of the putative $[FXN^{81-210}]_2 \bullet [NFS1]_2 \bullet [ISD11]_4 \bullet [ISCU]_2$ complex, one monomer of FXN^{81-210} was thought to bind in a pocket at the NFS1–ISCU interface through direct interactions between the conserved negatively charged iron-binding surface on FXN^{81-210} and a positively charged patch on NFS1 (Colin et al., 2013). Surprisingly, these interactions blocked the ability of FXN^{81-210} to bind iron and also blocked iron entry into ISCU, while stimulation of the cysteine desulfurase activity via addition of L-cysteine restored iron entry in the presence of bound FXN^{81-210} (Colin et al., 2013). These effects have remained difficult to explain, and in general, studies focusing on monomeric forms of frataxin orthologues have shed little light on how the iron donor might bind to the cysteine desulfurase and to the scaffold, and on how iron might be delivered to the site of Fe–S cluster assembly (Amela et al., 2013; Colin et al., 2013; Cook et al., 2010; Manicki et al., 2014; Prischi et al., 2010). Meanwhile, work from our laboratory showed that oligomerization is a mechanism by which both yeast and human frataxin can promote assembly of the core machinery for Fe–S cluster synthesis both in vitro and in cells, in such a manner that the scaffold protein and the cysteine desulfurase can simultaneously bind to frataxin, the iron donor, with high affinity (Gakh et al., 2010; Li, Gakh, Smith, & Isaya, 2009). We therefore reasoned that oligomeric frataxin might provide a stable platform to enable coordinate delivery of sulfur and iron needed for cluster assembly. As it turned out, this approach enabled us to shed light on the iron-donation mechanism (Gakh et al., 2016; Ranatunga et al., 2016).

Early on, FXN^{42-210} and its bacterial and yeast orthologues, CyaY, and Yfh1, had demonstrated a strong propensity to oligomerize in vitro (Adamec et al., 2000; Gakh et al., 2010; Layer, Ollagnier-de Choudens, Sanakis, &

Fontecave, 2006; Söderberg et al., 2013) and in vivo (Gakh, Smith, & Isaya, 2008; Gakh et al., 2010; Seguin et al., 2010). We knew that unlike CyaY or Yfh1, FXN^{42-210} oligomerized in an iron-independent manner via its N-terminal region upstream of Ser-81 (Gakh et al., 2010). We also knew that certain point mutations enabled Yfh1 to oligomerize in an iron-independent manner just like FXN^{42-210} (Gakh et al., 2008; Karlberg et al., 2006), and using one of these mutations, Y73A, we were able to uncouple oligomerization of Yfh1 from iron binding and to characterize the complex formed by the apo-forms of Yfh1 and Isu1 (Li et al., 2009), which revealed for the first time a potential path for iron donation from Yfh1 to Isu1 (Ranatunga et al., 2016). By similarity, we assumed that oligomeric FXN^{42-210} might help us illuminate the path for iron donation in the human system. We therefore set out to characterize the complexes and subcomplexes formed by FXN^{42-210} with ISCU, NFS1, and ISD11. Rather than reconstituting these complexes from individually purified components in vitro as we had previously done (Gakh et al., 2010), we developed a new method to coexpress the four proteins in *E. coli*. Since chaperones involved in protein folding and assembly in the bacterial cytoplasm and mitochondrial matrix are highly conserved (Voos & Rottgers, 2002), we relied on bacterial cells to fold and assemble the four human proteins. This approach enabled the isolation of mg amounts of a stable and active complex consisting of all four proteins with 1:1:1:1 stoichiometry. Negative staining transmission EM and single-particle analysis yielded a three-dimensional (3D) model of the complex at $\sim$14 Å resolution. Molecular dynamics flexible fitting (MDFF) of protein structures docked into the EM map of the model further revealed a $[FXN^{42-210}]_{24}$•$[NFS1]_{24}$•$[ISD11]_{24}$•$[ISCU]_{24}$ complex, consistent with the measured 1:1:1:1 stoichiometry of its four-protein components. Chemical cross-linking and limited proteolysis of the complex coupled with tandem mass spectrometry (MS/MS) identification of cross-linked peptides confirmed the protein interfaces present in the 3D model, which were found to involve hydrogen bonds, salt bridges, or hydrophobic interactions between conserved residues. These approaches together revealed that the human Fe–S cluster assembly machinery consists of a central, roughly cubic $[FXN^{42-210}]_{24}$•$[ISCU]_{24}$ subcomplex where one symmetric ISCU trimer binds on top of one symmetric FXN^{42-210} trimer at each of the eight vertices of the cube (Fig. 2). Moreover, binding of 12 $[NFS1]_2$•$[ISD11]_2$ subcomplexes to the surface of the $[FXN^{42-210}]_{24}$•$[ISCU]_{24}$ subcomplex yields in a globular macromolecule with a diameter of $\sim$15 Å and creates 24 Fe–S cluster assembly centers

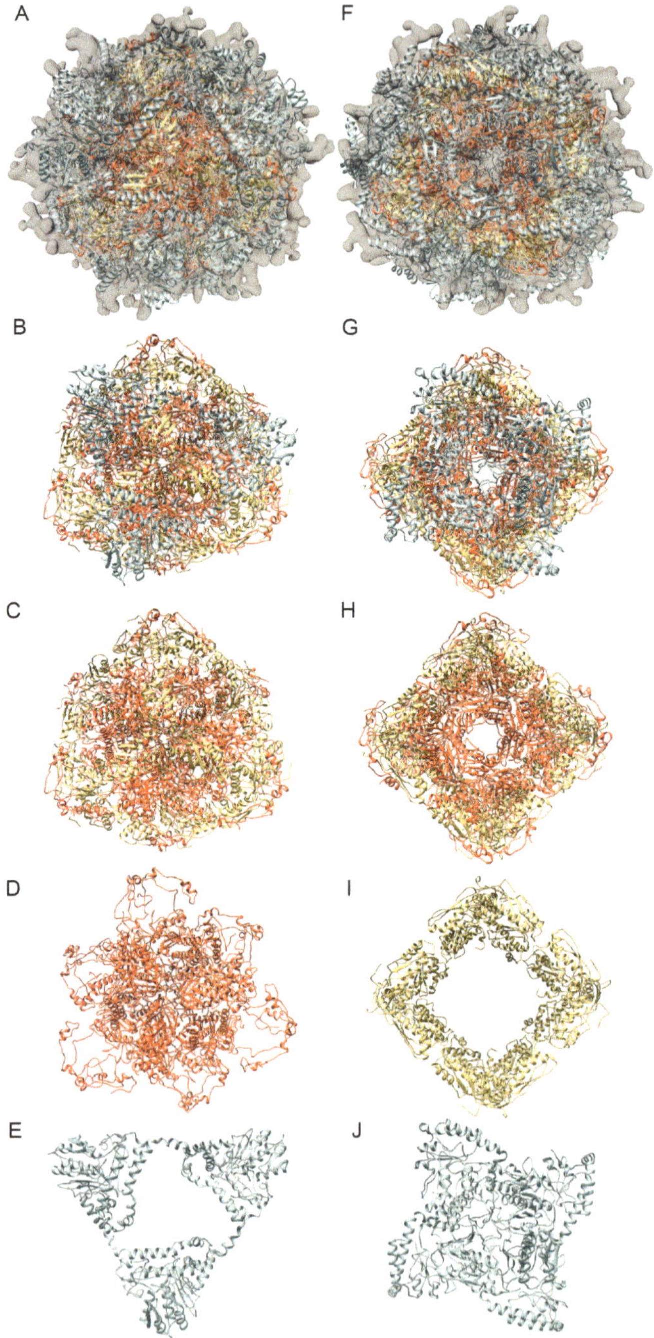

Fig. 2 See legend on opposite page.

(Fig. 3). The organization of each center recapitulates the previously proposed mechanism for sulfur donation from NFS1 to ISCU (Fig. 4) and reveals a path for iron donation from FXN^{42-210} to ISCU (Fig. 5) (Gakh et al., 2016).

2. MULTIPROTEIN COMPLEX EXPRESSION AND PURIFICATION STRATEGIES

Functional $[FXN^{42-210}]_n$•$[NFS1]_n$•$[ISD11]_n$•$[ISCU]_n$ complexes can be reconstituted in vitro from independently expressed and purified FXN^{42-210} oligomer, [NFS1]•[ISD11] subcomplex, and ISCU monomer/dimer (Gakh et al., 2010). In addition, complexes containing native FXN^{42-210}, NFS1, and ISCU can be isolated from human cell extracts by use of size-exclusion chromatography followed by coimmunoprecipitation (Gakh et al., 2010). Each of these two approaches has limitations. On one hand, the in vitro approach can be scaled up to yield substantial amounts of complex, but it is nonphysiological. On the other hand, complexes formed in physiological conditions in human cells are very difficult to purify in workable amounts. Given the high degree of conservation between prokaryotic and mitochondrial Fe–S cluster assembly systems (Bedekovics, Gajdos, Kispal, & Isaya, 2007; Blanc, Gerez, & Ollagnier de Choudens, 2015; Kim, Bothe, Alderson, & Markley, 2015), as well as between the molecular chaperones that assemble these systems in bacteria and mitochondria (Voos & Rottgers, 2002), we postulated that coexpression of the protein components of the human system in *E. coli* might provide us with complexes assembled under close to physiological conditions and in sufficient quantities to enable structural studies. We therefore coexpressed FXN^{42-210}, NFS1, ISD11, and ISCU in *E. coli* as described later.

Fig. 2 Overall architecture of the human $[FXN^{42-210}]_{24}$•$[ISCU]_{24}$•$[NFS1]_{24}$ complex. (A–F) In order to visualize the structure of the entire complex, the simulated half of the structure was aligned with itself into the EM density map of the refined 3D model. ISD11 was not included in the simulation and is not shown in this figure. The structure consists of 24 FXN^{42-210} (*salmon ribbon*), 24 ISCU (*yellow ribbon*), and 24 NFS1 (*light blue ribbon*) subunits. (A and F) Docked structure at the threefold (A) and fourfold (F) axes of the complex. The structural building blocks of the complex at the threefold axis (B–E) and the fourfold axis (G–J) of the complex are shown with EM density map removed. *This research was originally published in The Journal of Biological Chemistry. Gakh, O., et al. (2016). Architecture of the human mitochondrial iron-sulfur cluster assembly machinery.* The Journal of Biological Chemistry, 291, *21296–21321. © The American Society for Biochemistry and Molecular Biology.*

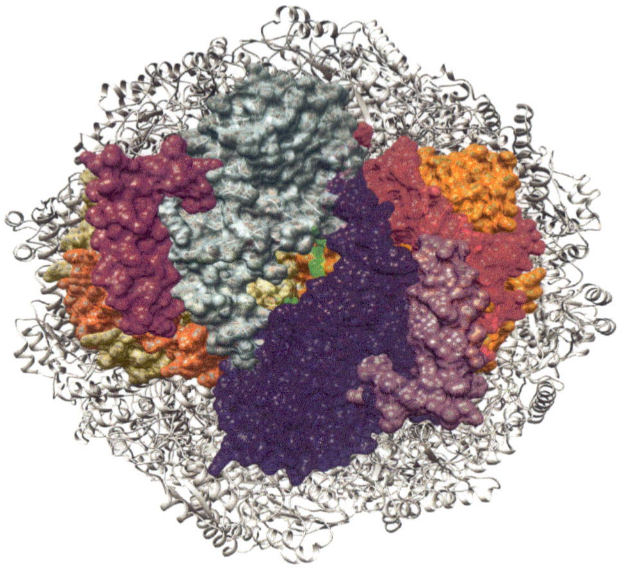

Fig. 3 Fe–S cluster assembly center at the twofold axis of the human $[FXN^{42-210}]_{24}$•$[ISCU]_{24}$•$[NFS1]_{24}$•$[ISD11]_{24}$ complex reveals a mechanism suitable for concomitant iron and sulfur transfer to the scaffold. Top view of two Fe–S cluster assembly sites formed by two adjacent $[FXN^{42-210}]_3$•$[ISCU]_3$ subcomplexes and one $[NFS1]_2$•$[ISD11]_2$ subcomplex at the twofold axis of the complex. The two adjacent ISCU trimers are shown as *yellow* and *orange ribbons* and surfaces with their Fe–S cluster coordinating residues shown as *green sticks*. The two adjacent FXN^{42-210} trimers are shown as *salmon* and *magenta ribbons* and surfaces. The two NFS1 subunits are shown as *blue* and *light blue ribbons* and surfaces, and ISD11 subunits as *purple* and *light purple ribbons* and surfaces. This configuration brings amino acids involved in iron and sulfur donation and coordination in close proximity, which is suitable for safe handling of potentially toxic iron and sulfur atoms. *This research was originally published in The Journal of Biological Chemistry. Gakh, O., et al. (2016). Architecture of the human mitochondrial iron-sulfur cluster assembly machinery.* The Journal of Biological Chemistry, 291, *21296–21321. © The American Society for Biochemistry and Molecular Biology.*

The purification procedure we developed relies on the presence of a six-histidine tag fused with the N terminus of NFS1 and a Strep•Tag® II fused with the N terminus of FXN^{42-210} (Fig. 6A). Initially, nickel affinity chromatography of the total bacterial cell extract separates a pool containing mostly NFS1 (which in our expression system is produced in excess of the other three proteins) from another pool containing all four proteins (Fig. 6B). The latter pool is first reisolated by size-exclusion chromatography (Fig. 6C) and further purified by StrepTrap™ affinity chromatography, which yields a pool that once again, contains all four proteins (Figs. 6D and 7A).

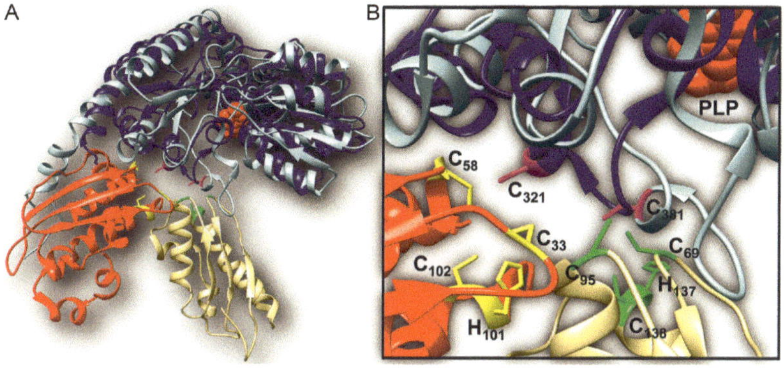

Fig. 4 Comparison of bacterial and human Fe–S cluster assembly complexes reveals conservation of the mechanism for sulfur donation from NFS1 to ISCU. (A) Alignment of the *E. coli* [IscS]•[IscU] heterodimer (PDB: 3LVL) with the [NFS1]•[ISCU] heterodimer shows the different positions of IscU (*red ribbon*) and ISCU (*yellow ribbon*) relative to the flexible loop and catalytic Cys of IscS (*blue ribbon*) and NFS1 (*light blue ribbon*). The PLP cofactor is shown in *red*. (B) Closer view of the flexible loops of IscS (*blue ribbon*) and NFS1 (*light blue ribbon*), with the catalytic Cys-321 of IscS and Cys-381 of NFS1 shown as *magenta sticks*, relative to the positions of the Fe–S cluster-coordinating residues of IscU (*red ribbon*) and ISCU (*yellow ribbon*), shown as *yellow* (IscU) and *green* (ISCU) *sticks*. Cys-69 and Cys-138 of ISCU are proposed acceptors of the persulfide formed on Cys-381 of NFS1 (Bridwell-Rabb et al., 2014; Marinoni et al., 2012; Parent et al., 2015). The fold of the NFS1 and ISCU monomers is very similar to the fold of *E. coli* IscS and IscU monomers. However, when the *A. fulgidus* [IscS]•[IscU] heterodimer is aligned with the [NFS1]•[ISCU] heterodimer, NFS1 and IscS overlap each other, whereas ISCU is rotated ~90° relative to IscU on the same plane. This brings the Fe–S cluster coordination site of ISCU closer to the NFS1 flexible loop. This configuration is supported by cross-links between NFS1 and ISCU as well as NFS1 and FXN^{42-210} (Gakh et al., 2016). *This research was originally published in The Journal of Biological Chemistry. Gakh, O., et al. (2016). Architecture of the human mitochondrial iron-sulfur cluster assembly machinery.* The Journal of Biological Chemistry, 291, *21296–21321. © The American Society for Biochemistry and Molecular Biology.*

Subsequent size-exclusion chromatography and SDS-PAGE of this pool demonstrate that the vast majority of the proteins in this pool are eluted together in one peak (Fig. 7B).

It is important to verify that the four proteins in the peak are indeed bound to each other, which can be assessed by pooling the fractions comprising the peak and incubating them with a cross-linker. We have used the dithiobis[sulfosuccinimidyl propionate] (DTSSP) cross-linker, a water-soluble cross-linker consisting of two amine-reactive NHS-ester groups at each end of an eight-atom spacer arm with a central disulfide bond that can be cleaved with reducing agents. DTSSP reacts primarily with the ϵ-amino group of the

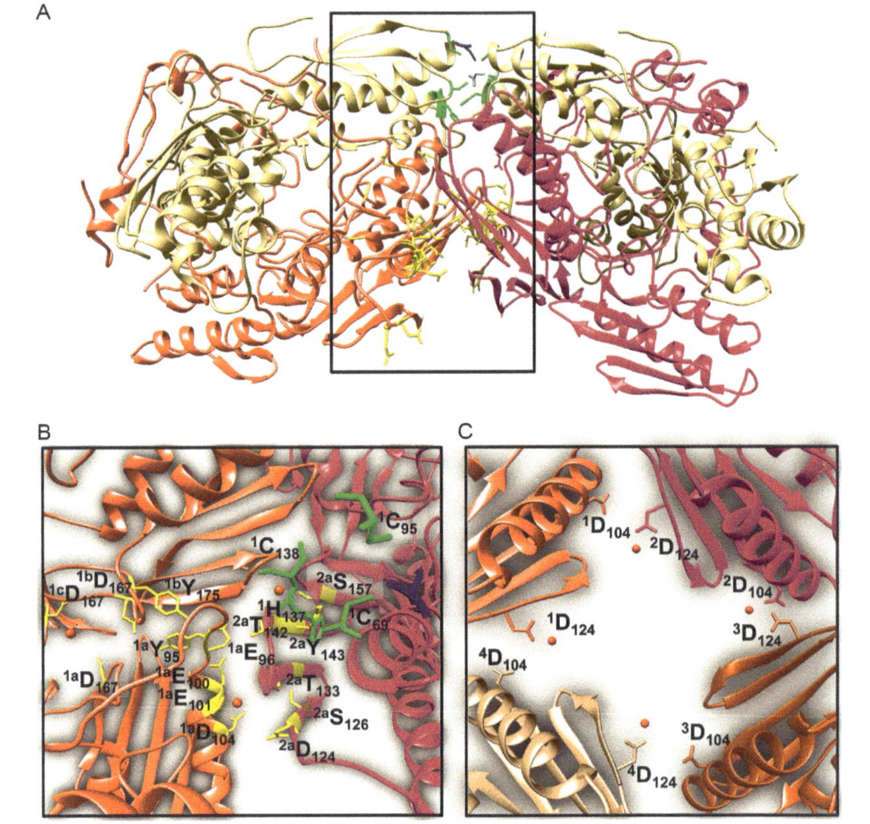

Fig. 5 The Fe–S cluster assembly center of the human $[FXN^{42-210}]_{24}$•$[ISCU]_{24}$•$[NFS1]_{24}$•$[ISD11]_{24}$ complex. (A) Ribbon representation of the Fe–S cluster assembly center formed by two adjacent $[FXN^{42-210}]_3$•$[ISCU]_3$ subcomplexes and one $[NFS1]_2$•$[ISD11]_2$ subcomplex at the twofold axis of the complex; for the two NFS1 subunits, only the Cys-381 residues are shown as a *light blue* and a *blue stick*; ISD11 subunits are not shown. Two ISCU subunits, one from each of the two ISCU trimers, face each other with their respective [2Fe–2S] coordinating sites formed by the invariant Cys-69, Cys-95, His-137, and Cys-138 residues. The two [2Fe–2S] cluster-coordinating sites are above the interface between the two FXN^{42-210} trimers, which is formed by two adjacent subunits from each of the two trimers. (B) Alignment with two available structures of iron-bound $Yfh1^{Y73A}$ trimer (PDB: 2FQL, 4EC2) (Karlberg et al., 2006; Söderberg et al., 2011) identifies potential iron-binding sites that may enable iron transfer to each of the two [2Fe–2S] cluster-coordinating sites. The numbers *1* and *2* denote two adjacent FXN^{42-210} or ISCU trimers; the letters *a*, *b*, and *c* denote different subunits of FXN^{42-210} or ISCU trimer *1* or *2*. For ISCU, only the Fe–S cluster coordinating residues are shown as *green sticks*. For NFS1, only the catalytic Cys-381 is shown as a *blue stick*. FXN^{42-210} residues involved in iron binding are shown as *yellow sticks*. One iron-binding site is in the channel at the threefold axis of each FXN^{42-210} trimer, formed by the invariant Asp-167, one from each subunit of the trimer (trimer *1* in this figure). From this site, iron could move to a second

side chain of lysine residues and the α-amino group of the N terminus of polypeptides and more weakly with the OH group of the side chains of tyrosine, threonine, or serine residues. We also used bis[sulfosuccinimidyl]suberate (BS^3) cross-linker (Thermo Fisher Scientific), which is very similar to DTSSP except that it is not cleavable. Complex cross-linked with DTSSP is isolated again by size-exclusion chromatography and analyzed by SDS-PAGE both in the presence and in the absence of β-mercaptoethanol (Fig. 7C vs D). In the presence of the reducing agent, one can detect the four individual protein components (Fig. 7C), while in the absence of the reducing agent there is a single band that migrates in the high molecular mass region of the gel (Fig. 7D), indistinguishable from complex treated with BS^3 and analyzed by SDS-PAGE in the presence of β-mercaptoethanol (Fig. 7E). These results are consistent with the isolation of a fully assembled $[FXN^{42–210}]_n$•$[NFS1]_n$•$[ISD11]_n$•$[ISCU]_n$ complex.

The fact that the BS^3 cross-linked complex remains intact during reducing SDS-PAGE (Fig. 7E) is consistent with an early observation that oligomeric $FXN^{42–210}$ forms stable contacts with ISCU and/or [NFS1]•[ISD11] under reducing conditions in vitro (Gakh et al., 2010; Li et al., 2009). To assess this further, purified complex can be treated with 5 m*M* β-mercaptoethanol and subsequently reisolated by size-exclusion chromatography in buffer also containing 5 m*M* β-mercaptoethanol. These conditions resolve high molecular weight species present in the earliest fractions eluted from the column (Fig. 7F, fractions 38–50) but do not otherwise affect the elution profile of the complex (Fig. 7F vs 7B). This and additional analyses not described here (Gakh et al., 2016) together indicate that

site formed by conserved residues Thr-142 and Ser-157 from one subunit of the opposite $FXN^{42–210}$ trimer (trimer *2* in this figure). In addition, alignment with the $[Yfh1]_{24}$•$[Isu1]_{24}$ complex (Ranatunga et al., 2016) suggests the presence of a third iron-binding site formed by conserved residues Glu-96, Glu-100, Glu-101, and Asp-104 of trimer *1*, together with residues Asp-124, Ser-126, and Thr-133 of trimer *2*. (C) Alignment of cobalt-bound CyaY monomer (PDB: 2EFF) identified a fourth potential iron-binding site formed by conserved residues Asp-104 from trimer *1* and Asp-124 from trimer *2*. Four of these sites are present at the fourfold axis of the complex and may form one of the iron mineralization sites of the protein, as described for the Yfh1 24-mer (Gakh et al., 2006; Söderberg et al., 2013). The numbers *1*, *2*, *3*, and *4* denote four adjacent $FXN^{42–210}$ trimers. *This research was originally published in The Journal of Biological Chemistry. Gakh, O., et al. (2016). Architecture of the human mitochondrial iron-sulfur cluster assembly machinery.* The Journal of Biological Chemistry, 291, *21296–21321.*

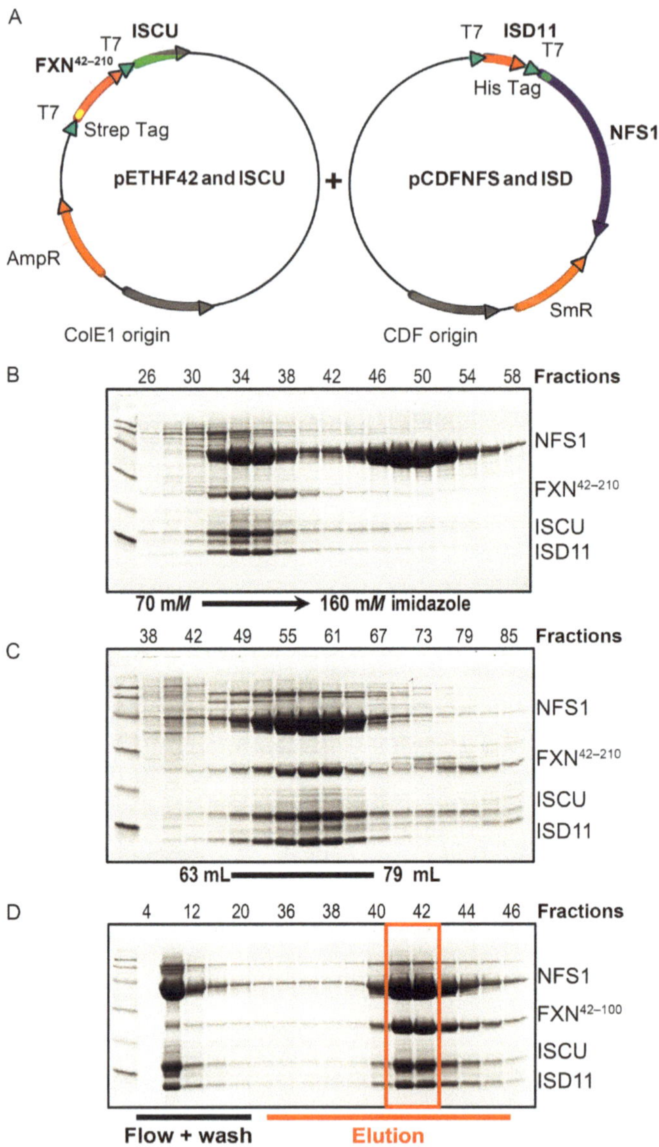

Fig. 6 Coexpression of human FXN$^{42-210}$, NFS1, ISD11, and ISCU proteins in *E. coli* yields a stable four-protein complex. (A) Maps of: (i) a vector bearing the cDNAs coding for Streptavidin-tagged FXN$^{42-210}$ and untagged ISCU, each cloned downstream of a T7 promoter as shown and (ii) a vector bearing the cDNAs coding for histidine-tagged NFS1 and untagged ISD11 each cloned downstream of a T7 promoter. (B–D) *E. coli* cells were cotransformed with the two vectors described earlier. At the end of the overexpression step, complex was isolated from total bacterial cell extract using four consecutive chromatography steps. (B) Complex was eluted from a nickel affinity

inter- or intramolecular disulfide bonds do not play a significant role in the formation or stability of the $[\mathrm{FXN}^{42-210}]_n \bullet [\mathrm{NFS1}]_n \bullet [\mathrm{ISD11}]_n \bullet [\mathrm{ISCU}]_n$ complex.

2.1 DNA Constructs Used for Four-Protein Coexpression in *E. coli*

DNA fragments encoding residues 35–168 of ISCU, which comprise the predicted mature form of this protein (Tong & Rouault, 2000), and residues 42–210 of FXN, which comprise the longest native isoform of FXN (Gakh et al., 2010), are PCR amplified and cloned into the pET-52b(+) 3C/LIC vector using the LIC Duet Adapter and LIC Cloning kit (Novagen) using a protocol provided by the manufacturer. This cloning enables the simultaneous expression of ISCU and FXN^{42-210}, the latter protein fused to an N-terminal Strep•Tag® II (Fig. 6A). It should be noted that the positioning of the Strep•Tag® II-FXN^{42-210} DNA fragment upstream of the ISCU DNA fragment results in robust expression of both proteins, whereas the opposite orientation yields very low levels of FXN^{42-210}. DNA fragments corresponding to residues 56–458 of NFS1 and 6–92 of ISD11, comprising the predicted mature forms of these proteins (Li et al., 2009; Marelja, Stocklein, Nimtz, & Leimkuhler, 2008), are, respectively, cloned into the *Bgl*II and *Hin*dIII sites and the *Nco*I and *Eco*RI sites of the pCDFDuet-1 vector (Novagen). This cloning enables the simultaneous expression of ISD11 and NFS1, the latter protein fused to an N-terminal six-histidine tag (Fig. 6A). The two plasmids encoding FXN^{42-210} and ISCU, and NFS1 and ISD11, are cotransformed in the *E. coli* strain BL21 (DE3) (Novagen) for four-protein coexpression.

2.2 Purification of the Four-Protein Complex From *E. coli*

The four-protein complex can be isolated from total bacterial cell extract using four consecutive chromatography steps, which includes nickel affinity chromatography, preparative size-exclusion chromatography, Streptavidin

chromatography column by use of an imidazole gradient, from 70 to 160 m*M*; uncomplexed NFS1 was more tightly bound to the column and eluted later at an imidazole concentration >200 m*M*. (C) Complex from the step described earlier was subjected to preparative Sephacryl S-300 size-exclusion chromatography and was eluted with an elution volume between 60 and 80 mL. (D) Complex from the step described earlier was subjected to Streptavidin Sepharose™ affinity chromatography and was eluted with 2.5 m*M* D-desthiobiotin.

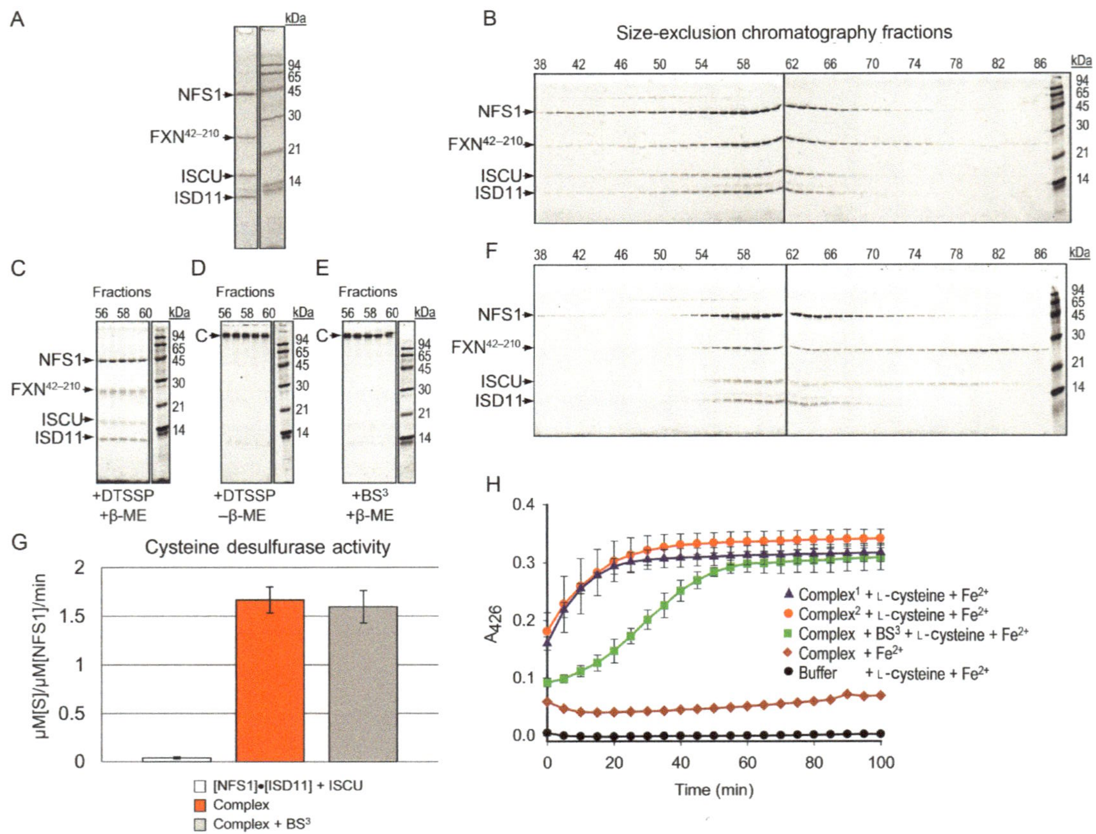

Fig. 7 See legend on opposite page.

Sepharose™ affinity chromatography, and preparative size-exclusion chromatography (Figs. 6B–D and 7B).

2.2.1 Equipment

- BioLogic Duo-Flow workstations for protein purification (Bio-Rad)
- Bio-Rad ChemiDoc XRS+, with Image Lab software (Bio-Rad)
- Refrigerated Centrifuge Avanti J-25 (Beckman)
- Refrigerated incubator shaker (New Brunswick)
- Nickel affinity column (5 mL HiTrap Ni-column, GE Healthcare)
- Preparative size-exclusion column (16 mm × 60 cm column packed with Sephacryl S-300 support, GE Healthcare)
- Streptavidin Sepharose™ affinity column (1 mL StrepTrap™ HP column, GE Healthcare)
- Amicon® Ultra Ultracel® 3K, Centrifugal Filters (Merk Millipore Ltd.)

2.2.2 Buffers and Reagents

- Streptomycin—25 mg/mL (1000 × stock solution)
- Ampicillin—100 mg/mL (1000 × stock solution)

Fig. 7 Biochemical characterization of human $[FXN^{42-210}]_{24}$•$[ISCU]_{24}$•$[NFS1]_{24}$•$[ISD11]_{24}$ complex. (A) Histidine-tagged NFS1 was coexpressed in *E. coli* with ISD11, ISCU, and Streptavidin-tagged FXN^{42-210}, and a complex containing all four proteins was purified as described in Fig. 6. (B) Sephacryl S-300 size-exclusion chromatography and SDS-PAGE analysis of purified complex. (C–E) Complex eluted from the StrepTrap™ HP affinity column was treated with DTSSP (C and D) or BS^3 (E) cross-linker and purified by size-exclusion chromatography, after which fractions 56–60 were analyzed by SDS-PAGE in the presence (C and E) or absence (D) of the reducing agent β-mercaptoethanol (β-ME). (F) Complex eluted from the StrepTrap™ HP affinity column was incubated for 1 h in the presence of 5 m*M* β-mercaptoethanol, purified by size-exclusion chromatography in TN150 buffer containing 5 m*M* β-mercaptoethanol, and analyzed by SDS-PAGE. (G) Cysteine desulfurase activity assays were performed anaerobically in the presence of 1 μ*M* $[FXN^{42-210}]_{24}$•$[ISCU]_{24}$•$[NFS1]_{24}$•$[ISD11]_{24}$ uncross-linked or cross-linked complex. Reactions containing independently expressed and purified [NFS1]•[ISD11] and ISCU (1 μ*M* each) were used as a negative control. Shown are the mean ± S.D. of at least two independent experiments with different protein preparations. (H) The indicated complex preparations were tested for their ability to catalyze Fe–S cluster assembly; $Complex^1$ and $Complex^2$ denote complex after the third and the last purification steps. Assays were performed as described in Section 3.2.2. Each of the plots shows the mean ± S.D. of three independent measurements with two different complex preparations. *This research was originally published in The Journal of Biological Chemistry. Gakh, O., et al. (2016). Architecture of the human mitochondrial iron-sulfur cluster assembly machinery.* The Journal of Biological Chemistry, 291, *21296–21321. © The American Society for Biochemistry and Molecular Biology.*

- 1 *M* Isopropyl β-D-1-thiogalactopyranoside (IPTG, Thermo Fisher Scientific)
- Complete Protease Inhibitor EDTA-free (Roche Diagnostics)
- PBI10—50 m*M* NaH_2PO_4, pH 8.0, 300 m*M* NaCl, 10 m*M* imidazole
- PBI400—50 m*M* NaH_2PO_4, pH 8.0, 300 m*M* NaCl, 400 m*M* imidazole (adjust pH to 8.0 after imidazole addition)
- TN150—50 m*M* Tris–HCl, pH 8.0, 150 m*M* NaCl
- PNK—20 m*M* NaH_2PO_4, pH 7.4, 150 m*M* NaCl, 6 m*M* KCl
- PNKD—20 m*M* NaH_2PO_4, pH 7.4, 150 m*M* NaCl, 6 m*M* KCl, 2.5 m*M* D-desthiobiotin

2.2.3 Procedure

1. The procedure below is for 1 L of bacterial culture; to obtain a greater yield (e.g., for chemical cross-linking studies), one can start with up to four 1 L cultures, process each culture separately, and then combine the four-protein samples together at step 7.
2. In an incubator shaker, grow cells overnight at 37°C in 50 mL of Luria broth containing 100 μg/mL ampicillin and 25 μg/mL streptomycin.
3. Inoculate the 50 mL culture into 1 L of Luria broth (prewarmed to 37°C) containing 100 μg/mL ampicillin and 25 μg/mL streptomycin and continue growth at 25°C until the culture reaches an OD_{600} of 0.6–0.8 (~2–3 h).
4. To induce protein expression, add IPTG (0.5 m*M* final concentration) and continue growth at 15°C for ~18 h. Overexpression is performed at a lower temperature in order to slow down protein expression and allow proteins to fold properly.
5. Harvest cells by centrifugation at 4500 × *g* for 10 min at 4°C, wash the cell pellet in PBI10 buffer, and harvest the cells again by repeating the centrifugation step.
6. Suspend cells in 10 mL of PBI10 buffer containing 1 tablet/50 mL of complete protease inhibitor cocktail EDTA-free (Roche Diagnostics) and disrupt them through five sonication cycles, each consisting of one 30 s pulse of sonication and 30 s rest on ice.
7. Remove cell debris by centrifugation at 25,000 × *g* for 20 min at 4°C, dilute the soluble fraction (~12 mL from one 1 L culture or ~48 mL from four 1 L cultures) with PBI10 buffer to reach a total volume of 60 mL, and load this sample on the affinity HiTrap Ni-column using the program described later. Analyze eluted fractions using SDS-PAGE (see Section 2.2.3.2) (Fig. 6B).

8. At 4°C, concentrate fractions containing all four proteins to a total volume of ~1 mL using an Amicon® Ultra unit and then load this 1 mL sample onto a Sephacryl S-300 size-exclusion column equilibrated with TN150 buffer using the program described later. The complex elutes as a broad peak, from ~60 to 80 mL elution volume, with the center of the peak at ~72 mL elution volume. Analyze fractions using SDS-PAGE (see Section 2.2.3.2) (Fig. 6C).
9. At 4°C, concentrate fractions containing all four proteins to ~1 mL in an Amicon® Ultra unit and load this 1 mL sample onto a StrepTrap™ HP column. Elute the complex with five column volumes of PNKD buffer. Analyze fractions using SDS-PAGE (see Section 2.2.3.2) (Fig. 6D).
10. Determine protein concentration using a BCA kit (Thermo Fisher Scientific) and take an appropriate amount of complex for the cross-linking reaction (see cross-linking procedure).
11. Load cross-linked or uncross-linked complexes on the Sephacryl S-300 size-exclusion column equilibrated with TN150 buffer as in step 8.
12. Analyze fractions using SDS-PAGE (see Section 2.2.3.2). Only fractions at the center of the peak containing the four-protein complex (fractions 56–60 in Fig. 7B) are used for further analyses.
13. This procedure yields ~1–1.5 mg of purified complex per L of bacterial culture. The yield can be doubled using two connected StrepTrap™ HP columns in the third purification step.
14. The purified complex is stable and soluble in TN150 buffer at 4°C for at least 1 week.
15. To assess the presence of intermolecular disulfide bonds in the complex, complex eluted from the StrepTrap™ HP affinity column can be incubated for 1 h in the presence of 5 m*M* β-mercaptoethanol (added freshly to the buffer from a 14.3 *M* stock solution), purified by size-exclusion chromatography in TN150 buffer containing freshly added 5 m*M* β-mercaptoethanol as earlier, and analyzed by SDS-PAGE (Fig. 6F).

2.2.3.1 Programs for Affinity and Size-Exclusion Chromatography

2.2.3.1.1 Step 1: HiTrap Ni-Column

1. Set the program to collect 5 mL fractions, starting at 25 mL and ending at 125 mL; then change to 1.5 mL fractions, starting at 125 mL and ending at 195 mL. The complex elutes between 70 and 160 m*M* imidazole (Fig. 6B).
2. Wash column with 25 mL PBI10 buffer at a flow rate of 2 mL/min.

3. Inject the 60 mL bacterial cell extract at a flow rate of 1 mL/min.
4. Wash the column with 40 mL of a mixture of 98% PBI10 and 2% PBI400 buffer at a flow rate of 1 mL/min.
5. Elute proteins from the column with a 70 mL linear gradient, from 2% to 90% PBI400 buffer, at the flow rate of 1 mL/min.
6. Wash the column with 25 mL PBI400 buffer at a flow rate of 1 mL/min.
7. Regenerate the column with 25 mL PBI10 buffer at a flow rate of 1 mL/min.

2.2.3.1.2 Steps 2 and 4: Sephacryl S-300 Column

1. Set the program to collect 1 mL fractions starting at the elution volume of 16 mL and ending at the elution volume of 120 mL.
2. Wash the column with 16 mL of TN150 buffer at a flow rate of 0.4 mL/min.
3. Load the 1 mL sample from the first purification step through the injection loop with 5 mL of TN150 buffer at a flow rate of 0.4 mL/min.
4. Elute proteins from the column with 240 mL of TN150 buffer at a flow rate of 0.4 mL/min (Fig. 6C).

2.2.3.1.3 Step 3: StrepTrap™ HP Column

1. Use two 1 mL columns connected in tandem. Set the program to collect 0.5 mL fractions starting at 5 mL and ending at 37 mL.
2. Wash the columns with 5 mL of PNK buffer at a flow rate of 1 mL/min.
3. Load the 1 mL sample from the second purification step through the injection loop with 5 mL of PNK buffer at a flow rate of 0.5 mL/min.
4. Wash the columns with 6 mL of PNK buffer at a flow rate of 0.5 mL/min.
5. Elute the complex from the columns using 5 mL of a mixture of 60% PNK and 40% PNKD buffer at a flow rate of 0.5 mL/min. The complex starts to elute as soon as PNKD buffer is added, and with this setup, it elutes around fractions 38–43 mL from the start of the program (Fig. 6D).
6. Wash the column with 3 mL of PNKD buffer at a flow rate of 0.5 mL/min.
7. Wash the column with 5 mL of PNK buffer at a flow rate of 0.5 mL/min.
8. Regenerate the column with 1 mL of 0.5 *M* NaOH at a flow rate of 0.5 mL/min (reverse flow).
9. Wash the column with 10 mL of 100% PNK buffer at a flow rate of 0.5 mL/min; this prepares the column for the next sample.

2.2.3.2 SDS-PAGE Analysis of the Complex

2.2.3.2.1 Materials and Buffers

- 18% Criterion™ Precast Gels (Bio-Rad)
- Criterion™ Gel Box (Bio-Rad)
- PowerPac 3000 (Bio-Rad)
- Plastic gel box with cover
- Bio-Rad ChemiDoc XRS+ imager
- 1 × running buffer (25 m*M* Tris–HCl, 190 m*M* glycine, 1% SDS; after mixing pH will be ~8.5)
- 2 × Laemmli buffer:
  - ➢ Add 6.25 mL of 1 *M* Tris–HCl, pH 6.8
  - ➢ Add 25 mL of 4% SDS
  - ➢ Add 10 mL of 100% glycerol
  - ➢ Add 2.5 mL of 14.2 *M* β-mercaptoethanol for reducing conditions
  - ➢ Add several crystals of bromophenol blue
  - ➢ Add H_2O to 50 mL
- 7.5% acetic acid
- SYPRO™ Orange Protein Gel Stain (Molecular Probes)

2.2.3.2.2 Procedure

1. Take a 10–20 μL aliquot from each of the column fractions to be analyzed.
2. Mix each sample with 2 × Laemmli buffer. For reducing SDS-PAGE, protein samples are mixed v/v with 2 × Laemmli buffer in the presence of β-mercaptoethanol (350 m*M* final concentration), which is added freshly to the buffer from a 14.3 *M* stock solution. For non-reducing SDS-PAGE, β-mercaptoethanol is excluded from the Laemmli buffer. Both types of samples are then processed as described earlier.
3. Heat at 90°C for 5 min and then centrifuge for 30 s at 14,000 × *g*.
4. Load samples on an 18% Criterion™ Precast Gel and run at 180 V for 70 min.
5. Remove the gel from the plastic case and place in a gel box containing 7.5% acetic acid, making sure that the gel is completely covered by the solution.
6. Microwave for 1 min or until the acetic acid solution starts boiling.
7. Place the box on a shaker for 10 min at room temperature.
8. Discard the solution and add 50 mL fresh 7.5% acetic acid.

9. Add 10 μL SYPRO Orange and place the box in a black plastic bag to prevent direct exposure to light. Incubate for 10 min on a shaker at room temperature.
10. Discard the solution and add 50 mL fresh 7.5% acetic acid. Place the box back in the black plastic bag and incubate for 5 min on the shaker.
11. Discard the solution and rinse the gel 2–3 × with distilled water.
12. Visualize protein bands on the Bio-Rad ChemiDoc XRS+ imager.

2.3 Chemical Cross-Linking of the Complex

2.3.1 Materials and Buffers

- Dithiobis[sulfosuccinimidyl propionate] cross-linker (DTSSP) (Thermo Fisher Scientific).
- Bis[sulfosuccinimidyl]suberate cross-linker (BS^3) (Thermo Fisher Scientific), a water-soluble nonhydrolyzable cross-linker (Kalkhof & Sinz, 2008; Madler, Gschwind, & Zenobi, 2010).
- 1 *M* Tris–HCl buffer, pH 7.5.
- 18% Criterion Tris–HCl precast gel (Bio-Rad).
- Benchtop centrifuge (Eppendorf, 5804R).

2.3.2 Cross-Linking Procedure

1. Measure protein concentration and calculate molar concentration. The protein sample should be in a buffer suitable for cross-linking like phosphate buffer (PNK or PNKD; see above), which does not contain primary amines that will otherwise affect the cross-linking reaction.
2. Freshly dissolve the DTSSP or BS^3 cross-linker in deionized water at a final concentration of 50 m*M*.
3. In a final volume of 1 mL, mix protein complex and cross-linker at a 1:60 (DTSSP) or 1:100 (BS^3) molar ratio and incubate the reaction at room temperature for 30 min on a shaker.
4. Stop the reaction by adding 20 m*M* Tris–HCl, pH 7.5 and incubate for an additional 5 min at room temperature.
5. Centrifuge the cross-linked mixture at 20,000 × *g* for 10 min at 4°C and load on the Sephacryl S-300 column.
6. Analyze fractions using SDS-PAGE; pool and concentrate fractions containing the cross-linked complex (fractions 56–60 in Fig. 7C–E) to ~1 mL for further analysis.

3. APPROACHES TO BIOCHEMICALLY CHARACTERIZE THE PURIFIED COMPLEX

As we will see later in this chapter, negative-staining transmission EM and limited proteolysis studies involve the use of BS^3-treated complex. It is therefore important to ensure that this treatment does not affect the physical and functional properties of the complex. As determined by size-exclusion chromatography and SDS-PAGE in the presence of β-mercaptoethanol, complex treated with BS^3 cross-linker has the same elution volume as uncross-linked or DTSSP-cross-linked complex (Fig. 7E vs C and D) and appears as a single band that migrates in the high molecular mass region of the gel, as observed with DTSSP cross-linker (Fig. 7E vs D). From a functional stand point, it is useful to measure the cysteine desulfurase activity of the uncross-linked and BS^3 cross-linked complex as well as their ability to catalyze Fe–S cluster formation. Complex cross-linked with BS^3 exhibits both activities (Fig. 7G and H). As compared to uncross-linked complex, BS^3 cross-linked complex catalyzes Fe–S cluster formation more slowly than the uncross-linked complex but with a similar final yield (Fig. 7H). Based on the cross-linking data, the slower time course of Fe–S cluster assembly exhibited by the cross-linked complex most likely result from cross-linking between flexible loops of NFS1 and ISCU involved in the Fe–S cluster assembly (Gakh et al., 2016).

Accurately assessing the stoichiometry of the four-protein components of the complex is a critical step in defining its architecture. This cannot be done simply by use of staining and densitometry of protein bands on SDS-PAGE gels owing to the fact that the four proteins are significantly different in size, especially NFS1 as compared to ISCU and ISD11 (~45 kDa vs ~15 kDa and ~11 kDa). Consequently, equimolar levels of NFS1, ISCU, and ISD11 may correspond to NFS1 having ~3 × and ~4 × higher local protein concentration inside the gel, which may result in enhanced stainability of NFS1 compared to ISCU and ISD11 (Neuhoff et al., 1990). To avoid this problem, the stoichiometry of the four proteins in the complex can be assessed by amino acid analysis. The experimentally measured amino acid composition of the complex can then be compared to different hypothetical amino acid compositions calculated based on the assumption that the complex might be made up of FXN^{42-210}, ISCU, NFS1, and ISD11 in a 1:1:1:1 or 1:1:2:1 or 1:1:1:2 or other molar ratios. For each hypothetical molar ratio, one can then calculate the variation for each individual amino acid as well as the total % variation from the experimental amino acid composition (Table 1). We found that the total % variation for the 1:1:1:1, 1:1:2:1, 1:1:1:2, 2:1:1:1, or 1:2:1:1 *M* ratio was 1.2%, 32.9%, 8.4% 17.6%, or 13.2%, respectively (Gakh et al., 2016), which supported a $[FXN^{42-210}]_n$•$[NFS1]_n$•$[ISD11]_n$•$[ISCU]_n$ complex containing stoichiometric amounts of its four-protein components in a 1:1:1:1 *M* ratio.

3.1 Amino Acid Analysis to Define the Subunit Stoichiometry

This analysis is performed by a commercial laboratory, e.g., the AAA Service Laboratory Inc., Damascus, OR. After measuring protein concentrations, aliquots of fractions containing complex eluted from the Sephacryl S-300 size-exclusion chromatography column are flash frozen and shipped in dry ice to the commercial laboratory. There, they will typically be hydrolyzed in 6 N HCl/2% (v/v) phenol for 22 h at 110°C. After hydrolysis, samples will be analyzed an amino acid analyzer, using postcolumn ninhydrin derivatization. Amino acids are quantified both individually and as total number of amino acids. Sixteen amino acids are normally quantified since tryptophan and cysteine are degraded by the hydrolysis process, and asparagine and glutamine are detected as aspartic and glutamic acid, respectively (Moore, 1963). This output represents the experimentally determined amino acid composition of the purified complex, which can be compared to the expected compositions based on different hypothetical FXN^{42-210}: ISCU:NFS1:ISD11 molar ratios. For each molar ratio, one can then easily calculate the % variation between the experimental and expected number of each amino acid and the total number of amino acids (Table 1).

3.2 Enzymatic Activity Assays

3.2.1 Cysteine Desulfurase Activity Assay

Sulfide production by NFS1 can be measured using the methylene blue assay essentially as described by Tsai and Barondeau (2010). All buffers and solutions needed for the reaction must be purged with argon gas (<0.2 ppm O_2) in vials tightly sealed with rubber septa (Kendall, Monoject, Blood Collection Tubes). Each reaction is assembled in a vial and purged with nitrogen gas before and after addition of each component, and finally sealed with a tight-fitting rubber septum.

3.2.1.1 Equipment

- Incubator set at 37°C
- Spectrophotometer (Beckman DU640B)
- Bench top centrifuge (Eppendorf, 5804R)

Table 1 Amino Acid Analysis of Human $[FXN^{42-210}]_{24} \bullet [ISCU]_{24} \bullet [NFS1]_{24} \bullet [ISD11]_{24}$ Complex

		FXN-ISCU-NFS1-ISD11 (1:1:1:1)		FXN-ISCU-NFS1-ISD11 (1:1:2:1)		FXN-ISCU-NFS1-ISD11 (1:1:1:2)	
Amino Acid	**Experimental Composition (Number of Residues)**	**Expected Composition (Number of Residues)**	**Variation**	**Expected Composition (Number of Residues)**	**Variation**	**Expected Composition (Number of Residues)**	**Variation**
Ala	70.8	63	−7.8	93	22.2	69	−1.8
Arg	51.0	49	−2.0	76	25.0	62	11.0
Asp	78.9	72	−6.9	102	23.1	82	3.1
Glu	96.8	85	−11.8	131	34.2	93	−3.8
Gly	62.3	60	−2.3	92	29.7	62	−0.3
His	25.8	28	2.2	47	21.2	29	3.2
Ile	44.6	46	1.4	76	31.4	52	7.4
Leu	75.7	76	0.3	111	35.3	83	7.3
Lys	55.1	61	5.9	82	26.9	67	11.9
Met	21.6	24	2.4	39	17.4	27	5.4
Phe	18.3	17	−1.3	24	5.7	19	0.7
Pro	30.4	35	4.6	55	24.6	37	6.6
Ser	60.9	67	6.1	104	43.1	71	10.1
Thr	46.6	45	−1.6	65	18.4	49	2.4

Continued

Table 1 Amino Acid Analysis of Human $[FXN^{42-210}]_{24}$•$[ISCU]_{24}$•$[NFS1]_{24}$•$[ISD11]_{24}$ Complex—cont'd

		FXN-ISCU-NFS1-ISD11 (1:1:1:1)		FXN-ISCU-NFS1-ISD11 (1:1:2:1)		FXN-ISCU-NFS1-ISD11 (1:1:1:2)	
Amino Acid	**Experimental Composition (Number of Residues)**	**Expected Composition (Number of Residues)**	**Variation**	**Expected Composition (Number of Residues)**	**Variation**	**Expected Composition (Number of Residues)**	**Variation**
Tyr	29.7	30	0.3	45	15.3	35	5.3
Val	58.5	59	0.5	91	32.5	66	7.5
Total amino acids	827.2	817	−10.2	1233	405.8	903	75.8
Total percent variation	—	—	1.20%	—	32.90%	—	8.40%

The experimental amino acid composition of the purified complex, as determined via amino acid analysis, is compared to three expected compositions based on the three potential FXN^{42-210}:ISCU:NFS1:ISD11 molar ratios shown. For each molar ratio, the variation between the experimental and expected number of each amino acid residue and the total % variation are shown. Sixteen amino acids were quantified; tryptophan and cysteine are degraded by the hydrolysis process, and asparagine and glutamine are detected as aspartic and glutamic acid, respectively (Moore, 1963).

This research was originally published in The Journal of Biological Chemistry: Gakh, O., et al. (2016). Architecture of the human mitochondrial iron-sulfur cluster assembly machinery. *The Journal of Biological Chemistry, 291*, 21296–21321.

3.2.1.2 Buffers and Solutions

- NANOpure water purged with argon gas (<0.2 ppm O_2)
- 5 × buffer (100 m*M* HEPES–KOH, pH 7.3, 750 m*M* NaCl) purged with argon gas (<0.2 ppm O_2)
- 1 *M* DTT
- 5 m*M* pyridoxal 5′-phosphate (PLP)
- 40 m*M* L-cysteine freshly prepared with the argon gas-purged NANOpure water
- 20 m*M* *N*,*N*-dimethyl-*p*-phenylenediamine [DNPD] (dissolve 27.2 mg of DNPD with 7.2 mL of 10 *M* HCl and 2.8 mL of NANOpure water)
- 30 m*M* $FeCl_3$ (dissolve 81.09 mg of $FeCl_3$ with 8.8 mL of NANOpure water and 1.2 mL of 10 *M* HCl)

3.2.1.3 Procedure

1. Add 308 μL of NANOpure water to vial.
2. Add 80 μL of 5 × buffer (final 1 ×).
3. Add 2 μL of DTT (final 5 m*M*).
4. Add 1 μL of PLP (final 12.5 μ*M*).
5. Add 1 μ*M* $[FXN^{42-210}]_{24}$•$[ISCU]_{24}$•$[NFS1]_{24}$•$[ISD11]_{24}$ complex.
6. Start reaction by adding 1 μL of L-cysteine (final 100 μ*M*).
7. Incubate reaction at 37°C for 10–30 min.
8. Stop reaction by adding 50 μL of DNPD (final 2 m*M*) and 50 μL of $FeCl_3$ (final 3 m*M*).
9. Incubate reaction at 37°C for 20 min.
10. Centrifuge tubes for 5 min at 20,000 × *g*.
11. Measure the absorbance of the soluble fraction at 670 nm.
12. Activity is determined from the conversion of the generated sulfide to methylene blue using a standard curve made with known increasing concentrations of sodium sulfide and extinction coefficient, $\varepsilon = 29.5 \times 10^3$ (Cline, 1969).

3.2.2 Fe–S Cluster Assembly Assay (Adapted From Kondapalli, Kok, Dancis, & Stemmler, 2008; Layer et al., 2006)

For Fe–S cluster synthesis assays, all buffers and solutions are purged with argon gas (<0.2 ppm O_2) in vials tightly sealed with rubber septa. Each reaction is assembled in a quartz cuvette and is purged with nitrogen gas before and after addition of each reaction component, and finally sealed with a tight-fitting rubber septum.

3.2.2.1 Equipment

- Spectrophotometer (Beckman DU640B)

3.2.2.2 Buffers and Solutions

- NANOpure water purged with argon gas (<0.2 ppm O_2)
- 5 × buffer (100 m*M* HEPES–KOH, pH 7.3, 750 m*M* NaCl) purged with argon gas (<0.2 ppm O_2)
- 1 *M* DTT
- 200 m*M* L-cysteine freshly prepared with the argon gas-purged NANOpure water
- 4 m*M* ferrous ammonium sulfite freshly prepared with argon gas-purged NANOpure water
- 100 μ*M* $[FXN^{42-210}]_{24}$•$[ISCU]_{24}$•$[NFS1]_{24}$•$[ISD11]_{24}$ complex

3.2.2.3 Procedure

1. To a quartz cuvette, add 433.5 μL water.
2. Add 120 μL of 5 × buffer (final 1 ×).
3. Add 3 μL of DTT (final 5 m*M*).
4. Add 30 μL of $[FXN^{42-210}]_{24}$•$[ISCU]_{24}$•$[NFS1]_{24}$•$[ISD11]_{24}$ complex (final 5 μ*M*) or buffer for a negative control.
5. Purge mixture with argon gas gently (i.e., blowing the argon gas on the surface while mixing the solution with magnetic stirrer) to prevent foam formation.
6. Add 7.5 μL of 4 m*M* ferrous ammonium sulfite (final 50 μ*M*) and incubate for 20 min at room temperature.
7. Start reaction by adding 6 μL of L-cysteine (final 2 m*M*).
8. Measure the reaction at 1 min intervals in the spectrophotometer at A_{426} using *Kinetic Mode.*
9. Plot A_{426} values vs time.
10. Measure absorption spectra from 300 to 800 nm at the end of each reaction.

4. APPROACHES TO DEFINE THE STRUCTURE OF THE COMPLEX

In preparation for EM studies, it is important to assess the dimensions of the complex and its homogeneity. Dynamic light scattering is a simple technique that can be applied to complex freshly eluted from the size-exclusion chromatography column in the last purification step. We consistently measured a hydrodynamic radius (R_h) of 7.4 ± 0.3 nm with unimodal distribution (Fig. 8A) indicative of a homogeneous sample suitable for EM studies. As a reference, it is also helpful to compare the R_h and elution profiles of the $[FXN^{42-210}]_n$•$[NFS1]_n$•$[ISD11]_n$•$[ISCU]_n$ complex with those of previously characterized globular complexes, e.g., the $Yfh1^{Y73A}$ 24-mer ($R_h = 7.6 \pm 0.5$) (Karlberg et al., 2006), the $[Yfh1^{Y73A}]_{24}$•$[Isu1]_{24}$ complex ($R_h = 7.2 \pm 0.2$) (Ranatunga et al., 2016), and the horse-spleen ferritin 24-mer ($R_h = 7.5 \pm 0.2$). When these complexes are analyzed by size-exclusion chromatography under the same solution conditions, they are eluted with similar retention volumes, between ~66 and 74 mL (measured between the ferritin and the human complex peaks in Fig. 8B). This result is consistent with the measured R_h of the four complexes (Andrews, 1965) and with the size-exclusion pore dimension of the Sephacryl S-300 gel matrix (~13 nm) (Hagel, Östberg, & Andersson, 1996), and indicated that the $[FXN^{42-210}]_n$•$[NFS1]_n$•$[ISD11]_n$•$[ISCU]_n$ complex most likely has a globular shape.

With the information above in hand, we can proceed to stain aliquots of purified complex with uranyl acetate. To initially assess if the stained complex can be visualized and to then optimize the conditions to prepare the grids (e.g., what is the optimal protein concentration to obtain a field of well-separated particles?), a medium-resolution microscope (e.g., a Tecnai G^2 12 Transmission Electron Microscope, FEI, Inc., Hillsboro, OR) is suitable. This initial assessment reveals roughly globular particles with a diameter of ~15 nm, consistent with the R_h measured by dynamic light scattering and with the behavior of the complex during size-exclusion chromatography as compared to known globular complexes.

The procedure for transmission EM and single-particle analysis of the complex involves (i) taking EM images of the uranyl acetate-stained particles at the best possible magnification and Å/pixel ratio; and (ii) collecting several thousand particles and using the best ~500–1000 particles to generate reference free class averages using the EMAN2 software package (Tang et al., 2007) (Fig. 8C). The resulting class averages should exhibit a similar globular shape and consistent diameter of ~15.3 ± 0.7 nm, but each class should have unique structural details since the complex is not perfectly spherical and can adsorb to the carbon film of the EM grids with different orientations corresponding to its four-, three-, and twofold symmetry axes (Gakh et al., 2016). When we first performed this analysis, the only clue we had was the high degree of conservation between the yeast and the human systems (Cavadini et al., 2000; Gakh et al., 2010), which led us to assume that

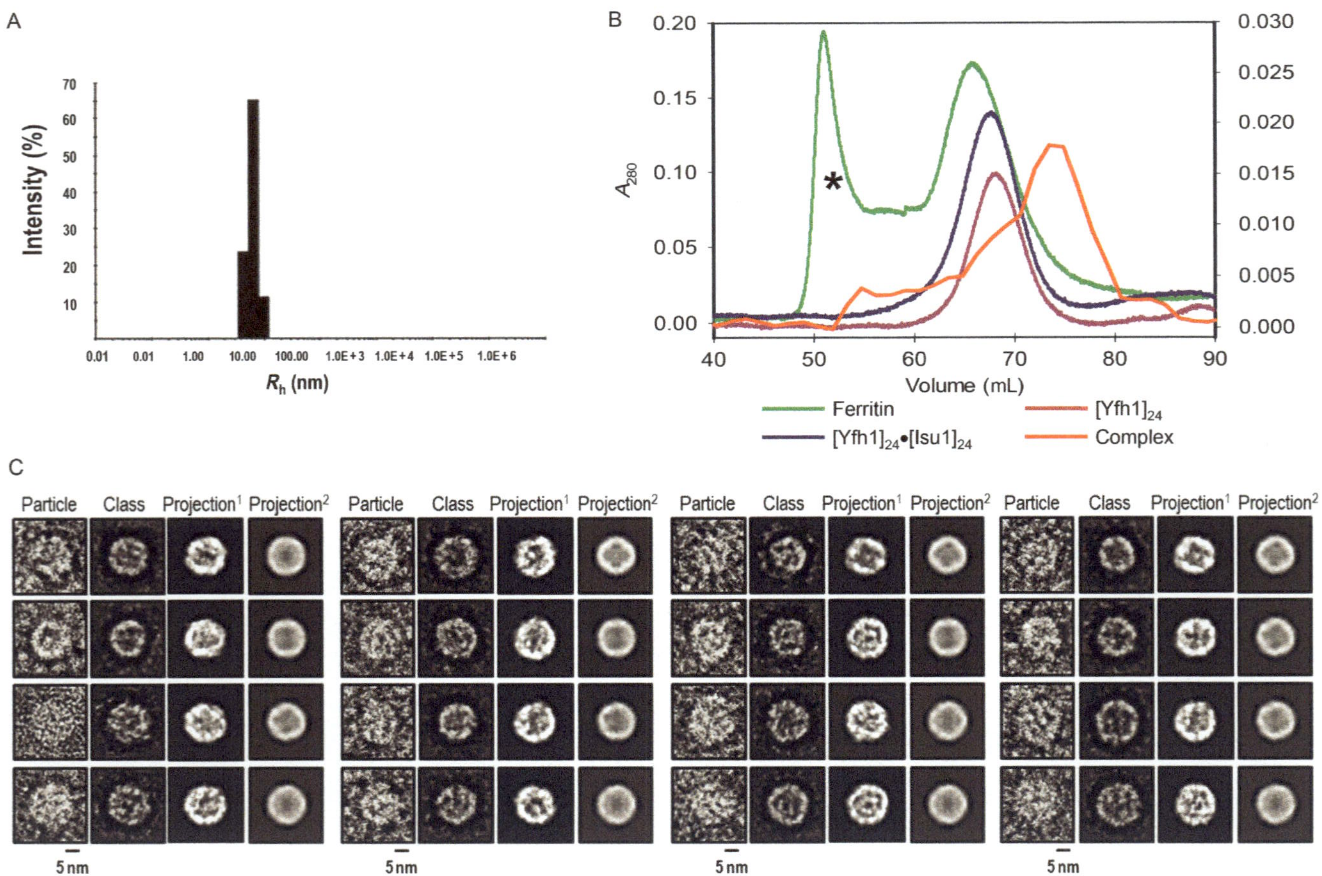

Fig. 8 See legend on opposite page.

the architecture of the human four-protein complex might be similar to the architecture of its cognate yeast complex, $[Yfh1]_{24}$•$[Isu1]_{24}$•$[Nfs1]_{24}$, which we knew had 432 symmetry (Ranatunga et al., 2016). Therefore, we generated initial and refined 3D models of the human complex without any imposed symmetry (Fig. 9A). Segmentation of these 3D models using the program Chimera (Pettersen et al., 2004) revealed that they contained what appeared as fourfold and threefold symmetry axes (Fig. 9B and C). The presence of symmetry in a 3D model can be assessed in an unbiased manner using the program POLARRFN of the CCP4 package https://services.mbi.ucla.edu/selfrot/, which calculates structure factors from the EM density map of the 3D model of interest and provides a self-rotation function (Winn et al., 2011). We calculated the self-rotation function for the EM density map of the refined 3D model without symmetry applied, and repeated the same analysis for a refined 3D model created with 432 symmetry applied as a positive control (Fig. 9D–F). This analysis revealed the presence of 432 symmetry in the 3D model created without symmetry applied (Fig. 9G vs H). We then used the PDBe Fourier Shell Correlation (FSC) server https://www.ebi.ac.uk/pdbe/emdb/validation/fsc/ to calculate the FSC curve for our

Fig. 8 Biophysical analysis and transmission EM of the $[FXN^{42-210}]_{24}$•$[ISCU]_{24}$•$[NFS1]_{24}$•$[ISD11]_{24}$ complex. (A) Dynamic light scattering measurements were performed on BS^3-cross-linked complex freshly eluted from the size-exclusion chromatography column, and the hydrodynamic radius (R_h) was obtained. (B) Elution volumes of the human complex (Complex) and the indicated protein complexes analyzed by Sephacryl S-300 size-exclusion chromatography. *Asterisk* denotes ferritin dimer. Although the predicted molecular mass of the human complex (~2 MDa) is greater than that of ferritin (700–800 kDa due to iron content), these two complexes are eluted according to their similar dimensions (Andrews, 1965). The fact that the human complex elutes slightly later than ferritin may also be due to retardation through weak adsorption to the gel. (C) Electron micrographs of purified, uranyl acetate-stained complex particles were obtained, and images processed with the EMAN2 software package. Shown is a gallery of class averages, with one representative particle from each class and the corresponding projection of the initial 3D reconstructions without symmetry applied (*Projection*[1] columns) and with 432 symmetry applied (*Projection*[2] columns). Particles, class averages, and projections representing the two-, three-, and fourfold axis of the complex and intermediate orientations are shown sequentially from the left to the right starting with the *top row*. The particle diameter was 15.2 ± 0.8 nm (average of 103 particles with twofold orientations), 15.1 ± 0.7 nm (average of 123 particles with threefold orientations), 15.4 ± 0.7 nm (average of 91 particles with fourfold orientations). *This research was originally published in The Journal of Biological Chemistry. Gakh, O., et al. (2016). Architecture of the human mitochondrial iron-sulfur cluster assembly machinery.* The Journal of Biological Chemistry, 291, *21296–21321. © The American Society for Biochemistry and Molecular Biology.*

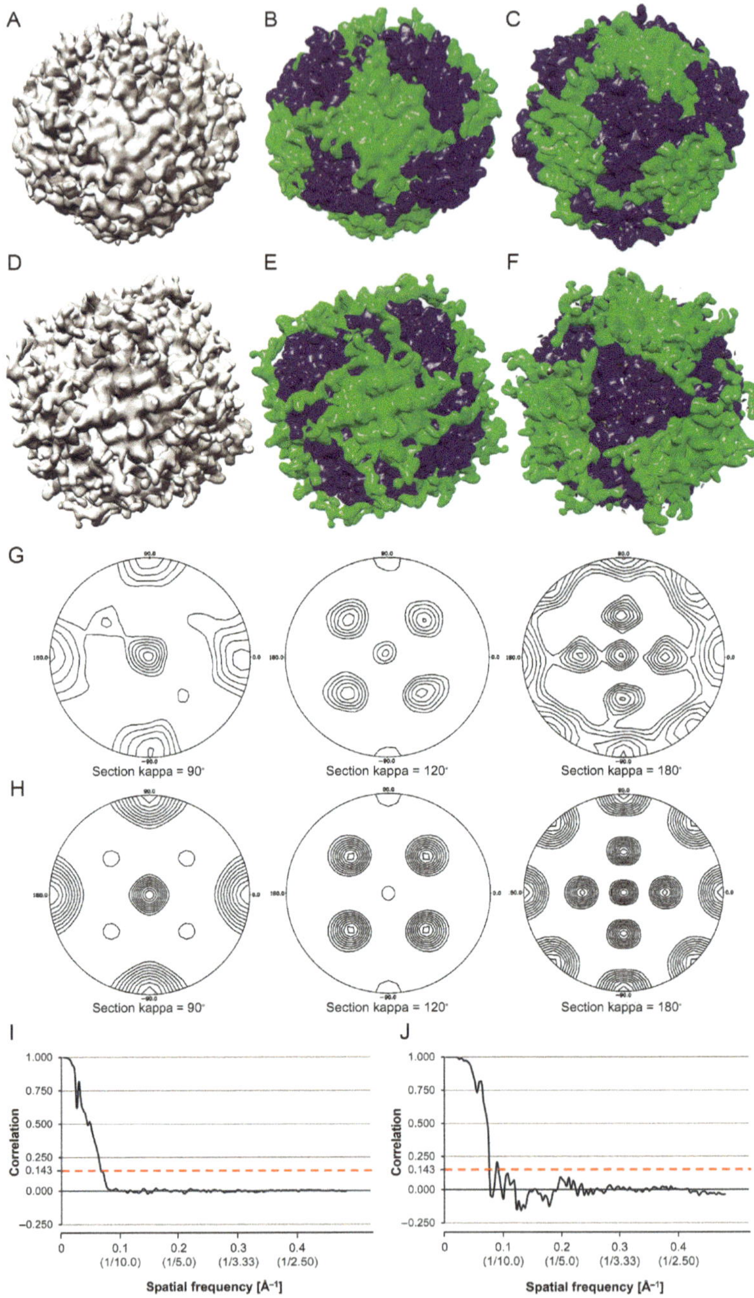

Fig. 9 Single-particle analysis of human four-protein complex. (A–F) Refined 3D models were generated without imposed symmetry (A) or with 432 symmetry applied (D). Both models were segmented using Chimera, and the fourfold (B and E) and threefold

refined 3D models. For any given model, the program provides the resolution according to three complementary ways of measuring it, including (i) where the FSC curve crosses the correlation value of 0.143, as well as (ii) the resolution based on the 3-sigma criterion, and (iii) the resolution based on the 1/2 bit criterion (van Heel & Schatz, 2005). Using the 0.143 cutoff (Ludtke, 2010), our FSC plots indicated resolutions of 15.0 and 13.5 Å, respectively, for the model without and with symmetry applied (Fig. 9I and J). For the two models respectively, the 3-sigma criterion indicated that at resolutions of 13.7 and 15.5 Å we had collected information significantly above the noise level; in addition, the 1/2 bit criterion indicated that at resolutions of 16.0 and 14.9 Å we had collected information sufficient for interpretation (van Heel & Schatz, 2005). The process described earlier is required to evaluate the quality of the EM reconstruction.

We noted that volumes around the threefold symmetry axes of the 3D models had triangular shapes (Fig. 9C and F), which was reminiscent of the shape of the yeast $[Yfh1]_3$•$[Isu1]_3$ subcomplex, the building block of the $[Yfh1]_{24}$•$[Isu1]_{24}$ complex (Ranatunga et al., 2016). Based on this observation, we modeled $FXN^{42–210}$ and ISCU trimers, but without the extended N-terminal regions of the two proteins (residues 42–93 of $FXN^{42–210}$ and 36–50 of ISCU), which were predicted to have a high degree of flexibility. We initially docked the $FXN^{42–210}$ and ISCU trimers into the segmented EM density map of the refined 3D model with 432 symmetry applied both separately (cross-correlation coefficients = 0.58 and 0.65, respectively) or together as a $[FXN^{42–210}]_3$•$[ISCU]_3$ subcomplex (cross-correlation coefficient 0.62), and selected the latter mode of docking to fit eight $[FXN^{42–210}]_3$•$[ISCU]_3$ subcomplexes into the EM map (cross-correlation coefficient ranged from 0.57 to 0.62). After the N-termini of $FXN^{42–210}$

(C and F) axes were identified and colored in *green* and *blue*, respectively. (G and H) Stereographic projection plots of the kappa = 90° (fourfold), kappa = 120° (threefold), and kappa = 180° (twofold) sections of the self-rotation function of the EM density map of the refined 3D model of the complex without symmetry applied (G), and with 432 symmetry applied (H), obtained using POLARRFN. (I and J) The PDBe Fourier Shell Correlation (FSC) server was used to calculate and plot the FSC curve for the refined 3D model without (I) and with 432 symmetry (J) applied. The *dotted red line* shows where the FSC curve crosses the correlation value of 0.143. *This research was originally published in The Journal of Biological Chemistry. Gakh, O., et al. (2016). Architecture of the human mitochondrial iron-sulfur cluster assembly machinery.* The Journal of Biological Chemistry, 291, *21296–21321.* *© The American Society for Biochemistry and Molecular Biology.*

and ISCU were modeled into nearby unoccupied densities (based on cross-linking data described later), there were 24 unoccupied volumes with a similar shape to a homology model of the NFS1 monomer, which could be docked into each of the 24 volumes with a cross-correlation coefficient ranged from 0.57 to 0.63. We repeated the same process for the segmented EM density map of the refined 3D model generated without symmetry applied and were able to fit eight $[FXN^{42-210}]_3$•$[ISCU]_3$ subcomplexes (cross-correlation coefficient ranged from 0.37 to 0.54) and 24 NFS1 monomers (cross-correlation coefficient ranged from 0.41 to 0.56).

After fitting 24 FXN^{42-210}, 24 ISCU and 24 NFS1 subunits into the EM density map, the question remained of where ISD11 might fit. The cross-linking data and the presence of small unoccupied volumes just above the docked NFS1 subunits suggested that ISD11 might be on the surface of the complex. However, the unoccupied volumes at the surface of the model were not clearly distinguished from the background to enable reliable docking of ISD11 monomers. Therefore, we had to rely on the amino acid analysis of the complex, indicating that ISD11 was present in equimolar levels as the other three proteins, and on the cross-linking data. This information enabled us to model 24 ISD11 subunits on the surface of the complex. However, we could not include the 24 ISD11 subunits in the molecular dynamic simulations described later.

We were unable to simulate the entire docked structure of the $[FXN^{42-210}]_{24}$•$[NFS1]_{24}$•$[ISCU]_{24}$ complex due to the very large number of atoms involved (>100,000). Instead, we subjected one-half of the structure, consisting of six NFS1 dimers docked on top of four $[FXN^{42-210}]_3$•$[ISCU]_3$ subcomplexes, to mild simulation using Molecular Dynamics Flexible Fitting (Chan et al., 2011; Trabuco, Villa, Schreiner, Harrison, & Schulten, 2009). This mild simulation improved the fitting of the structure into the EM density map, with cross-correlation coefficients of 0.46 and 0.62 before and after the simulation, respectively.

4.1 Dynamic Light Scattering

4.1.1 Equipment

- DynaPro Molecular Sizing Instrument with Dynamics V6 software (Protein Solutions, Inc.) equipped with an 824.7 nm laser diode and detector that measures scattered light at a 90° angle to the incident beam.

4.1.2 Procedure

1. Fractions containing the complex freshly eluted from the size-exclusion chromatography column are centrifuged at 20,000 × *g* for 10 min at 4°C, after which the supernatant is transferred to an appropriate cuvette.
2. Several consecutive measurements are performed for each sample, each consisting of 50 counts of 5 s each.
3. From the measurements, the software calculates the average autocorrelation function and the particle translational diffusion coefficient. The software then calculates the R_h using the diffusion coefficient and Stokes–Einstein equation and displays the data in both numerical and graphic forms.

4.2 Comparative Size-Exclusion Chromatography

4.2.1 Proteins

- Holo horse-spleen ferritin (F4503, Sigma)
- Yeast Yfh1^{Y73A} 24-mer (Karlberg et al., 2006; Schagerlöf et al., 2008)
- Yeast [Yfh1^{Y73A}]$_{24}$•[Isu1]$_{24}$ complex (Ranatunga et al., 2016)
- Human [FXN$^{42–210}$]$_{24}$•[ISCU]$_{24}$•[NFS1]$_{24}$•[ISD11]$_{24}$ complex (Gakh et al., 2016)

4.2.2 Procedure

- Each of the protein complexes listed earlier (40 μ*M*) is independently analyzed by size-exclusion chromatography on a Sephacryl S-300 column as described in the purification protocol.
- The resulting chromatograms are overlaid for comparison of the respective elution volumes.

4.3 Electron Microscopy

4.3.1 Grid Preparation and Negative Staining of Purified Complex

4.3.1.1 Equipment

- Carbon-coated copper grids, 400 mesh (Electron Microscopy Sciences, EMS)
- Glow discharger (Denton Vacuum Inc.)
- Parafilm M® (Bemis Company, Inc.)
- PELCO reverse fine tweezers (TED PELLA Inc.)

4.3.1.2 Buffers and Solutions

- NANOpure water
- TN150 buffer (50 m*M* Tris–HCl, pH 8.0, 150 m*M* NaCl)
- 1% uranyl acetate freshly prepared in NANOpure water

4.3.1.3 Procedure

1. Glow-discharge carbon-coated copper grids for 30 s using a DV-502A vacuum evaporator within a 1–2 h time frame before applying the protein sample.
2. Preincubate grid for 1 min by placing it on a 20-μL TN150 buffer drop on Parafilm M®.
3. Remove excess buffer with blotting paper, and place a 10-μL drop of protein sample at a concentration between 0.2 and 0.3 mg/mL on the grid for ~2 min.
4. Remove excess protein sample with blotting paper and wash the grid by placing a 7-μL drop of sterile water on the grid for 1 min. Repeat this step once.
5. Remove excess water with blotting paper and stain the grid with 5 μL of 1% (w/v) uranyl acetate for 5 and 30 s by successively placing two separate drops of uranyl acetate on the grid, with excess stain drawn off after each application.
6. Dry the grid on the tweezer for at least 30 min and store grids at room temperature.

4.3.2 Transmission Electron Microscopy

4.3.2.1 Equipment

- FEI Tecnai G^2 F30 field emission gun cryo-transmission electron microscope (FEI, Inc., Hillsboro, OR) with an acceleration voltage at 300 kV equipped with a 4k × 4k UltraScan CCD camera (4096 × 4096 pixels) with an Electron Energy-Loss Gatan imaging filter
- Grid-holder to carry carbon-coated copper grids with negatively stained particles
- PELCO reverse fine tweezers (TED PELLA Inc.)

4.3.2.2 Procedure

1. Prepare microscope as per Tecnai Operating Instructions
2. Collect images with a defocus range from −0.2 to −3.0 μm at a magnification of 115,000-fold with an Å/pixel ratio of 1.034

4.4 Single-Particle Reconstruction

4.4.1 Equipment

High-speed/performance computer such as a MacPro with 12 Core Intel Xeon and 64 GB memory for fast data processing time.

4.4.2 Procedure (Adapted From the EMAN2 Tutorial http://blake.bcm.edu/emanwiki/eman2/tutorials)

1. Collect EM micrographs in "*.dm3*" format.
2. On the EMAN2 program window, select "*Raw Data,*" and import your micrographs into your project.
3. Select "*Interactive Boxing*" and choose the appropriate box size for your particles, which should be 1.5 × larger than each particle (use 256 box size for the human complex).
4. Pick particles in manual mode making sure that the selected particles are well separated from the surrounding particles. Several thousand particles are needed.
5. Generate the boxed particles output using "*Generate Output.*"
6. To achieve "*CTF (contrast transfer function) Fitting and Correction,*" start with an automated fitting using "*Auto High Pass*" and "*Amplitude Contrast*" set at 60%–70%.
7. Next, select "*Interactive Tuning*" to check the fitting and do manual corrections if improvement is necessary.
8. Select "*Generate Structure Factor*" and repeat the "*Automated Fitting*" and "*Interactive Tuning*" steps and eliminate particles that do not achieve a spectral signal-to-noise ratio (SSNR) of at least 0.5.
9. When the particle fits are satisfactory based on SSNR and a defocus range -0.2 to $-3.0\,\mu m$, select "*Generate Output.*"
10. Go to "*Particle Sets*" and select "*Wiener Filtered Particles.*" This allows you to manually identify "*bad*" particles and generate "*sets.*"
11. Most of the "*bad*" particles should be automatically eliminated during the "*CTF Fitting and Correction*" step. However, at this point it is helpful to also evaluate the particles manually and eliminate those that are difficult to distinguish from the background.
12. Next, in the "*Build Particle Sets*" dialog highlight all particle images you want to use, ideally 4000–5000, name the set "*set-all,*" and click *OK.*
13. Open "*Build Particle Sets*" again, select particle images that represent the best particles, ideally the best ~1000 particles; name this set "*set-small,*" and click *OK.*
14. In "*Reference Free Class Averages*" select "*set-small.*" From this ~1000 best particle set, the program will generate 8–10 reference free class averages, called "*classes_01, classes_02,* etc." Since each iteration improves the data, select the last two classes.
15. Eliminate obviously "bad" class averages, such as those that are not centered, too big or too small or indistinguishable from the background,

and save the remaining "good" classes with a different name, like "*good-classes.hdf*."

16. Go to the "*Initial Model*" dialog, choose "*make model*" and select the "*good-classes.hdf*" file created in the previous step. Select the parameters indicated in the EMAN2 tutorial. At this point, we recommend generating several models, both without any imposed symmetry as well as with 432 symmetry applied.
17. Analyze the models, called "*model_01_01_aptcl, model_01_02_aptcl*, etc.," and select the "best" initial models based on (i) the degree of similarity between each projection of the model and the corresponding class-average; and (ii) the presence of symmetry, which can be calculated by POLARRFN of the CCP4 package using a self-rotation function sever https://services.mbi.ucla.edu/selfrot/ (Winn et al., 2011) (Fig. 9).
18. Go to the "*3D Refinement*" dialog and select "*Run e2refine,*" and refine the best "*initial model*" against both the "*set-small*" particles and the "*set-all*" particles, once again both without any imposed symmetry as well as with 432 symmetry applied.
19. Go to the https://services.mbi.ucla.edu/selfrot/ server and check the generated models for the presence of symmetry.
20. Go to the "*Resolution*" dialog window and create "*even*" and "*odd*" models; these models can be uploaded to the FSC server (https://www.ebi.ac.uk/pdbe/emdb/validation/fsc/) to calculate and plot FSC curves and obtain resolution values.

4.5 Docking of Structures Into the EM Density Maps of the 3D Models

4.5.1 Procedure

1. Generate a homology model of the FXN^{42-210} monomer based on the X-ray crystal structure of one subunit of the $Yfh1^{Y73A}$ trimer (Söderberg et al., 2011) (PDB: 3OEQ) using the I-TASSER (http://zhanglab.ccmb.med.umich.edu/I-TASSER/) web resource (Yang et al., 2015).
2. Next, generate a model of the FXN^{42-210} trimer by aligning three FXN^{42-210} monomers with the N-terminal region (residues 42–93) removed, with the X-ray crystal structure of the $Yfh1^{Y73A}$ trimer.
3. Generate homology models of ISCU and NFS1 monomers based on available crystal and NMR structures (ISCU PDB structures: 2Z7eB, 1wfzA, 2kqkA, 4eb5C, 1q48A; NFS1 PDB structures: 3lvmB,

4eb5C, 3a9yA, 3qzcA) of bacterial orthologues using the I-TASSER web resource, which will automatically perform a search for all available homology structures for any given protein of interest.

4. Next, generate a model of the ISCU trimer by aligning three ISCU monomers with the N-terminal region (residues 35–50) removed with the structure of the Isu1 trimer from the 3D model of the yeast $[Yfh1]_{24}\bullet[Isu1]_{24}$ complex (Ranatunga et al., 2016) (PDB: 5T0V).
5. Both the model generation and the subsequent docking into the EM density map are facilitated if the extended N-terminal regions of $FXN^{42–210}$ and ISCU (residues 42–93 and 36–50, respectively) are excluded from the homology models as these regions are predicted to be largely unstructured or highly flexible.
6. Since, to our knowledge, there are as yet no structures available for ISD11, a model of the ISD11 monomer can be generated based on the protein's primary amino acid sequence using the I-TASSER web resource.
7. The pyridoxal 5′-phosphate (PLP) cofactor can be modeled into NFS1 based on the *Archaeoglobus fulgidus* $[IscS]_2\bullet[IscU]_2$ complex structure (PDB: 4EB5).
8. Verify the quality of each model based on the Ramachandran plot and ProQ-Protein Quality Predictor measures (http://www.sbc.su.se/~bjornw/ProQ/ProQ.cgi).
9. Start with the refined 3D model generated with 432 symmetry applied; open the model *.mrc* file in the program *Chimera*, and using the "*Volume Viewer*" dialog, change the "*Level Field*" to the suggested contour level of 0.8.
10. In the main window, select *Tools* → *Volume Data* → *Segment Map*
11. Select the EM map of the refined 3D model and then execute the "*Segment*" command.
12. Visually analyze the segmented map to identify segments that together may represent structural units of the complex, and then execute the "*Group*" command.
13. In the "*Segment Map*" dialog, select "*Regions*" → "*Show only selected.*" This will hide the 3D model except for the selected volume.
14. In the main window, select *Tools* → *Volume Data* → *Fit to Segments*, then select "*Structure to Be Fitted*," and select the "*Fit*" command. Begin by docking the $FXN^{42–210}$ and ISCU trimers. They can be docked both sequentially or together as a $[FXN^{42–210}]_3\bullet[ISCU]_3$ subcomplex. Since both modes of docking give similar cross-correlation functions of ~0.6,

it is convenient to use the latter to sequentially fit eight $[FXN^{42-210}]_3$•$[ISCU]_3$ subcomplexes into the EM map.

15. Use cross-linking data to model the N-termini of FXN^{42-210} and ISCU into nearby unoccupied densities.
16. Dock the homology model of the NFS1 monomer into the 24 remaining unoccupied volumes.
17. Repeat steps 9–16 for the refined 3D model generated without symmetry applied.
18. Adjust the fitting of the protein structures into the EM density maps using "*Fitting Options.*"
19. The closer the cross-correlation function is to 1, the better the fitting.

4.6 MDFF for Docked Structures

4.6.1 Programs

- Install the Situs program package (http://situs.biomachina.org).
- Install the NAMD 2.10 program package (http://www.ks.uiuc.edu/Development/Download/download.cgi?PackageName=NAMD).
- Install the program Coot (https://www2.mrc-lmb.cam.ac.uk/personal/pemsley/coot/).

4.6.2 Procedure

1. Prior to performing molecular dynamics simulations and energy minimizations, it is necessary to convert the EM density map of the 3D model to Situs format and then perform rigid-body docking of the complex structure into the EM map. Ideally one would want to simulate the entire structure of the $[FXN^{42-210}]_{24}$•$[NFS1]_{24}$•$[ISCU]_{24}$ complex. However, depending on the computational power of the available setup, this may not be feasible. In this case, one-half of the structure, consisting of six NFS1 dimers docked on top of four $[FXN^{42-210}]_3$•$[ISCU]_3$ subcomplexes, can be subjected to molecular dynamic simulations and energy minimizations using *NA*noscale *M*olecular *D*ynamics, NAMD 2.10 (Trabuco, Villa, Mitra, Frank, & Schulten, 2008) (http://www.ks.uiuc.edu/Training/Tutorials/science/mdff/tutorial_mdff.pdf).
2. Use NAMD to perfect the complex structure while simultaneously improving its fit to the EM map (Chan et al., 2011; Trabuco et al., 2009).
3. Use the CHARMM22 force field in a vacuum with a dielectric constant of 80, a constant simulation temperature of 50K, and Langevin dynamics coupled with a damping constant of $5\,ps^{-1}$.

4. Apply harmonic restraints for amine N, carbonyl C, and α-C atoms to maintain the model's 432 symmetry with a symmetry force constant of 20 for 100 ps (200 steps) (Chan et al., 2011).
5. Apply harmonic restraints to maintain chirality and secondary structure.
6. Perform iterations of model improvement using the program Coot (Emsley, Lohkamp, Scott, & Cowtan, 2010), followed by molecular dynamics simulations and energy minimizations using NAMD 2.10 against the EM map, until a plausible model of the complex is obtained.
7. This process should yield a structure that closely matches the $[FXN^{42-210}]_3$•$[ISCU]_3$ and NFS1 homology models used for the initial docking. A step-by-step protocol for molecular dynamic simulations of the human complex is available from the authors upon request.
8. Submit the simulated structure of the complex fitted to the EM map of the 3D model with the PDB. The PDB Summary Validation Report will provide parameters that together indicate if the structural model has reasonable geometry as compared to previously deposited EM-based structures. The parameters include the Ramachandran plot, MolProbity score (http://molprobity.biochem.duke.edu/), β-carbon deviations >0.25 Å (Chen et al., 2010), and *Z* score (Emsley et al., 2010). In addition, the wwPDB Annotation Staff will highlight issues that will have to be resolved prior to the release of the file. These issues may include atomic clashes that can be minimized by *Molecular Dynamic Flexible Fitting* using NAMD 2.10. There may also be chirality errors and peptide linkage issues that can be corrected using Coot.
9. For the purpose of visualizing the entire complex structure, the simulated one half can be aligned with itself into the EM density map, followed by structure improvement iterations with the program Coot to resolve steric clashes mainly between NFS1 subunits at the interface between the two halves of the structure.

5. APPROACHES TO VALIDATE THE PROTEIN–PROTEIN INTERFACES OF THE COMPLEX

Chemical cross-linking coupled with limited proteolysis and MS/MS identification of cross-linked peptides is another approach to obtain information about protein–protein interactions in a multiprotein complex such as the $[FXN^{42-210}]_{24}$•$[ISCU]_{24}$•$[NFS1]_{24}$•$[ISD11]_{24}$ complex (Leitner et al., 2010). This approach was feasible for our complex because we had found that cross-linking with the uncleavable cross-linker BS^3 did not alter the physical and enzymatic properties of the complex (Fig. 7E, G, and H). The program StavroX 3.5.1 (Gotze et al., 2012) (www.stavrox.com/help.php) was instrumental for the analysis of the MS/MS output given its complexity. Based on the amino acid sequences of FXN, ISCU, NFS1, and ISD11, the molecular weight and amino acid specificity of BS^3, and the cleavage specificity of two different peptidases, GluC and AspN, the program calculated all theoretical possible combinations of cross-linked peptides and compared them to the precursor peptide masses in the MS file (Gotze et al., 2012). Then, the program generated a list of identified cross-linked peptides between FXN and ISCU, FXN and NFS1, FXN and ISD11, NFS1 and ISCU, NFS1 and ISD11, as well as cross-linked peptides within FXN, ISCU, NFS1, or ISD11. Each peptide was assigned a score based on a comparison between the theoretical fragmentation and the actual MS/MS spectrum of the peptide (Gotze et al., 2012), and a false discovery rate (FDR) based on a comparison between the number of candidates in the experimental dataset and the number of false-positive candidates from a decoy dataset created from the inverted amino acid sequences of FXN, ISCU, NFS1, and ISD11. This procedure (described in detail later) led to the identification of 124 unique cross-linked peptides, 43% and 90% of which had a false FDR of $\leq$5% and $\leq$25%, respectively (Table 2). The vast majority of Lys residues present in FXN^{42-210}, ISCU, NFS1, and ISD11 as well as the N-terminal amino groups and several Ser, Thr, and Tyr residues of the four proteins were involved. This confirmed that we were dealing with a tight complex in which the four-protein components were in close contact with one another. To assess the degree of agreement between the structure of the complex and the cross-linking data we measured, in the simulated half-structure of the complex, the distances between all possible pairs of cross-linked residues within any given cross-linked peptide. Next, we identified the mean distances ($\pm$S.D.) that were equal to or lower than the distance constraints or the maximum allowable distance constraints; or greater than the maximum allowable distance constraints (defined later). We assigned agreement between the simulated structure and any given cross-linked peptide if at least one of the mean distances measured in the simulated half-structure was within the distance constraints or the maximum allowable distance constraints. We found that the simulated structure was in agreement with 90% of the cross-linked peptides identified (Table 2). There was apparent disagreement between the simulated half-structure of the complex and $\sim$10% of the cross-linked peptides identified (Table 2), which could be explained by the fact that these peptides involved flexible regions of the four proteins.

Table 2 Summary of Cross-Linking Results

Cross-Linked Partners	Total Cross-Linked Peptides	Peptides With FDR ≤5%	Peptides With FDR ≤10%	Peptides With FDR ≤15%	Peptides With FDR ≤20%	Peptides With FDR ≤25%	Peptides With FDR ≤35%	Peptides With FDR ≤45%	Peptides With FDR ≤60%	Cross-Linked Peptides That Support Complex Structure Based on Distance Constraints	Cross-Linked Peptides That Support Complex Structure Based on Maximum Allowable Distance Constraints	Cross-Linked Peptides in Disagreement With Complex Structure
FXN^{42-210}–FXN	6	2	—	—	—	3	—	1	—	5	1	—
FXN^{42-210}–ISCU	13	4	—	—	4	1	1	1	2	10	3	—
FXN^{42-210}–ISD11	4	2	—	—	—	1	—	1	—	1	—	3
FXN^{42-210}–NFS1	12	3	1	—	4	2	—	1	1	4	7	1
ISCU–ISCU	26	16	3	1	3	3	—	—	—	26	—	—
ISCU–ISD11	4		—	—	—	1	2	—	1	2	1	1
ISCU–NFS1	22	10	2	—	8	2	—	—	—	9	8	5
ISD11–ISD11	4	2	—	—	—	1	1	—	—	2	2	—

Continued

Table 2 Summary of Cross-Linking Results—cont'd

Cross-Linked Partners	Total Cross-Linked Peptides	Peptides With FDR ≤5%	Peptides With FDR ≤10%	Peptides With FDR ≤15%	Peptides With FDR ≤20%	Peptides With FDR ≤25%	Peptides With FDR ≤35%	Peptides With FDR ≤45%	Peptides With FDR ≤60%	Cross-Linked Peptides That Support Complex Structure Based on Distance Constraints	Cross-Linked Peptides That Support Complex Structure Based on Maximum Allowable Distance Constraints	Cross-Linked Peptides in Disagreement With Complex Structure
ISD11–NFS1	11	3	—	—	5	3	—	—	—	6	4	1
NFS1–NFS1	22	11	3	—	6	1	1	—	—	15	5	2
Total	124	53	9	1	30	18	5	4	4	80	31	13
% of Total	*100%*	43%	7%	1%	24%	15%	4%	3%	3%	65%	25%	10%

Cross-linked peptides were identified and categorized based on their respective FDR. In evaluating the FDR of each cross-linked peptide, it is important to consider that we are not dealing with a complex mixture of unknown proteins. Rather, we are dealing with a complex consisting of stoichiometric amounts of four known proteins that generated a relatively small number of cross-linked peptides by single or double digestion. In light of this, the fact that 75% of the peptides have overall FDR ≤20% (corresponding to ≥80% confidence) and the fact that 90% of the peptides are in agreement with the distance constraints set by the complex structure at multiple different interfaces together indicate that this analysis and the conclusions drawn from it are robust. The distance constraints and maximum allowable distance constraints between different pairs of cross-linked residues were calculated as described in Gakh et al. (2016). Agreement between the simulated three-dimensional structure and any given cross-linked peptide was established as described in Gakh et al. (2016).

When we physically mapped the cross-links in the simulated structure of the complex, it became evident that the cross-linking data supported the existence of several recurring protein–protein interaction surfaces both within and between subunits of the complex. Owing to the symmetrical arrangement of the complex, some FXN^{42-210}–FXN^{42-210} or ISCU–ISCU cross-links were possible within the same monomer or between two monomers of the same trimer or between monomers from two adjacent trimers. For example, in the case of FXN^{42-210}, cross-links Lys-135–Lys-147 and Lys-135–Lys-164 are possible within one FXN^{42-210} subunit (Fig. 10A) or between subunits of two adjacent trimers (Fig. 10A). However, the distances measured in the simulated structure indicate that these particular cross-links are more likely to form between subunits of two adjacent FXN^{42-210} trimers. Also, cross-link Tyr-93–Lys-116 is only possible between subunits of two adjacent trimers (Fig. 10A). Similarly, in the case of ISCU, cross-links Lys-121–Lys-135 and Lys-127–Lys-135, which involve Lys-135 (part of the highly conserved PVK (Pro-133, Val-134, Lys-135) motif of ISCU) (Manicki et al., 2014), are possible both within the same ISCU subunit and between ISCU subunits of two adjacent ISCU trimers (Fig. 10B). However, cross-links Lys-160–Lys-160 and Lys-160–Lys-161 are only possible between two adjacent ISCU subunits of the same trimer (Fig. 10B). Most of the NFS1–NFS1 cross-links could only form within the same NFS1 monomer (Gakh et al., 2016). However, there were cross-links that could only be explained by the presence of adjacent NFS1 subunits. For example, the Lys-157–Ser-437 cross-link was possible only in the context of the NFS1 dimer at the twofold axis of the complex (Fig. 10C). Similarly, the Ser-437–Lys-320 cross-link was only possible in the context of the NFS1 trimer at the threefold axis of the complex (Fig. 10D). Moreover, the Ser-99–Lys-157 cross-link was only possible in the context of the NFS1 tetramer at the fourfold axis of the complex (Fig. 10E). Overall, the cross-links supported the fold of the modeled proteins and were also in excellent agreement with protein–protein interaction surfaces at the two-, three-, and fourfold axes of the complex (Gakh et al., 2016).

5.1 Limited Proteolysis of Cross-Linked Complex and Tandem Mass Spectrometry Identification of Cross-Linked Peptides

5.1.1 Equipment

- Thermo Ultimate 3000 RSLCnano HPLC system
- Orbitrap Elite Mass Spectrometer (Thermo Fisher Scientific)
- 250 nL OPTI-PAK trap (Optimize Technologies) custom packed with Michrom Magic C18 solid phase (Michrom Bioresources)
- PicoFrit Magic C18, 3 μm, 75 μm × 250 mm column (New Objective)

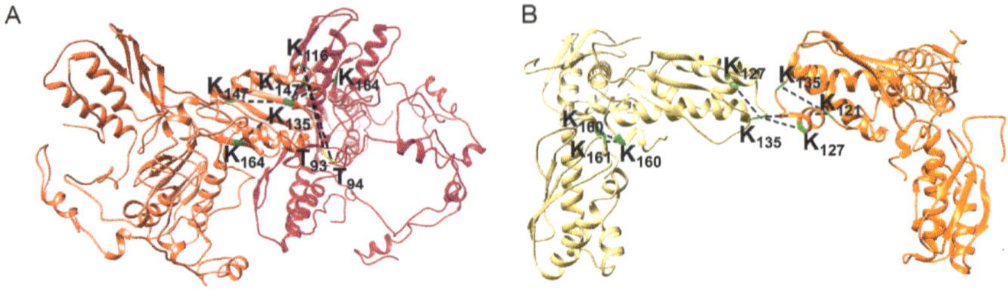

Cross-Linked Peptides		Distances (Å) Measured in Structure		
FXN^{42-210}	FXN^{42-210}	FXN^{42-210} Intra monomer	FXN^{42-210}–FXN^{42-210} Intra trimer	FXN^{42-210}–FXN^{42-210} Inter trimers
DVSFGSGVLTV K_{135}LGG	DLGTYVINK_{147}QTP NK_{152}QIWLSSPSS GPK_{164}RY			
K_{135}	K_{147}	17.6 ± 0.7	33.0 ± 1.8	17.3 ± 3.9
K_{135}	K_{164}	18.4 ± 0.8	34.7 ± 1.1	14.3 ± 2.2
FFEDLADK_{116}PY TFE	$T_{93}T_{94}Y_{95}$E			
K_{116}	T_{93}	35.8 ± 1.5	31.2 ± 1.1	28.1 ± 0.8
K_{116}	T_{94}	33.3 ± 0.9	32.9 ± 1.1	26.3 ± 1.1

Cross-Linked Peptides		Distances (Å) Measured in Structure		
ISCU	ISCU	ISCU Intra monomer	ISCU–ISCU Intratrimer	ISCU–ISCU Intertrimers
LCLPPVK_{135}LHC S_{139}MLAE	ALT_{119}IK_{121}NT_{123}DI AK_{127}E			
K_{135}	K_{121}	13.5 ± 0.8	44.8 ± 1.3	17.2 ± 1.6
K_{135}	K_{127}	12.0 ± 0.7	46.4 ± 1.5	12.9 ± 1.5
P$K_{160}K_{161}$GE	P$K_{160}K_{161}$GE			
K_{160}	K_{160}	NA	11.8 ± 0.7	69.1 ± 0.2
K_{160}	K_{161}	NA	10.5 ± 0.5	67.6 ± 0.3

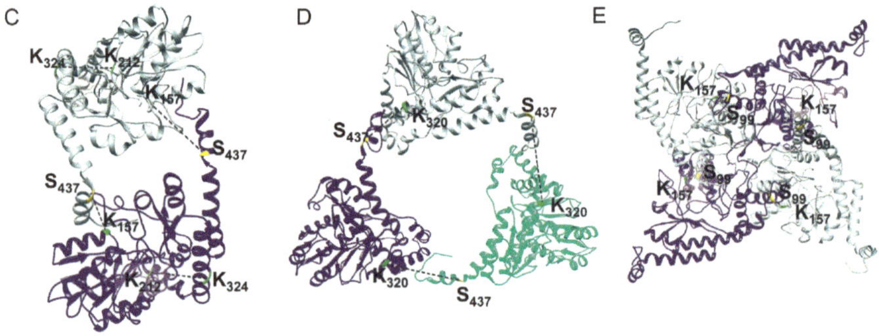

Cross-Linked Peptides		Distances (Å) Measured in Structure			
NFS1	NFS1	Monomer	Dimer	Trimer	Tetramer
IGVK_{212}QPIAE	Y_{317}DHK_{320}RI$S_{323}K_{324}$LSE				
K_{212}	K_{324}	16.0 ± 0.3	66.4 ± 2.3	66.4 ± 2.2	50.2 ± 1.8
MS_{437}PLWE	HK_{157}CVLDS_{162}CRS_{165}LE				
S_{437}	K_{157}	40.2 ± 1.1	19.0 ± 0.9	55.5 ± 3.7	67.6 ± 1.2
MS_{437}PLWE	Y_{317}DHK_{320}RI$S_{323}K_{324}$LSE				
S_{437}	K_{320}	44.8 ± 0.8	47.2 ± 1.4	27.7 ± 3.9	65.0 ± 2.9
S_{99}EAAME	HK_{157}CVL				
S_{99}	K_{157}	39.5 ± 1.1	64.3 ± 1.4	48.4 ± 2.9	18.7 ± 4.9

Fig. 10 Architecture of FXN^{42-210} and ISCU trimers and NFS1 monomer, dimer, trimer and tetramer analyzed using chemical cross-linking. Mapping of representative cross-links (*dotted lines*; Lys residues are in *green* and non-Lys residues in *yellow*) in the simulated structures of FXN^{42-210}, ISCU, and NFS1. Distances were measured between all possible pairs of cross-linked residues within each of the cross-linked peptides in the simulated half-structure of the complex, and were expressed as mean ± S.D. Distances equal to or lower than the distance constraints are highlighted in *light gray*; distances equal to or lower than the maximum allowable distance constraints are highlighted in *dark gray*, and distances greater than the maximum allowable distance constraints are highlighted in *yellow*. (A and B) *Intramonomer* distances are measured within individual FXN^{42-210} or ISCU subunits, *intratrimer* distances between FXN^{42-210} or ISCU subunits of the same $[FXN^{42-210}]_3$•$[ISCU]_3$ subcomplex, and *intertrimers* distances

5.1.2 Buffers and Solutions

- Endoproteinase GluC (New England BioLabs)
- Endoproteinase AspN (Roche/Sigma-Aldrich)
- Zwittergent 3-16 detergent
- 1 *M* Tris–HCl, pH 8.1
- Trifluroacetic acid
- 500 m*M* Tris-carboxyethyl phosphine (Hampton Research)
- Solvent A [98% water/2% acetonitrile (v/v) and 0.2% (v/v) formic acid]
- Solvent B [80% acetonitrile/10% isopropanol/10% water (v/v) and 0.2% (v/v) formic acid]

5.1.3 Procedure (Adapted From Ranatunga et al., 2016)

1. Incubate 20 μg of cross-linked or uncross-linked complex freshly eluted from the Sephacryl S-300 column with 1 μg of the endoproteinase GluC at 37°C.
2. Using SDS-PAGE, analyze a time course of digestion keeping in mind that over 60 h of incubation may be needed to achieve ≥95% digestion of the cross-linked complex.
3. Terminate the digestion reaction by the addition of 0.5% (v/v) trifluroacetic acid.
4. At this point, the reaction can be frozen and used for MS analysis later.
5. For double digestion with GluC and AspN endoproteinase, do the following prior to step 4.

---

between FXN^{42-210} or ISCU subunits of two adjacent $[FXN^{42-210}]_3$•$[ISCU]_3$ subcomplexes. For FXN^{42-210}, cross-links between K_{135} and K_{147} or K_{164} are possible both within one FXN^{42-210} subunit or between two FXN^{42-210} subunits of two adjacent trimers. Cross-links between K_{116} and T_{93} or T_{94} are only possible between two FXN^{42-210} subunits of two adjacent trimers. For ISCU, cross-links between K_{135} and K_{121} or K_{127} are possible within one ISCU subunit as well as between ISCU subunits of adjacent trimers but not between subunits of the same trimer. Conversely, the K_{160}–K_{160} and K_{160}–K_{161} cross-links are only possible between subunits of the same trimer. *NA*, not applicable. (C–E) *Monomer, dimer, trimer*, and *tetramer* distances are measured, respectively, within the same NFS1 subunit or between two NFS1 subunits at the two-, three- and fourfold axes of the complex. The table shows cross-linked peptides that are uniquely compatible for NFS1 monomer, dimer, trimer, or tetramer. *This research was originally published in The Journal of Biological Chemistry. Gakh, O., et al. (2016). Architecture of the human mitochondrial iron-sulfur cluster assembly machinery.* The Journal of Biological Chemistry, 291, *21296–21321.*

6. Add 0.0002% Zwittergent 3-16 detergent.
7. Add 0.25 μg (w/v) endoproteinase AspN (Roche/Sigma-Aldrich) and incubate at 37°C overnight.
8. Terminate the digestion reaction by the addition of 0.5% (v/v) trifluroacetic acid.
9. Incubate samples with 50 m*M* reducing agent, Tris-carboxyethyl phosphine, at 60°C for 30 min and use the sample for MS/MS analysis.
10. Load the digested peptide mixture onto a 250 nL OPTI-PAK trap (Optimize Technologies) custom packed with Michrom Magic C18 solid phase (Michrom Bioresources).
11. Perform chromatography on a Thermo Ultimate 3000 RSLCnano HPLC system using solvent A and solvent B, with a 2% B to 40% B gradient over 60 min at 325 nL/min through a PicoFrit Magic C18, 3 μm, 75 μm × 250 mm column (New Objective).
12. Perform a Fourier transform full scan from 340 to1600 *m/z* at a resolution of 120,000 (400 *m/z* resolving power), followed by higher energy collisional dissociation Orbitrap MS/MS scans on the top 15 ions at a resolution of 15,000 on the mass spectrometer Orbitrap Elite Mass Spectrometer (Thermo Fisher Scientific).
13. Set the MS scan 1 automatic gain control target to $1e^6$ and the MS scan 2 target to $5e^5$ with maximum ion inject times of 50 and 100 ms, respectively.

5.2 Analysis of Cross-Linked Peptides

5.2.1 Equipment

- High-speed/performance computer for fast data processing time

5.2.2 Programs

- Install StavroX 3.5.1 program (Gotze et al., 2012) (www.stavrox.com/help.php)
- Microsoft Excel 2010

5.2.3 Procedure (Adapted From http://www.Stavrox.Com/Help.Php)

1. Open StavroX and provide the information needed for the project by clicking *Settings* as follows:
   1.1 In the *Protease Sites* tab: provide the protease cleavage sites (for example, ↓*E* for GluC).
   1.2 In the *Modifications* tab, use default.

1.3 In the *Cross-Linker* tab, select the cross-linker used and its specificity.

1.4 In the *Mass Comparison* tab, keep the default values for *Precursor Precision* and *Fragment Ion Precision* at 3.0 ppm and 0.8 Da, respectively; select a *Lower Mass Limit* of 147 Da (molecular mass of lysine); select a *Upper Mass Limit* between 5000 and 30,000 Da to account for possible peptide oligomers; for the *S/N Ratio* use default (*2.0*); select *Ion Types b and y*; select *Apply a Prescore > 10%* intensity; select defaults for *Losses* (*neutral loss of identified fragments*) and *Max. Losses (1)*.

1.5 In the *Graphical User Interface* tab, use default settings (*Digits of Deviations (ppm)* = *2*; *Digits of MS Data* = 3; *Digits of MSMS Data* = 1). Select *Automatically Check for Updates* and *Site Numbering as in Peptides*. Select *Ions colored by Different Peptides* to visualize different types of ions in the MS spectra.

Save the settings with your project's name.

2. Click *Load Fasta*. This will direct you to load the file containing the amino acid sequences of the four-protein components of the complex in Fasta format.
3. Click on *Load MS Data* and upload the MS/MS data file (in *.mgf format).
4. Click *Start*.

4.1 It will take a few minutes to upload the data and display the results. A table will show the identified cross-linked peptides, and their respective score, *m/z*, and other information.

4.2 Click *Show Decoy Analysis*. This will open a new window showing the FDR range with the corresponding score range of all identified cross-linked peptides. For example, FDR < 0.05 (38) will indicate that cross-linked peptides with a score ≥ 38 are identified with 95% confidence.

4.3 Select the peptides based on their FDR and transfer the data into an Excel spreadsheet and organize the data (Fig. 10; see also Supplemental Table S2 in Gakh et al., 2016).

5. Measure protein–protein distances between each possible pair of cross-linked residues within each of the identified cross-linked peptides in the structure using a script (distance: resi#1.subunit#1@CA:resi#2.subunit#2@CA) in the command line of Chimera (Pettersen et al., 2004) and calculate the mean distance ± S.D. for each possible pair of cross-linked residues. Add this information to the Excel spreadsheet.

6. Since BS^3 has a space arm of 11.4 Å, assume that the distance constraint between the backbone α-carbons of two cross-linked residues is equal to the length of the BS^3 spacer arm + the lengths of the side chains of the two cross-linked residues. Based on this, the distance constraints for all of the possible pairs of residues that can be cross-linked by BS^3 can be calculated as reported (Gakh et al., 2016).
7. Assume that the maximum allowable distance constraint between the backbone α-carbons of two cross-linked residues is equal to the distance constraint plus the estimated resolution-dependent errors (Blow, 2002) in the positions of FXN^{42-210}, ISCU, NFS1, or ISD11 protein atoms, as reported (Gakh et al., 2016).
8. In the Excel spreadsheet mark, with different colors, the mean distances measured in the structure that are equal or lower than the distance constraints, equal or lower than the maximum allowable distance constraints, or greater than the maximum allowable distance constraint (Fig. 10).
9. Map the cross-links on the structure as shown in Fig. 10.

5.3 Analysis of Complex Interfaces

5.3.1 Equipment

- High-speed/performance computer for fast data processing time

5.3.2 Programs

- PISA program (http://www.ebi.ac.uk/msd-srv/prot_int/cgi-bin/piserver).
- PDB2PQR Server (http://nbcr-222.ucsd.edu/pdb2pqr_2.0.0/).
- Adaptive Poisson–Boltzmann Solver (APBS) software package (http://pymolwiki.org/index.php/APBS).
- PyMol (https://www.pymol.org/).

5.3.3 Procedure

1. Upload a coordinate file of the four-protein complex (it is easier to work with a subcomplex, e.g., $[FXN^{42-210}]_6$•$[NFS1]_2$•$[ISCU]_6$, that can be extracted from the simulated half-structure of the complex), into the PISA program.
2. Use the PISA Interfaces algorithm to identify FXN-FXN, ISCU-ISCU, NFS1-NFS1, FXN-ISCU, FXN-NFS1, and ISCU-NFS1 interfaces, their buried surface areas, and conserved residues involved in each interface as described (Gakh et al., 2016).

3. Upload a coordinate file of the four-protein complex into the PDB2PQR Server and calculate the electrostatic surface potential using the Poisson–Boltzmann equation.
4. The APBS software package can be used to visualize electrostatic protein surfaces in PyMol.

ACKNOWLEDGMENTS

The research described here was originally published in *The Journal of Biological Chemistry*. Gakh, O., et al. (2016). Architecture of the human mitochondrial iron-sulfur Cluster assembly machinery. *The Journal of Biological Chemistry, 291*(21), 21296–21321. © The American Society for Biochemistry and Molecular Biology. The authors acknowledge the supported by a grant from the National Institutes of Health, National Institute on Aging, AG15709-20.

REFERENCES

Adamec, J., Rusnak, F., Owen, W. G., Naylor, S., Benson, L. M., Gacy, A. M., et al. (2000). Iron-dependent self-assembly of recombinant yeast frataxin: Implications for Friedreich ataxia. *American Journal of Human Genetics, 67*(3), 549–562. http://dx.doi.org/10.1086/303056.

Amela, I., Delicado, P., Gomez, A., Querol, E., & Cedano, J. (2013). A dynamic model of the proteins that form the initial iron-sulfur cluster biogenesis machinery in yeast mitochondria. *The Protein Journal, 32*(3), 183–196. http://dx.doi.org/10.1007/s10930-013-9475-4.

Andrews, P. (1965). The gel-filtration behaviour of proteins related to their molecular weights over a wide range. *The Biochemical Journal, 96*(3), 595–606.

Bedekovics, T., Gajdos, G. B., Kispal, G., & Isaya, G. (2007). Partial conservation of functions between eukaryotic frataxin and the *Escherichia coli* frataxin homolog CyaY. *FEMS Yeast Research*, 7(8), 1276–1284. http://dx.doi.org/10.1111/j.1567-1364.2007.00296.x.

Blanc, B., Gerez, C., & Ollagnier de Choudens, S. (2015). Assembly of Fe/S proteins in bacterial systems: Biochemistry of the bacterial ISC system. *Biochimica et Biophysica Acta, 1853*(6), 1436–1447. http://dx.doi.org/10.1016/j.bbamcr.2014.12.009.

Blow, D. M. (2002). Rearrangement of Cruickshank's formulae for the diffraction-component precision index. *Acta Crystallographica. Section D, Biological Crystallography, 58*(Pt. 5), 792–797.

Bridwell-Rabb, J., Fox, N. G., Tsai, C. L., Winn, A. M., & Barondeau, D. P. (2014). Human frataxin activates Fe-S cluster biosynthesis by facilitating sulfur transfer chemistry. *Biochemistry, 53*(30), 4904–4913. http://dx.doi.org/10.1021/bi500532e.

Cavadini, P., Adamec, J., Taroni, F., Gakh, O., & Isaya, G. (2000). Two-step processing of human frataxin by mitochondrial processing peptidase. Precursor and intermediate forms are cleaved at different rates. *The Journal of Biological Chemistry, 275*(52), 41469–41475. http://dx.doi.org/10.1074/jbc.M006539200.

Chan, K. Y., Gumbart, J., McGreevy, R., Watermeyer, J. M., Sewell, B. T., & Schulten, K. (2011). Symmetry-restrained flexible fitting for symmetric EM maps. *Structure, 19*(9), 1211–1218. http://dx.doi.org/10.1016/j.str.2011.07.017.

Chen, V. B., Arendall, W. B., 3rd, Headd, J. J., Keedy, D. A., Immormino, R. M., Kapral, G. J., et al. (2010). MolProbity: All-atom structure validation for macromolecular crystallography. *Acta Crystallographica. Section D, Biological Crystallography, 66*(Pt. 1), 12–21. http://dx.doi.org/10.1107/S0907444909042073.

Cline, J. D. (1969). Spectrophotometric determination of hydrogen sulfide in natural waters. *Limnology and Oceanography*, *14*(3), 454–458. http://dx.doi.org/10.4319/lo.1969.14.3.0454.

Colin, F., Martelli, A., Clemancey, M., Latour, J. M., Gambarelli, S., Zeppieri, L., et al. (2013). Mammalian frataxin controls sulfur production and iron entry during de novo Fe_4S_4 cluster assembly. *Journal of the American Chemical Society*, *135*(2), 733–740. http://dx.doi.org/10.1021/ja308736e.

Cook, J. D., Kondapalli, K. C., Rawat, S., Childs, W. C., Murugesan, Y., Dancis, A., et al. (2010). Molecular details of the yeast frataxin-Isu1 interaction during mitochondrial Fe-S cluster assembly. *Biochemistry*, *49*(40), 8756–8765. http://dx.doi.org/10.1021/bi1008613.

Emsley, P., Lohkamp, B., Scott, W. G., & Cowtan, K. (2010). Features and development of Coot. *Acta Crystallographica. Section D, Biological Crystallography*, *66*(Pt. 4), 486–501. http://dx.doi.org/10.1107/S0907444910007493.

Fox, N. G., Das, D., Chakrabarti, M., Lindahl, P. A., & Barondeau, D. P. (2015). Frataxin accelerates [2Fe-2S] cluster formation on the human Fe-S assembly complex. *Biochemistry*, *54*(25), 3880–3889. http://dx.doi.org/10.1021/bi5014497.

Gakh, O., Bedekovics, T., Duncan, S. F., Smith, D. Y. I. V., Berkholz, D. S., & Isaya, G. (2010). Normal and Friedreich ataxia cells express different isoforms of frataxin with complementary roles in iron-sulfur cluster assembly. *The Journal of Biological Chemistry*, *285*(49), 38486–38501. http://dx.doi.org/10.1074/jbc.M110.145144.

Gakh, O., Park, S., Liu, G., Macomber, L., Imlay, J. A., Ferreira, G. C., et al. (2006). Mitochondrial iron detoxification is a primary function of frataxin that limits oxidative damage and preserves cell longevity. *Human Molecular Genetics*, *15*(3), 467–479.

Gakh, O., Ranatunga, W., Smith, D. Y. I. V., Ahlgren, E. C., Al-Karadaghi, S., Thompson, J. R., et al. (2016). Architecture of the human mitochondrial iron-sulfur cluster assembly machinery. *The Journal of Biological Chemistry*, *291*(40), 21296–21321. http://dx.doi.org/10.1074/jbc.M116.738542.

Gakh, O., Smith, D. Y., IV, & Isaya, G. (2008). Assembly of the iron-binding protein frataxin in *Saccharomyces cerevisiae* responds to dynamic changes in mitochondrial iron influx and stress level. *The Journal of Biological Chemistry*, *283*(46), 31500–31510. http://dx.doi.org/10.1074/jbc.M805415200.

Gotze, M., Pettelkau, J., Schaks, S., Bosse, K., Ihling, C. H., Krauth, F., et al. (2012). StavroX—A software for analyzing crosslinked products in protein interaction studies. *Journal of the American Society for Mass Spectrometry*, *23*(1), 76–87. http://dx.doi.org/10.1007/s13361-011-0261-2.

Hagel, L., Östberg, M., & Andersson, T. (1996). Apparent pore size distributions of chromatography media. *Journal of Chromatography A*, *743*(1), 33–42.

Kalkhof, S., & Sinz, A. (2008). Chances and pitfalls of chemical cross-linking with amine-reactive N-hydroxysuccinimide esters. *Analytical and Bioanalytical Chemistry*, *392*(1–2), 305–312.

Karlberg, T., Schagerlof, U., Gakh, O., Park, S., Ryde, U., Lindahl, M., et al. (2006). The structures of frataxin oligomers reveal the mechanism for the delivery and detoxification of iron. *Structure*, *14*(10), 1535–1546. http://dx.doi.org/10.1016/j.str.2006.08.010.

Kim, J. H., Bothe, J. R., Alderson, T. R., & Markley, J. L. (2015). Tangled web of interactions among proteins involved in iron-sulfur cluster assembly as unraveled by NMR, SAXS, chemical crosslinking, and functional studies. *Biochimica et Biophysica Acta*, *1853*(6), 1416–1428. http://dx.doi.org/10.1016/j.bbamcr.2014.11.020.

Kondapalli, K. C., Kok, N. M., Dancis, A., & Stemmler, T. L. (2008). Drosophila frataxin: An iron chaperone during cellular Fe-S cluster bioassembly. *Biochemistry*, *47*(26), 6917–6927. http://dx.doi.org/10.1021/bi800366d.

Layer, G., Ollagnier-de Choudens, S., Sanakis, Y., & Fontecave, M. (2006). Iron-sulfur cluster biosynthesis: Characterization of *Escherichia coli* CyaY as an iron donor for the assembly of [2Fe-2S] clusters in the scaffold IscU. *The Journal of Biological Chemistry*, *281*(24), 16256–16263.

Leitner, A., Walzthoeni, T., Kahraman, A., Herzog, F., Rinner, O., Beck, M., et al. (2010). Probing native protein structures by chemical cross-linking, mass spectrometry, and bioinformatics. *Molecular & Cellular Proteomics*, *9*(8), 1634–1649. http://dx.doi.org/10.1074/mcp.R000001-MCP201.

Li, H., Gakh, O., Smith, D. Y. I. V., & Isaya, G. (2009). Oligomeric yeast frataxin drives assembly of core machinery for mitochondrial iron-sulfur cluster synthesis. *The Journal of Biological Chemistry*, *284*(33), 21971–21980. http://dx.doi.org/10.1074/jbc.M109.011197.

Li, H., Gakh, O., Smith, D. Y. I. V., Ranatunga, W. K., & Isaya, G. (2013). Missense mutations linked to friedreich ataxia have different but synergistic effects on mitochondrial frataxin isoforms. *The Journal of Biological Chemistry*, *288*(6), 4116–4127. http://dx.doi.org/10.1074/jbc.M112.435263.

Ludtke, S. J. (2010). 3-D structures of macromolecules using single-particle analysis in EMAN. *Methods in Molecular Biology*, *673*, 157–173. http://dx.doi.org/10.1007/978-1-60761-842-3_9.

Madler, S., Gschwind, S., & Zenobi, R. (2010). Role of arginine in chemical cross-linking with N-hydroxysuccinimide esters. *Analytical Biochemistry*, *398*(1), 123–125. http://dx.doi.org/10.1016/j.ab.2009.11.020.

Manicki, M., Majewska, J., Ciesielski, S., Schilke, B., Blenska, A., Kominek, J., et al. (2014). Overlapping binding sites of the frataxin homologue assembly factor and the heat shock protein 70 transfer factor on the Isu iron-sulfur cluster scaffold protein. *The Journal of Biological Chemistry*, *289*(44), 30268–30278. http://dx.doi.org/10.1074/jbc.M114.596726.

Marelja, Z., Stocklein, W., Nimtz, M., & Leimkuhler, S. (2008). A novel role for human Nfs1 in the cytoplasm: Nfs1 acts as a sulfur donor for MOCS3, a protein involved in molybdenum cofactor biosynthesis. *The Journal of Biological Chemistry*, *283*(37), 25178–25185. http://dx.doi.org/10.1074/jbc.M804064200.

Marinoni, E. N., de Oliveira, J. S., Nicolet, Y., Raulfs, E. C., Amara, P., Dean, D. R., et al. (2012). $(IscS\text{-}IscU)_2$ complex structures provide insights into Fe_2S_2 biogenesis and transfer. *Angewandte Chemie*, *51*(22), 5439–5442. http://dx.doi.org/10.1002/anie.201201708.

Moore, S. (1963). On the determination of cysteine and cysteic acid. *The Journal of Biological Chemistry*, *238*(1), 235–237.

Neuhoff, V., Stamm, R., Pardowitz, I., Arold, N., Ehrhardt, W., & Taube, D. (1990). Essential problems in quantification of proteins following colloidal staining with coomassie brilliant blue dyes in polyacrylamide gels, and their solution. *Electrophoresis*, *11*(2), 101–117. http://dx.doi.org/10.1002/elps.1150110202.

Pandey, A., Gordon, D. M., Pain, J., Stemmler, T. L., Dancis, A., & Pain, D. (2013). Frataxin directly stimulates mitochondrial cysteine desulfurase by exposing substrate-binding sites, and a mutant Fe-S cluster scaffold protein with frataxin-bypassing ability acts similarly. *The Journal of Biological Chemistry*, *288*(52), 36773–36786. http://dx.doi.org/10.1074/jbc.M113.525857.

Parent, A., Elduque, X., Cornu, D., Belot, L., Le Caer, J. P., Grandas, A., et al. (2015). Mammalian frataxin directly enhances sulfur transfer of NFS1 persulfide to both ISCU and free thiols. *Nature Communications*, *6*, 5686. http://dx.doi.org/10.1038/ncomms6686.

Pettersen, E. F., Goddard, T. D., Huang, C. C., Couch, G. S., Greenblatt, D. M., Meng, E. C., et al. (2004). UCSF chimera—A visualization system for exploratory research and analysis. *Journal of Computational Chemistry*, *25*(13), 1605–1612. http://dx.doi.org/10.1002/jcc.20084.

Prischi, F., Konarev, P. V., Iannuzzi, C., Pastore, C., Adinolfi, S., Martin, S. R., et al. (2010). Structural bases for the interaction of frataxin with the central components of iron-sulphur cluster assembly. *Nature Communications*, *1*, 95. http://dx.doi.org/10.1038/ncomms1097.

Ranatunga, W., Gakh, O., Galeano, B. K., Smith, D. Y. I. V., Soderberg, C. A., Al-Karadaghi, S., et al. (2016). Architecture of the yeast mitochondrial iron-sulfur cluster assembly machinery: The sub-complex formed by the iron donor, Yfh1, and the scaffold, Isu1. *The Journal of Biological Chemistry*, *291*(19), 10378–10398. http://dx.doi.org/10.1074/jbc.M115.712414.

Schagerlöf, U., Elmlund, H., Gakh, O., Nordlund, G., Hebert, H., Lindahl, M., et al. (2008). Structural basis of the iron storage function of Frataxin from single-particle reconstruction of the iron-loaded oligomer. *Biochemistry*, *47*, 4948–4954.

Schmucker, S., Argentini, M., Carelle-Calmels, N., Martelli, A., & Puccio, H. (2008). The *in vivo* mitochondrial two-step maturation of human frataxin. *Human Molecular Genetics*, *17*(22), 3521–3531. http://dx.doi.org/10.1093/hmg/ddn244.

Seguin, A., Sutak, R., Bulteau, A. L., Garcia-Serres, R., Oddou, J. L., Lefevre, S., et al. (2010). Evidence that yeast frataxin is not an iron storage protein *in vivo*. *Biochimica et Biophysica Acta*, *1802*(6), 531–538. http://dx.doi.org/10.1016/j.bbadis.2010.03.008.

Shi, R., Proteau, A., Villarroya, M., Moukadiri, I., Zhang, L., Trempe, J. F., et al. (2010). Structural basis for Fe-S cluster assembly and tRNA thiolation mediated by IscS protein-protein interactions. *PLoS Biology*, *8*(4), e1000354. http://dx.doi.org/10.1371/journal.pbio.1000354.

Söderberg, C. A., Rajan, S., Shkumatov, A. V., Gakh, O., Schaefer, S., Ahlgren, E. C., et al. (2013). The molecular basis of iron-induced oligomerization of frataxin and the role of the ferroxidation reaction in oligomerization. *The Journal of Biological Chemistry*, *288*(12), 8156–8167. http://dx.doi.org/10.1074/jbc.M112.442285.

Söderberg, C. A., Shkumatov, A. V., Rajan, S., Gakh, O., Svergun, D. I., Isaya, G., et al. (2011). Oligomerization propensity and flexibility of yeast frataxin studied by X-ray crystallography and small-angle X-ray scattering. *Journal of Molecular Biology*, *414*(5), 783–797. http://dx.doi.org/10.1016/j.jmb.2011.10.034.

Tang, G., Peng, L., Baldwin, P. R., Mann, D. S., Jiang, W., Rees, I., et al. (2007). EMAN2: An extensible image processing suite for electron microscopy. *Journal of Structural Biology*, *157*(1), 38–46. http://dx.doi.org/10.1016/j.jsb.2006.05.009.

Tong, W. H., & Rouault, T. (2000). Distinct iron-sulfur cluster assembly complexes exist in the cytosol and mitochondria of human cells. *The EMBO Journal*, *19*(21), 5692–5700.

Trabuco, L. G., Villa, E., Mitra, K., Frank, J., & Schulten, K. (2008). Flexible fitting of atomic structures into electron microscopy maps using molecular dynamics. *Structure*, *16*(5), 673–683. http://dx.doi.org/10.1016/j.str.2008.03.005.

Trabuco, L. G., Villa, E., Schreiner, E., Harrison, C. B., & Schulten, K. (2009). Molecular dynamics flexible fitting: A practical guide to combine cryo-electron microscopy and X-ray crystallography. *Methods*, *49*(2), 174–180. http://dx.doi.org/10.1016/j.ymeth.2009.04.005.

Tsai, C. L., & Barondeau, D. P. (2010). Human frataxin is an allosteric switch that activates the Fe-S cluster biosynthetic complex. *Biochemistry*, *49*(43), 9132–9139. http://dx.doi.org/10.1021/bi1013062.

van Heel, M., & Schatz, M. (2005). Fourier shell correlation threshold criteria. *Journal of Structural Biology*, *151*(3), 250–262. http://dx.doi.org/10.1016/j.jsb.2005.05.009.

Vaubel, R. A., & Isaya, G. (2013). Iron-sulfur cluster synthesis, iron homeostasis and oxidative stress in Friedreich ataxia. *Molecular and Cellular Neurosciences*, *55*, 50–61. http://dx.doi.org/10.1016/j.mcn.2012.08.003.

Voos, W., & Rottgers, K. (2002). Molecular chaperones as essential mediators of mitochondrial biogenesis. *Biochimica et Biophysica Acta*, *1592*(1), 51–62.

Winn, M. D., Ballard, C. C., Cowtan, K. D., Dodson, E. J., Emsley, P., Evans, P. R., et al. (2011). Overview of the CCP4 suite and current developments. *Acta Crystallographica. Section D, Biological Crystallography*, *67*(Pt. 4), 235–242. http://dx.doi.org/10.1107/S0907444910045749.

Yang, J., Yan, R., Roy, A., Xu, D., Poisson, J., & Zhang, Y. (2015). The I-TASSER suite: Protein structure and function prediction. *Nature Methods*, *12*(1), 7–8. http://dx.doi.org/10.1038/nmeth.3213.

CHAPTER SIX

Fe–S Cluster Hsp70 Chaperones: The ATPase Cycle and Protein Interactions

Rafal Dutkiewicz*, Malgorzata Nowak*, Elizabeth A. Craig†,[1], Jaroslaw Marszalek*,†,[1]

*Intercollegiate Faculty of Biotechnology, University of Gdansk and Medical University of Gdansk, Gdansk, Poland

†University of Wisconsin–Madison, Madison, WI, United States

[1]Corresponding authors: e-mail address: ecraig@wisc.edu; jaroslaw.marszalek@biotech.ug.edu.pl

Contents

Abstract

Hsp70 chaperones and their obligatory J-protein cochaperones function together in many cellular processes. Via cycles of binding to short stretches of exposed amino acids on substrate proteins, Hsp70/J-protein chaperones not only facilitate protein folding but also drive intracellular protein transport, biogenesis of cellular structures, and disassembly of protein complexes. The biogenesis of iron–sulfur (Fe–S) clusters is one of the critical cellular processes that require Hsp70/J-protein action. Fe–S clusters are ubiquitous cofactors critical for activity of proteins performing diverse functions in, for example, metabolism, RNA/DNA transactions, and environmental sensing. This biogenesis process can be divided into two sequential steps: first, the assembly of an Fe–S cluster on a conserved scaffold protein, and second, the transfer of the cluster from the scaffold to a recipient protein. The second step involves Hsp70/J-protein chaperones. Via

Methods in Enzymology, Volume 595
ISSN 0076-6879
http://dx.doi.org/10.1016/bs.mie.2017.07.004

binding to the scaffold, chaperones enable cluster transfer to recipient proteins. In eukaryotic cells mitochondria have a key role in Fe–S cluster biogenesis. In this review, we focus on methods that enabled us to dissect protein interactions critical for the function of Hsp70/J-protein chaperones in the mitochondrial process of Fe–S cluster biogenesis in the yeast *Saccharomyces cerevisiae.*

1. INTRODUCTION

As an Hsp70/J-protein system is key to the process of Fe–S cluster biogenesis, the cycle of interaction of this class of chaperones with substrate proteins is described first. A brief overview of how the chaperone system fits into the process of Fe–S cluster biogenesis and a description of the complex evolutionary history of the chaperones involved in cluster biogenesis follows.

1.1 Substrate-Binding Cycle of Hsp70/J-Protein Chaperones

The ability of Hsp70s to perform their cellular functions depends on their cyclic interaction with substrates—short, hydrophobic stretches of residues in full-length proteins (Clerico, Tilitsky, Meng, & Gierasch, 2015; Kim, Hipp, Bracher, Hayer-Hartl, & Hartl, 2013). This cycle is allosterically regulated by nucleotide binding, hydrolysis, and exchange (Mayer, 2013). All members of the Hsp70 protein family have two domains, an N-terminal nucleotide-binding domain (NBD) of ~44 kDa, with two lobes forming a deep pocket where adenine nucleotide binds; and a C-terminal substrate-binding domain (SBD) of ~26 kDa, which consists of a β-sandwich subdomain (SBDβ) forming a peptide-binding pocket and an α-helical subdomain (SBDα) forming a "lid" that covers the peptide-binding pocket (Fig. 1) (Kampinga & Craig, 2010). The two domains are joined by a flexible linker, which is important for interdomain allosteric communication.

In the ATP-bound state, the SBD and interdomain linker are docked onto the NBD. The α-helical lid (SBDα) and SBDβ are restrained such that the peptide-binding pocket is not covered by the lid (Fig. 1). This "open" conformation allows for rapid binding and releasing of a substrate. Upon ATP hydrolysis to ADP, the conformation of Hsp70 changes dramatically. The SBD dissociates from the NBD and only the extended linker connects the two domains. Unrestrained, the α-helical lid covers the peptide-binding pocket of the SBDβ, preventing dissociation of substrate. Most Hsp70s bind

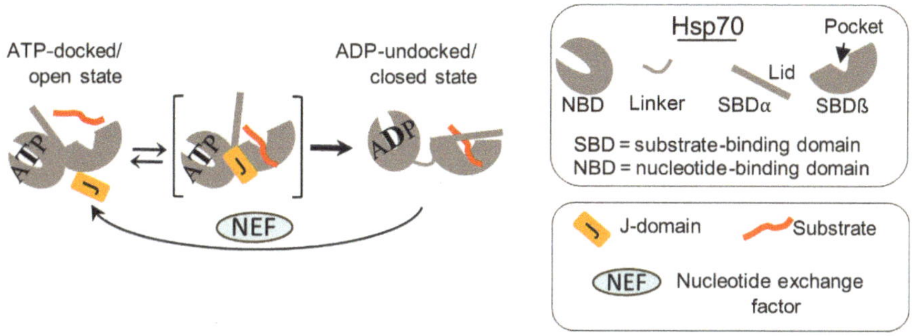

Fig. 1 Overview of Hsp70–substrate interaction cycle. When ATP is bound to Hsp70 (*left*), substrate has easy access to the peptide-binding pocket in β-subdomain of SBD (SBDβ). This is called the open or docked state, because the lid (SBDα), which traps the substrate by covering the binding pocket when ADP is bound (*right*), is restrained through interaction with the NBD, as are SBDβ and interdomain linker. In concert with substrate in the binding pocket, J-domain, binding at the interface of NBD and SBDβ, stimulates hydrolysis of ATP to ADP. Due to resulting conformational changes, the domains disengage, forming the undocked/closed state, stabilizing substrate interaction. Nucleotide exchange factor (NEF) completes the cycle (*bottom*). Brackets (*middle*) indicate dynamic transitions between ATP- and ADP-bound conformations. Note disengagement of the SBDα lid, but not SBDβ, from NBD.

adenine nucleotides with high affinity. Therefore, a nucleotide exchange factor (NEF), which stimulates ADP release, is required to complete the substrate interaction cycle. Upon rebinding of ATP, which is normally in higher abundance than ADP in the cell, substrate is released, allowing the start of a new substrate interaction cycle (Mayer, 2013).

ATP hydrolysis is key to this cycle of Hsp70–substrate interaction. Because the intrinsic ATPase activity of Hsp70 is very low, a J-protein cochaperone, which stimulates ATPase activity, is obligatory. J-proteins are defined by the presence of a "J-domain" (Kampinga & Craig, 2010). This highly conserved domain consists of four α-helices, with two of them forming a finger-like structure connected by a loop region. This loop invariantly contains a His-Pro-Asp (HPD) tripeptide, critical for interaction with Hsp70. The J-domain interacts with ATP-bound Hsp70 and stimulates its ATPase activity. Although the molecular details of this transient interaction are not well understood, evidence suggest that the J-domain-binding interface is formed in the open conformation when the SBD and interdomain linker are docked onto the ATP-bound NBD (Mayer, 2013). Other domains of J-proteins are highly variable, in many cases displaying no sequence or structure similarities. Many J-proteins have a J-domain at their

N-terminus followed by different types of domains that bind substrates, which allow them to interact with particular substrate proteins on their own, and as consequence to target substrate for Hsp70 binding (Craig & Marszalek, 2017). While J-proteins are obligatory, they only modestly stimulate Hsp70's ATPase activity. Interaction with substrate results in the conformational change in the NBD that also, on its own, modestly increases ATPase activity. It is the synchronized interaction of the J-domain and the substrate protein that result in rapid hydrolysis, and thus the major conformational change that leads to the SBD dissociation from the NBD and closing of the α-helical lid over the substrate associated with the SBDβ (Mayer, 2013).

1.2 An Overview of Mitochondrial Iron–Sulfur Cluster Biogenesis

Proteins involved in mitochondrial Fe–S cluster biogenesis were inherited from bacterial ancestors (Lill et al., 2015; Roche et al., 2013). The central component of both bacterial and mitochondrial pathways is a scaffold protein, termed Isu, on which an Fe–S cluster is assembled, before it is transferred to a recipient apo-protein. *Saccharomyces cerevisiae* contains two paralogous, highly conserved and functionally redundant, scaffold proteins termed Isu1 and Isu2. The homologous scaffold in bacteria is termed IscU.

The process of Fe–S cluster biogenesis involves two sequential steps: cluster synthesis on the Isu scaffold and cluster transfer from Isu onto recipient apo-proteins (Fig. 2A). Each step engages a dedicated set of proteins forming multiprotein complexes (Beilschmidt & Puccio, 2014; Kim, Bothe, Alderson, & Markley, 2015; Rouault, 2015). Fe–S cluster assembly requires formation of the assembly complex composed of cysteine desulfurase interacting with the Isu scaffold and iron-binding protein, Yfh1, the yeast frataxin homolog (Prischi et al., 2010; Tsai & Barondeau, 2010). In mitochondria, cysteine desulfurase functions as a heterooligomer composed of the enzymatic subunit Nfs1, which provides sulfur needed for Fe–S cluster synthesis, and the tightly associated small accessory protein Isd11 (Richards & van der Giezen, 2006; Wiedemann et al., 2006). We refer to this complex as Nfs1 (Isd11) throughout. The biological role of Yfh1, which interacts with both Nfs1 and Isu, remains obscure (Pastore & Puccio, 2013; Stemmler, Lesuisse, Pain, & Dancis, 2010; Vaubel & Isaya, 2013). It has been implicated as an iron donor and/or as a regulatory subunit of the Fe–S cluster assembly complex (Bridwell-Rabb, Fox, Tsai, Winn, & Barondeau, 2014). Synthesis of the Fe–S cluster on the Isu scaffold also requires sulfur reduction by an

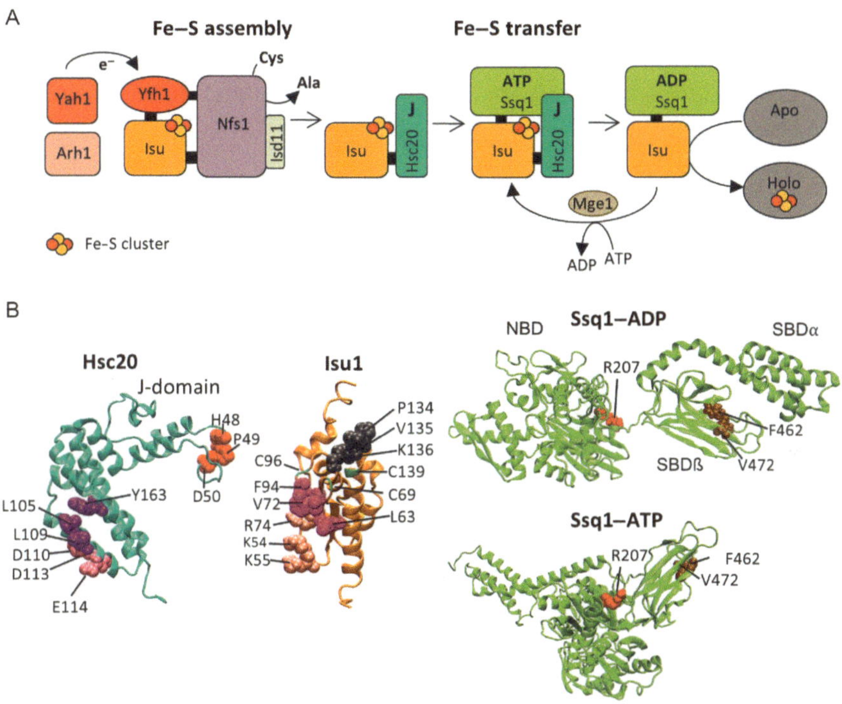

Fig. 2 Fe–S cluster biogenesis in mitochondria of *Saccharomyces cerevisiae*. (A) Overview. *Assembly step*: Fe–S cluster assembly requires interaction of the Isu scaffold with cysteine desulfurase Nfs1(Isd11) and frataxin Yfh1. Together they form the assembly complex that catalyzes Fe–S cluster synthesis. Electrons needed for sulfur reduction are provided by the electron transfer chain via ferredoxin (Yah1) and ferredoxin reductase (Arh1). *Transfer step*: Following its assembly, the Fe–S cluster is transferred from the Isu scaffold to a recipient protein. Transfer requires the Hsp70 chaperone system consisting of Hsp70 Ssq1, J-protein cochaperone Hsc20, and nucleotide exchange factor Mge1. Hsc20 binds Isu forming the Hsc20–Isu complex; via its ability to compete with Nfs1 for Isu binding, Hsc20 has a key role in the transition between the assembly and the transfer steps. The Isu–Hsc20 complex interacts with ATP-bound Hsp70; orchestrated interaction of Isu and J-domain of Hsc20 with Hsp70 results in formation of a transient tripartite complex Ssq1–Isu–Hsc20 and stimulation of Ssq1's ATPase activity. Hydrolysis of ATP to ADP induces conformational changes of Hsp70 (Fig. 1), resulting in the formation of a stable Ssq1–Isu complex. As a result, the Fe–S cluster is transferred to a recipient protein. Isu is released from Ssq1, due to the nucleotide exchange activity of Mge1 protein, upon rebinding of ATP. A new cycle of Isu interaction then ensues. (B) Molecular models of Isu1, Hsc20, and Ssq1. Protein crystal structure of Hsc20 (Jac1 from *S. cerevisiae*, PDB code 3UO3, *cyan*). Residues involved in Isu1 interaction L^{105}, L^{106}, D^{110}, D^{113}, E^{114}, Y^{163} and H^{48} P^{49} D^{50} motif of the J-domain critical for Hsp70 interaction are highlighted. Homology model of the Isu1 structure (*orange*, based on PDB code 3LVL). Residues involved in Hsc20 interaction K^{54}, K^{55}, L^{63}, V^{72}, R^{74}, F^{94}, and residues of the LPPVK motif (P^{134}, V^{135}, K^{136}) involved in Hsp70 binding are highlighted. Three cysteine residues (C^{69}, C^{96}, C^{139}) that coordinate

(Continued)

electron transfer chain from NAD(P)H via ferredoxin reductase (Arh1) to ferredoxin (Yah1) (Webert et al., 2014; Yan, Adinolfi, & Pastore, 2015) (Fig. 2A). But, how these additional factors interact with the Fe–S assembly complex is not well understood.

The Fe–S cluster transfer from Isu scaffold onto recipient apo-proteins requires a dedicated Hsp70/J-protein chaperone system (Vickery & Cupp-Vickery, 2007). In mitochondria of *S. cerevisiae*, this system is composed of the Hsp70 Ssq1, its J-protein cochaperone called Hsc20 in most organisms (often called Jac1 in *S. cerevisiae*), and an NEF (Mge1 in *S. cerevisiae*) (Dutkiewicz et al., 2003). It is clear that the Isu scaffold is a substrate of Hsp70, and that the ATP-dependent formation of the Hsp70–Isu complex is directly responsible for Fe–S cluster transfer from Isu onto recipient protein (Fig. 2A) (Bonomi, Iametti, Morleo, Ta, & Vickery, 2008; Chandramouli & Johnson, 2006; Dutkiewicz et al., 2006). However, the exact molecular mechanism behind Hsp70 action during the Fe–S cluster transfer is not well understood.

1.3 Complex Evolutionary History of Mitochondrial Hsp70s Involved in Fe–S Cluster Biogenesis

Most proteins functioning in mitochondrial Fe–S cluster biogenesis were inherited from bacterial ancestors. Yet, the evolutionary history of the Hsp70s involved in this pathway is not simple (Schilke et al., 2006). In bacteria, including *Escherichia coli*, the Hsp70/J-protein chaperones functioning in Fe–S cluster biogenesis, termed HscA and Hsc20, respectively, are highly specialized and distantly related to other Hsp70/J-protein systems (Vickery & Cupp-Vickery, 2007). While the major bacterial Hsp70, DnaK, binds many different substrates and interacts with several J-protein cochaperones (Genevaux, Georgopoulos, & Kelley, 2007), HscA interacts exclusively with the Fe–S cluster scaffold and Hsc20 cochaperone. Moreover, in contrast to DnaK, HscA does not require an NEF to complete the substrate-binding cycle, as its low affinity for adenine nucleotide enables spontaneous ADP to ATP exchange (Vickery & Cupp-Vickery, 2007).

Fig. 2—Cont'd the Fe–S cluster are marked in *green*. Homology models of the Ssq1 (*green*) in ADP-bound (*top*, based on PDB code 2KHO) and ATP-bound (*bottom*, based on PDB code 4JN4) conformations. Residues of the peptide-binding pocket (F^{462}, V^{472}) that interact with the LPPVK sequence of Isu, and residue R^{207} of the nucleotide-binding domain (NBD) involved in the interaction with J-domain HPD motif of Hsc20 are highlighted.

Mitochondria inherited the Fe–S cluster scaffold and the Hsc20 cochaperone from their bacterial ancestors, but specialized Hsp70 HscA was lost during evolution (Kominek, Marszalek, Neuveglise, Craig, & Williams, 2013; Schilke et al., 2006). As a result, many eukaryotic species, including humans, have only a single mitochondrial (mt)Hsp70, a descendant of DnaK, which binds the Isu scaffold and functions in the Fe–S cluster biogenesis with Hsc20 as cochaperone (Delewski et al., 2016; Pukszta et al., 2010).

Yet, this evolutionary story does not end here. A duplication of the mtHsp70 gene in a common ancestor of *S. cerevisiae* and closely related yeast species resulted in formation of an Hsp70 paralog, which became specialized in Fe–S cluster biogenesis (Kominek et al., 2013; Schilke et al., 2006). In the postduplication species one copy of the mtHsp70, termed Ssc1, remained multifunctional, while the other copy, termed Ssq1, functions exclusively in Fe–S cluster biogenesis. Therefore, although both *E. coli* and *S. cerevisiae* have specialized Hsp70s dedicated to Fe–S cluster biogenesis, their evolutionary origins are very different and therefore they can be considered as an example of convergent evolution at the biochemical level. Another interesting aspect, illustrated by this evolutionary story, is that both specialized and multifunctional Hsp70s can efficiently function in Fe–S cluster biogenesis, provided the specialized J-protein Hsc20 is available to target the Isu scaffold for Hsp70 binding. It is also important to stress that all key protein–protein interactions that are critical for the Hsp70/J-protein–Isu-binding cycle are conserved across phylogeny, regardless of whether a specialized or multifunctional Hsp70 is involved. Below we discuss these conserved interactions in more detail.

2. Hsc20–Isu INTERACTION

The Hsp70/Hsc20–Isu interaction cycle begins with formation of the Hsc20–Isu complex (Fig. 2A). Like many other J-proteins, Hsc20 interacts with substrate protein independently from its Hsp70 partner (Craig & Marszalek, 2017). Yet, the specificity of this interaction is quite unique, with Isu and Hsc20 forming a large binding interface. The interaction between Hsc20 and Isu is critical in vivo. Studies in yeast revealed that, when Hsc20 is severely defective in Isu-binding, cells are inviable; the presence of Hsc20 variants with diminished Isu affinity results in slow growth and reduced activity of mitochondrial enzymes (aconitase and succinate dehydrogenase) requiring Fe–S clusters for activity (Ciesielski et al., 2012).

On the C-terminal domain of Hsc20 (Fig. 2B), three highly conserved hydrophobic residues Leu^{105}, Leu^{109}, Tyr^{163} and a patch of less conserved negatively charged residues (Asp^{110}, Asp^{113}, Glu^{114}) are involved in the interaction with Isu (Andrew, Dutkiewicz, Knieszner, Craig, & Marszalek, 2006; Ciesielski et al., 2012; Fuzery, Oh, Ta, Vickery, & Markley, 2011; Fuzery et al., 2008). As revealed by amino acid replacement experiments, the hydrophobic residues play a key role, contributing greatly to the stability of the Hsc20–Isu complex. Despite having a lesser role in stability, the charged residues may have a role in providing specificity and orienting the interacting proteins. The complementary binding surface (Fig. 2B), on a three-stranded antiparallel β-sheet of the Isu scaffold, also consists of highly conserved hydrophobic residues (Leu^{63}, Val^{72}, Phe^{94}), as well as a positively charged patch (Lys^{54}, Lys^{55}, Arg^{74}) (Kim et al., 2009; Majewska et al., 2013).

The interaction site of Hsc20 does not overlap with that of Hsp70. The Hsp70 binding site is the Leu^{132} Pro^{133} Pro^{134} Val^{135} Lys^{136} (LPPVK) pentapeptide on a distant flexible loop (Dutkiewicz et al., 2004; Hoff, Ta, Tapley, Silberg, & Vickery, 2002) (Fig. 2B). However, residues involved in the Isu–Hsc20 interaction are also critical for interaction of Isu with the cysteine desulfurase (i.e., *E. coli* IscS and *S. cerevisiae* Nfs1) present in the Fe–S cluster assembly complex (Majewska et al., 2013; Prischi et al., 2010). This raises a question of the possible physiological importance of overlapping binding sites for proteins having roles in the two different steps of Fe–S cluster biogenesis (i.e., Nfs1 in cluster assembly and Hsc20 in cluster transfer) (Fig. 2A). It is possible that sequential Isu interactions, first with Nfs1 and then with Hsc20, drive the ordered transition from assembly to transfer. One factor may be the relative affinities of Nfs1 and Hsc20 for the apo- and holo-form of Isu, with Nfs1 having higher affinity for the apo-form and Hsc20 for the holo-form. Hsc20 could also have a more direct role. The charged patch of Isu (K^{54} K^{55} R^{74}) (Fig. 2B) involved in the Hsc20–Isu interaction is not a part of the Nfs1-binding interface, suggesting that Hsc20 interaction might be initiated even when Isu is in the complex with Nfs1 (Majewska et al., 2013). This might allow Hsc20 to play a role of a kinetic switch between cluster assembly and transfer steps. Recent studies of bacterial Hsc20 indicating that it directly interacts with cysteine desulfurase (Puglisi, Yan, Adinolfi, & Pastore, 2016) provide further support for this attractive scenario. Moreover, Hsc20, as well as Nfs1, interaction protects Isu from proteolytic degradation by mitochondrial Lon protease (Pim1 in *S. cerevisiae*) both in vivo and in vitro (Ciesielski, Schilke, Marszalek, &

Craig, 2016; Song, Marszalek, & Craig, 2012). These results underscore the importance of the Hsc20–Isu interaction, and indeed the complex network of Isu interactions, for both driving and coordinating Fe–S cluster biogenesis.

2.1 Glutathione-*S*-Transferase Pull-Down Assay

Many different techniques have been used to detect the Hsc20–Isu interaction, including: microcalorimetry, surface plasmon resonance (SPR) (Dutkiewicz et al., 2003; Hoff, Silberg, & Vickery, 2000), glycerol gradient centrifugation (Dutkiewicz et al., 2003), nuclear magnetic resonance spectroscopy (Kim et al., 2009), and glutathione-*S*-transferase (GST) pull-down assay (Ciesielski et al., 2012). In our hands the latter assay has worked very effectively. These assays are easy to set up and provide semi-quantitative results (Fig. 3), allowing us to test how replacements of residues involved in a given protein–protein interaction affect protein complex formation. We have also used this assay to analyze the competition between different proteins for binding to the same interaction partner (Fig. 3B). Below we present the procedure utilizing Isu1–GST fusion protein to detect proteins interacting with the Isu scaffold.

2.1.1 Equipment

No highly specialized equipment is necessary to perform GST pull-down assays. The specific equipment we typically use is listed below.

- Rotilabo-Block-Heater H250 for reaction mixture incubation
- Sigma 2-16K microcentrifuge
- Grant-bio PTR-30 rotator for bead incubation with rotation
- Standard SDS-PAGE equipment, 15% SDS-PAGE gels
- Epson Perfection 4990 PHOTO film scanner
- ImageJ quantification software

2.1.2 Buffers and Reagents

- Binding buffer (PB): 40 m*M* HEPES (4-(2-hydroxyethyl)piperazine-1-ethanesulfonic acid potassium salt) (HEPES–KOH, pH 7.5), 10% (v/v) glycerol, 100 m*M* KCl, 1 m*M* dithiothreitol, 10 m*M* $MgCl_2$
- Laemmli sample buffer fourfold concentrated (SB): 0.125 *M* Tris (hydroxymethyl)-aminomethane hydrochloride (Tris–HCl, pH 6.8), 20% (v/v) glycerol, 10% 2-mercaptoethanol, 5% SDS (sodium dodecyl sulfate), 0.004% bromophenol blue

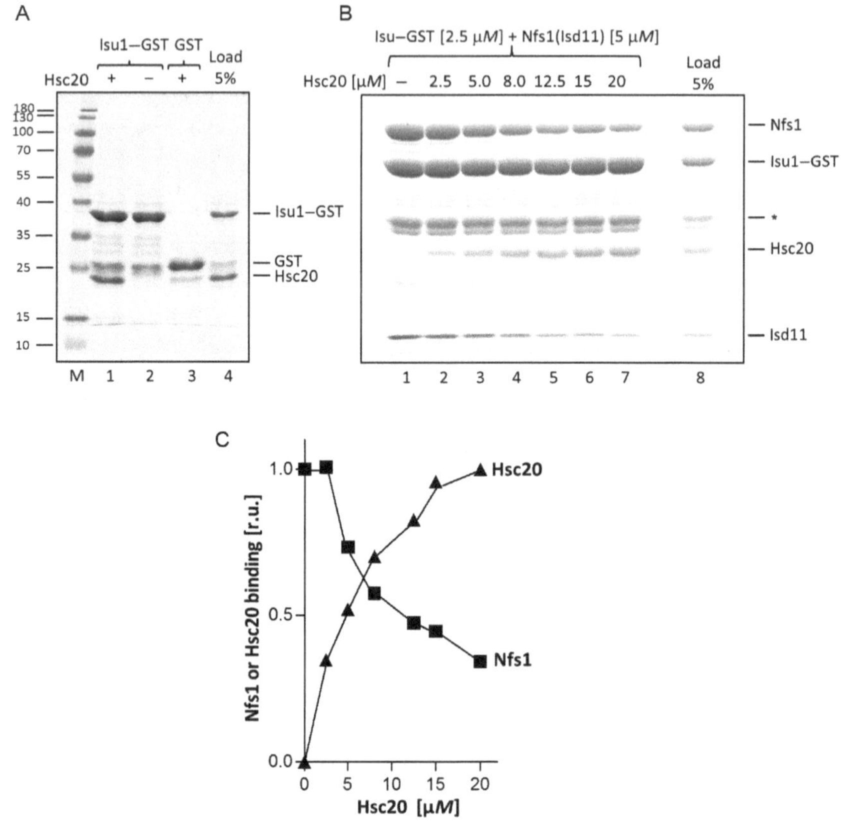

Fig. 3 Hsc20–Isu interaction. (A) Hsc20–Isu1 interaction detected by the GST pull-down assay. (Lanes I–3) Isu1–GST (2.5 μM) or GST (2.5 μM) were incubated in the presence (+) or absence (−) of Hsc20 (5 μM), as indicated. Glutathione resin was added to pull-down GST and associated proteins, which were separated by SDS-PAGE and stained with Coomassie Blue. (Lane 4) Five percent of the reaction mixture run on lane 1 was run as a loading control (load 5%). Notice strong Hsc20 signal (lane 1) when compared to the background signal (lane 3), which was obtained when Hsc20 was incubated with GST. "M" indicates the lane that has molecular weight markers. (B) Hsc20 competes with Nfs1(Isd11) for Isu1 binding. (Lanes 1–7) Isu1–GST (2.5 μM) was mixed with Nfs1(Isd11) (5 μM) and incubated for 15 min to allow Nfs1(Isd11)–Isu1–GST complex formation. Incubation was continued for another 15 min after addition of the Hsc20 competitor at the indicated concentrations. Glutathione resin was added to pull-down Isu1–GST complexes. (Lane 8) Five percent of the reaction mixture run in lane 7 was run as a loading control (load 5%). "*" indicates GST band that contaminates Isu1–GST protein preparation. Proteins were separated by SDS-PAGE, visualized by Coomassie Blue staining, and quantitated by densitometry. (C) Data from B were plotted as relative units (r.u.) with maximum binding of Nfs1 or Hsc20 set at 1. Notice decrease of the Nfs1 signal and increase of the Hsc20 signal with increasing concentrations of Hsc20 present in the reaction mixtures.

2.1.3 Protein Preparations

- Isu1–GST (2.3 mg/mL) protein purification was performed according to the procedure described in Ciesielski et al. (2012)
- Hsc20/Jac1 (4.9 mg/mL) purified as described previously in Voisine et al. (2001), except that *E. coli* strain C41(DE3) was used for Hsc20/Jac1 expression
- Ssq1 (2.1 mg/mL) protein purification was performed according to the procedure described in Manicki et al. (2014)

2.1.4 Procedure

1a. Equilibrate the reduced glutathione-immobilized agarose beads (Fluka, Sigma-Aldrich) by adding 210 mg of dry beads to 40 mL of buffer PB and incubate overnight at 4°C with rotation ~15 rpm. Centrifuge swollen beads for 5 min at 1000 × *g*.

1b. Wash beads three times with buffer PB (double volume of the beads), centrifuge between wash steps for 2 min at 1000 × *g*.

1c. Incubate beads with PB buffer with 0.1% (v/v) Triton X-100 for 2.5 h at 4°C with rotation ~15 rpm. Centrifuge beads for 5 min at 1000 × *g*.

1d. Wash beads three times with buffer PB (triple volume of the beads), centrifuge between wash steps for 2 min at 1000 × *g*.

1e. Incubate beads with PB buffer with 0.1% (w/v) BSA (bovine serum albumin, Fluka, Sigma-Aldrich) added for 2.5 h at 4°C with rotation ~15 rpm. Wash beads three times with buffer PB (triple volume of the beads), centrifuge between wash steps for 2 min 1000 × *g*. After extensive wash steps incubate resin in a refrigerator overnight.

1f. Finally, wash beads three times in buffer PB (double volume of the beads) and suspend in equal volume of buffer PB. Store refrigerated for, at most, 30 days.

2. In Eppendorf tube (500 μL) set up a reaction mixture (150 μL) consisting of buffer PB, Isu1–GST (2.5 μ*M*), Hsc20 (5 μ*M*). Incubate at 25°C for 15–30 min in a thermo-block without rotation.

3. Take an aliquot of the reaction mixture (7.5 μL) as a 5% "loading control," add 7.5 μL of twofold concentrated buffer SB, and store for further use.

4. To the remaining of the reaction mixture add 40 μL of glutathione-beads (~20 μL bead volume) and incubate at 4°C for 1 h with rotation ~15 rpm.

5. Wash beads one time with 500 μL of buffer PB and then three times with 200 μL of buffer PB. Centrifuge between wash steps at 1000 × *g* for 2 min.
6. Suspend beads in 20 μL of fourfold concentrated SB buffer, incubate for 10 min at 90°C, load 15 μL on SDS-PAGE 15% gel, and visualize by Coomassie Blue staining.
7. Image gel on film scanner and quantify bands by ImageJ or similar software.

2.1.5 Notes

1. As a control for nonspecific background binding, we substitute purified GST for Isu–GST in the reaction mixture. We process the control samples the same way as we do the experimental samples.
2. To control for both the amount of protein in the reaction mixture and the efficiency of the pull-down procedure we run "loading control" samples (see Section 2.1.4, step 3) on the same gel as we run the pull-down samples.

3. Hsp70/Hsc20–Isu-BINDING CYCLE

Isu scaffold interaction with the Hsp70/J-protein chaperones is critical for Fe–S cluster transfer from the scaffold to recipient apo-proteins, both in vivo and in vitro (Chandramouli & Johnson, 2006; Dutkiewicz et al., 2006; Vickery & Cupp-Vickery, 2007). For example, depletion of either Ssq1 or Hsc20 results in Fe–S cluster accumulation on the Isu scaffold and reduced activities of mitochondrial enzymes that require clusters for catalysis (aconitase, succinate dehydrogenase) (Dong et al., 2017; Muhlenhoff, Gerber, Richhardt, & Lill, 2003). These results demonstrate that the Hsp70/J-protein system is required for cluster transfer, but not for cluster synthesis on Isu. Moreover, biochemical experiments with purified bacterial HscA/Hsc20 chaperones demonstrated that efficient cluster transfer from the scaffold to recipient proteins such as apo-ferredoxin (Fdx) or apo-monothiol glutaredoxin (Grx5) requires the presence of the chaperones (Bonomi et al., 2008; Bonomi, Iametti, Morleo, Ta, & Vickery, 2011; Chandramouli & Johnson, 2006; Shakamuri, Zhang, & Johnson, 2012). Consistent with the canonical Hsp70/J-protein substrate-binding cycle (Fig. 1) cluster transfer is strictly dependent on ATP hydrolysis. Below, we describe individual steps of the Hsp70/Hsc20–Isu-binding cycle and biochemical techniques that enabled us to study individual protein–protein interactions, as well as overall progression of the reaction cycle.

3.1 Hsp70–Isu Interaction

The binding site of Hsp70 on the Isu scaffold is the LPPVK pentapeptide localized on a flexible loop in close proximity (~8 Å) away from the Fe–S cluster-binding site (Fig. 2B) (Dutkiewicz et al., 2004; Hoff et al., 2002). The LPPVK motif is evolutionary conserved; it is present in all orthologs of Isu, both prokaryotic and eukaryotic, which utilize an Hsp70/J-protein system for Fe–S cluster transfer onto recipient proteins. Biochemical evidence suggests that Hsp70s specialized in Fe–S cluster biogenesis (i.e., *E. coli* HscA and *S. cerevisiae* Ssq1) recognize only this sequence. Yet, binding is via the same molecular mechanism as that of a typical, more promiscuous, Hsp70–peptide interactions (Cupp-Vickery, Peterson, Ta, & Vickery, 2004). The same conserved residues, in the peptide-binding pocket of the SBDβ subdomain, are critical for HscA/Ssq1 interaction with Isu (Phe^{462}, Val^{472}; see Fig. 2B) and DnaK interaction with substrate proteins (e.g., DnaK–σ^{32} interaction) (Hoff, Cupp-Vickery, & Vickery, 2003; Knieszner et al., 2005; Mayer et al., 2000; Tapley, Cupp-Vickery, & Vickery, 2005, 2006). In fact, the LPPVK sequence is very similar to the consensus peptide sequence recognized by DnaK (Rudiger, Germeroth, Schneider-Mergener, & Bukau, 1997).

Although the importance of the Hsp70–Isu interaction for the progression of the Fe–S cluster biogenesis is well established, the molecular mechanism directly responsible for the cluster transfer is not well understood (Bonomi et al., 2008; Chandramouli & Johnson, 2006; Shakamuri et al., 2012). One model suggests that the formation of the Hsp70–Isu complex affects the conformation of the Isu scaffold in such a way that it can no longer coordinate cluster binding, thus promoting cluster transfer onto recipient protein (Bonomi et al., 2011). Such a mechanism would be consistent with "remodeling" of protein structures by Hsp70 chaperones, such as Hsp70/J-protein-dependent disassembly of phage lambda DNA replication complex (Liberek, Georgopoulos, & Zylicz, 1988) or disassembly of clathrin-coated vesicles (De Los Rios & Goloubinoff, 2016; Sousa et al., 2016).

3.1.1 Detection of Ssq1–Isu Interaction Using GST Pull-Down Assay

Interaction between Hsp70 and Isu has been detected previously (Cai et al., 2013; Dutkiewicz et al., 2003; Silberg, Hoff, Tapley, & Vickery, 2001). However, complex formation is very inefficient. This inefficiency is consistent with the canonical molecular mechanism of the Hsp70 substrate-binding cycle (Fig. 1) (Mayer, 2013). In the presence of ATP an open conformation of Hsp70 allows for rapid association, but also for rapid dissociation of Isu, while the closed lid of the ADP-bound conformation prevents efficient binding of Isu. However, very efficient formation of the Hsp70–Isu complex is observed in the presence of substoichiometric (catalytic) amounts of Hsc20 (Dutkiewicz et al., 2003; Hoff et al., 2000).

As depicted in Fig. 2, Hsc20 directly binds Isu and targets it for Hsp70 binding. The Hsc20–Isu complex interacts with ATP-bound Hsp70 allowing orchestration of the two critical interaction events: Isu interaction with the peptide-binding pocket in the SBDβ subdomain, and J-domain of Hsc20 interaction with the ATP-bound Hsp70. The two interactions synergistically stimulate ATP hydrolysis to ADP, resulting in conformational changes of Hsp70 (Fig. 2B) that promote its stable interaction with Isu. Because Hsc20 dissociates from the tripartite complex, upon ATP hydrolysis, it can interact with another Isu molecule. In consequence, over time, Hsp70–Isu complex accumulates in the reaction mixture. Hsc20-dependent formation of the Hsp70–Isu complex has been monitored by various biochemical assays: SPR and glycerol gradient centrifugation (Dutkiewicz et al., 2003; Hoff et al., 2000). Here we present results obtained using pull-down assay utilizing Isu1–GST fusion protein (Manicki et al., 2014). The procedure is essentially the same as described in Section 2.1 for Isu–GST interaction with Hsc20; however, the binding reaction is initiated by addition of ATP. This assay has also allowed us to test how individual amino acid substitutions in the intermolecular binding interfaces involved in the Hsp70/Hsc20–Isu interaction cycle affect the efficiency of Hsp70–Isu complex formation (Figs. 4 and 5).

3.2 ATPase Activity as a Measure of the Progression of the Hsp70/Hsc20–Isu Interaction Cycle

Similar to other members of the Hsp70 family chaperones specialized in Fe–S cluster biogenesis, HscA/Ssq1 possesses low, but measurable, ATPase activity, which is mildly stimulated in the presence of either Hsc20 or Isu (Dutkiewicz et al., 2003; Silberg, Tapley, Hoff, & Vickery, 2004; Silberg & Vickery, 2000). However, in the presence of both Hsc20 and Isu, this activity is stimulated over a hundred-fold. Because this stimulation depends on intermolecular interactions involved in the Hsp70/Hsc20–Isu-binding cycle, measurements of ATPase stimulation can be used to monitor the progression of the Isu-binding cycle. In this sense, the ATPase

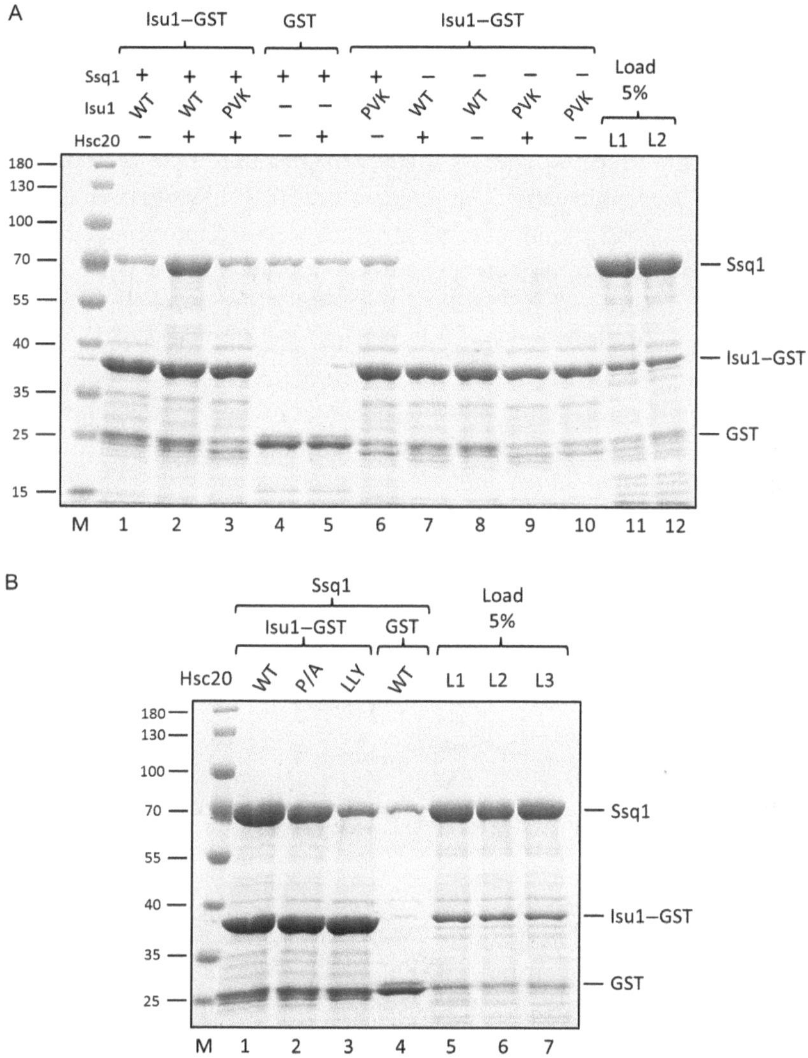

Fig. 4 Hsp70–Isu interaction. (A) Detection of Ssq1–Isu1 interaction via GST pull-down assay. (Lanes 1–10) Isu1–GST (2.5 μ*M*) or GST (2.5 μ*M*) was incubated in the presence of ATP (2 m*M*) with Ssq1 (5 μ*M*) and/or Hsc20 (0.25 μ*M*) as indicated. Glutathione resin was added to pull-down GST and associated proteins, which were separated by SDS-PAGE and stained with Coomassie Blue. (Lanes 11–12) Five percent of the reaction mixtures, which were run on lanes 3 and 2, were run as loading control (load 5%, L1 and L2, respectively). "M" indicates lane that has molecular weight markers. Notice strong Ssq1 signal (lane 2) when Isu1–GST wild type (WT) and Hsc20 were present in the reaction mixture. The Ssq1 signal was at the background level (lanes 4–5) when Hsc20 was absent (lane 1) or when Isu–GST WT was replaced by the Isu1$^{P134A/V135A/K136A}$–GST (PVK) variant (lane 3), which is defective in Ssq1 interaction. (B) Hsc20 variants defective in *(Continued)*

stimulation assay is complementary to the GST pull-down assays described above. In both cases, amino acid replacements that affect individual protein–protein interactions reduce assay output. It is worth noting, however, that for a given amino acid substitution variant a stronger negative effect is usually observed for the protein complex formation measured by GST pull-down than for ATPase stimulation (Ciesielski et al., 2012; Delewski et al., 2016; Majewska et al., 2013; Manicki et al., 2014). Therefore, although both assays can be applied for monitoring progression of the Hsp70/J-protein–Isu-binding cycle, ATPase stimulation is less sensitive to minor disruption of the protein–protein interactions.

Various assays have been used to measure the ATPase activity of Hsp70s involved in Fe–S cluster biogenesis (Dutkiewicz et al., 2003; Silberg et al., 2004; Silberg & Vickery, 2000). Here we present a method utilizing radioactive ATP (Fig. 6).

3.2.1 Equipment

No highly specialized equipment is needed to perform ATPase assay. The specific equipment we used is listed below.

- Rotilabo-Block-Heater H250 for sample incubation
- Sigma 1-15 microcentrifuge
- Beckman LS6000 scintillation counter
- Vortex

3.2.2 Buffers and Reagents

- ATPase reaction buffer (RB): 40 m*M* HEPES–KOH, pH 7.5, 100 m*M* KCl, 10% (v/v) glycerol, 1 m*M* DTT, 10 m*M* $MgCl_2$
- Stop buffer (ST): 1 *M* perchloric acid, 1 m*M* sodium phosphate

Fig. 4—Cont'd Ssq1 interaction, $Hsc20^{P134A}$ (P/A), or Isu1 binding, $Hsc20^{L105A/L109A/Y163A}$ (LLY), inhibit formation of the Ssq1–Isu1 complex. (Lanes 1–4) Isu1–GST (2.5 μ*M*) or GST (2.5 μ*M*) was incubated in the presence of ATP (2 m*M*) with Ssq1 (5 μ*M*) and Hsc20 (0.25 μ*M*) either wild type (WT), substitution variant (P/A), or (LLY), as indicated. Glutathione resin was added to pull-down GST and associated proteins, which were separated by SDS-PAGE and stained with Coomassie Blue. (Lanes 5–7) Five percent of the reaction mixtures, which were run on lanes 1, 2, and 3, were run as loading control (load 5%, L1, L2, L3, respectively). "M" indicates lane that has molecular weight markers. Notice reduced Ssq1 signal when WT Hsc20 was replaced by substitution variants defective in interaction with Ssq1 (P/A) or with Isu1 (LLY).

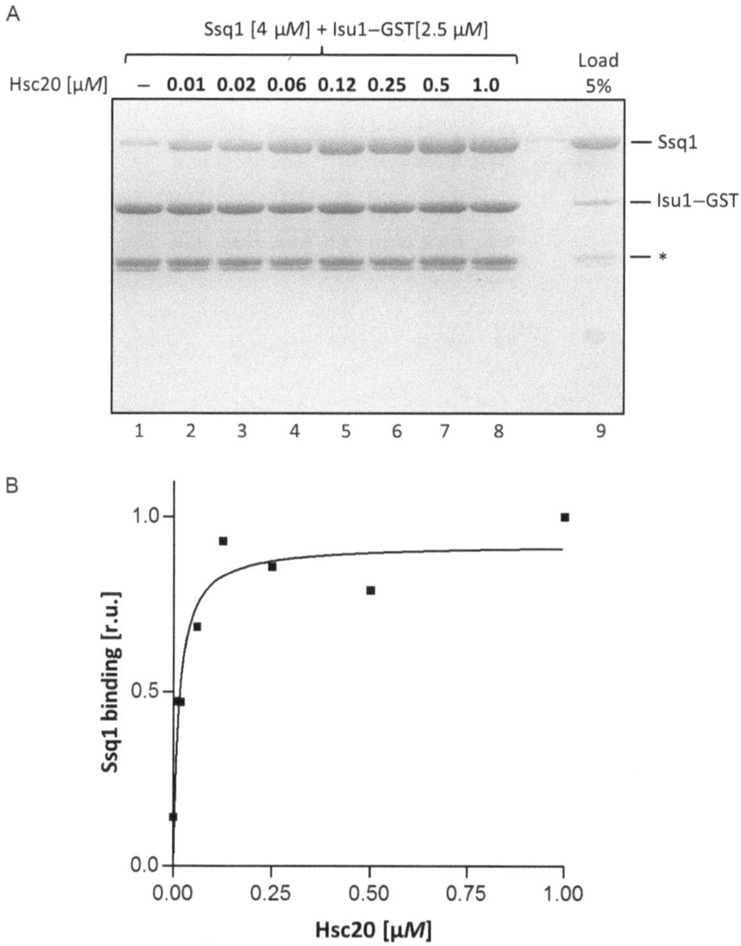

Fig. 5 Substoichiometric amounts of Hsc20 are sufficient for Hsp70–Isu complex formation. (A) (Lanes 1–8) Isu1–GST (2.5 μM) was mixed with Ssq1 (5 μM) and varying concentrations of Hsc20, as indicated. Glutathione resin was added to pull-down GST and associated proteins, which were separated by SDS-PAGE and stained. (Lane 9) Five percent of the reaction mixture, which was run on lane 8, was run as loading control (load 5%). "*" indicates GST band that contaminates Isu1–GST protein preparation. (B) Densitometry values plotted as relative units (r.u.) with maximum binding of Ssq1 set at 1. Curve represents best fit to the data of Michaelis–Menten hyperbolic equation estimated by nonlinear regression using GraphPad Prism software. Notice that the Ssq1 signal increases with increasing concentrations of Hsc20 in the reaction mixture. Yet, the reaction is saturated at the substoichiometric concentrations of Hsc20 (0.125 μM), consistent with the mechanism of the Hsp70/Hsc20–Isu interaction cycle (see Fig. 1).

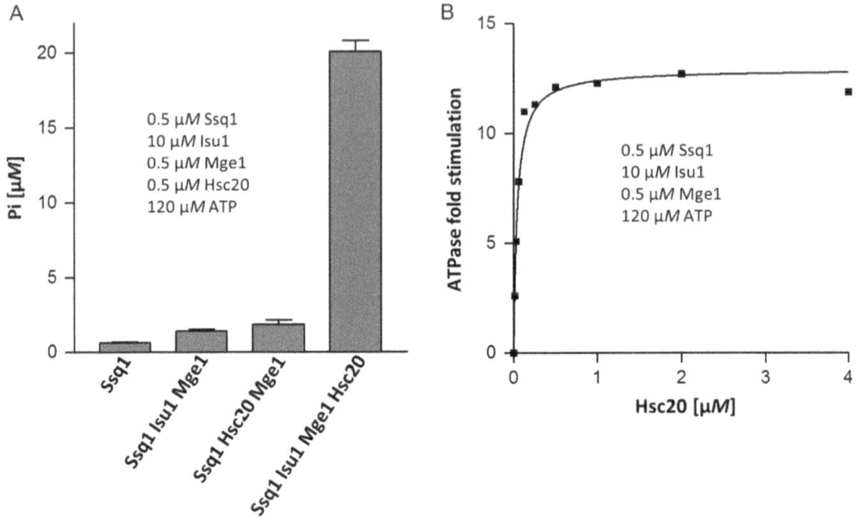

Fig. 6 ATPase activity of Hsp70 is stimulated in the presence of Isu and cochaperones. (A) Steady-state ATPase activity of Ssq1 was measured alone, or in the presence of indicated proteins. *Bars* represent average value for three repeated measurements with presented error bars as SD. (B) Stimulation of Ssq1 ATPase activity by Hsc20 in reaction containing Ssq1, Isu1, and Mge1, at indicated concentrations of Hsc20. *Curve* represents best fit to the data of Michaelis–Menten hyperbolic equation estimated by nonlinear regression using GraphPad Prism software. Notice synergistic stimulation of the ATPase activity of Ssq1 in the presence of cochaperones (Hsc20 and Mge1) and substrate (Isu1). Hsc20 stimulates the ATPase activity of Ssq1 at substoichiometric (0.125 μ*M*) concentration consistent with the mechanism of the Hsp70/Hsc20 cyclic interaction with Isu (see Fig. 1). Also, notice that similar concentration of Hsc20 is needed to saturate both the ATPase of Ssq1 and formation of the Ssq1–Isu–GST complex (Fig. 5). Both reactions can be used to monitor the progression of the Hsp70/Hsc20–Isu interaction cycle.

3.2.3 Protein Preparations

- Isu1 (1 mg/mL) protein purification was performed according to the procedure described in Dutkiewicz et al. (2003)
- Mge1 (2.5 mg/mL) purified as described in Azem et al. (1997)

3.2.4 Procedure

1. In an Eppendorf tube (1.5 mL), set up a reaction mixture (12.5 μL) consisting of buffer RB, Ssq1 (1 μ*M*), and other proteins as needed: Hsc20 (0.5 μ*M*), Isu1 (10 μ*M*), and Mge1 (0.5 μ*M*). Start reaction by addition of radioactive ATP to a final concentration of 120 μ*M* [0.7 μCi [γ^{33}P]ATP (Hartmann Analytic GmbH, SCF-301, 2200 Ci/mmol)]; [γ^{32}]ATP can also be used.

2. Incubate reaction mixture at 25°C in the thermo-block. Terminate reaction after 15 min by the addition of 100 μL of buffer ST. Keep the sample on ice.
3. Determine the total radioactivity in 5 μL of sample after stopping reaction. Add aliquot to the scintillation cocktail (EcoLite™(+), ICN) and quantify radioactivity by liquid scintillation counting (Beckman LS6000).
4. Next, add 0.4 mL of 20 m*M* ammonium molybdate and 0.4 mL of isopropyl acetate and mix the solution vigorously using vortex.
5. Separate the phases by centrifugation using microcentrifuge (1 min at 15,000 × *g*).
6. Remove 150 μL of the organic phase, containing the radioactive orthophosphate-molybdate complex, and determine the ^{33}P radioactivity by liquid scintillation counting, as described in step 3.

4. CONCLUDING REMARKS

Biogenesis of iron–sulfur clusters, similar to other cellular processes, is driven by multiprotein complexes that require specific protein–protein interactions for their assembly and function. Therefore, to gain mechanistic insight, one needs to dissect and validate those interactions. In this review, we have presented biochemical assays (Isu1–GST pull-down and ATPase stimulation of Hsp70) that allowed us to dissect interactions between the Isu scaffold and the Hsp70/J-protein chaperones involved in the Fe–S cluster biogenesis. However, besides the examples presented in this review, we have used GST pull-down assays to study other protein–protein interactions involved in the Fe–S cluster biogenesis. For example, by using Isu1–GST fusion protein, described here, we have also detected formation of the Fe–S assembly complex composed of Nfs1(Isd11)–Yfh1–Isu–GST. Moreover, we have used other GST fusion constructs, e.g., Yfh1–GST, Jac1–GST, to test their interactions with protein partners. In all cases, results obtained by the GST pull-down assays enabled semi-quantitative interpretations of the effects. Moreover, recent technical developments allowed us to use GST fusion constructs to measure protein interactions in real time. We have recently accomplished this using the GST constructs described above immobilized to a biosensor of the Bio-layer Interferometer (BLI-Pall Corporation, USA) to measure interactions with their protein partners (data not shown).

ACKNOWLEDGMENTS

We thank Milena Stolarska for help with preparing Fig. 2B. This work was supported by the Polish National Science Center Grant DEC-2012/06/A/NZ1/00002 and the Foundation for Polish Science Master-6/2014 programme (J.M.) and by NIH grant GM27870 (E.A. C.). The work of R.D. and M.N. was additionally supported by Polish National Science Center Grant DEC-2015/19/B/NZ1/00237 (R.D.).

REFERENCES

Andrew, A. J., Dutkiewicz, R., Knieszner, H., Craig, E. A., & Marszalek, J. (2006). Characterization of the interaction between the J-protein Jac1p and the scaffold for Fe-S cluster biogenesis, Isu1p. *The Journal of Biological Chemistry*, *281*(21), 14580–14587. http://dx.doi.org/10.1074/jbc.M600842200.

Azem, A., Oppliger, W., Lustig, A., Jeno, P., Feifel, B., Schatz, G., et al. (1997). The mitochondrial hsp70 chaperone system. Effect of adenine nucleotides, peptide substrate, and mGrpE on the oligomeric state of mhsp70. *The Journal of Biological Chemistry*, *272*(33), 20901–20906.

Beilschmidt, L. K., & Puccio, H. M. (2014). Mammalian Fe-S cluster biogenesis and its implication in disease. *Biochimie*, *100*, 48–60. http://dx.doi.org/10.1016/j.biochi.2014.01.009.

Bonomi, F., Iametti, S., Morleo, A., Ta, D., & Vickery, L. E. (2008). Studies on the mechanism of catalysis of iron-sulfur cluster transfer from IscU[2Fe2S] by HscA/HscB chaperones. *Biochemistry*, *47*(48), 12795–12801. http://dx.doi.org/10.1021/bi801565j.

Bonomi, F., Iametti, S., Morleo, A., Ta, D., & Vickery, L. E. (2011). Facilitated transfer of IscU-[2Fe2S] clusters by chaperone-mediated ligand exchange. *Biochemistry*, *50*(44), 9641–9650. http://dx.doi.org/10.1021/bi201123z.

Bridwell-Rabb, J., Fox, N. G., Tsai, C. L., Winn, A. M., & Barondeau, D. P. (2014). Human frataxin activates Fe-S cluster biosynthesis by facilitating sulfur transfer chemistry. *Biochemistry*, *53*(30), 4904–4913. http://dx.doi.org/10.1021/bi500532e.

Cai, K., Frederick, R. O., Kim, J. H., Reinen, N. M., Tonelli, M., & Markley, J. L. (2013). Human mitochondrial chaperone (mtHSP70) and cysteine desulfurase (NFS1) bind preferentially to the disordered conformation, whereas co-chaperone (HSC20) binds to the structured conformation of the iron-sulfur cluster scaffold protein (ISCU). *The Journal of Biological Chemistry*, *288*(40), 28755–28770. http://dx.doi.org/10.1074/jbc.M113.482042.

Chandramouli, K., & Johnson, M. K. (2006). HscA and HscB stimulate [2Fe-2S] cluster transfer from IscU to apoferredoxin in an ATP-dependent reaction. *Biochemistry*, *45*(37), 11087–11095. http://dx.doi.org/10.1021/bi061237w.

Ciesielski, S. J., Schilke, B., Marszalek, J., & Craig, E. A. (2016). Protection of scaffold protein Isu from degradation by the Lon protease Pim1 as a component of Fe-S cluster biogenesis regulation. *Molecular Biology of the Cell*, *27*(7), 1060–1068. http://dx.doi.org/10.1091/mbc.E15-12-0815.

Ciesielski, S. J., Schilke, B. A., Osipiuk, J., Bigelow, L., Mulligan, R., Majewska, J., et al. (2012). Interaction of J-protein co-chaperone Jac1 with Fe-S scaffold Isu is indispensable in vivo and conserved in evolution. *Journal of Molecular Biology*, *417*(1–2), 1–12. http://dx.doi.org/10.1016/j.jmb.2012.01.022.

Clerico, E. M., Tilitsky, J. M., Meng, W., & Gierasch, L. M. (2015). How hsp70 molecular machines interact with their substrates to mediate diverse physiological functions. *Journal of Molecular Biology*, *427*(7), 1575–1588. http://dx.doi.org/10.1016/j.jmb.2015.02.004.

Craig, E. A., & Marszalek, J. (2017). How do J-proteins get Hsp70 to do so many different things? *Trends in Biochemical Sciences*, *42*, 355–368. http://dx.doi.org/10.1016/j.tibs.2017.02.007.

Cupp-Vickery, J. R., Peterson, J. C., Ta, D. T., & Vickery, L. E. (2004). Crystal structure of the molecular chaperone HscA substrate binding domain complexed with the IscU recognition peptide ELPPVKIHC. *Journal of Molecular Biology*, *342*(4), 1265–1278. http://dx.doi.org/10.1016/j.jmb.2004.07.025.

Delewski, W., Paterkiewicz, B., Manicki, M., Schilke, B., Tomiczek, B., Ciesielski, S. J., et al. (2016). Iron-sulfur cluster biogenesis chaperones: Evidence for emergence of mutational robustness of a highly specific protein-protein interaction. *Molecular Biology and Evolution*, *33*(3), 643–656. http://dx.doi.org/10.1093/molbev/msv254.

De Los Rios, P., & Goloubinoff, P. (2016). Hsp70 chaperones use ATP to remodel native protein oligomers and stable aggregates by entropic pulling. *Nature Structural & Molecular Biology*, *23*(9), 766–769. http://dx.doi.org/10.1038/nsmb.3283.

Dong, Y., Zhang, D., Yu, Q., Zhao, Q., Xiao, C., Zhang, K., et al. (2017). Loss of Ssq1 leads to mitochondrial dysfunction, activation of autophagy and cell cycle arrest due to iron overload triggered by mitochondrial iron-sulfur cluster assembly defects in Candida albicans. *The International Journal of Biochemistry & Cell Biology*, *85*, 44–55. http://dx.doi.org/10.1016/j.biocel.2017.01.021.

Dutkiewicz, R., Marszalek, J., Schilke, B., Craig, E. A., Lill, R., & Muhlenhoff, U. (2006). The Hsp70 chaperone Ssq1p is dispensable for iron-sulfur cluster formation on the scaffold protein Isu1p. *The Journal of Biological Chemistry*, *281*(12), 7801–7808. http://dx.doi.org/10.1074/jbc.M513301200.

Dutkiewicz, R., Schilke, B., Cheng, S., Knieszner, H., Craig, E. A., & Marszalek, J. (2004). Sequence-specific interaction between mitochondrial Fe-S scaffold protein Isu and Hsp70 Ssq1 is essential for their in vivo function. *The Journal of Biological Chemistry*, *279*(28), 29167–29174. http://dx.doi.org/10.1074/jbc.M402947200.

Dutkiewicz, R., Schilke, B., Knieszner, H., Walter, W., Craig, E. A., & Marszalek, J. (2003). Ssq1, a mitochondrial Hsp70 involved in iron-sulfur (Fe/S) center biogenesis. Similarities to and differences from its bacterial counterpart. *The Journal of Biological Chemistry*, *278*(32), 29719–29727. http://dx.doi.org/10.1074/jbc.M303527200.

Fuzery, A. K., Oh, J. J., Ta, D. T., Vickery, L. E., & Markley, J. L. (2011). Three hydrophobic amino acids in Escherichia coli HscB make the greatest contribution to the stability of the HscB-IscU complex. *BMC Biochemistry*, *12*, 3. http://dx.doi.org/10.1186/1471-2091-12-3.

Fuzery, A. K., Tonelli, M., Ta, D. T., Cornilescu, G., Vickery, L. E., & Markley, J. L. (2008). Solution structure of the iron-sulfur cluster cochaperone HscB and its binding surface for the iron-sulfur assembly scaffold protein IscU. *Biochemistry*, *47*(36), 9394–9404. http://dx.doi.org/10.1021/bi800502r.

Genevaux, P., Georgopoulos, C., & Kelley, W. L. (2007). The Hsp70 chaperone machines of Escherichia coli: A paradigm for the repartition of chaperone functions. *Molecular Microbiology*, *66*(4), 840–857. http://dx.doi.org/10.1111/j.1365-2958.2007.05961.x.

Hoff, K. G., Cupp-Vickery, J. R., & Vickery, L. E. (2003). Contributions of the LPPVK motif of the iron-sulfur template protein IscU to interactions with the Hsc66-Hsc20 chaperone system. *The Journal of Biological Chemistry*, *278*(39), 37582–37589. http://dx.doi.org/10.1074/jbc.M305292200.

Hoff, K. G., Silberg, J. J., & Vickery, L. E. (2000). Interaction of the iron-sulfur cluster assembly protein IscU with the Hsc66/Hsc20 molecular chaperone system of Escherichia coli. *Proceedings of the National Academy of Sciences of the United States of America*, *97*(14), 7790–7795. http://dx.doi.org/10.1073/pnas.130201997.

Hoff, K. G., Ta, D. T., Tapley, T. L., Silberg, J. J., & Vickery, L. E. (2002). Hsc66 substrate specificity is directed toward a discrete region of the iron-sulfur cluster template protein

IscU. *The Journal of Biological Chemistry, 277*(30), 27353–27359. http://dx.doi.org/10.1074/jbc.M202814200.

Kampinga, H. H., & Craig, E. A. (2010). The HSP70 chaperone machinery: J proteins as drivers of functional specificity. *Nature Reviews. Molecular Cell Biology, 11*(8), 579–592. http://dx.doi.org/10.1038/nrm2941.

Kim, J. H., Bothe, J. R., Alderson, T. R., & Markley, J. L. (2015). Tangled web of interactions among proteins involved in iron-sulfur cluster assembly as unraveled by NMR, SAXS, chemical crosslinking, and functional studies. *Biochimica et Biophysica Acta, 1853*(6), 1416–1428. http://dx.doi.org/10.1016/j.bbamcr.2014.11.020.

Kim, J. H., Fuzery, A. K., Tonelli, M., Ta, D. T., Westler, W. M., Vickery, L. E., et al. (2009). Structure and dynamics of the iron-sulfur cluster assembly scaffold protein IscU and its interaction with the cochaperone HscB. *Biochemistry, 48*(26), 6062–6071. http://dx.doi.org/10.1021/bi9002277.

Kim, Y. E., Hipp, M. S., Bracher, A., Hayer-Hartl, M., & Hartl, F. U. (2013). Molecular chaperone functions in protein folding and proteostasis. *Annual Review of Biochemistry, 82*, 323–355. http://dx.doi.org/10.1146/annurev-biochem-060208-092442.

Knieszner, H., Schilke, B., Dutkiewicz, R., D'Silva, P., Cheng, S., Ohlson, M., et al. (2005). Compensation for a defective interaction of the hsp70 ssq1 with the mitochondrial Fe-S cluster scaffold isu. *The Journal of Biological Chemistry, 280*(32), 28966–28972. http://dx.doi.org/10.1074/jbc.M503031200.

Kominek, J., Marszalek, J., Neuveglise, C., Craig, E. A., & Williams, B. L. (2013). The complex evolutionary dynamics of Hsp70s: A genomic and functional perspective. *Genome Biology and Evolution, 5*(12), 2460–2477. http://dx.doi.org/10.1093/gbe/evt192.

Liberek, K., Georgopoulos, C., & Zylicz, M. (1988). Role of the Escherichia coli DnaK and DnaJ heat shock proteins in the initiation of bacteriophage lambda DNA replication. *Proceedings of the National Academy of Sciences of the United States of America, 85*(18), 6632–6636.

Lill, R., Dutkiewicz, R., Freibert, S. A., Heidenreich, T., Mascarenhas, J., Netz, D. J., et al. (2015). The role of mitochondria and the CIA machinery in the maturation of cytosolic and nuclear iron-sulfur proteins. *European Journal of Cell Biology, 94*(7-9), 280–291. http://dx.doi.org/10.1016/j.ejcb.2015.05.002.

Majewska, J., Ciesielski, S. J., Schilke, B., Kominek, J., Blenska, A., Delewski, W., et al. (2013). Binding of the chaperone Jac1 protein and cysteine desulfurase Nfs1 to the iron-sulfur cluster scaffold Isu protein is mutually exclusive. *The Journal of Biological Chemistry, 288*(40), 29134–29142. http://dx.doi.org/10.1074/jbc.M113.503524.

Manicki, M., Majewska, J., Ciesielski, S., Schilke, B., Blenska, A., Kominek, J., et al. (2014). Overlapping binding sites of the frataxin homologue assembly factor and the heat shock protein 70 transfer factor on the Isu iron-sulfur cluster scaffold protein. *The Journal of Biological Chemistry, 289*(44), 30268–30278. http://dx.doi.org/10.1074/jbc.M114.596726.

Mayer, M. P. (2013). Hsp70 chaperone dynamics and molecular mechanism. *Trends in Biochemical Sciences, 38*(10), 507–514. http://dx.doi.org/10.1016/j.tibs.2013.08.001.

Mayer, M. P., Schroder, H., Rudiger, S., Paal, K., Laufen, T., & Bukau, B. (2000). Multistep mechanism of substrate binding determines chaperone activity of Hsp70. *Nature Structural Biology*, 7(7), 586–593. http://dx.doi.org/10.1038/76819.

Muhlenhoff, U., Gerber, J., Richhardt, N., & Lill, R. (2003). Components involved in assembly and dislocation of iron-sulfur clusters on the scaffold protein Isu1p. *The EMBO Journal, 22*(18), 4815–4825. http://dx.doi.org/10.1093/emboj/cdg446.

Pastore, A., & Puccio, H. (2013). Frataxin: A protein in search for a function. *Journal of Neurochemistry, 126*(Suppl. 1), 43–52. http://dx.doi.org/10.1111/jnc.12220.

Prischi, F., Konarev, P. V., Iannuzzi, C., Pastore, C., Adinolfi, S., Martin, S. R., et al. (2010). Structural bases for the interaction of frataxin with the central components of iron-sulphur cluster assembly. *Nature Communications*, *1*, 95. http://dx.doi.org/10.1038/ncomms1097.

Puglisi, R., Yan, R., Adinolfi, S., & Pastore, A. (2016). A New Tessera into the interactome of the isc operon: A novel interaction between HscB and IscS. *Frontiers in Molecular Biosciences*, *3*, 48. http://dx.doi.org/10.3389/fmolb.2016.00048.

Pukszta, S., Schilke, B., Dutkiewicz, R., Kominek, J., Moczulska, K., Stepien, B., et al. (2010). Co-evolution-driven switch of J-protein specificity towards an Hsp70 partner. *EMBO Reports*, *11*(5), 360–365. http://dx.doi.org/10.1038/embor.2010.29.

Richards, T. A., & van der Giezen, M. (2006). Evolution of the Isd11-IscS complex reveals a single alpha-proteobacterial endosymbiosis for all eukaryotes. *Molecular Biology and Evolution*, *23*(7), 1341–1344. http://dx.doi.org/10.1093/molbev/msl001.

Roche, B., Aussel, L., Ezraty, B., Mandin, P., Py, B., & Barras, F. (2013). Iron/sulfur proteins biogenesis in prokaryotes: Formation, regulation and diversity. *Biochimica et Biophysica Acta*, *1827*(3), 455–469. http://dx.doi.org/10.1016/j.bbabio.2012.12.010.

Rouault, T. A. (2015). Mammalian iron-sulphur proteins: Novel insights into biogenesis and function. *Nature Reviews. Molecular Cell Biology*, *16*(1), 45–55. http://dx.doi.org/10.1038/nrm3909.

Rudiger, S., Germeroth, L., Schneider-Mergener, J., & Bukau, B. (1997). Substrate specificity of the DnaK chaperone determined by screening cellulose-bound peptide libraries. *The EMBO Journal*, *16*(7), 1501–1507. http://dx.doi.org/10.1093/emboj/16.7.1501.

Schilke, B., Williams, B., Knieszner, H., Pukszta, S., D'Silva, P., Craig, E. A., et al. (2006). Evolution of mitochondrial chaperones utilized in Fe-S cluster biogenesis. *Current Biology*, *16*(16), 1660–1665. http://dx.doi.org/10.1016/j.cub.2006.06.069.

Shakamuri, P., Zhang, B., & Johnson, M. K. (2012). Monothiol glutaredoxins function in storing and transporting [Fe2S2] clusters assembled on IscU scaffold proteins. *Journal of the American Chemical Society*, *134*(37), 15213–15216. http://dx.doi.org/10.1021/ja306061x.

Silberg, J. J., Hoff, K. G., Tapley, T. L., & Vickery, L. E. (2001). The Fe/S assembly protein IscU behaves as a substrate for the molecular chaperone Hsc66 from Escherichia coli. *The Journal of Biological Chemistry*, *276*(3), 1696–1700. http://dx.doi.org/10.1074/jbc.M009542200.

Silberg, J. J., Tapley, T. L., Hoff, K. G., & Vickery, L. E. (2004). Regulation of the HscA ATPase reaction cycle by the co-chaperone HscB and the iron-sulfur cluster assembly protein IscU. *The Journal of Biological Chemistry*, *279*(52), 53924–53931. http://dx.doi.org/10.1074/jbc.M410117200.

Silberg, J. J., & Vickery, L. E. (2000). Kinetic characterization of the ATPase cycle of the molecular chaperone Hsc66 from Escherichia coli. *The Journal of Biological Chemistry*, *275*(11), 7779–7786.

Song, J. Y., Marszalek, J., & Craig, E. A. (2012). Cysteine desulfurase Nfs1 and Pim1 protease control levels of Isu, the Fe-S cluster biogenesis scaffold. *Proceedings of the National Academy of Sciences of the United States of America*, *109*(26), 10370–10375. http://dx.doi.org/10.1073/pnas.1206945109.

Sousa, R., Liao, H. S., Cuellar, J., Jin, S., Valpuesta, J. M., Jin, A. J., et al. (2016). Clathrin-coat disassembly illuminates the mechanisms of Hsp70 force generation. *Nature Structural & Molecular Biology*, *23*(9), 821–829. http://dx.doi.org/10.1038/nsmb.3272.

Stemmler, T. L., Lesuisse, E., Pain, D., & Dancis, A. (2010). Frataxin and mitochondrial FeS cluster biogenesis. *The Journal of Biological Chemistry*, *285*(35), 26737–26743. http://dx.doi.org/10.1074/jbc.R110.118679.

Tapley, T. L., Cupp-Vickery, J. R., & Vickery, L. E. (2005). Sequence-dependent peptide binding orientation by the molecular chaperone DnaK. *Biochemistry*, *44*(37), 12307–12315. http://dx.doi.org/10.1021/bi051145r.

Tapley, T. L., Cupp-Vickery, J. R., & Vickery, L. E. (2006). Structural determinants of HscA peptide-binding specificity. *Biochemistry*, *45*(26), 8058–8066. http://dx.doi.org/10.1021/bi0606187.

Tsai, C. L., & Barondeau, D. P. (2010). Human frataxin is an allosteric switch that activates the Fe-S cluster biosynthetic complex. *Biochemistry*, *49*(43), 9132–9139. http://dx.doi.org/10.1021/bi1013062.

Vaubel, R. A., & Isaya, G. (2013). Iron-sulfur cluster synthesis, iron homeostasis and oxidative stress in Friedreich ataxia. *Molecular and Cellular Neurosciences*, *55*, 50–61. http://dx.doi.org/10.1016/j.mcn.2012.08.003.

Vickery, L. E., & Cupp-Vickery, J. R. (2007). Molecular chaperones HscA/Ssq1 and HscB/Jac1 and their roles in iron-sulfur protein maturation. *Critical Reviews in Biochemistry and Molecular Biology*, *42*(2), 95–111. http://dx.doi.org/10.1080/10409230701322298.

Voisine, C., Cheng, Y. C., Ohlson, M., Schilke, B., Hoff, K., Beinert, H., et al. (2001). Jac1, a mitochondrial J-type chaperone, is involved in the biogenesis of Fe/S clusters in Saccharomyces cerevisiae. *Proceedings of the National Academy of Sciences of the United States of America*, *98*(4), 1483–1488. http://dx.doi.org/10.1073/pnas.98.4.1483.

Webert, H., Freibert, S. A., Gallo, A., Heidenreich, T., Linne, U., Amlacher, S., et al. (2014). Functional reconstitution of mitochondrial Fe/S cluster synthesis on Isu1 reveals the involvement of ferredoxin. *Nature Communications*, *5*, 5013. http://dx.doi.org/10.1038/ncomms6013.

Wiedemann, N., Urzica, E., Guiard, B., Muller, H., Lohaus, C., Meyer, H. E., et al. (2006). Essential role of Isd11 in mitochondrial iron-sulfur cluster synthesis on Isu scaffold proteins. *The EMBO Journal*, *25*(1), 184–195. http://dx.doi.org/10.1038/sj.emboj.7600906.

Yan, R., Adinolfi, S., & Pastore, A. (2015). Ferredoxin, in conjunction with NADPH and ferredoxin-NADP reductase, transfers electrons to the IscS/IscU complex to promote iron-sulfur cluster assembly. *Biochimica et Biophysica Acta*, *1854*(9), 1113–1117. http://dx.doi.org/10.1016/j.bbapap.2015.02.002.

CHAPTER SEVEN

B. subtilis as a Model for Studying the Assembly of Fe–S Clusters in Gram-Positive Bacteria

Patricia C. Dos Santos[1]
Wake Forest University, Winston-Salem, NC, United States
[1]Corresponding author: e-mail address: dossanpc@wfu.edu

Contents

Abstract

Complexes of iron and sulfur (Fe–S clusters) are widely distributed in nature and participate in essential biochemical reactions. The biological formation of Fe–S clusters involves dedicated pathways responsible for the mobilization of sulfur, the assembly of Fe–S clusters, and the transfer of these clusters to target proteins. Genomic analysis of *Bacillus subtilis* and other Gram-positive bacteria indicated the presence of only one Fe–S cluster biosynthesis pathway, which is distinct in number of components and organization from previously studied systems. *B. subtilis* has been used as a model system for the characterization of cysteine desulfurases responsible for sulfur mobilization reactions in the biogenesis of Fe–S clusters and other sulfur-containing cofactors. Cysteine desulfurases catalyze the cleavage of the C–S bond from the amino acid cysteine and subsequent transfer of sulfur to acceptor molecules. These reactions can be monitored by the rate of alanine formation, the first product in the reaction, and sulfide formation, a byproduct of reactions performed under reducing conditions. The assembly of Fe–S clusters on protein scaffolds and the transfer of these clusters to target acceptors are determined through a combination of spectroscopic methods probing the rate of cluster assembly and transfer. This chapter provides a description of reactions

Methods in Enzymology, Volume 595
ISSN 0076-6879
http://dx.doi.org/10.1016/bs.mie.2017.07.009

promoting the assembly of Fe–S clusters in bacteria as well as methods used to study functions of each biosynthetic component and identify mechanistic differences employed by these enzymes across different pathways.

1. INTRODUCTION

Fe–S clusters are ubiquitous inorganic cofactors critical for existence of life on Earth (Beinert, 2000; Johnson, Dean, Smith, & Johnson, 2005). Complexes of Fe and S can adopt minimal configurations such as rhombic [2Fe–2S] or cubane [4Fe–4S] clusters. Variations of these simple structures with higher nuclearity, incorporation of certain 4D and 5D transition metals, and coordination with a variety of organic ligands expand the inventory and functionalities of Fe–S clusters found in nature (Dos Santos et al., 2005; Doukov, Iverson, Seravalli, Ragsdale, & Drennan, 2002; Shepard et al., 2011; Suess, Kuchenreuther, De La Paz, Swartz, & Britt, 2016). The chemical nature of cluster ligands, the protein environment surrounding these metal centers, and their location relative to the solvent shell modulates redox potentials and restricts reactivities of Fe–S clusters, thus guaranteeing specificity in reactions performed by enzymes utilizing these metal cofactors (Bak & Elliott, 2014; Langen, Jensen, Jacob, Stephens, & Warshel, 1992). The association of Fe–S clusters to proteins through covalent and noncovalent interactions bestows these inorganic molecules with an array of biological activities ranging from simple one-electron transfer to complex redox reactions, formation of radicals, and substrate activation through Lewis acid–base catalysis (Broderick, Duffus, Duschene, & Shepard, 2014; Esakova et al., 2016; Rees & Howard, 2003). Fe–S clusters have also been proposed to serve in sacrificial roles in reactions involving sulfur and molybdenum donation (Dos Santos & Dean, 2008; Hernandez et al., 2008; McLaughlin et al., 2016) and in noncatalytic roles when serving as structural elements or mediating signal transduction events that are driven by protein conformational changes upon chemical and/or electronic changes to the metal center (Bartels et al., 2017; Fuss, Tsai, Ishida, & Tainer, 2015; Mettert & Kiley, 2015). Because of the versatility of Fe–S clusters, their associated proteins participate in a wide range of biochemical pathways, and are considered critical components of several life sustaining reactions (Py & Barras, 2015). As a result, the biological formation of Fe–S clusters and their subsequent insertion into protein partners is essential to most forms of life (Huet, Daffe, & Saves, 2005; Johnson, Unciuleac, &

Dean, 2006; Kobayashi et al., 2003; Lill et al., 1999; Rocha & Dancis, 2016; Rouault & Tong, 2008).

Despite the well-acknowledged importance of Fe–S clusters in biology for many decades (Nicholas, Wilson, Heinen, Palmer, & Beinert, 1962), the first clues of how simple Fe–S clusters are formed in vivo only began to emerge in the 1990s (Kispal, Csere, Prohl, & Lill, 1999; Takahashi & Nakamura, 1999; Zheng, Cash, Flint, & Dean, 1998; Zheng, White, Cash, Jack, & Dean, 1993). Several factors have confounded the scientific community regarding the requirement and existence of pathways responsible for the synthesis of these cofactors. First, the chemical synthesis of simple clusters can be readily achieved in a test tube in the presence of free iron, sulfide, and suitable ligands under anoxic and reducing conditions (Lee, Lo, & Holm, 2014). This observation supported the assumption that Fe–S clusters would also be spontaneously assembled in cellular environments. Second, this proposal was further supported by the absence of reports associating genetic defects with inactivation of biosynthetic genes. It is important to highlight that Fe–S clusters are essential in nearly all model systems studied to date (Huet et al., 2005; Johnson et al., 2006; Kobayashi et al., 2003; Lill et al., 1999; Rocha & Dancis, 2016; Rouault & Tong, 2008), thus inactivation of cellular capacity to synthesize Fe–S clusters is lethal. Not surprisingly, the first biosynthetic system to be discovered was not a system involved in the general assembly of Fe–S clusters, but instead an accessory system specific to the supply of Fe–S clusters for nitrogen fixation enzymes (Dos Santos & Dean, 2014). The identification of the Nif system, composed of NifU and NifS, involved in the initial assembly of Fe–S clusters to nitrogenase in *Azotobacter vinelandii*, led to the subsequent discovery of the Isc system involved in the housekeeping formation of Fe–S clusters in the same organism and later in many other species (Dos Santos, 2014; Iwasaki, 2010; Takahashi & Nakamura, 1999).

To date, several Fe–S cluster biosynthetic pathways have been identified in species from diverse ecological and phylogenetic groups (Dos Santos, 2014; Ellis, Clough, Saldanha, & Wilson, 2001; Iwasaki, 2010; LaGier, Tachezy, Stejskal, Kutisova, & Keithly, 2003; Takahashi & Nakamura, 1999). These pathways vary in number and types of biosynthetic components, and are subjected to distinct regulatory mechanisms. Nevertheless, the unifying feature common to all Fe–S cluster biosynthetic pathways studied so far involves a sulfur activation reaction, the preformation of Fe–S clusters, and the insertion of these clusters into the final acceptor proteins (Fig. 1). Thus, a paradigm for the biological formation of Fe–S clusters involves the requirement of a sulfur-activating enzyme and a protein that

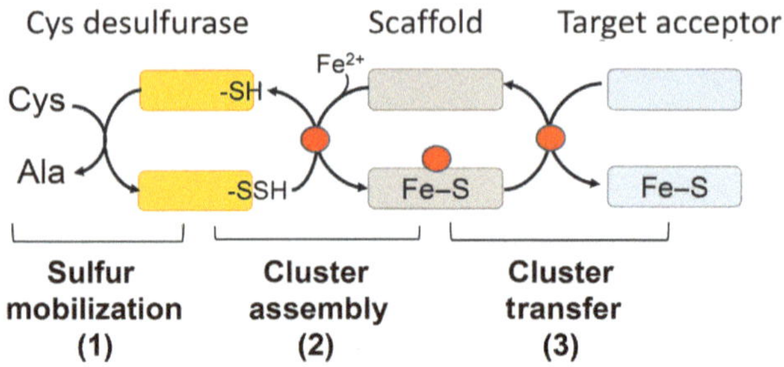

Fig. 1 Basic steps in Fe–S cluster biosynthesis. The formation of Fe–S clusters in biological systems include (1) sulfur activation by a cysteine desulfurase, (2) a sulfur transfer reaction to a protein scaffold for the formation of Fe–S clusters, and (3) the transfer of preformed Fe–S units to final target acceptor proteins. *Red circles* indicate the biosynthetic steps potentially involving redox reactions.

serves as the platform for the construction of Fe–S clusters, denominated as the scaffold. Once formed, clusters are then transferred to final targets where they exert their biological functions. This chapter provides an overview of the three general steps involving the biological assembly of Fe–S clusters in bacteria (Fig. 1). Here, we describe the experimental methods used to characterize these discrete steps in *Bacillus subtilis* to establish the functions of each biosynthetic component and identify mechanistic differences employed by these enzymes.

1.1 Sulfur-Activating Enzyme

In bacteria and eukaryotic species, the sulfur-activating enzymes are designated as cysteine desulfurases (Black & Dos Santos, 2015b; Mihara & Esaki, 2002). These enzymes utilize the cofactor pyridoxal-5′-phosphate (PLP) to activate and transfer sulfur from cysteine to an acceptor protein molecule. They are found as homodimers harboring two active sites each containing a PLP cofactor tethered to the enzyme via a Schiff base with a strictly conserved lysine residue. The reaction performed by cysteine desulfurases is proposed to be a sulfur transfer reaction in which the sulfur from cysteine is transferred to an acceptor molecule via a persulfide enzyme covalent intermediate. The mandatory formation of a persulfide bond at a conserved cysteine residue is considered a hallmark of cysteine desulfurase reactions (Mueller, 2006). These enzymes in their persulfurated forms are known to interact with acceptor molecules during persulfide sulfur transfer. While all cysteine desulfurases studied to date are assumed to employ the same sequence of chemical steps leading to the formation of the persulfide intermediate (Behshad & Bollinger, 2009; Kaiser et al., 2000; Zheng, White, Cash, & Dean, 1994), the subsequent transfer event is distinct and varies according to the enzyme and associated acceptor molecule, ultimately dictating their physiological functions.

Sulfur transfer in the biogenesis of Fe–S clusters can occur directly to a Fe–S cluster scaffold or indirectly via a sulfur carrier protein. There are two classes of cysteine desulfurases known to participate in sulfur donation during Fe–S cluster biogenesis (Mihara & Esaki, 2002). Members of both Class I and Class II cysteine desulfurases share a high degree of sequence similarity and unified mechanisms involving persulfide formation intermediate. However, representatives of each class exhibit distinct reaction profiles in enzymatic assays and display characteristic sequence elements allowing their in silico classification. Members of Class I, like IscS and NifS, contain a characteristic amino acid sequence insertion following the active site cysteine that provides a structural loop proposed to adopt distinct conformations during catalysis and further mediate sulfur transfer (Kaiser et al., 2000; Shi et al., 2010). Class II cysteine desulfurases, SufS and CsdA, lack this structural element, and sulfur transfer typically involves the participation of a dedicated sulfur acceptor protein that serves as an intermediate S-carrier to the scaffold protein (Blauenburg et al., 2016; Kim & Park, 2013; Tirupati, Vey, Drennan, & Bollinger, 2004). In bacterial systems, the gene coding the sulfur acceptor protein is typically adjacent to the cysteine desulfurase coding sequence as exemplified by *Escherichia coli sufS–sufE* and *csdA–csdE and B. subtilis sufS–sufU* genes pairs.

1.2 Scaffold Protein as the Platform for the Synthesis of Fe–S Clusters

The assembly of simple 2Fe–2S and 4Fe–4S clusters is proposed to occur remotely at the scaffold protein (Blanc, Gerez, & Ollagnier de Choudens, 2015; Dos Santos & Dean, 2014; Outten, 2015). The exact mechanism of sulfur transfer to the scaffold protein during the synthesis of Fe–S clusters remains unresolved. Labeling studies and mass spectrometry analysis showed that both scaffold and sulfur acceptor carrier proteins are capable of receiving multiple sulfur atoms in reactions using cysteine and cysteine desulfurases (Dai & Outten, 2012; Layer et al., 2007; Loiseau et al., 2005; Parent et al., 2015; Selbach et al., 2014; Selbach, Pradhan, & Dos Santos, 2013; Smith et al., 2001). In the absence of Fe, scaffold proteins can receive sulfur through persulfide sulfur transfer onto one or more of the cluster-ligating cysteine residues. The ability of scaffold proteins to accumulate persulfides and polysulfides provides support for a "sulfur-first model," in which the construction of Fe–S clusters would require sulfur binding preceding the iron-binding event. Alternatively, an "iron-first model" is supported through experimental evidence that scaffold proteins bind iron independently of their sulfuration state (Agar, Yuvaniyama, et al., 2000; Nuth, Yoon, & Cowan, 2002). In both models, the catalytic formation of Fe–S clusters onto the scaffold protein requires a source of iron and a reductant in addition to the sulfur source supplied by a cysteine desulfurase (Fig. 1). In vitro, ferrous salts and artificial reducing agents such as dithiothreitol (DTT) serve these roles, whereas in vivo the identity of these sources has not been clearly defined. Iron-binding proteins like frataxin and its bacterial orthologs (CyaY and Fra) as well as A-type cluster carriers/scaffolds (IscA and SufA) have been proposed to be candidates for iron delivery (Albrecht et al., 2011; Ding, Smith, & Ding, 2005; Layer, Ollagnier-de Choudens, Sanakis, & Fontecave, 2006; Lu, Yang, Tan, & Ding, 2008). Coupled-redox reactions through conversion of Fe^{2+} to Fe^{3+} or oxidation of cysteine to cystine have also been suggested, albeit less efficiently, as potential sources of electrons during formation of Fe–S clusters (Parent et al., 2015). Physiological redox agents such as the Isc ferredoxin have been shown to directly interact with active forms of cysteine desulfurases (Kim, Bothe, Alderson, & Markley, 2015), to enhance the rate of Fe–S cluster formation (Fox, Chakrabarti, McCormick, Lindahl, & Barondeau, 2015; Yan, Adinolfi, & Pastore, 2015), and to provide electrons for the conversion of 2Fe–2S clusters into 4Fe–4S clusters (Chandramouli et al., 2007). Depending on the type of scaffold protein, the appearance of 2Fe–2S cluster species precedes the formation of 4Fe–4S clusters. This suggests that in some cases the sequential assembly of two 2Fe–2S clusters followed by the formation of a 4Fe–4S cluster via a reductive coupling reaction (Smith et al., 2005). Typically, Fe–S clusters formed on the scaffold are labile and transiently associated with the protein. These nascent clusters show increased rates of degradation upon air exposure or incubation with chelating agents or strong reductants (Raulfs, O'Carroll, Dos Santos, Unciuleac, & Dean, 2008). Thus, spectroscopic characterization of nascent metallocenters constitutes an experimental challenge.

1.3 Transfer of Preformed Clusters to Final Acceptors

Once formed, transient Fe–S clusters are destined to be transferred to final apo-target proteins. Kinetic studies showed that the transfer of preformed

Fe–S clusters is faster and more efficient than in situ construction of Fe–S clusters onto apo-proteins, providing additional support for the scaffold hypothesis (Dos Santos et al., 2004; Unciuleac et al., 2007; Wollers et al., 2010; Wu, Wu, Surerus, & Cowan, 2002). The rate of cluster transfer can then be accelerated several orders of magnitude by the cluster carriers, which are proteins that can receive Fe–S clusters from scaffolds and deliver them to final acceptors (Bandyopadhyay et al., 2008; Mapolelo et al., 2013). The involvement of cluster carriers is dispensable under in vivo standard conditions, however, these proteins become essential under conditions known to increase the demand for efficient Fe–S cluster synthesis and repair such as oxidative stress and iron depletion (Angelini et al., 2008; Bandyopadhyay et al., 2008; Tan, Lu, Bitoun, Huang, & Ding, 2009). Chaperone proteins have also been shown to assist cluster transfer reactions. The Isc chaperone HscA and cochaperone HscB are known to interact with the scaffold protein IscU and facilitate the cluster transfer in reaction dependent of ATP-hydrolysis (Bonomi, Iametti, Morleo, Ta, & Vickery, 2008; Chandramouli & Johnson, 2006; Iametti, Barbiroli, & Bonomi, 2015). The exact role of ATP hydrolysis in this process is not fully understood, but is proposed to drive conformational change on the scaffold protein, allowing distortion of transient cluster and therefore promote facile transfer. Interestingly, the involvement of ATP-dependent reactions is also employed by the Suf system. In this case the activity of the ATPase SufC, which is phylogenetically unrelated to the Hsc chaperones, is necessary for the functionality of the Suf Fe–S cluster biosynthetic pathway (Outten, 2015; Saini, Mapolelo, Chahal, Johnson, & Outten, 2010). The synthesis of polypeptides in their apo-forms followed by subsequent insertion of preformed clusters by scaffold or carrier proteins constitutes a second paradigm in Fe–S cluster biosynthesis that is a bonafide posttranslational event. However, it is thermodynamically and mechanistically difficult to conceive that the folding of translated Fe–S proteins has to await cluster insertion. Therefore, it is anticipated that chaperone proteins may also be involved in final stage of this pathway by interacting with the apo-proteins, assisting the cluster insertion, and completing the folding of newly formed holo-structures.

2. FE–S CLUSTER ASSEMBLY IN *B. SUBTILIS*

The formation of Fe–S clusters in Gram-positive bacteria is far less understood than other model bacterial systems such as Gram-negative bacteria *E. coli*, *Salmonella enterica*, and *A. vinelandii* (Dos Santos, 2014;

Outten, 2015; Riboldi, de Mattos, & Frazzon, 2013). Nevertheless, this group of bacteria is also proposed to follow the same three sequential biosynthetic steps: sulfur activation, Fe–S cluster assembly, and transfer to apo-proteins (Fig. 1). Initial identification of proposed pathways was based on bioinformatic analysis using protein sequences known to participate in this process in studied species (Fig. 2). There have been three types of Fe–S cluster pathways identified across taxonomic class groups of Gram-positive species named after the similarities to the Gram-negative systems: Clostridia-ISC, Actinobacteria-SUF, and Bacilli-SUF (Dos Santos, 2014). Interestingly, only one set of biosynthetic genes has been identified in the vast majority of Gram-positive species analyzed (>100), and these predicted pathways varied in number and genomic organization when compared to studied Gram-negative species. These computational predictions were supported by experimental evidence demonstrating that these biosynthetic genes are essential, and supporting the notion that these species lack backup systems for the assembly of Fe–S clusters (Huet et al., 2005; Kobayashi et al., 2003). Another important difference is that Gram-positive bacterial genomes contain multiple cysteine desulfurase genes, and the chromosome location of these coding sequences suggests that sulfur mobilization for the

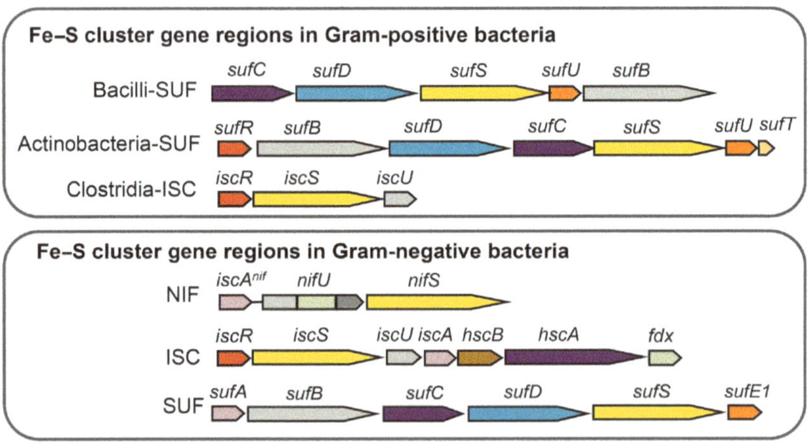

Fig. 2 Bacterial Fe–S cluster biosynthetic gene regions. Diagram representation shows the gene regions coding for the main Fe–S cluster biosynthetic components. Gene whose products are proposed to perform equivalent biosynthetic functions are color-coded accordingly: sulfur-activating enzymes in *yellow*, sulfur acceptors in *orange*, scaffold protein in *gray*, iron donor/cluster carrier in *pink*, ATPase in *purple*, cochaperone in *brown*, electron donor ferredoxin or ferredoxin-domain in *green*, transcription regulators in *red*, and SufD of yet-unidentified function in *blue*.

biosynthesis of other sulfur-containing cofactors is partially decoupled from Fe–S cluster formation (Fig. 3) (Black & Dos Santos, 2015b). This lies in stark contrast with the Gram-negative species in which sulfur activation for the biosynthesis of other thio-cofactors is provided by the same cysteine desulfurase involved in Fe–S cluster formation (Lauhon & Kambampati, 2000; Schwartz, Djaman, Imlay, & Kiley, 2000).

Mechanistic insights into the biological formation of Fe–S clusters in Gram-positive bacteria have only begun to emerge in the last few years. *Mycobacterium tuberculosis*, *Staphylococcus aureus*, *Streptococcus pneumoniae*, and *B. subtilis* are proposed to rely on the Suf system for the general synthesis of Fe–S clusters (Albrecht et al., 2010; Huet et al., 2005; Riboldi, Larson, & Frazzon, 2011; Rosario-Cruz, Chahal, Mike, Skaar, & Boyd, 2015; Selbach, Earles, & Dos Santos, 2010). Among these species, *B. subtilis* has been the best Gram-positive bacterium studied so far. This nonpathogenic bacterium has been extensively investigated in high-throughput analyses, is amenable to genetic modification, and has a rapid generation time (<15 min). Genes encoding Fe–S cluster biosynthetic components display high sequence similarity with other species of Gram-positive bacteria of pathogenetic relevance, thus making *B. subtilis* an excellent model system to study the biosynthesis of Fe–S clusters. In *B. subtilis*, the Suf system is encoded by the suf operon, *sufCDSUB* genes. Additional coding sequences at remote chromosome locations have also been proposed to code for accessory proteins assisting in the formation of Fe–S clusters and to be involved in other aspects of sulfur metabolism (Fig. 3) (Dos Santos, 2014).

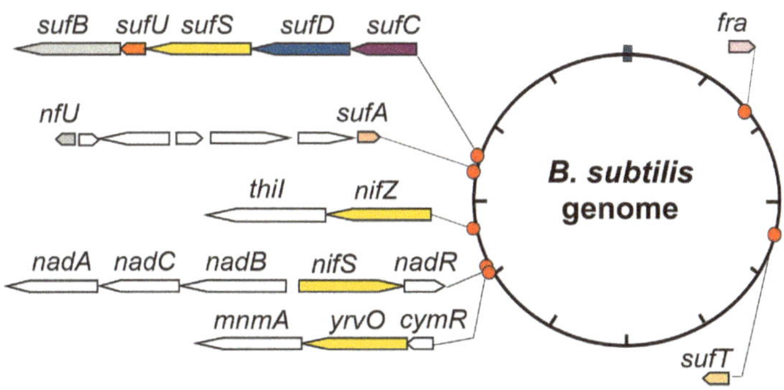

Fig. 3 Genomic location of *B. subtilis* genes known or suspect to be involved in S-cofactor and Fe–S cluster biogenesis. Diagram representation shows the relative chromosomal location, size and orientation of genes, and relevant genomic neighborhood.

3. METHODS TO PROBE INDIVIDUAL BIOSYNTHETIC STEPS IN FE–S CLUSTER ASSEMBLY

3.1 Cysteine Desulfurase Reactions

SufS has been identified as the sulfur-activating enzyme during the general formation of Fe–S clusters in *B. subtilis* (Albrecht et al., 2010; Selbach et al., 2010). This enzyme catalyzes the Cys:SufU sulfurtransferase reaction via a double displacement (Ping-Pong) mechanism (Fig. 4). The reaction scheme is marked by the formation of an enzyme persulfide intermediate and release of alanine prior the binding of the second substrate SufU. The kinetic characterization of the SufS reaction, first described regarding the *B. subtilis* enzyme, is believed to be universal to all cysteine desulfurases involved in Fe–S cluster biogenesis. The sulfur acceptor substrate SufU is a zinc-dependent sulfurtransferase and is the active participant in the second half of the SufS reaction (Selbach et al., 2014). The presence of SufU enhances the catalytic turnover rate of enzyme reaction by nearly 200-fold. The pH-dependent reaction profile along with labeling studies indicates that at least one cysteine thiol group within SufU, in its deprotonated form, conducts the nucleophilic attack onto SufS persulfide sulfur, regenerating the enzyme for the next turnover. Kinetic analyses of cysteine desulfurase

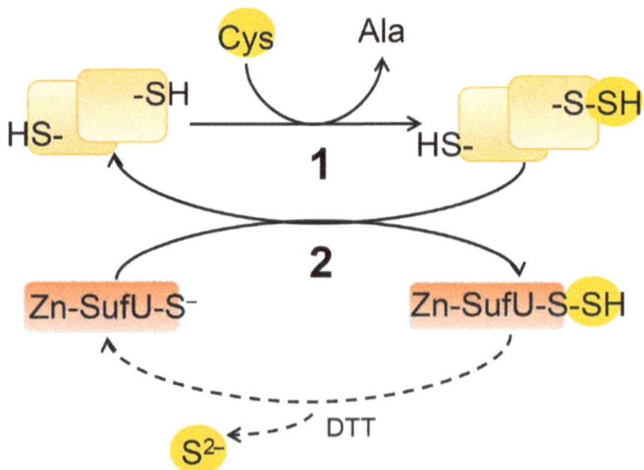

Fig. 4 Cysteine:SufU sulfurtransferase reaction of *B. subtilis* SufS. The reaction proceeds via a ping-pong mechanism. In the first half of the reaction (1) the cleavage of the carbon sulfur bond from cysteine leads to the release of alanine and the formation of a covalent S-enzyme intermediate at the active site cysteine residue within SufS (*yellow box*). In the second half of the reaction (2) the persulfide sulfur is transferred to a cysteine residue within the zinc-bound SufU. In vitro reactions in the presence of DTT, the second product, SufU-SSH, is recycled through the release of sulfide (*dashed lines*).

reactions can be measured through the formation of the first product, alanine and/or through quantification of sulfide generated when reactions are performed under reducing conditions. Critical aspects to consider when studying the kinetic profile of these enzymes are the pH of assay reactions, the presence of sulfur acceptor molecules, and the interference of reducing agents.

3.1.1 Ala-NDA Assay

The rate of alanine formation in SufS or any other cysteine desulfurase assays can be determined through a derivatization reaction that forms a fluorescence adduct (Selbach et al., 2010), or continuously in coupled assays with NADH-dependent alanine dehydrogenase reactions (Sigma-Aldrich®). Probing the formation of Ala is particularly interesting when determining the effect of the presence of the second substrate in cysteine desulfurase reactions and dispenses the need of reducing agents, which often skew the kinetic behavior of these enzymes. Here, we provide a description of the naphthalene dialdehyde (NDA)-derivatization method used to quantify Ala formed in the cysteine desulfurase reactions through fluorescence detection.

In the presence of cyanide, the amino group of Ala reacts with NDA to form a N-substituted highly fluorescent 1-cyano-2-alkyl-benz[*f*]isoindole adduct (Fig. 5). The NDA reaction is not specific to Ala but is reactive to any primary amine (Cho et al., 2002). Thus, standard assays must be performed in buffers that are not primary amines (e.g., Tris buffer reacts with NDA generating high background signals). The substrate cysteine also contains a primary amine; however, because of the nucleophilicity of the thiol group of cysteine, equivalent reaction of NDA with Cys leads to an unstable alternate adduct displaying lower fluorescence intensity (Zhang & Sun, 2005). Although reactions performed with high concentrations of cysteine generate background of fluorescence, it does not affect the rate of alanine formation over time. In order to correct for the variation in background, Ala standard curves can be prepared with the same concentration of cysteine present in the assay (Selbach et al., 2010). While the kinetic profile of cysteine desulfurases is affected by the presence of reducing agents, the detection of Ala-NDA is not. So far, we have determined that Ala derivatization is not affected by reducing agents such as DTT, β-mercaptoethanol, glutathione, or Tris (2-carboxyethyl) phosphine hydrochloride (TCEP). The rates of Ala formation are determined by taking reaction aliquots (100 μL) at each time point, quenching with 20 μL of 10% trichloroacetic acid, and reacting

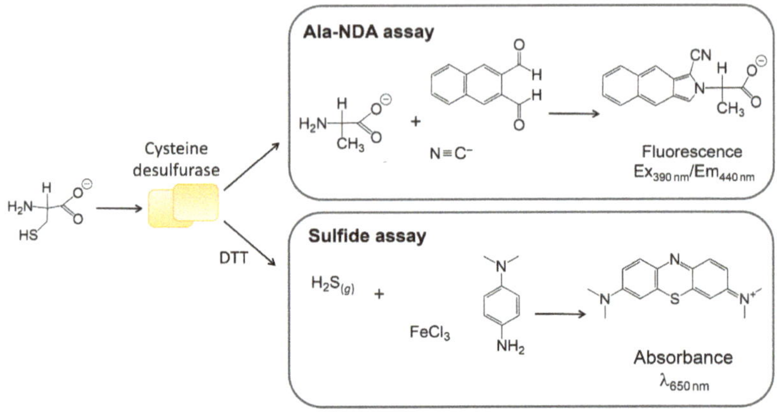

Fig. 5 Methods of detection in cysteine desulfurase reactions. The substrate cysteine reacts with the enzyme forming one equivalent of alanine which can be measured using the Ala-NDA assay through fluorescence detection. The enzyme in its persulfurated form can transfer the sulfur to a sulfur acceptor molecule to complete the reaction cycle. Under reducing conditions, the persulfide sulfur bound to the enzyme and/or sulfur acceptor is released in the form of sulfide gas that under acid conditions can be detected through formation of methylene blue.

with 1 mL of freshly prepared NDA mix (0.1 m*M* NDA and 1 m*M* KCN in 100 m*M* borate buffer pH 9). While NDA, cyanide, and borate solution can be prepared as far as 1 week in advance, NDA mix should be prepared immediately before the derivatization reaction. Upon incubation in the dark, Ala-NDA adducts can be quantified directly in a fluorimeter or subjected to chromatography using a Zorbax SB C18 column (Agilent) column coupled with a fluorescence detector set at Ex_{390nm}/Em_{440nm} (Selbach et al., 2010, 2013). The latter is recommended for assays generating lower rates of Ala since this method allows detections as low as 0.1 pmol of Ala-NDA adduction in each sample. The fluorescence intensity directly measured or quantified through the integrated peak area is converted into concentration of Ala from a standard curve generated under the same conditions.

3.1.2 Sulfide Assays

Sulfide can be generated upon reduction of persulfide intermediates formed in cysteine desulfurase reactions (Fig. 5). The rate of sulfide formation in reactions performed in the presence of artificial reducing agents such as DTT has been widely used as a readout of cysteine desulfurase activity (Chen & Mortenson, 1977; Siegel, 1965). In standard assays, reaction times points (800 μL) are first stopped upon addition of 25 μL of 12% NaOH and

50 μL 10% zinc acetate and subsequently reacted 100 μL 20 mM *N,N*-dimethyl-*p*-phenylenediamine sulfate (DMPD) in 7.2 *M* HCl and 100 μL 30 m*M* $FeCl_3$ in 1.2 *M* HCl. The formation of methylene blue is limited by the concentration of sulfide, thus the concentration of the chromophore with absorption maxima at 650 nm is directly proportional to the concentration of sulfide generated in the enzymatic assay. While assays can be performed in open tubes, formation of methylene blue occurs under acidic conditions in which hydrogen sulfide gas is formed. Therefore, the reaction development is performed in sealed vials or closed reaction containers. This protocol can be adapted to a 96-well plate format and smaller reaction volumes; in this case, wells are sealed with parafilm prior to the addition of DMPD and $FeCl_3$ with a syringe. Important to consider that certain chemicals interfere with the reaction development or can convert methylene blue to its colorless reduced form. Therefore, S^{2-} standard curves should be prepared under the same conditions used in the assay (buffer, reducing agent, and cysteine concentrations).

3.1.3 Sulfur Acceptors and Reducing Agents

Discrete chemical steps in the first half of the cysteine desulfurase reaction involving the binding of cysteine and formation of the persulfide intermediate have been characterized in detail (Behshad & Bollinger, 2009; Kaiser et al., 2000; Tirupati et al., 2004; Zheng et al., 1994). The proposed mechanism, recently reviewed elsewhere (Black & Dos Santos, 2015b), is based on spectroscopic and labeling studies probing the formation and consumption of reaction intermediates associated with the PLP cofactor. The mechanism of sulfur activation by PLP is initiated through displacement of the lysine residue and formation of a cysteine-PLP aldimine intermediate (external Schiff base). Sequential chemical events in this reaction lead to changes in the absorption spectrum of PLP allowing for characterization of reaction intermediates in the first turnover of the cysteine desulfurase reaction through rapid kinetic measurements. The formation of the enzyme persulfide and subsequent release of alanine marks the end of the first half of the cysteine desulfurase reaction (Black & Dos Santos, 2015b; Mihara & Esaki, 2002).

Steady-state reaction measurements, however, have been mainly analyzed in the presence of artificial reducing agents, a typical component of assay mixtures (Black & Dos Santos, 2015a; Ikeuchi, Shigi, Kato, Nishimura, & Suzuki, 2006; Kato et al., 2002; Lauhon, Erwin, & Ton, 2004; Mihara, Kurihara, Yoshimura, & Esaki, 2000; Park et al., 2003;

Zheng & Dean, 1994). Under these conditions, DTT itself can directly interact with the PLP cofactor disturbing the overall absorption spectrum. Furthermore, reducing agents can directly react with the persulfide bond, releasing sulfide and resetting the enzyme for the next catalytic cycle. This presents an experimental problem. In some cases, reduction of the reaction S-intermediate skews the kinetic behavior of the sulfurtransferase reactions and bypasses the involvement of sulfur acceptor molecules, which are critical to the function of these pathways in vivo (Black & Dos Santos, 2015a; Ikeuchi et al., 2006; Kato et al., 2002; Lauhon et al., 2004; Mihara et al., 2000; Park et al., 2003; Zheng & Dean, 1994). For example, we found that the *B. subtilis* cysteine desulfurase NifZ transfers sulfur to the thiouridylase ThiI in the biosynthesis of 4-thiouridine in tRNA, but in vitro, the sulfur-transfer reaction involving these proteins can be substituted by sulfide under reducing conditions (Rajakovich, Tomlinson, & Dos Santos, 2012). Furthermore, the sulfur acceptor ThiI can enhance the catalytic turnover rate of NifZ only in reactions absent of synthetic reducing agents displaying a reaction profile similar to the one shown in Fig. 6. This suggests that in vitro DTT bypasses ThiI's role as a sulfurtransferase. However, this feature may not be universal in reactions involving sulfurtransferases. In the *E. coli* cysteine desulfurase SufS and its sulfur acceptor SufE, both persulfide intermediates are resistant to the action of certain reducing agents, protecting the sulfur transfer path to the final acceptor SufBCD during Fe–S cluster assembly under oxidative stress conditions (Dai & Outten, 2012; Selbach et al., 2013). In this cysteine desulfurase-sulfur acceptor reaction, a stronger reducing agent (e.g., TCEP) can effectively reduce the persulfide bond at the sulfur acceptor protein leading to higher turnover number. In the *B. subtilis* SufS, the persulfide bond is protected from the action of DTT (Selbach et al., 2014), which helps to explain the low turnover number of the enzyme in the presence of reducing agents. Thus, SufU can effectively abstract the persulfide sulfur from SufS. However, unlike SufE, DTT can effectively reduce the sulfur acceptor persulfide. In this assay, the transfer of the persulfide sulfur from SufS to SufU is the rate-limiting reaction step in the overall reaction. While the kinetic behavior of select members of this class of enzymes has been investigated, no comparative analyses have been reported on their cross-reactivity as experimental evidence of their structural and functional specificities. It is anticipated, however, that specific protein–protein interactions between cysteine desulfurase and sulfur acceptor must occur to promote effective and rapid sulfur transfer under physiological settings.

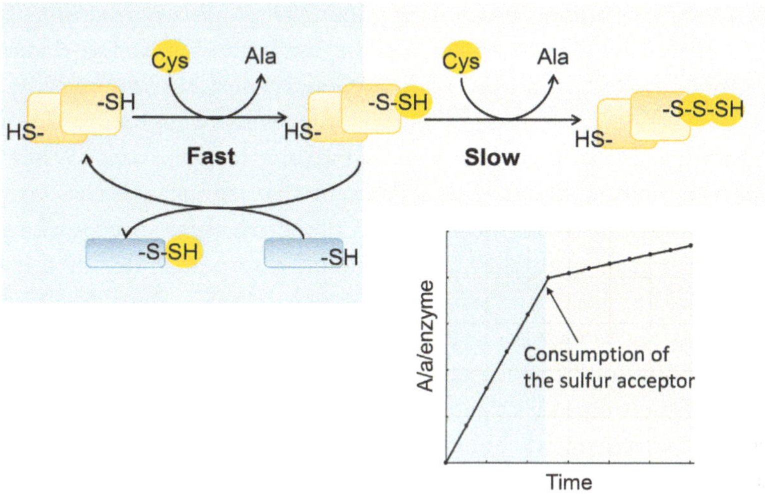

Fig. 6 Cysteine desulfurase reaction under nonreducing conditions. The formation of alanine and the subsequent sulfur transfer to the acceptor molecule proceeds at a fast rate until the consumption of the sulfur acceptor substrate (represented with a *shaded blue background*). The number of turnovers of the fast phase of the reaction is proportional to the amount of sulfur acceptor substrate present in the reaction. Once all the substrate is persulfurated, the reaction proceeds through a slow steady-state phase (*shaded in gray*). The persulfurated (showed in the diagram) or polysulfurated form of the enzyme can undergo subsequent reactions with cysteine adding one additional sulfur at a time to a growing polysulfide intermediate that will eventually lead to the formation of cyclic polysulfide (colloidal sulfur, S_8).

Direct sulfur-transfer reactions between cysteine desulfurases and sulfur acceptors are readily achieved under nonreducing conditions. The sulfur transfer event can be tracked in reactions performed with ^{35}S-Cys and both cysteine desulfurase and sulfur acceptor and analyzed through nonreducing SDS-PAGE and imaged with phosphorimager (Selbach et al., 2013). When performing this type of experiment, reactions can be stopped upon addition of an alkylating agent like *N*-ethylmaleimide or iodoacetamide prior electrophoretic separation. Alkylation of thiols including cysteine substrate protein thiols, and persulfides prevents unspecific ^{35}S-labeling of proteins and the formation of mixed disulfides after denaturation.

Multiple turnover reactions of cysteine desulfurases are also observed under nonreducing condition (Fig. 6). That is, following the release of the Ala, another molecule of Cys substrate can bind and complete the first half of the reaction through formation of a polysulfurated form of the enzyme. The activity of the enzyme measured through the rate of alanine formation shows much lower activity (Fig. 6, *gray shaded area*). Thus, the rate of polysulfide formation is much slower than the rate of sulfur transfer. Titration of sulfur acceptor proteins in kinetic assays can be used to determine rates of sulfur transfer reactions. Enzymatic assays performed with short time points allow for determination of initial sulfur transfer rates before consumption of sulfur acceptor protein substrate (Fig. 6, *blue shaded area*). When performing this type of assays, it is important that sulfur acceptors are fully reduced so they can assume their fully competent form in promoting the nucleophilic attack onto the cysteine desulfurase persulfide. In this case, these proteins are pretreated with reducing agents following a desalting step to remove excess reductants. Alternatively, sulfur acceptors may be reacted with TCEP-immobilized to sepharose for 30 min prior to assay reactions. Solutions of the substrate cysteine and standard sodium sulfide are also susceptible to oxidation and should always be freshly prepared.

3.2 Fe–S Cluster Assembly Reactions

The formation of Fe–S clusters in *B. subtilis* is proposed to occur on the scaffold protein SufB when in complex with SufD and SufC. This proposal, although lacking experimental validation, is based on the established function of the *E. coli* orthologous proteins (Chahal, Dai, Saini, Ayala-Castro, & Outten, 2009; Layer et al., 2007). Heterologous expression of the *suf* operon under the pBAD promoter leads to the isolation of protein complexes of SufS–SufU, as well as SufD–SufC. Although several methods of expression of SufB have been explored by our laboratory thus far, accumulation of soluble forms of SufB has not been achieved. The isolation strategy involving expression of SufB with a C-terminus His-tag using a pET28 vector in *E. coli* BL21DE3 cells has been optimized and is our current standard of practice. Using this system, SufB is found as inclusion bodies which are then isolated from Ni-IMAC purification under 6 *M* urea denaturing conditions (Rangachari et al., 2002). Refolding of the isolated protein is then performed through dialysis against 25 m*M* Tris pH 8, 0.3 *M* NaCl, 0.15 *M* KCl, 2 m*M* $MgCl_2$, 2 m*M* DTT, 2 m*M* EDTA, twice, for 4 h each. Both DTT and EDTA were found to be critical for proper refolding. As-isolated SufD or refolded SufB is equally competent in stimulating the ATPase activity of SufC in similar fashion to what has been described for *Thermatoga maritima, Erwinia chrysanthemi* enzymes (Nachin, Loiseau, Expert, & Barras, 2003; Rangachari et al., 2002).

In vitro, clusters can be formed on a scaffold protein upon incubation of ferrous ammonium sulfate and a source of sulfur under anoxic conditions.

While sodium sulfide is a suitable source of sulfur, reactions using a cysteine desulfurase and cysteine display faster rates and higher levels of Fe–S cluster accumulation onto the scaffold (Agar, Zheng, Cash, Dean, & Johnson, 2000; Bandyopadhyay et al., 2008; Dos Santos et al., 2004; Raulfs et al., 2008; Selbach et al., 2014; Smith et al., 2005; Yuvaniyama, Agar, Cash, Johnson, & Dean, 2000; Zheng & Dean, 1994). When performing these reactions, it is important to keep solutions devoid of oxygen as possible. For that, buffers should be degassed for at least 30 min prior introduction into the anaerobic chamber. Ideally, all reactants, pipettes, and glassware can be placed inside the glove box several hours or even on the day before experimentation. Freshly prepared and degassed cysteine and iron solutions along with still-frozen aliquots of enzymes that can then be introduced into the anaerobic chamber prior to the reaction. To avoid accumulation and the subsequent undesired formation of free iron-sulfide precipitates, limiting amounts of iron source can be used in these reactions. For example, cluster assembly reactions can be performed in the presence of a 50 μM of the scaffold protein, 0.25 mM $Fe(NH_4)_2(SO_4)_2$, 0.5 mM Cys, 2 mM DTT, 1 μM of the cysteine desulfurase and incubated at room temperature for 30 min. When using sealed cuvettes (Starna®), the reaction can be followed by tracking the increase in absorbance over time with characteristic peaks at 320 and 400 nm regions. However, the UV/Vis absorption spectrum should not be used as the sole indicator of cluster assembly since Fe–S species spontaneously formed in these reactions may not be associated to the scaffold. Therefore, complementary methods to detect accumulation of Fe–S clusters should be used to confirm the association of newly synthesized clusters to the scaffold protein.

The protein environment surrounding Fe–S clusters leads to induced chirality of these metal centers and they can be detected through circular dichroism (CD) within visible wavelength region (350–700 nm) (Chahal et al., 2009; Iametti et al., 2015; Selbach et al., 2014; Urbina, Silberg, Hoff, & Vickery, 2001). Kinetics of cluster assembly can then be monitored and traced back to changes in the signal intensity of a specific wavelength overtime. The overall inducible CD (ICD) spectrum is taken after the cluster assembly is completed. This allows for the selection of a specific wavelength that is associated with the formation of a unique protein-bound cluster. Since the protein environment of Fe–S clusters changes with the scaffold and final targets, ICD has also been used as a method to detect cluster transfer reactions. This is possible because the overall spectrum is unique to the type of cluster and the protein environment associated with that cluster.

Kinetics of cluster formation and transfer can be followed by continuous measurements over time at a selected wavelength that is characteristic of the cluster of interest. It is important to mention that not all Fe–S clusters associated to proteins result in a strong CD signal. In some cases, the signal falls below the limit of detection and limits the use of this technique to only certain types of scaffolds.

To determine the stoichiometry and types of Fe–S clusters formed on scaffold proteins, additional sample treatment is necessary. Following cluster assembly, the scaffold protein loaded with Fe–S clusters can be separated from reaction ingredients by a size exclusion desalting column. When considering the use of this approach, it is critical to optimize the amount of desalting resin to the reaction volumes loaded onto these columns to maximize the effectiveness of removing unbound Fe–S species which may form high-molecular-weight aggregates and free forms of Fe–S of low molecular weight from the protein fraction containing cluster bound to the scaffold. Alternatively, affinity chromatography can be employed when the scaffold is tagged with a His-tag or Strep-tag. This type of purification is often advantageous because it does not dilute the sample and allows for rapid purification of the scaffold loaded with transient clusters. A pitfall of this technique, however, is observed when using His-tagged proteins. The affinity resin may chelate the cluster and purified fractions will contain imidazole that have the potential to destabilize the labile cluster associated with the scaffold through direct interaction and/or may interfere with activation experiments. Once purified, the cluster-loaded scaffold can be analyzed through direct UV/Vis absorption spectroscopy and elemental analysis of acid labile sulfur and iron content (Fortune & Mellow, 1938; Siegel, 1965). When comparing the spectra of the cluster-bound and apo-forms of the scaffold, it is ideal to measure the protein concentration of both samples through an indirect method such as a Bradford assay (Bradford, 1976), because Fe–S clusters also display absorption at 280 nm. Characterization of cluster species can also be supplemented with additional spectroscopic methods (Bandyopadhyay et al., 2008; Chandramouli et al., 2007; Gupta et al., 2009; Smith et al., 2005). The electron paramagnetic resonance spectrum of cluster-bound proteins after titration with a strong reductant like dithionite can inform about the type of cluster associated with the scaffold. This technique, however, has not been successful in characterizing the metal centers associated with certain scaffolds, because dithionite reduction led to destruction of labile clusters. Alternatively, magnetic circular dichroism (MCD) and Raman spectroscopies can inform about the types of protein and nonprotein ligands associated with the nascent Fe–S clusters. Proteins reconstituted with ^{57}Fe can be examined through Mössbauer spectroscopy, in this case cluster-bound samples need to be generated at high concentrations (>300 μ*M*) and purified from free iron added to reconstitution reactions (Pandelia, Lanz, Booker, & Krebs, 2015). As mentioned earlier, transient clusters formed on Fe–S scaffold proteins are labile toward the action of strong reducing agents and easily susceptible to oxygen damage making an experimental challenge the characterization of cluster species.

3.3 Fe–S Cluster Transfer Reactions

The transfer of preformed Fe–S cluster onto the protein scaffold to final target proteins constitutes the last step of the biosynthetic process. This biosynthetic step can be monitored by spectroscopic methods that allow for detection of a signature signal associated with the binding of the cluster to the final acceptor and/or disappearance of signal associated with the cluster bound to the scaffold. As aforementioned, ICD has been an invaluable tool in accessing the rates of cluster transfer from the scaffold to a final target (Fox et al., 2015; Iametti et al., 2015; Mapolelo et al., 2013). Determining the activity of enzymes utilizing Fe–S clusters is also an effective method in discerning the efficiency of cluster transfer. In both cases, in vitro reactions involved the use of apo-proteins devoid of their cluster partners. Methods of isolation of proteins and removal of Fe–S clusters involve incubation of as-isolated holo-enzymes with chelating agents such as dipyridyl or EDTA and the addition of cyanide or treatment with reducing agents (Dos Santos et al., 2004; Kennedy & Beinert, 1988). This helps to avoid the formation of persulfides or disulfide protein intermediates. Dropwise titration with chelating agents results in higher yields of apo-proteins. Chelation reactions are then subjected to gel filtration purification to remove chelated metal and free sulfide species along with unfolded protein aggregates resulted from this type of treatment. Since the formation of Fe–S clusters and the transfer of intact clusters are both processes sensitive to oxygen, it is important that the final step of purification of apo-target proteins is performed under anoxic conditions. Well-studied targets to quantify cluster transfer events are the 2Fe–2S cluster ferredoxin because of characteristic ICD spectrum (Chandramouli & Johnson, 2006) and the 4Fe–4S cluster aconitase and isoleucine dehydratase because its activity can be monitored in continuous assays using a standard spectrophotomer (Kennedy & Beinert, 1988; Unciuleac et al., 2007). Once a method to detect cluster on final targets is established, characterization of the last step on this pathway can be determined in time-course experiments and ratio of cluster-loaded scaffold to final acceptor. Typically, a bonafide scaffold is capable of (1) coordinating a defined stoichiometry of Fe–S clusters and (2) transferring clusters at stoichiometric ratio (cluster to target protein) within minutes. In these experiments, cluster assembly and transfer are performed in the same day to maximize the occupancy of intact clusters bound to the scaffold site.

Accessory proteins are known to facilitate the transfer from scaffold to final acceptors leading to faster rates of cluster transfer. Additional proteins, mainly cluster carriers can participate in this final stage by enhancing the rate of cluster transfer and efficiency of this biosynthetic step. The *B. subtilis* genome encodes for Nfu and SufA proteins, both single domain proteins with similarities to cluster carriers in other species. While inactivation of *sufA* led to no defects in Fe–S metabolism (Albrecht et al., 2010), *nfu* gene could not be deleted (our unpublished results) suggesting a critical role of this protein for survival of *B. subtilis*. Direct cluster assembly on Nfu shows that this protein can harbor an Fe–S cluster within a dimer and this cluster can be easily transferred to apo-aconitase (unpublished results), this result is in agreement with biochemical data indicating the similar function of close ortholog *S. aureus* Nfu (Rosario-Cruz & Boyd, 2015).

4. CONCLUDING REMARKS

In *B. subtilis* Fe–S cluster assembly is proposed to be performed by the Suf system. In this system, sulfur is mobilized by SufS and transferred to SufU. The assembly of Fe–S clusters is then proposed to occur at the scaffold SufB when in a complex with SufC and SufD. Assembled clusters can then be transferred directly to target proteins or this transfer step can be expedited by carrier proteins such as Nfu. Many steps of this biosynthetic process remain yet uncharacterized. What are the sulfur acceptor molecules of SufU? How does SufB assemble and transfer Fe–S clusters? What is the involvement of SufC and SufD in this process? What is the physiological reductant in this pathway? When is the Nfu carrier recruited? These and additional questions are currently being explored by our group and others through investigation using the approaches for the characterization of the three main Fe–S cluster biosynthetic steps. These studies include not only investigation of individual steps (Fig. 1) using the methods described in this chapter, but also through investigation of the complete pathway in reactions including all biosynthetic proteins in one reaction. These conditions allow for characterization of turnover of the Fe–S cluster biosynthetic system, rather than analysis of scaffold proteins preloaded with clusters and quantification transfer in separate events.

The presence of additional cysteine desulfurases apart from the Suf system, however, does not exclude the participation of these enzymes in the formation of Fe–S cluster to certain proteins (Dos Santos, 2014). Although cysteine desulfurases can bypass scaffolds and assist the direct in vitro assembly of Fe–S clusters, in vivo all cluster biosynthetic systems described so far have involved a scaffold protein. Interestingly, the *B. subtilis* NifS cysteine desulfurase is required in the biosynthesis of nicotinamide adenine dinucleotide (NAD) (Sun & Setlow, 1993) possibly activating the [4Fe–4S] cluster-containing quinolinate synthase, NadA enzyme (Marinoni et al., 2008). It is possible that NifS is able to provide sulfur atoms for the direct assembly of [4Fe–4S] cluster on NadA in a novel system that bypasses the need for preassembly on a scaffold. The validation of this hypothesis will represent a *paradigm shift* in our understanding of Fe–S cluster metabolism.

ACKNOWLEDGMENT

The research in the laboratory of P.C.D.S. is supported by National Science Foundation MCB-1054623.

REFERENCES

Agar, J. N., Yuvaniyama, P., Jack, R. F., Cash, V. L., Smith, A. D., Dean, D. R., et al. (2000). Modular organization and identification of a mononuclear iron-binding site within the NifU protein. *Journal of Biological Inorganic Chemistry*, *5*(2), 167–177.

Agar, J. N., Zheng, L., Cash, V. L., Dean, D. R., & Johnson, M. K. (2000). Role of the IscU protein in iron-sulfur cluster biosynthesis: IscS-mediated assembly of a [2Fe-2S] cluster in IscU. *Journal of the American Chemical Society*, *122*, 2136–2137.

Albrecht, A. G., Landmann, H., Nette, D., Burghaus, O., Peuckert, F., Seubert, A., et al. (2011). The frataxin homologue Fra plays a key role in intracellular iron channeling in *Bacillus subtilis*. *Chembiochem: A European Journal of Chemical Biology*, *12*(13), 2052–2061.

Albrecht, A. G., Netz, D. J., Miethke, M., Pierik, A. J., Burghaus, O., Peuckert, F., et al. (2010). SufU is an essential iron-sulfur cluster scaffold protein in *Bacillus subtilis*. *Journal of Bacteriology*, *192*(6), 1643–1651. http://dx.doi.org/10.1128/JB.01536-09.

Angelini, S., Gerez, C., Ollagnier-de Choudens, S., Sanakis, Y., Fontecave, M., Barras, F., et al. (2008). NfuA, a new factor required for maturing Fe/S proteins in *Escherichia coli* under oxidative stress and iron starvation conditions. *The Journal of Biological Chemistry*, *283*(20), 14084–14091. http://dx.doi.org/10.1074/jbc.M709405200.

Bak, D. W., & Elliott, S. J. (2014). Alternative FeS cluster ligands: Tuning redox potentials and chemistry. *Current Opinion in Chemical Biology*, *19*, 50–58. http://dx.doi.org/10.1016/j.cbpa.2013.12.015.

Bandyopadhyay, S., Naik, S. G., O'Carroll, I. P., Huynh, B. H., Dean, D. R., Johnson, M. K., et al. (2008). A proposed role for the *Azotobacter vinelandii* NfuA protein as an intermediate iron-sulfur cluster carrier. *The Journal of Biological Chemistry*, *283*(20), 14092–14099. http://dx.doi.org/10.1074/jbc.M709161200.

Bartels, P. L., Zhou, A., Arnold, A. R., Nunez, N. N., Crespilho, F. N., David, S. S., et al. (2017). Electrochemistry of the [4Fe4S] cluster in base excision repair proteins: Tuning the redox potential with DNA. *Langmuir*, *33*(10), 2523–2530. http://dx.doi.org/10.1021/acs.langmuir.6b04581.

Behshad, E., & Bollinger, J. M., Jr. (2009). Kinetic analysis of cysteine desulfurase CD0387 from *Synechocystis sp.* PCC 6803: Formation of the persulfide intermediate. *Biochemistry*, *48*(50), 12014–12023. http://dx.doi.org/10.1021/bi802161u.

Beinert, H. (2000). A tribute to sulfur. *European Journal of Biochemistry*, *267*(18), 5657–5664. http://dx.doi.org/10.1046/j.1432-1327.2000.01637.x.

Black, K. A., & Dos Santos, P. C. (2015a). Abbreviated pathway for biosynthesis of 2-thiouridine in *Bacillus subtilis*. *Journal of Bacteriology*, *197*(11), 1952–1962.

Black, K. A., & Dos Santos, P. C. (2015b). Shared-intermediates in the biosynthesis of thio-cofactors: Mechanism and functions of cysteine desulfurases and sulfur acceptors. *Biochimica et Biophysica Acta*, *1853*(6), 1470–1480.

Blanc, B., Gerez, C., & Ollagnier de Choudens, S. (2015). Assembly of Fe/S proteins in bacterial systems: Biochemistry of the bacterial ISC system. *Biochimica et Biophysica Acta*, *1853*(6), 1436–1447. http://dx.doi.org/10.1016/j.bbamcr.2014.12.009.

Blauenburg, B., Mielcarek, A., Altegoer, F., Fage, C. D., Linne, U., Bange, G., et al. (2016). Crystal structure of *Bacillus subtilis* cysteine desulfurase SufS and its dynamic interaction with frataxin and Scaffold protein SufU. *PLoS One*, *11*(7), e0158749. http://dx.doi.org/10.1371/journal.pone.0158749.

Bonomi, F., Iametti, S., Morleo, A., Ta, D., & Vickery, L. E. (2008). Studies on the mechanism of catalysis of iron-sulfur cluster transfer from IscU[2Fe2S] by HscA/HscB chaperones. *Biochemistry*, *47*(48), 12795–12801. http://dx.doi.org/10.1021/bi801565j.

Bradford, M. M. (1976). A rapid and sensitive method for the quantitation of microgram quantities of protein utilizing the principle of protein-dye binding. *Analytical Biochemistry*, *72*, 248–254.

Broderick, J. B., Duffus, B. R., Duschene, K. S., & Shepard, E. M. (2014). Radical S-adenosylmethionine enzymes. *Chemical Reviews*, *114*(8), 4229–4317. http://dx.doi.org/10.1021/cr4004709.

Chahal, H. K., Dai, Y., Saini, A., Ayala-Castro, C., & Outten, F. W. (2009). The SufBCD Fe-S scaffold complex interacts with SufA for Fe-S cluster transfer. *Biochemistry*, *48*(44), 10644–10653. http://dx.doi.org/10.1021/bi901518y.

Chandramouli, K., & Johnson, M. K. (2006). HscA and HscB stimulate [2Fe-2S] cluster transfer from IscU to apoferredoxin in an ATP-dependent reaction. *Biochemistry*, *45*(37), 11087–11095.

Chandramouli, K., Unciuleac, M. C., Naik, S., Dean, D. R., Huynh, B. H., & Johnson, M. K. (2007). Formation and properties of [4Fe-4S] clusters on the IscU scaffold protein. *Biochemistry*, *46*(23), 6804–6811.

Chen, J. S., & Mortenson, L. E. (1977). Inhibition of methylene blue formation during determination of the acid- labile sulfide of iron-sulfur protein samples containing dithionite. *Analytical Biochemistry*, *79*(1–2), 157–165.

Cho, Y. H., Yoo, H. S., Min, J. K., Lee, E. Y., Hong, S. P., Chung, Y. B., et al. (2002). Comparative study of naphthalene-2,3-dicarboxaldehyde and o-phthalaldehyde fluorogenic reagents for chromatographic detection of sphingoid bases. *Journal of Chromatography A*, *977*(1), 69–76.

Dai, Y., & Outten, F. W. (2012). The E. coli SufS-SufE sulfur transfer system is more resistant to oxidative stress than IscS-IscU. *FEBS Letters*, *586*(22), 4016–4022.

Ding, B., Smith, E. S., & Ding, H. (2005). Mobilization of the iron centre in IscA for the iron-sulphur cluster assembly in IscU. *The Biochemical Journal, 389*(Pt. 3), 797–802.

Dos Santos, P. C. (2014). Fe-S assembly in Gram-positive bacteria. In T. Rouault (Ed.), *Iron sulfur clusters in chemistry and biology* (pp. 347–366). Germany: Verlag Walter de Gruyter.

Dos Santos, P. C., & Dean, D. R. (2008). A newly discovered role for iron-sulfur clusters. *Proceedings of the National Academy of Sciences of the United States of America, 105*(33), 11589–11590. http://dx.doi.org/10.1073/pnas.0805713105.

Dos Santos, P. C., & Dean, D. R. (2014). A retrospective on the discovery of [Fe-S] cluster biosynthetic machineries in *Azotobacter vinelandii*. In T. Rouault (Ed.), *Iron-sulfur clusters in chemistry and biology* (pp. 267–296). Verlag Walter de Gruyter.

Dos Santos, P. C., Igarashi, R. Y., Lee, H. I., Hoffman, B. M., Seefeldt, L. C., & Dean, D. R. (2005). Substrate interactions with the nitrogenase active site. *Accounts of Chemical Research, 38*(3), 208–214.

Dos Santos, P. C., Smith, A. D., Frazzon, J., Cash, V. L., Johnson, M. K., & Dean, D. R. (2004). Iron-sulfur cluster assembly: NifU-directed activation of the nitrogenase Fe protein. *The Journal of Biological Chemistry, 279*(19), 19705–19711.

Doukov, T. I., Iverson, T. M., Seravalli, J., Ragsdale, S. W., & Drennan, C. L. (2002). A Ni-Fe-Cu center in a bifunctional carbon monoxide dehydrogenase/acetyl-CoA synthase. *Science, 298*(5593), 567–572.

Ellis, K. E., Clough, B., Saldanha, J. W., & Wilson, R. J. (2001). Nifs and Sufs in malaria. *Molecular Microbiology, 41*(5), 973–981.

Esakova, O. A., Silakov, A., Grove, T. L., Saunders, A. H., McLaughlin, M. I., Yennawar, N. H., et al. (2016). Structure of quinolinate synthase from *Pyrococcus horikoshii* in the presence of its product, quinolinic acid. *Journal of the American Chemical Society, 138*(23), 7224–7227. http://dx.doi.org/10.1021/jacs.6b02708.

Fortune, W. B., & Mellow, M. G. (1938). Determination of iron with o-phenanthroline. *Industrial and Engineering Chemistry, Analytical Edition, 10*, 61–64.

Fox, N. G., Chakrabarti, M., McCormick, S. P., Lindahl, P. A., & Barondeau, D. P. (2015). The human iron-sulfur assembly complex catalyzes the synthesis of [2Fe-2S] clusters on ISCU2 that can be transferred to acceptor molecules. *Biochemistry, 54*(25), 3871–3879. http://dx.doi.org/10.1021/bi5014485.

Fuss, J. O., Tsai, C. L., Ishida, J. P., & Tainer, J. A. (2015). Emerging critical roles of Fe-S clusters in DNA replication and repair. *Biochimica et Biophysica Acta, 1853*(6), 1253–1271. http://dx.doi.org/10.1016/j.bbamcr.2015.01.018.

Gupta, V., Sendra, M., Naik, S. G., Chahal, H. K., Huynh, B. H., Outten, F. W., et al. (2009). Native *Escherichia coli* SufA, coexpressed with SufBCDSE, purifies as a [2Fe-2S] protein and acts as an Fe-S transporter to Fe-S target enzymes. *Journal of the American Chemical Society, 131*(17), 6149–6153. http://dx.doi.org/10.1021/ja807551e.

Hernandez, J. A., Curatti, L., Aznar, C. P., Perova, Z., Britt, R. D., & Rubio, L. M. (2008). Metal trafficking for nitrogen fixation: NifQ donates molybdenum to NifEN/NifH for the biosynthesis of the nitrogenase FeMo-cofactor. *Proceedings of the National Academy of Sciences of the United States of America, 105*(33), 11679–11684. http://dx.doi.org/10.1073/pnas.0803576105.

Huet, G., Daffe, M., & Saves, I. (2005). Identification of the *Mycobacterium tuberculosis* SUF machinery as the exclusive mycobacterial system of [Fe-S] cluster assembly: Evidence for its implication in the pathogen's survival. *Journal of Bacteriology, 187*(17), 6137–6146.

Iametti, S., Barbiroli, A., & Bonomi, F. (2015). Functional implications of the interaction between HscB and IscU in the biosynthesis of FeS clusters. *Journal of Biological Inorganic Chemistry, 20*(6), 1039–1048. http://dx.doi.org/10.1007/s00775-015-1285-z.

Ikeuchi, Y., Shigi, N., Kato, J., Nishimura, A., & Suzuki, T. (2006). Mechanistic insights into sulfur relay by multiple sulfur mediators involved in thiouridine biosynthesis at tRNA wobble positions. *Molecular Cell, 21*(1), 97–108.

Iwasaki, T. (2010). Iron-sulfur world in aerobic and hyperthermoacidophilic archaea Sulfolobus. *Archaea, 2010,* 1–14. http://dx.doi.org/10.1155/2010/842639.
Johnson, D. C., Dean, D. R., Smith, A. D., & Johnson, M. K. (2005). Structure, function, and formation of biological iron-sulfur clusters. *Annual Review of Biochemistry, 74,* 247–281.
Johnson, D. C., Unciuleac, M. C., & Dean, D. R. (2006). Controlled expression and functional analysis of iron-sulfur cluster biosynthetic components within *Azotobacter vinelandii*. *Journal of Bacteriology, 188*(21), 7551–7561.
Kaiser, J. T., Clausen, T., Bourenkow, G. P., Bartunik, H. D., Steinbacher, S., & Huber, R. (2000). Crystal structure of a NifS-like protein from Thermotoga maritima: Implications for iron sulphur cluster assembly. *Journal of Molecular Biology, 297*(2), 451–464.
Kato, S., Mihara, H., Kurihara, T., Takahashi, Y., Tokumoto, U., Yoshimura, T., et al. (2002). Cys-328 of IscS and Cys-63 of IscU are the sites of disulfide bridge formation in a covalently bound IscS/IscU complex: Implications for the mechanism of iron-sulfur cluster assembly. *Proceedings of the National Academy of Sciences of the United States of America, 99*(9), 5948–5952.
Kennedy, M. C., & Beinert, H. (1988). The state of cluster SH and S2- of aconitase during cluster interconversions and removal. A convenient preparation of apoenzyme. *The Journal of Biological Chemistry, 263*(17), 8194–8198.
Kim, J. H., Bothe, J. R., Alderson, T. R., & Markley, J. L. (2015). Tangled web of interactions among proteins involved in iron-sulfur cluster assembly as unraveled by NMR, SAXS, chemical crosslinking, and functional studies. *Biochimica et Biophysica Acta, 1853*(6), 1416–1428. http://dx.doi.org/10.1016/j.bbamcr.2014.11.020.
Kim, S., & Park, S. (2013). Structural changes during cysteine desulfurase CsdA and sulfur acceptor CsdE interactions provide insight into the trans-persulfuration. *The Journal of Biological Chemistry, 288*(38), 27172–27180. http://dx.doi.org/10.1074/jbc.M113.480277.
Kispal, G., Csere, P., Prohl, C., & Lill, R. (1999). The mitochondrial proteins Atm1p and Nfs1p are essential for biogenesis of cytosolic Fe/S proteins. *The EMBO Journal, 18*(14), 3981–3989.
Kobayashi, K., Ehrlich, S. D., Albertini, A., Amati, G., Andersen, K. K., Arnaud, M., et al. (2003). Essential *Bacillus subtilis* genes. *Proceedings of the National Academy of Sciences of the United States of America, 100*(8), 4678–4683.
LaGier, M. J., Tachezy, J., Stejskal, F., Kutisova, K., & Keithly, J. S. (2003). Mitochondrial-type iron-sulfur cluster biosynthesis genes (IscS and IscU) in the apicomplexan *Cryptosporidium parvum*. *Microbiology, 149*(Pt. 12), 3519–3530.
Langen, R., Jensen, G. M., Jacob, U., Stephens, P. J., & Warshel, A. (1992). Protein control of iron-sulfur cluster redox potentials. *The Journal of Biological Chemistry, 267*(36), 25625–25627.
Lauhon, C. T., Erwin, W. M., & Ton, G. N. (2004). Substrate specificity for 4-thiouridine modification in *Escherichia coli*. *The Journal of Biological Chemistry, 279*(22), 23022–23029.
Lauhon, C. T., & Kambampati, R. (2000). The iscS gene in *Escherichia coli* is required for the biosynthesis of 4-thiouridine, thiamin, and NAD. *The Journal of Biological Chemistry, 275*(26), 20096–20103.
Layer, G., Gaddam, S. A., Ayala-Castro, C. N., Ollagnier-de Choudens, S., Lascoux, D., Fontecave, M., et al. (2007). SufE transfers sulfur from SufS to SufB for iron-sulfur cluster assembly. *The Journal of Biological Chemistry, 282*(18), 13342–13350.
Layer, G., Ollagnier-de Choudens, S., Sanakis, Y., & Fontecave, M. (2006). Iron-sulfur cluster biosynthesis: Characterization of *Escherichia coli* CYaY as an iron donor for the assembly of [2Fe-2S] clusters in the scaffold IscU. *The Journal of Biological Chemistry, 281*(24), 16256–16263.

Lee, S. C., Lo, W., & Holm, R. H. (2014). Developments in the biomimetic chemistry of cubane-type and higher nuclearity iron-sulfur clusters. *Chemical Reviews*, *114*(7), 3579–3600. http://dx.doi.org/10.1021/cr4004067.

Lill, R., Diekert, K., Kaut, A., Lange, H., Pelzer, W., Prohl, C., et al. (1999). The essential role of mitochondria in the biogenesis of cellular iron-sulfur proteins. *Biological Chemistry*, *380*(10), 1157–1166.

Loiseau, L., Ollagnier-de Choudens, S., Lascoux, D., Forest, E., Fontecave, M., & Barras, F. (2005). Analysis of the heteromeric CsdA-CsdE cysteine desulfurase, assisting Fe-S cluster biogenesis in *Escherichia coli*. *The Journal of Biological Chemistry*, *280*(29), 26760–26769.

Lu, J., Yang, J., Tan, G., & Ding, H. (2008). Complementary roles of SufA and IscA in the biogenesis of iron-sulfur clusters in *Escherichia coli*. *The Biochemical Journal*, *409*(2), 535–543. http://dx.doi.org/10.1042/BJ20071166.

Mapolelo, D. T., Zhang, B., Randeniya, S., Albetel, A. N., Li, H., Couturier, J., et al. (2013). Monothiol glutaredoxins and A-type proteins: Partners in Fe-S cluster trafficking. *Dalton Transactions*, *42*(9), 3107–3115. http://dx.doi.org/10.1039/c2dt32263c.

Marinoni, I., Nonnis, S., Monteferrante, C., Heathcote, P., Hartig, E., Bottger, L. H., et al. (2008). Characterization of L-aspartate oxidase and quinolinate synthase from *Bacillus subtilis*. *The FEBS Journal*, *275*(20), 5090–5107.

McLaughlin, M. I., Lanz, N. D., Goldman, P. J., Lee, K. H., Booker, S. J., & Drennan, C. L. (2016). Crystallographic snapshots of sulfur insertion by lipoyl synthase. *Proceedings of the National Academy of Sciences of the United States of America*, *113*(34), 9446–9450. http://dx.doi.org/10.1073/pnas.1602486113.

Mettert, E. L., & Kiley, P. J. (2015). Fe-S proteins that regulate gene expression. *Biochimica et Biophysica Acta*, *1853*(6), 1284–1293. http://dx.doi.org/10.1016/j.bbamcr.2014.11.018.

Mihara, H., & Esaki, N. (2002). Bacterial cysteine desulfurases: Their function and mechanisms. *Applied Microbiology and Biotechnology*, *60*(1-2), 12–23. http://dx.doi.org/10.1007/s00253-002-1107-4.

Mihara, H., Kurihara, T., Yoshimura, T., & Esaki, N. (2000). Kinetic and mutational studies of three NifS homologs from *Escherichia coli*: Mechanistic difference between L-cysteine desulfurase and L- selenocysteine lyase reactions. *Journal of Biochemistry*, *127*(4), 559–567.

Mueller, E. G. (2006). Trafficking in persulfides: Delivering sulfur in biosynthetic pathways. *Nature Chemical Biology*, *2*(4), 185–194.

Nachin, L., Loiseau, L., Expert, D., & Barras, F. (2003). SufC: An unorthodox cytoplasmic ABC/ATPase required for [Fe-S] biogenesis under oxidative stress. *The EMBO Journal*, *22*(3), 427–437.

Nicholas, D. J., Wilson, P. W., Heinen, W., Palmer, G., & Beinert, H. (1962). Use of electron paramagnetic resonance spectroscopy in investigations of functional metal components in micro-organisms. *Nature*, *196*, 433–436.

Nuth, M., Yoon, T., & Cowan, J. A. (2002). Iron-sulfur cluster biosynthesis: Characterization of iron nucleation sites for assembly of the [2Fe-2S]2+ cluster core in IscU proteins. *Journal of the American Chemical Society*, *124*(30), 8774–8775.

Outten, F. W. (2015). Recent advances in the Suf Fe-S cluster biogenesis pathway: Beyond the Proteobacteria. *Biochimica et Biophysica Acta*, *1853*(6), 1464–1469. http://dx.doi.org/10.1016/j.bbamcr.2014.11.001.

Pandelia, M. E., Lanz, N. D., Booker, S. J., & Krebs, C. (2015). Mossbauer spectroscopy of Fe/S proteins. *Biochimica et Biophysica Acta*, *1853*(6), 1395–1405. http://dx.doi.org/10.1016/j.bbamcr.2014.12.005.

Parent, A., Elduque, X., Cornu, D., Belot, L., Le Caer, J. P., Grandas, A., et al. (2015). Mammalian frataxin directly enhances sulfur transfer of NFS1 persulfide to both ISCU and free thiols. *Nature Communications*, *6*, 5686. http://dx.doi.org/10.1038/ncomms6686.

Park, J. H., Dorrestein, P. C., Zhai, H., Kinsland, C., McLafferty, F. W., & Begley, T. P. (2003). Biosynthesis of the thiazole moiety of thiamin pyrophosphate (vitamin B1). *Biochemistry*, *42*(42), 12430–12438.

Py, B., & Barras, F. (2015). Genetic approaches of the Fe-S cluster biogenesis process in bacteria: Historical account, methodological aspects and future challenges. *Biochimica et Biophysica Acta*, *1853*(6), 1429–1435. http://dx.doi.org/10.1016/j.bbamcr.2014.12.024.

Rajakovich, L. J., Tomlinson, J., & Dos Santos, P. C. (2012). Functional analysis of *Bacillus subtilis* genes involved in the biosynthesis of 4-thiouridine in tRNA. *Journal of Bacteriology*, *194*(18), 4933–4940.

Rangachari, K., Davis, C. T., Eccleston, J. F., Hirst, E. M., Saldanha, J. W., Strath, M., et al. (2002). SufC hydrolyzes ATP and interacts with SufB from Thermotoga maritima. *FEBS Letters*, *514*(2–3), 225–228.

Raulfs, E. C., O'Carroll, I. P., Dos Santos, P. C., Unciuleac, M. C., & Dean, D. R. (2008). In vivo iron-sulfur cluster formation. *Proceedings of the National Academy of Sciences of the United States of America*, *105*(25), 8591–8596. http://dx.doi.org/10.1073/pnas.0803173105.

Rees, D. C., & Howard, J. B. (2003). The interface between the biological and inorganic worlds: Iron-sulfur metalloclusters. *Science*, *300*(5621), 929–931.

Riboldi, G. P., de Mattos, E. P., & Frazzon, J. (2013). Biogenesis of [Fe-S] cluster in Firmicutes: An unexploited field of investigation. *Antonie Van Leeuwenhoek*, *104*(3), 283–300.

Riboldi, G. P., Larson, T. J., & Frazzon, J. (2011). *Enterococcus faecalis sufCDSUB* complements *Escherichia coli sufABCDSE*. *FEMS Microbiology Letters*, *320*(1), 15–24.

Rocha, A. G., & Dancis, A. (2016). Life without Fe-S clusters. *Molecular Microbiology*, *99*(5), 821–826. http://dx.doi.org/10.1111/mmi.13273.

Rosario-Cruz, Z., & Boyd, J. M. (2016). Physiological roles of bacillithiol in intracellular metal processing. *Current Genetics*, *62*(1), 59–65. http://dx.doi.org/10.1007/s00294-015-0511-0.

Rosario-Cruz, Z., Chahal, H. K., Mike, L. A., Skaar, E. P., & Boyd, J. M. (2015). Bacillithiol has a role in Fe-S cluster biogenesis in *Staphylococcus aureus*. *Molecular Microbiology*, *98*(2), 218–242. http://dx.doi.org/10.1111/mmi.13115.

Rouault, T. A., & Tong, W. H. (2008). Iron-sulfur cluster biogenesis and human disease. *Trends in Genetics*, *24*(8), 398–407. http://dx.doi.org/10.1016/j.tig.2008.05.008.

Saini, A., Mapolelo, D. T., Chahal, H. K., Johnson, M. K., & Outten, F. W. (2010). SufD and SufC ATPase activity are required for iron acquisition during in vivo Fe-S cluster formation on SufB. *Biochemistry*, *49*(43), 9402–9412.

Schwartz, C. J., Djaman, O., Imlay, J. A., & Kiley, P. J. (2000). The cysteine desulfurase, IscS, has a major role in in vivo Fe-S cluster formation in *Escherichia coli*. *Proceedings of the National Academy of Sciences of the United States of America*, *97*(16), 9009–9014.

Selbach, B. P., Chung, A. H., Scott, A. D., George, S. J., Cramer, S. P., & Dos Santos, P. C. (2014). Fe-S cluster biogenesis in Gram-positive bacteria: SufU is a zinc-dependent sulfur transfer protein. *Biochemistry*, *53*(1), 152–160. http://dx.doi.org/10.1021/bi4011978.

Selbach, B., Earles, E., & Dos Santos, P. C. (2010). Kinetic analysis of the bisubstrate cysteine desulfurase SufS from *Bacillus subtilis*. *Biochemistry*, *49*(40), 8794–8802.

Selbach, B. P., Pradhan, P. K., & Dos Santos, P. C. (2013). Protected sulfur transfer reactions by the *Escherichia coli* Suf system. *Biochemistry*, *52*(23), 4089–4096. http://dx.doi.org/10.1021/bi4001479.

Shepard, W., Soutourina, O., Courtois, E., England, P., Haouz, A., & Martin-Verstraete, I. (2011). Insights into the Rrf2 repressor family—The structure of CymR, the global cysteine regulator of *Bacillus subtilis*. *The FEBS Journal*, *278*(15), 2689–2701.

Shi, R., Proteau, A., Villarroya, M., Moukadiri, I., Zhang, L., Trempe, J. F., et al. (2010). Structural basis for Fe-S cluster assembly and tRNA thiolation mediated by IscS protein-protein interactions. *PLoS Biology, 8*(4), e1000354. http://dx.doi.org/10.1371/journal.pbio.1000354.
Siegel, L. M. (1965). A direct microdetermination for sulfide. *Analytical Biochemistry, 11*, 126–132.
Smith, A. D., Agar, J. N., Johnson, K. A., Frazzon, J., Amster, I. J., Dean, D. R., et al. (2001). Sulfur transfer from IscS to IscU: The first step in iron-sulfur cluster biosynthesis. *Journal of the American Chemical Society, 123*(44), 11103–11104.
Smith, A. D., Jameson, G. N., Dos Santos, P. C., Agar, J. N., Naik, S., Krebs, C., et al. (2005). NifS-mediated assembly of [4Fe-4S] clusters in the N- and C-terminal domains of the NifU scaffold protein. *Biochemistry, 44*(39), 12955–12969.
Suess, D. L., Kuchenreuther, J. M., De La Paz, L., Swartz, J. R., & Britt, R. D. (2016). Biosynthesis of the [FeFe] hydrogenase H cluster: A central role for the radical SAM enzyme HydG. *Inorganic Chemistry, 55*(2), 478–487. http://dx.doi.org/10.1021/acs.inorgchem.5b02274.
Sun, D., & Setlow, P. (1993). Cloning, nucleotide sequence, and regulation of the *Bacillus subtilis* nadB gene and a nifS-like gene, both of which are essential for NAD biosynthesis. *Journal of Bacteriology, 175*(5), 1423–1432.
Takahashi, Y., & Nakamura, M. (1999). Functional assignment of the ORF2-iscS-iscU-iscA-hscB-hscA-fdx-ORF3 gene cluster involved in the assembly of Fe-S clusters in Escherichia coli. *Journal of Biochemistry, 126*(5), 917–926.
Tan, G., Lu, J., Bitoun, J. P., Huang, H., & Ding, H. (2009). IscA/SufA paralogues are required for the [4Fe-4S] cluster assembly in enzymes of multiple physiological pathways in *Escherichia coli* under aerobic growth conditions. *The Biochemical Journal, 420*(3), 463–472. http://dx.doi.org/10.1042/BJ20090206.
Tirupati, B., Vey, J. L., Drennan, C. L., & Bollinger, J. M., Jr. (2004). Kinetic and structural characterization of Slr0077/SufS, the essential cysteine desulfurase from *Synechocystis sp.* PCC 6803. *Biochemistry, 43*(38), 12210–12219.
Unciuleac, M. C., Chandramouli, K., Naik, S., Mayer, S., Huynh, B. H., Johnson, M. K., et al. (2007). In vitro activation of apo-aconitase using a [4Fe-4S] cluster-loaded form of the IscU [Fe-S] cluster scaffolding protein. *Biochemistry, 46*(23), 6812–6821.
Urbina, H. D., Silberg, J. J., Hoff, K. G., & Vickery, L. E. (2001). Transfer of sulfur from IscS to IscU during Fe/S cluster assembly. *The Journal of Biological Chemistry, 276*(48), 44521–44526.
Wollers, S., Layer, G., Garcia-Serres, R., Signor, L., Clemancey, M., Latour, J. M., et al. (2010). Iron-sulfur (Fe-S) cluster assembly: The SufBCD complex is a new type of Fe-S scaffold with a flavin redox cofactor. *The Journal of Biological Chemistry, 285*(30), 23331–23341. http://dx.doi.org/10.1074/jbc.M110.127449.
Wu, S. P., Wu, G., Surerus, K. K., & Cowan, J. A. (2002). Iron-sulfur cluster biosynthesis. Kinetic analysis of [2Fe-2S] cluster transfer from holo ISU to apo Fd: Role of redox chemistry and a conserved aspartate. *Biochemistry, 41*(28), 8876–8885.
Yan, R., Adinolfi, S., & Pastore, A. (2015). Ferredoxin, in conjunction with NADPH and ferredoxin-NADP reductase, transfers electrons to the IscS/IscU complex to promote iron-sulfur cluster assembly. *Biochimica et Biophysica Acta, 1854*(9), 1113–1117. http://dx.doi.org/10.1016/j.bbapap.2015.02.002.
Yuvaniyama, P., Agar, J. N., Cash, V. L., Johnson, M. K., & Dean, D. R. (2000). NifS-directed assembly of a transient [2Fe-2S] cluster within the NifU protein. *Proceedings of the National Academy of Sciences of the United States of America, 97*(2), 599–604.
Zhang, L. Y., & Sun, M. X. (2005). Selective determination of gamma-aminobutyric acid, glutamate and alanine by mixed micellar electrokinetic chromatography and fluorescence detection. *Journal of Chromatography A, 1095*(1-2), 185–188. http://dx.doi.org/10.1016/j.chroma.2005.09.050.

Zheng, L., Cash, V. L., Flint, D. H., & Dean, D. R. (1998). Assembly of iron-sulfur clusters. Identification of an iscSUA-hscBA-fdx gene cluster from *Azotobacter vinelandii*. *The Journal of Biological Chemistry*, *273*(21), 13264–13272.

Zheng, L., & Dean, D. R. (1994). Catalytic formation of a nitrogenase iron-sulfur cluster. *The Journal of Biological Chemistry*, *269*(29), 18723–18726.

Zheng, L., White, R. H., Cash, V. L., & Dean, D. R. (1994). Mechanism for the desulfurization of L-cysteine catalyzed by the nifS gene product. *Biochemistry*, *33*(15), 4714–4720.

Zheng, L., White, R. H., Cash, V. L., Jack, R. F., & Dean, D. R. (1993). Cysteine desulfurase activity indicates a role for NIFS in metallocluster biosynthesis. *Proceedings of the National Academy of Sciences of the United States of America*, *90*(7), 2754–2758.

CHAPTER EIGHT

Structural Characterization of Poised States in the Oxygen Sensitive Hydrogenases and Nitrogenases

Jacob H. Artz*, Oleg A. Zadvornyy*, David W. Mulder†, Paul W. King†, John W. Peters*[,1]
*Institute of Biological Chemistry, Washington State University, Pullman, WA, United States
†National Renewable Energy Laboratory, Biosciences Center, Golden, CO, United States
[1]Corresponding author: e-mail address: jw.peters@wsu.edu

Contents

Abstract

The crystallization of FeS cluster-containing proteins has been challenging due to their oxygen sensitivity, and yet these enzymes are involved in many critical catalytic reactions. The last few years have seen a wealth of innovative experiments designed to elucidate not just structural but mechanistic insights into FeS cluster enzymes. Here, we focus on the crystallization of hydrogenases, which catalyze the reversible reduction of protons to hydrogen, and nitrogenases, which reduce dinitrogen to ammonia. A specific focus is given to the different experimental parameters and strategies that are used to trap distinct enzyme states, specifically, oxidants, reductants, and gas

Methods in Enzymology, Volume 595
ISSN 0076-6879
http://dx.doi.org/10.1016/bs.mie.2017.07.005

treatments. Other themes presented here include the recent use of Cryo-EM, and how coupling various spectroscopies to crystallization is opening up new approaches for structural and mechanistic analysis.

1. INTRODUCTION TO FeS CLUSTERS

FeS clusters are widely distributed in biology and are critical for many oxidation–reduction reactions and for structural stability. Specialized FeS clusters, such as the nitrogenase FeMo cofactor, P-cluster, and hydrogenase H-cluster (Fig. 1), have been formed by evolution to catalyze otherwise difficult reactions, which are of increasing importance to understand to drive and develop new energy technologies. Structural characterization provides a platform for understanding many aspects of these challenging chemistries. While some enzymes (for example, the hydrogenase from *Chlamydomonas reinhardtii*) consist of a minimal catalytic subunit (Kamp et al., 2008), many hydrogenases contain complex arrays of FeS cluster electron transfer pathways, with possible integration of other cofactors (Adams, 1990; Wang, Huang, Kahnt, & Thauer, 2013). Often the FeS clusters are of the [4Fe–4S] type (Lubitz, Ogata, Rüdiger, & Reijerse, 2014), but [3Fe–4S], [4Fe–3S], and [2Fe–2S] clusters are found as well, and can have the canonical 4 Cys coordination or can be site differentiated, for example [4Fe–4S] with a single His, or in the case of [4Fe–3S] have supernumerary Cys coordination. Hydrogenase complexes can be quite large, >1 MDa (Vitt et al., 2014),

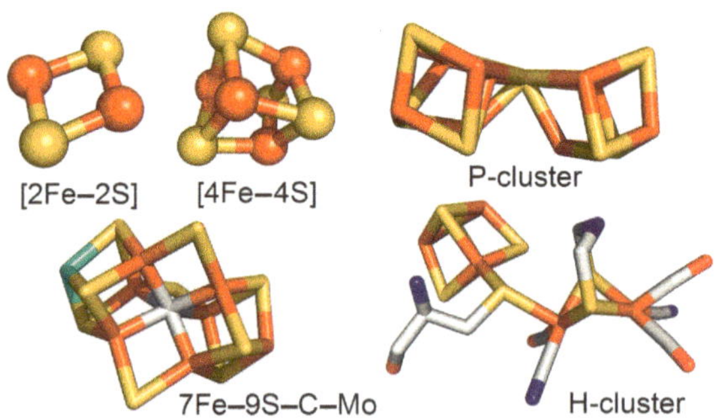

Fig. 1 Examples of basic and specialized FeS clusters from hydrogenase and nitrogenase. [2Fe–2S] and [4Fe–4S] clusters serve as basic building blocks for more specialized variations, including the P-cluster of nitrogenase that can be described as two fused [4Fe–4S] clusters, and the catalytic core (7Fe–9S–C–Mo) of nitrogenase and catalytic H-cluster site of [FeFe]-hydrogenase. Color scheme: C, *gray*; N, *blue*; O, *red*; S, *yellow*; Fe, *orange*; Mo, *cyan*.

and can be further complicated in characterization by being membrane associated (Tersteegen & Hedderich, 1999).

The growth of protein crystals is a complex physical and chemical process. Many factors affect crystallization of proteins, such as temperature, purity, and concentration of the protein, type and concentration of precipitants and stabilizers, concentration and pH value of different buffers, gravitation, method for crystallization, and technical bias (McPherson, 1976).

An additional difficulty in the study of hydrogenases and nitrogenases is that the FeS clusters are, in general, highly oxygen sensitive (Cohen et al., 2005; Erbes, King, & Gibbs, 1979; Lambertz et al., 2011; Liebgott et al., 2011; Stripp et al., 2009; Swanson et al., 2015). For example, exposure to molecular oxygen (Imlay, 2006), hydrogen peroxide (Lambertz et al., 2011), or superoxide (Lambertz et al., 2011) directly oxidizes a [4Fe–4S] cluster from the 2^+ to the 3^+ state, releasing one of the Fe atoms, and forming an unstable [3Fe–4S] cluster (Imlay, 2006). While the exact degradation pathway of these clusters is still subject to uncertainty, the damaged clusters are known to be easily subjected to further degradation (Imlay, 2006). The activity of certain [FeFe]-hydrogenases and nitrogenases has been shown to have a half-life of mere seconds when exposed to atmospheric oxygen (Stiebritz & Reiher, 2012). Thus, the purification, crystallization, and harvesting of crystals are best carried under the strictest of anaerobic conditions, often in conjunction with a reductant such as sodium dithionite or dithiothreitol in order to scrub out exogenous oxygen prior to the protein becoming exposed (Chen & Mortenson, 1974; Mulder et al., 2011; Peters et al., 2015). While these limitations are broadly true, certain [NiFe]-hydrogenases have been found to be oxygen tolerant (Brooke et al., 2016; Frielingsdorf et al., 2014; Liebgott et al., 2011; Volbeda et al., 2012), and certain [FeFe]-hydrogenases may simply become oxidized to an "inactive" state that can be rescued through reduction (Morra, Arizzi, Valetti, & Gilardi, 2016; van Dijk, van Berkel-Arts, & Veeger, 1983). In general, the FeS cluster-containing enzymes should be treated under as strictly anaerobic conditions as possible unless they have been demonstrated to be oxygen tolerant, or when the crystallization experiment is specifically targeting oxygen exposure.

After a protein has been purified, it is often assayed for iron content (Adams, 1987). If this number is lower than the amount predicted for the protein, a reconstitution is typically performed, wherein ammonium iron sulfate and sodium sulfide (Na_2S) are slowly titrated into a solution containing the protein, allowing the FeS clusters to form properly. The reconstitution may improve stability and help ensure homogenous protein samples.

Hydrogenases and nitrogenases are both redox-active proteins with complicated catalytic cycles, and crystallography is often used as a tool to examine the structural composition of the different catalytic intermediates. Poising the crystals under varying conditions is therefore critical. Often, different oxidants, reductants, or redox-mediating dyes are used to help stabilize the enzyme in the appropriate state. The addition of other molecules, such as hydrogen, oxygen, or CO, may also be of use. Noble gas derivatization with xenon or krypton is another tool that can be used to reveal gas channels in the enzyme.

Crystallographic experiments also benefit from the use of associated spectroscopies and activity assays. This can be crucial in order to verify protein activity and homogeneity prior to setting up crystal screens, and again once the crystallization has occurred. Some crystals may be specifically harvested for spectroscopic or activity assays. Without this type of secondary experiment it is often impossible in practice to verify that the crystallization experiment has trapped a desired state, and the resulting X-ray data are subject to ambiguous interpretations. In this chapter, we will review methods of structural characterization of the hydrogenases and nitrogenases performed by different research groups.

2. CRYSTALLIZATION OF HYDROGENASES

2.1 Introduction to Hydrogenases

Hydrogenases, which reversibly catalyze the oxidation of hydrogen (H_2) to protons and electrons (Chen & Blanchard, 1978; Peters et al., 2015; Vignais & Billoud, 2007), are broadly classified as either [FeFe]-, [NiFe]-, or [Fe]-only on the basis of the metal composition in the active site (Lubitz et al., 2014; Vignais, Billoud, & Meyer, 2001). The [Fe]-only hydrogenases do not contain FeS clusters and, as such, are not the focus of this chapter (Shima & Thauer, 2007). The other two types of hydrogenases, [FeFe] and [NiFe] (Fig. 2), are evolutionarily unrelated, although similar features of their structures and functions represent a profound case of convergent evolution (Peters et al., 2015). The [FeFe]-hydrogenases are distributed in the bacteria and lower eukarya, while the [NiFe]-hydrogenases are found in bacteria and archaea (Peters et al., 2015). These hydrogenases provide a wide breadth of functions in vivo, including disposal of excess reducing equivalents (Schut & Adams, 2009; Tamagnini et al., 2007;

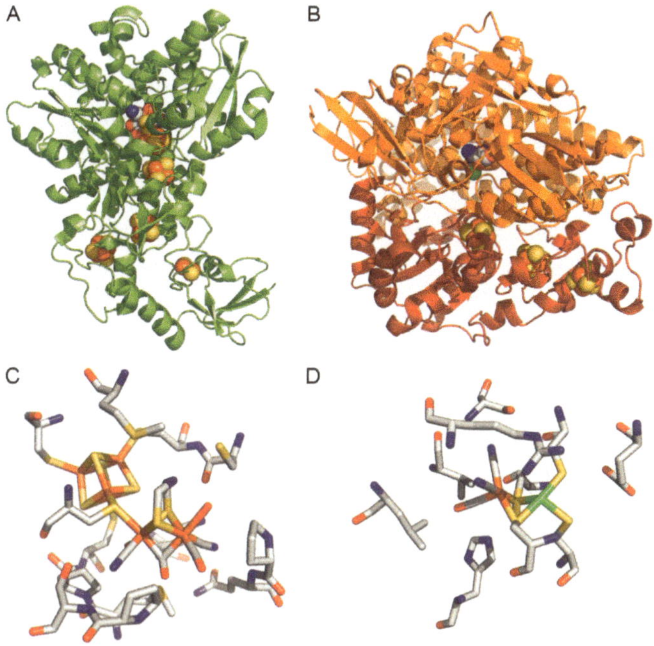

Fig. 2 [FeFe]- and [NiFe]-hydrogenase structures and active sites. (A) The structure of the monomeric [FeFe]-hydrogenase CpI (PDB ID 3C8Y) (Pandey, Harris, Giles, Peters, & Szilagyi, 2008). (B) The structure of [NiFe]-hydrogenase from *D. vulgaris* (PDB ID 4U9H) (Ogata, Nishikawa, & Lubitz, 2015). (C and D) The coordination environments around the active sites of [FeFe]- and [NiFe]-hydrogenases, respectively. Color scheme: C, *grey*; N, *blue*; O, *red*; S, *yellow*; Fe, *orange*; Ni, *green*.

Verhagen, O'Rourke, & Adams, 1999), hydrogen sensing (Brecht, van Gastel, Buhrke, Friedrich, & Lubitz, 2003; Lenz, Bernhard, Buhrke, Schwartz, & Friedrich, 2002), ion translocation (Hedderich, 2004), and oxidizing H_2 (Dementin et al., 2004; Kanai et al., 2011) to provide reducing equivalents for other substrates. Accordingly, they have been found to play key metabolic roles in methanogens (Thauer et al., 2010), sulfate reducers (Baltazar et al., 2011), nitrogen fixers (Zhang, Sherman, & Sherman, 2014), and photosynthesizers (Ghirardi et al., 2006; Melis, Seibert, & Happe, 2004; Tamagnini et al., 2007).

2.2 Methods of Structural Characterization of [FeFe]-Hydrogenases

Crystallization of the [FeFe]-hydrogenases has proved to be critical in the development of a functional understanding. Prior to crystallization, it was known that the hydrogenases contain an array of FeS clusters (Adams, 1990; Adams, Eccleston, & Howard, 1989), but one of them, predicted to be the active site, had been shown to have unusual spectroscopic features in comparison to standard FeS clusters. The 1.8Å crystal structure of [FeFe]-hydrogenase from *Clostridium pasteurianum*, CpI, ushered in a new era of hydrogenase structural and functional characterization (Peters, Lanzilotta, Lemon, & Seefeldt, 1998). This structure revealed that the active site, termed the H-cluster, contains a standard [4Fe–4S] cluster cubane ($[4Fe–4S]_H$) with a cysteine linkage to the unusual 2Fe subcluster ($[2Fe]_H$). This subcluster has two Fe atoms, termed the proximal and distal Fe atoms based on their proximity to the $[4Fe–4S]_H$ cubane. These Fe atoms are further coordinated by carbon monoxide ligands, cyanide ligands, and a dithiomethyl bridging ligand, though the bridgehead atom could not be conclusively identified. It was therefore hypothesized to be an S, N, or C atom (Nicolet, Piras, Legrand, Hatchikian, & Fontecilla-Camps, 1999; Pandey et al., 2008; Peters et al., 1998), though the latter two cases were more compelling, as they could leave an open coordination site that may help facilitate the catalytic cycle. This initial crystallization success was rapidly followed by a CO inhibition study, wherein concentrated CpI protein (52mg/mL) was exposed to CO in a sealed serum vial in the presence of 2m*M* sodium dithionite, and incubated for 5min. Activity assays were used to monitor the inhibition of hydrogenase activity by CO. The microcapillary batch diffusion method was then used as in the previous study, although, interestingly, a different crystal form was observed. Solving the structure through molecular replacement demonstrated CO bound in the active site (Lemon & Peters, 1999).

The CpI structure was soon followed up by the 1.6Å structure from the periplasmic, heterodimeric *Desulfovibrio desulfuricans* [FeFe]-hydrogenase (DdHase) (Nicolet et al., 1999). Crystallization was observed with either PEG 6000, 100m*M* Na acetate, pH 5.0, or ammonium sulfate in 50m*M* Tris pH 8. To set up the hanging drops, the protein solution was initially stirred under an atmosphere of 90% nitrogen, 10% hydrogen, and allowed to crystallize in a glovebox. Crystals were harvested by flash cooling anaerobically in liquid propane inside the glovebox. Interestingly, though the DdHase protein may be purified in an inactive form on the benchtop, it was found to be unstable in the crystallization drops under aerobic conditions. The crystal structure was solved by a MAD experiment, with four data sets collected at or near the Fe absorption edge. The structure revealed the H-cluster active site to be identical to that of CpI, suggesting that the

H-cluster is a conserved feature of [FeFe]-hydrogenases. While the H-cluster organometallic cluster has been found to be strictly conserved in all known [FeFe]-hydrogenases (Poudel et al., 2016), the protein framework around the H-cluster has been shown to profoundly influence catalytic function (Knörzer et al., 2012), and it is probable that natural variation in the secondary coordination sphere plays a role in fine-tuning the hydrogenase for its specific evolutionary niche (Boyd, Hamilton, et al., 2014; Boyd, Schut, Adams, & Peters, 2014; Peters et al., 2015; Poudel et al., 2016).

2.2.1 Structural Characterization of "Apo" Hydrogenases

Another major breakthrough in the crystallization of [FeFe]-hydrogenases came from our crystallization of the hydrogenase from *C. reinhardtii*, CrHydA (Mulder et al., 2010). This was done without the use of the maturases, HydE, HydF, and HydG, that had been shown to be critical in forming the fully active hydrogenase (Broderick et al., 2014; Mulder et al., 2009; Shepard et al., 2014). Previous crystal structures had relied upon native protein expression with the maturases present. Thus, in order to elucidate the effect of the maturases on the hydrogenase structure, CrHydA was purified heterologously in *Escherichia coli*, which lacked the necessary maturases. The microcapillary batch diffusion method was used for crystal screens inside a glovebox under nitrogen atmosphere, with favorable crystallization conditions being 25.5% PEG 8000, 0.085 *M* sodium cacodylate, pH 6.5, 0.17 *M* sodium acetate trihydrate, and 1 m*M* dithionite as a reductant. The structure was solved by molecular replacement with CpI. The structure demonstrates a high degree of homology to CpI, but, without the maturase genes, the $[2Fe]_H$ is absent, resulting in an "apo" form CrHydA being in an "open" conformation (Fig. 3). Thus, there is a solvent access channel from the solution to the active site, which contains the $[4Fe–4S]_H$ cubane portion of the H-cluster. This demonstrates that the formation of active hydrogenase proceeds in a stepwise fashion, with the general FeS cluster machinery first working to form the standard [4Fe–4S] clusters, and subsequently allowing the maturases to insert the $[2Fe]_H$ into the active site cavity completing the assembly of the H-cluster (Mulder et al., 2010). The open confirmation of CrHydA, compared to the closed conformation of CpI, suggests a general mechanism in [FeFe]-hydrogenases to generate the fully active protein. Indeed, conserved loop regions have been identified that have been hypothesized to gate the conformation, and specific conserved glycine residues are thought to act as hinges (Esselborn et al., 2016). The closed conformation of the

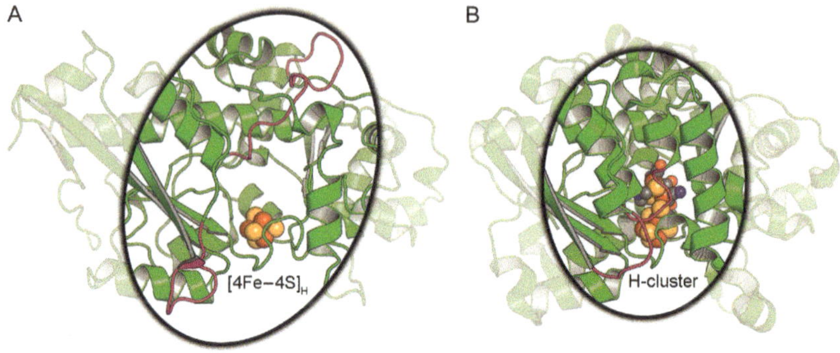

Fig. 3 Comparison of "apo" CrHydA1 [FeFe]-hydrogenase to the catalytic domain of holo CpI [FeFe]-hydrogenase. (A) The $[4Fe–4S]_H$ of "apo" CrHydA1 (PDB ID 3LX4) is solvent accessible due to the open conformation created by two conserved loop regions (*magenta*) which provides a pathway for $[2Fe]_H$ insertion. (B) The two homologous loop regions (*magenta*) in the catalytic domain of holo CpI (from PDB ID 3C8Y) exist in a closed conformation protecting the fully intact, catalytic H-cluster after completion of the $[2Fe]_H$ cluster insertion process of maturation.

protein is likely critical in order to block the active site from the bulk solvent, preventing degradation of the H-cluster.

2.2.2 Characterization of Oxygen Susceptibility in [FeFe]-Hydrogenases

Another crystal study performed on the CrHydA enzyme involved an examination of O_2 susceptibility in conjunction with spectroscopic methods (Swanson et al., 2015). Given that O_2 is known to inactivate [FeFe]-hydrogenases, trapping an O_2-inactivated form of the enzyme is clearly an attractive target. To approach this problem, we first heterologously expressed and purified the fully maturated and active enzyme. During the purification 5 m*M* dithionite was used as a reductant in order to prepare high-quality enzyme, though the dithionite was removed through serial dilution-concentration steps, resulting in the oxidized form of the enzyme as verified by Fourier transform infrared spectroscopy (FTIR). This protein was exposed to O_2 at a concentration of 0.01% for 2 h in a 4.5 mL Wheaton vial, and loss of activity was measured by gas chromatography using a dye-based assay. The O_2-damaged enzyme was then concentrated to 30 mg mL^{-1} and crystallized in a glovebox using the microcapillary batch diffusion method, under the conditions of 22.5% PEG 8000, 0.085 *M* sodium cacodylate pH 6.5, 0.17 *M* sodium acetate trihydrate, and 1 m*M* dithionite. The structure was able to be solved using molecular replacement with the previous CrHydA structure.

The O_2 damaged CrHydA enzyme closely resembled the undamaged protein in the "apo" form, with an intact $[4Fe–4S]_H$ cubane of the H-cluster, and the absence of the $2[Fe]_H$, suggesting that the oxidative damage to the H-cluster initially occurs at the $2[Fe]_H$. As in the earlier study, the enzyme was in the "open" conformation, suggesting release of the $[2Fe]_H$. In addition, the electron density maps revealed extra electron density around three cysteine residues, one of which, Cys169, functions to transfer protons to the H-cluster. Cys169 was modeled as a sulfenic acid, which would form under oxidative conditions. Mass spectrometry was used to confirm the assignment of this amino acid as a sulfenic acid. Subsequent studies corroborated that the activity of the CrHydA could be restored through the addition of the $[2Fe]_H$ from the maturase HydF, showing conclusively that the $[2Fe]_H$ is the site of O_2 damage under these conditions. While other studies have suggested that O_2 damage may first occur at the [4Fe–4S] cluster of the H-cluster (Stripp et al., 2009), this study illustrates how crystallography can be used as a powerful tool to trap differing enzymatic states.

2.2.3 X-Ray Damage-Free Structure of CpI Hydrogenase

Recently, we have solved a structure of CpI to 1.6 Å in a radiation damage-free form (Neutze, Wouts, van der Spoel, Weckert, & Hajdu, 2000) using an X-ray free electron laser (XFEL) at the Linac Coherent Light Source (Cohen et al., 2014). With this technique extremely bright femtosecond X-ray pulses are used to generate a diffraction pattern from the crystal faster than electronic or atomic rearrangements are possible (Barty et al., 2012). XFEL therefore preserves the structural information from the protein without causing the degradation of sample that is commonly observed for metalloenzymes (Alonso-Mori et al., 2012), and could, in principle, lead to experiments that are capable of analyzing the structure of [FeFe]-hydrogenases over a short time domain. In this case minor differences were observed in the structure of the XFEL-solved structure in comparison to the standard synchrotron-based structure, and these differences were attributed to the protein not having any X-ray damage in the case of the experiment (Cohen et al., 2014). Despite these initial success, XFEL still requires large, high-quality crystals for goniometer-based experiments, or, alternatively, a large supply of nanocrystals flowed through a liquid injector. This area of research is likely to reap great rewards in the metalloenzyme field as the technology continues to mature.

2.2.4 Synthetic Reconstitution of the [FeFe]-Hydrogenase Active Site

A major innovation in the field of [FeFe]-hydrogenase research has been the ability to synthetically build the $[2Fe]_H$ or close analogs (Berggren et al., 2013), and insert these into the "apo" form of hydrogenase, which lacks only the $[2Fe]_H$ but otherwise contains a full complement of FeS clusters (Fig. 4). These synthetic $[2Fe]_H$ compounds have been generated in a variety of forms, differing at the bridgehead position of the dithiomethyl bridge, and may be synthesized with an O, S, N, or C. Incorporation of these analogs, in conjunction with activity assays and various spectroscopies, has finally demonstrated that the native enzyme contains a bridging amine (Adamska-Venkatesh et al., 2015). Other experiments have made use of the synthetic clusters (Artz et al., 2017; Birrell et al., 2016; Caserta et al., 2016), as forming a catalytically inactive but otherwise normal enzyme offers many opportunities.

Thus, CpI has recently been crystallized both in the apo form of the enzyme, and in each of the states with the differing bridgehead atoms (Esselborn et al., 2016). The exact crystallization conditions varied from one crystal form to the next, but were overall similar to those reported for the native-like enzyme (CpI^{PDT}) with the bridging amine, with the hanging drop vapor diffusion method being used to grow crystals in 11% PEG 4000, 0.1 *M* MES pH 7.0, 0.2 *M* $MgCl_2$. Crystals were obtained under anaerobic conditions, protected from ice crystals with paraffin oil, and flash-frozen in liquid nitrogen. Structures were solved using molecular replacement to CpI. An interesting finding from this study is that the "apo" CpI crystallized in a conformation essentially identical to that of the CpI^{PDT},

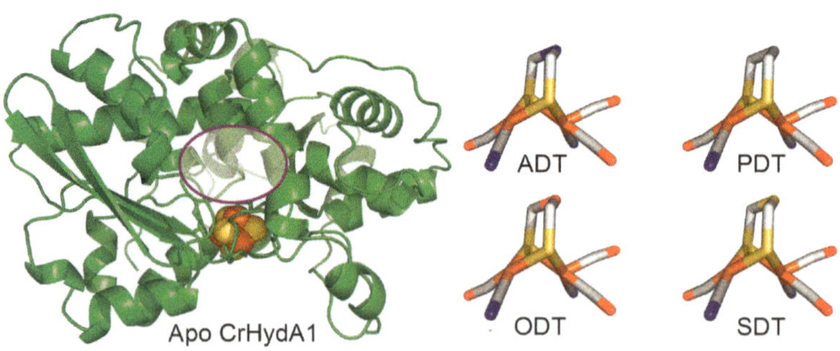

Fig. 4 The catalytic domain of CpI is shown in *green* in the "apo" form, that is, lacking the $[2Fe]_H$, with the $[2Fe]_H$ cavity highlighted by a *purple circle*. Various possible $[2Fe]_H$ configurations are displayed on the *right*. Color scheme: C, *grey*; N, *blue*; O, *red*; S, *yellow*. ADT, azide dithiolate (the wild type); ODT, oxygen dithiolate; PDT, propane dithiolate; SDT, sulfur dithiolate.

with the enzyme in the "closed" state, differentiating it from CrHydA in the apo form in an open state, suggesting an equilibrium between the two states in which a given state is more favorable for crystallization depending on the enzyme. The main outcome from this study was that the nonnative $[2Fe]_H$ forms had crystal structures virtually identical to that of the native-type CpI^{PDT}. Thus, differences in activity can be explained on the basis of ineffective active site chemistry, rather than any overarching structural deficiencies in the nonnative forms.

The [FeFe]-hydrogenase crystallography field as a whole is beginning to piece together a series of mechanistic insights. Further studies on more phylogenetically distinct enzymes, different redox states, and with creative use of nonnative cofactors are poised to reveal even more about the hydrogenase mechanism (Table 1).

2.3 Methods of Structural Characterization of [NiFe]-Hydrogenases

2.3.1 Background on [NiFe]-Hydrogenase Crystallization and Catalytic Cycle

The [NiFe]-hydrogenases have thus far received much more attention than the [FeFe]-hydrogenases in terms of structural characterization, with >50 solved structures to date (For a full list, please see Lubitz et al., 2014). As such, the objective here is not to exhaustively review each of these structures, but to describe, in general, the types of methods that are used, and some of the emerging advancements in the field. [NiFe]-hydrogenases represent a wide variety of structural and functional variation (Vignais & Billoud, 2007); however, the minimal subunit required for catalysis is closely conserved and consists of a binuclear metal center which has a Ni atom proximal to a [4Fe4S] cubane, and an Fe atom distal. The Fe atom is further coordinated by carbon monoxide and cyanide ligands that, like the [FeFe]-hydrogenases, poises the Fe in a low-spin state that facilitates heterolytic cleavage of H_2 (Nicolet et al., 1999).

The [NiFe]-hydrogenases have had their catalytic cycle worked out through many different techniques (EPR, FTIR, Raman, Mossbauer, protein film electrochemistry, DFT), including a variety of solved structures, corresponding to many different states in the catalytic cycle (Fig. 5) (Abou-Hamdan et al., 2015; Flanagan & Parkin, 2016; Horch et al., 2014; Jones et al., 2003; Lubitz et al., 2014; Pandelia et al., 2011; Pardo et al., 2006). Two inactive, oxidized states termed Ni-A and Ni-B, correspond to "unready" and "ready" forms of the enzyme, respectively, and

Table 1 Solved [FeFe]-Hydrogenase Structures

Protein	Crystallization Method	Phasing	Experiment	Resolution (Å)	PDB #	References
Clostridium pasteurianum [FeFe]-hydrogenase 1 (CpI)	Microcapillary batch diffusion	MAD[a]	First solved [FeFe]-hydrogenase	1.8	1FEH	Peters et al. (1998)
CpI	Microcapillary batch diffusion	MR[b]	Isolate CO-inhibited state	2.4	1C4C	Lemon and Peters (1999)
Desulfovibrio desulfuricans [FeFe]-hydrogenase (DDHase)	Hanging drop vapor diffusion	MAD		1.6	1HFE	Nicolet et al. (1999)
Chlamydomonas reinhardtii HydA1 (CrHydA)	Microcapillary batch diffusion	MR	Structure of apo protein	1.97	3LX4	Mulder et al. (2010)
CrHydA	Microcapillary batch diffusion	MR	Structure of O_2-damaged protein	2.29	4ROV	Swanson et al. (2015)
CpI	Microcapillary batch diffusion	MR	X-ray damage-free structure	1.6	NA	Cohen et al. (2014)
CpI	Hanging drop vapor diffusion	MR	Apo form, incorporation of variety of synthetic cofactors into active site	1.6–1.93	4XDC, 4XDD, 5BYQ, 5BYR, 5BYS	Esselborn et al. (2016)

[a]MAD, multiple anomalous diffraction.
[b]MR, molecular replacement.

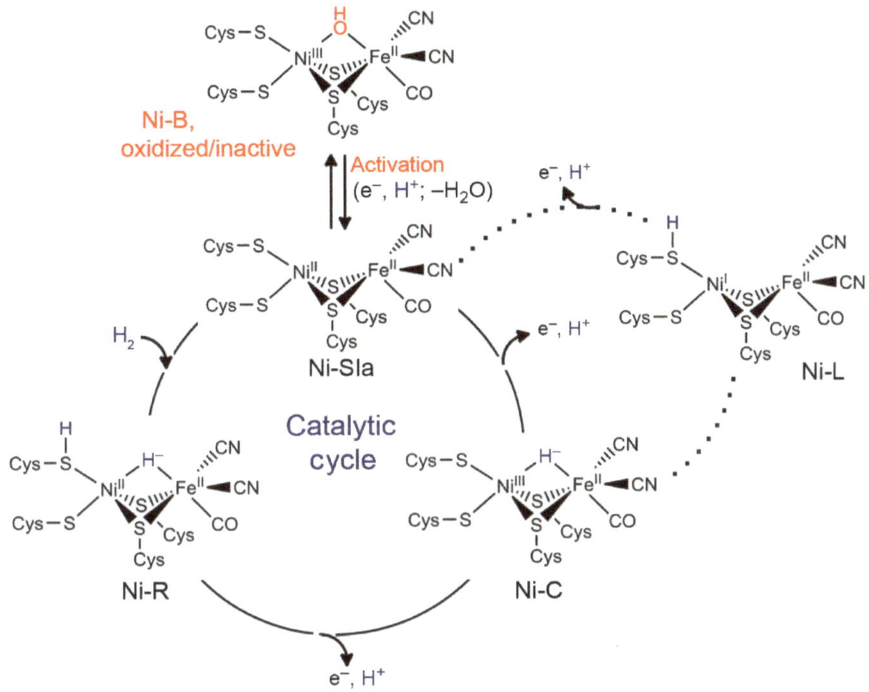

Fig. 5 Basic catalytic cycle for H_2 activation by [NiFe]-hydrogenase. Heterolytic bond cleavage of H_2 is accompanied by various oxidation, protonation, and hydride bound states of the [NiFe] active site during catalysis.

contain an additional bridging ligand between Ni and Fe sites such as OOH− or OH−. The Ni-B is rapidly reductively reactivated to the Ni-SIa state. Further reduction with H_2 brings the hydrogenase to the Ni-R state. Breaking the H_2 bond and releasing both a proton and an electron brings the hydrogenase to the Ni-C state. Further loss of a proton and an electron return the hydrogenase to the Ni-SIa state. Another enzyme state, the Ni-L state, has been proposed as an intermediate between the Ni-C and Ni-SIa states based on FTIR spectroscopy (Greene, Wu, McTernan, Adams, & Dyer, 2015; Hidalgo, Ash, Healy, & Vincent, 2015; Murphy et al., 2015). While neither the Ni-C nor the Ni-SIa state has a solved structure, a CO-inhibited structure of the Ni-SIa state has been solved, helping to fill in the catalytic cycle. Additional catalytic states of the enzyme may be reached through light illumination or inhibition with other small molecules that act as competitors with H_2 (O_2, NO, CN^-) (Ogata, Lubitz, & Higuchi, 2016). Though a full discussion of these states is beyond the scope of this review, many of these states represent attractive targets for future crystallographic investigations. This structural information is essential for establishing the ligand coordination and bonding interactions, for example, that complement biophysical and electrochemical-based approaches.

The first [NiFe]-hydrogenase crystallization was reported in 1987 (Higuchi et al., 1987). Protein of the *Desulfovibrio vulgaris* Miyazaki F hydrogenase was treated anaerobically and precipitated in PEG via the hanging drop vapor diffusion method. It was reported that the crystals were very small, and thus these were used to seed further crystal growth. Despite promising diffraction patterns, it was nearly a decade before the structure of the [NiFe]-hydrogenase was solved through multiple isomorphous replacement (MIR), that of the *Desulfovibrio gigas* enzyme (Volbeda et al., 1995). This set the stage for future crystallization experiments.

In 2005 the structures of the oxidized, inactivated Ni-A and Ni-B states were reported (Ogata et al., 2005). Enzyme was prepared in the Ni-B form by treating the purified form with methyl viologen and hydrogen, then oxidizing the protein with exposure to air. To trap the Ni-A state, the protein was treated with 500 m*M* sodium sulfide dissolved in 25 m*M* Tris, pH 7.4. This solution was added to the enzyme at a final concentration of 50 m*M*. Protein for both the Ni-A and Ni-B states were further purified by anion exchange chromatography in order to remove any denatured sample. The crystals were later analyzed by EPR to verify the appropriate assignment of the enzyme state, illustrating that the structure alone is often insufficient data to draw firm mechanistic conclusions. A contemporary approach to examining these states has relied upon catalytically inactivated mutants, which has allowed for a reexamination of the Ni-A state as containing a peroxide ligand, when it may in fact possess an oxidized cysteine (Volbeda et al., 2015).

The reduced, active, Ni-R state has been crystallized on a variety of occasions (Higuchi, Yagi, & Yasuoka, 1997; Ogata et al., 2005; Volbeda et al., 2005), most notably the *D. vulgaris* Miyazaki F structure has been solved to the subatomic resolution of 0.89 Å, rendering many of the protons visible for the first time (Ogata et al., 2015). This high-quality data unambiguously allowed for the assignment of a protonated cysteine residue near the active site, as well as a bridging hydride between the Ni and Fe atoms of the active site, indicative of heterolytic cleavage of the hydrogen. Data of this resolution were able to be collected by the purification and isolation of homogeneous, high-quality enzyme, which was monitored by FTIR spectroscopy to ensure the presence of only one catalytic intermediate. All samples were treated under strictly anoxic conditions. "Rather large" crystals were formed with sitting drop vapor diffusion, in buffer containing 33% 2-methyl-2,4-pentanediolin 25 m*M* Tris–HCl pH 7.4, with 0.27 m*M* protein. Eight diffraction sets from one crystal were merged.

2.3.2 Crystallization of [NiFe]-Hydrogenases With CO-Inhibition

The CO-inhibited Ni-SCO state has been structurally characterized in the *D. vulgaris* Miyazaki F enxyme (Ogata et al., 2002). Crystals were grown under aerobic conditions via sitting-drop vapor diffusion, and then reduced with methyl viologen in the presence of hydrogen, reducing the protein. Crystals were treated with pure CO in a glass capillary tube. These crystals were solved using molecular replacement and demonstrated CO binding to the Ni of the active site. As CO is a competitive inhibitor with H_2, this strongly suggests that the activation site of H_2 is also the Ni atom. The presence of CO in the crystals was verified by Resonance Raman spectroscopy.

2.3.3 Redox Poising of [NiFeSe]-Hydrogenase Crystals

Some [NiFe]-hydrogenases have swapped one of the Ni-coordinating cysteine for a selenocysteine. These [NiFeSe]-hydrogenases generally have high activity and rapid reactivation after O_2 exposure, which makes them an attractive target for technological innovation (Baltazar et al., 2011). To date they have been crystallized in several states, including the oxidized, reduced, and active states (Garcin et al., 1999; Marques, Coelho, Pereira, & Matias, 2013; Volbeda et al., 2013). The experimental approach of Marques et al. was to grow crystals aerobically at room temperature to generate an oxidized, as-isolated form of the crystals. After cryoprotecting with 20% glycerol, 20.5% PEG 1500, and 0.1 mm Tris–HCl pH 7.6, crystals were reduced in a new drop of cryoprotecting solution, into which 1 μL of a 10-m*M* methyl viologen and 50 m*M* Tris–HCl pH 7.6 buffer was added. The drops were then incubated under a 15% H_2/85% N_2 atmosphere until the crystals and solution turned blue, indicating reduction. Some of these crystals were further exposed to aerobic conditions for 24 or 48 h. This experiment was able to demonstrate certain reversible and irreversible processes of oxidative inactivation. Specifically, the [4Fe–4S] accessory cluster proximal to the active site was seen to be reversibly superoxidized in the presence of O_2, while in the active site it was discovered that the terminal Ni-bonded cysteine irreversibly oxidizes, likely partially explaining why the hydrogenase in susceptible to O_2 (Marques et al., 2013). This series of experiments provides a convenient demonstration of how to sequentially trap multiple enzyme states in order to correlate the structure with the function of the enzyme.

2.3.4 Cryo-Electron Microscopy as a Structural Characterization Technique of Hydrogenases

Cryo-electron microscopy (Cryo-EM) has recently emerged as an experimental compliment or replacement to traditional X-ray-based structural characterization methods, as it has the advantages of intrinsically overcoming the phase problem, and of performing well with large protein complexes that are often extremely difficult to characterize using traditional methods. Furthermore, protein solution may be applied directly to a grid without any need for an intermediate crystallization step (Milne et al., 2013; Orlova & Saibil, 2011; Shi et al., 2016). This technique is even more powerful when combined with traditional X-ray crystallographic methods. Recently, Cryo-EM has been employed on the 1.2 MDa F_{420}-reducing hydrogenase (Frh), an archaeal cytosolic group 3 [NiFe]-hydrogenase (Allegretti, Mills, McMullan, Kühlbrandt, & Vonck, 2014). The Cryo-EM data were used to solve the structure to a resolution of 3.36 Å, which is a high enough resolution to show electron density for secondary structure, the FAD cofactor, and many of the side chains. Protein purification was carried out anaerobically with 2 m*M* DTT and 25 μ*M* FAD over a size exclusion column. Active fractions of column eluate were concentrated and applied to a grid, and plunge-frozen under oxygen-free conditions into liquid ethane (Mills, Vitt, Strauss, Shima, & Vonck, 2013). This Cryo-EM structure allowed for the model building of the 1.7 Å crystal structure of the Frh complex (Vitt et al., 2014), which was able to reveal H_2 channels and possible proton relay pathways to a much higher degree than the Cryo-EM structure.

2.3.5 Structural Characterization of a Group 4 [NiFe]-Hydrogenases

We have recently characterized the first group 4 [NiFe]-hydrogenase, a membrane-bound, ion-translocating, ferredoxin-oxidizing, proton-reducing hydrogenase (Schut et al., 2016). This enzyme is large and membrane-associated, making it a challenge to crystallize. In order to overcome this difficulty, several of the membrane-associated subunits were not purified, allowing the cytosolic portion of the enzyme to remain stabilized. The crystal structure was solved by molecular replacement with analogous subunits of Complex I, to which the hydrogenase is closely evolutionarily related. It was found that in comparison to the [NiFe]-hydrogenase active site, the Complex I active site has undergone a remodeling process in order to accommodate a larger substrate. However, the core electron transfer pathway in the [NiFe]-hydrogenase is closely matched in Complex I.

2.3.6 Gas Derivatization as a Tool to Probe Gas Channels in Hydrogenases Structures

Gas diffusion through hydrogenases has been proposed as one of the factors regulating their oxygen sensitivity. In order to address this question more quantitatively, Kalms and coworkers have krypton-derivatized gas channels in an oxygen-sensitive and an oxygen-tolerant [NiFe]-hydrogenase, and compared the result (Kalms et al., 2016). Following crystallization by the sitting-drop vapor diffusion method, crystals were harvested into pluggable sample supports and transferred into a high-pressure sample cell, and then exposed to 80 bar of krypton gas for 10 min. Those crystals were directly flash-frozen. Several native and single anomalous data sets were collected, with the resulting krypton-derivatized structures strongly resembling the WT enzyme in the Ni-B state. The data reveal a complete overview of gas tunnels in both types of hydrogenase and show that tunnel size and quantity are both higher in the O_2-sensitive enzyme, suggesting that gas tunnel size is an important feature in determining O_2 tolerance. These findings are largely in agreement with a previous xenon derivation study that examined only an O_2-sensitive hydrogenase (Montet et al., 1997) (Table 2).

2.4 Conclusions and General Outlook on Hydrogenase Structural Characterization

Overall, the experiments discussed here capture the current state of the hydrogenase crystallographic field and illustrate many of the key methods that are typically employed, including treatment under varying redox conditions and combining data with many other spectroscopic methods. A critical element, especially with the [FeFe]-hydrogenases, is a strict anaerobic treatment of highly pure protein. Many exciting opportunities for crystallization remain in the field, including trapping more catalytic intermediate states, such as the Ni-C and Ni-SIa of [NiFe]-hydrogenases, and probing a variety of phylogenetically distinct enzymes. Emerging techniques such as XFEL, Cryo-EM, and perhaps even neutron diffraction will aid in future experiments and data collection.

3. CRYSTALLIZATION OF NITROGENASES

3.1 Introduction to Nitrogenases

Nitrogen fixation in the biological systems is performed by the nitrogenase enzyme, which catalyzes the reduction of atmospheric nitrogen gas to

Table 2 Selection of Solved [NiFe]-Hydrogenase Structures

Protein	Crystallization Method	Phasing	Experiment	Resolution (Å)	PDB #	References
Desulfovibrio gigas [NiFe]-hydrogenase	Hanging drop vapor diffusion	MIR[a]	First solved [NiFe]-hydrogenase	2.85	1FRV	Volbeda et al. (1995)
Desulfovibrio vulgaris [NiFe]-hydrogenase	Hanging drop vapor diffusion	MR[b]	Trap Ni-A and Ni-B states with methyl viologen, hydrogen, and oxygen exposure	1.04–1.4	1WUI, 1WUJ, 1WUK	Ogata et al. (2005)
D. vulgaris Miyazaki F [NiFe]-hydrogenase	Hanging drop vapor diffusion	MR	Ni-R state, subatomic resolution	0.89	4U9H	Ogata et al. (2015)
D. vulgaris Hildenborough [NiFeSe]-hydrogenase	Sitting drop vapor diffusion	MR	Trap multiple catalytic sites sequentially	1.33–1.95	3ZEA, 3ZE6, 3ZE7, 3ZE8, 3ZE9	Marques et al. (2013)
F_{420}-reducing hydrogenase (group 3)	Cryo-EM, MR	Cryo-EM, MR	Couple Cryo-EM of large H2ase with crystallography	3.3, 1.7	4CI0, 4OMF	Allegretti et al. (2014), Mills et al. (2013) and Vitt et al. (2014)
Membrane-bound hydrogenase (group 4)	Hanging drop vapor diffusion	MR	First group 4 hydrogenase			Schut et al. (2016)
Ralstonia eutropha membrane-bound hydrogenase	Sitting drop vapor diffusion	SAD[c]	Krypton derivatization	1.47	5D51	Kalms et al. (2016)

[a]Multiple isomorphous replacement.
[b]Molecular replacement.
[c]Single anomalous diffraction.

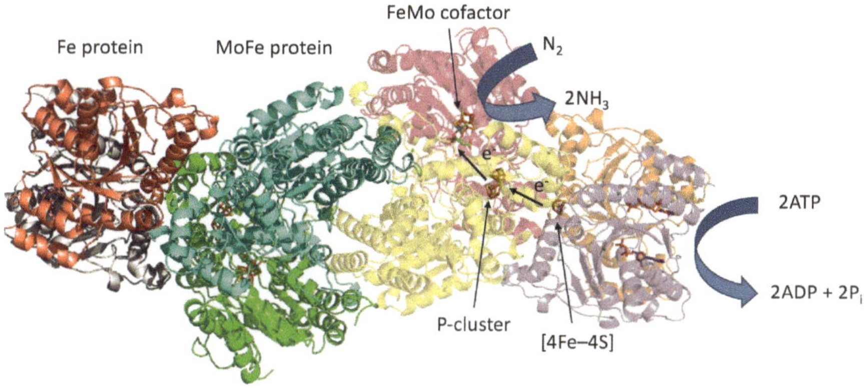

Fig. 6 The structure of the nitrogenase complex. The overall organization of the nitrogenase complex consists of two Fe proteins and two $\alpha_2\beta_2$ MoFe proteins. The transparent representation of one half of the complex shows the MgATP-binding sites, and the electron transfer pathway from the [4Fe–4S] cluster of the Fe protein to the active site FeMo cofactor via the P-cluster intermediate. *Figure generated from PDB file 1N2C: Schindelin, H., Kisker, C., Schlessman, J. L., Howard, J. B., & Rees, D. C. (1997). Structure of ADP x AlF4(-)-stabilized nitrogenase complex and its implications for signal transduction.* Nature, *387(6631), 370–376. http://dx.doi.org/10.1038/387370a0.*

ammonia (Fig. 6), and is found in certain bacteria and archaea (Burgess & Lowe, 1996; Eady, 1996; Peters, Fisher, & Dean, 1995).

This energy demanding process involves the hydrolysis of two MgATP molecules per electron transferred in the reaction and eight reducing equivalents, and can be represented by the simplified equation (Eq. 1) (Howard & Rees, 1996; Seefeldt, Hoffman, & Dean, 2009):

$$N_2 + 8H^+ + 16MgATP + 8e^- \rightarrow 2NH_3 + H_2 + 16MgADP + 16P_i \quad (1)$$

Industrial ammonia production used to make fertilizers is achieved by the Haber–Bosh process, which uses iron catalysts, pressures of around 250 atm, and temperatures of more than 400°C to reduce dinitrogen. In contrast, the nitrogenase as a biological catalyst performs this process at ambient temperature and atmospheric pressure (Erisman & Sutton, 2008; Kitano et al., 2012). Nitrogen-fixing organisms usually have a nitrogenase system containing Mo and Fe at the active site. In addition, some organisms have alternative nitrogenase systems containing V and Fe, or "Fe only" proteins (Eady, 1996; Hales, 1990; Howard & Rees, 1996).

The structural characterization of the nitrogenase complex has been an extensive research area for several decades (Benton & Peters, 2004; Kim & Rees, 1994; Lawson & Smith, 2002; Peters et al., 1995; Spatzal, Perez,

Einsle, Howard, & Rees, 2014). The molybdenum-dependent nitrogenase represents the most extensively studied nitrogenase system. This system consists of two complex metalloprotein components designated as the iron protein (Fe protein) and the molybdenum-iron protein (MoFe protein). Fe protein is a homodimer (~60 kDa) bridged by a single [4Fe–4S] cluster and has two sites for MgATP binding and hydrolysis, one in each subunit (Georgiadis et al., 1992). The MoFe protein is an $\alpha_2\beta_2$ heterotetramer (~250 kDa) that contains two [8Fe–7S] P-clusters and two active site [7Fe–9S–C–Mo–homocitrate] iron-molybdenum cofactors (FeMo cofactors). Conformational changes of Fe protein derived by multiple cycles of MgATP-dependent association and dissociation events are used to deliver electrons from the [4Fe–4S] cluster of the Fe protein to the site of substrate reduction at the FeMo cofactor through the P-cluster located in the MoFe protein (Fig. 6).

3.2 Methods of Structural Characterization of Mo-Nitrogenases

The first method for crystallization of the oxygen-sensitive MoFe protein was performed by Burns, Holsten, and Hardy (1970), then was modified by Shah and Brill (1973). Later, Weininger and Mortenson (1982) obtained first X-ray diffraction patterns of MoFe protein crystals. Robust nitrogenase crystals are difficult to obtain robust nitrogenase crystals are due to high flexibility and conformational changes within the MoFe and Fe proteins upon reduction of the substrate.

In 1992 Rees' group succeeded in growing the large crystals of MoFe and Fe proteins from *Azotobacter vinelandii* using the microcapillary batch diffusion method (Georgiadis et al., 1992; Kim & Rees, 1994). Later in 1997, after using ADP-tetrafluoroaluminate to stabilize the nitrogenase complex from *A. vinelandii* containing both MoFe and Fe protein components, the structure of the complete nitrogenase complex was successfully solved (Schindelin, Kisker, Schlessman, Howard, & Rees, 1997). In 2011 Einsle's and in 2015 Rees' groups have published structures of MoFe protein from *A. vinelandii* and *C. pasteurianum* at near atomic resolution (Spatzal et al., 2011; Zhang, Morrison, Kaiser, & Rees, 2015), respectively. In total, over the course of several decades there have been 25 MoFe proteins, 14 Fe proteins, and 9 nitrogenase complexes crystallographically determined and deposited into the PDB (Protein Data Bank, Tables 3–5).

Table 3 Selected List of Nitrogenase Fe Protein Structures

Protein	Crystallization Method	Method	Phasing[a]	Resolution (Å)	PDB Code	References
Nitrogenase Fe protein from *A. vinelandii*	Microcapillary batch	X-ray diffraction	MIR	2.9	1NIP	Georgiadis et al. (1992)
Nitrogenase Fe protein from *C. pasteurianum*	Microcapillary batch	X-ray diffraction	MR	1.93	1CP2	Schlessman, Woo, Joshua-Tor, Howard, and Rees (1998)
Nitrogenase Fe protein from *A. vinelandii*	Microcapillary batch	X-ray diffraction	MR	2.2	2NIP	Schlessman et al. (1998)
Nitrogenase Fe protein from *A. vinelandii* complexed with Mg ADP	Microcapillary batch	X-ray diffraction	MR	2.15	1FP6	Jang et al. (2000a) and Jang, Seefeldt, and Peters (2000b)
Modulating the midpoint potential of the [4Fe–4S] cluster of the nitrogenase Fe protein *A. vinelandii*	Microcapillary batch	X-ray diffraction	MR	2.4	1DE0	Jang et al. (2000b)
All-ferrous nitrogenase Fe protein from *A. vinelandii*	Microcapillary batch	X-ray diffraction	MR	2.25	1G1M	Strop et al. (2001)
Nitrogenase Fe protein from *A. vinelandii*	Microcapillary batch	X-ray diffraction	MR	2.2	1G5P	Strop et al. (2001)
Nitrogenase Fe protein l127 deletion variant *A. vinelandii*	Microcapillary batch	X-ray diffraction	MR	2.5	1RW4	Sen et al. (2004)
Crystal structure of the nitrogenase Fe protein Asp39Asn *A. vinelandii*	Microcapillary batch	X-ray diffraction	MR	2.7	1XD8	Jang, Jeong, Seefeldt, and Peters (2004)

Continued

Table 3 Selected List of Nitrogenase Fe Protein Structures—cont'd

Protein	Crystallization Method	Method	Phasing	Resolution (Å)	PDB Code	References
Crystal structure of the nitrogenase Fe protein Asp39Asn with MgADP bound *A. vinelandii*	Microcapillary batch	X-ray diffraction	MR	2.8	1XD9	Jang et al. (2004)
Crystal structure of the nitrogenase Fe protein Asp129Glu *A. vinelandii*	Microcapillary batch	X-ray diffraction	MR	2.8	1XDB	Jang et al. (2004)
Crystal structure of the nitrogenase Fe protein Phe135Trp with MgADP bound *A. vinelandii*	Microcapillary batch	X-ray diffraction	MR	3.2	1XCP	Jeong and Jang (2004)
Structural characterization of the nitrogenase Fe protein Leu127 deletion variant with bound MgATP *A. vinelandii*	Microcapillary batch	X-ray diffraction	MR	2.5	2C8V	Sen et al. (2006)
Fe protein from *A. vinelandii* independent substrate reduction by nitrogenase variants altered in intramolecular electron transfer	Microcapillary batch	X-ray diffraction	MR	1.97	4XPI	Danyal et al. (2015)

[a]MIR, multiple isomorphous replacement; MR, molecular replacement.

Table 4 Selected List of Nitrogenase MoFe Protein Structures

Protein	Crystallization Method	Method	Phasing[a]	Resolution (Å)	PDB Code	Reference
Nitrogenase MoFe protein from *A. vinelandii*	Microcapillary batch	X-ray diffraction	MIR	2.7		Kim and Rees (1992a)
Nitrogenase MoFe protein from *C. pasteurianum*	Microcapillary batch	X-ray diffraction	MIR	3.0	1MIO	Kim and Rees (1992a)
Nitrogenase MoFe protein from *A. vinelandii* (oxidized state)	Microcapillary batch	X-ray diffraction	MR	2.03	2MIN	Peters et al. (1997)
Nitrogenase MoFe protein from *A. vinelandii* (reduced state)	Microcapillary batch	X-ray diffraction	MR	2.03	3MIN	Peters et al. (1997)
Nitrogenase MoFe protein from *K. pneumoniae* (dithionite reduced state)	Microcapillary batch	X-ray diffraction	MR	1.6	1QGU	Mayer, Lawson, Gormal, Roe, and Smith (1999)
Nitrogenase MoFe protein from *K. pneumoniae* (phenosafranin oxidized state)	Microcapillary batch	X-ray diffraction	MR	1.6	1QH1	Mayer et al. (1999)
Nitrogenase MoFe protein from *K. pneumonia*, as crystallized (mixed oxidation state)	Microcapillary batch	X-ray diffraction	MR	1.6	1QH8	Mayer et al. (1999)
FeMo cofactor deficient nitrogenase MoFe protein	Microcapillary batch	X-ray diffraction	MR	2.3	1L5H	Schmid et al. (2002)
Nitrogenase MoFe protein from *K. pneumonia* NifV mutant	Microcapillary batch	X-ray diffraction	MR	1.9	1H1L	Mayer, Gormal, Smith, and Lawson (2002)

Continued

Table 4 Selected List of Nitrogenase MoFe Protein Structures—cont'd

Protein	Crystallization Method	Method	Phasing	Resolution (Å)	PDB Code	Reference
Nitrogenase MoFe protein from *A. vinelandii*	Vapor diffusion method	X-ray diffraction	MR	1.16	1M1N	Einsle et al. (2002)
Crystal structure of the α-H195Q mutant of nitrogenase	Microcapillary batch	X-ray diffraction	MR	2.5	1FP4	Sorlie et al. (2001)
Structure of the core domain of NafY from *A. vinelandii* determined at 1.8 Å resolution	Vapor diffusion, hanging drop	X-ray diffraction	MIR, MR	1.8	1P90	Dyer et al. (2003)
Insights into substrate binding at FeMo cofactor in nitrogenase from the structure of an alpha-70Ile MoFe protein variant	Microcapillary batch	X-ray diffraction	MR	2.23	3K1A	Sarma et al. (2010)
n-NafY from *A. vinelandii*. N-terminal domain of NafY		Nuclear magnetic resonance solution			2KIC	PDB 2KIC
Precursor bound NifEN from *A. vinelandii*	n/a	X-ray diffraction	MR	2.4	3PDI	Kaiser, Hu, Wiig, Rees, and Ribbe (2011)
A. vinelandii nitrogenase MoFe protein at atomic resolution	Vapor diffusion	X-ray diffraction	MR	1.0	3U7Q	Spatzal et al. (2011)

A. vinelandii Nitrogenase MoFe protein High pH form	Vapor diffusion	X-ray diffraction	MR	2.0	4ND8	Yang, Haynes, Spatzal, Rees, and Howard (2014)
Reactivated nitrogenase MoFe protein from *A. vinelandii*	Vapor diffusion sitting drop	X-ray diffraction	MR	1.43	4TKU	Spatzal et al. (2014)
CO-bound nitrogenase MoFe protein from *A. vinelandii*	Vapor diffusion sitting drop	X-ray diffraction	MR	1.5	4TKV	Spatzal et al. (2014)
Nitrogenase molybdenum-iron protein from *C. pasteurianum* at 1.08 Å resolution	Vapor diffusion	X-ray diffraction	MAD	1.08	4WES	Zhang et al. (2015)
Structure of the nitrogenase MoFe protein from *C. pasteurianum* pressurized with xenon	Vapor diffusion	X-ray diffraction	MAD	1.9	4WN9	Morrison, Hoy, Zhang, Einsle, and Rees (2015)
Structure of the nitrogenase MoFe protein from *A. vinelandii* pressurized with xenon	Vapor diffusion	X-ray diffraction	MAD	2.0	4WNA	Morrison et al. (2015)
Nitrogenase molybdenum-iron protein beta-K400E mutant *A. vinelandii*	Vapor diffusion	X-ray diffraction	MR	1.75	5CX1	Owens, Katz, Carter, Oswald, and Tezcan (2016)
Selenium incorporated nitrogenase MoFe protein (Av1-Se2B) from *A. vinelandii*	Vapor diffusion	X-ray diffraction	MR	1.6	5BVG	Spatzal, Perez, Howard, and Rees (2015)

Continued

Table 4 Selected List of Nitrogenase MoFe Protein Structures—cont'd

Protein	Crystallization Method	Method	Phasing	Resolution (Å)	PDB Code	Reference
CO-bound form of selenium incorporated nitrogenase MoFe protein (Av1-Se-CO) from *A. vinelandii*	Vapor diffusion	X-ray diffraction	MR	1.53	5BVH	Spatzal et al. (2015)
Nitrogenase MoFeP protein in the IDS oxidized state *Gluconacetobacter diazotrophicus*	Vapor diffusion	X-ray diffraction	MR	2.59	5KOJ	Owens et al. (2016)
Nitrogenase MoFeP from *G. diazotrophicus* in dithionite reduced state	Vapor diffusion	X-ray diffraction	MR	1.83	5KOH	Owens et al. (2016)

[a] MIR, multiple isomorphous replacement; MR, molecular replacement.

Table 5 Selected List of Nitrogenase MoFe and Fe Protein Complex Structures

Protein	Crystallization Method	Method	Phasing[a]	Resolution (Å)	PDB Code	References
Nitrogenase complex from *A. vinelandii* stabilized by ADP-tetrafluoroaluminate	Microcapillary batch	X-ray diffraction	MR	3.0	1N2C	Schindelin et al. (1997)
Nucleotide-free structure of nitrogenase protein complex between LEU127DEL-Fe protein and MoFe protein	Microcapillary batch 1	X-ray diffraction	MR	2.2	1G20	Chiu et al. (2001)
MGATP-bound structure of nitrogenase protein complex between LEU127DEL-Fe protein and MoFe protein	Microcapillary batch 1	X-ray diffraction	MR	3.0	1G21	Chiu et al. (2001)
Chemical crosslink of nitrogenase MoFe protein and Fe protein	Microcapillary batch 1	X-ray diffraction	MR	3.2	1M1Y	Schmid et al. (2002)
Nitrogenase complex from *A. vinelandii* stabilized by ADP-tetrafluoroaluminate	Microcapillary batch 1	X-ray diffraction	MR	2.3	1M34	Schmid et al. (2002)
Crystal structure of nucleotide-free Av2-Av1 complex	Vapor diffusion, sitting drops	X-ray diffraction	MR	2.1	2AFH	Tezcan et al. (2005)
Crystal structure of MgAMPPCP-bound Av2-Av1 complex	Vapor diffusion, sitting drops	X-ray diffraction	MR	2.3	4WZB	Tezcan et al. (2005)
Crystal structure of MgADP bound Av2-Av1 complex	Vapor diffusion, sitting drops	X-ray diffraction	MR	3.1	2AFI	Tezcan et al. (2005)
Asymmetric nucleotide binding in the nitrogenase complex	Vapor diffusion, sitting drops	X-ray diffraction	MR	1.9	4WZA	Tezcan, Kaiser, Howard, and Rees (2015)

[a]MIR, multiple isomorphous replacement; MR, molecular replacement.

3.2.1 Structural Characterization of the Nitrogenase Fe-Protein

Fe protein plays different roles in the nitrogenase enzyme system. In addition to serving as the physiological electron donor to MoFe protein, it is involved in FeMo-co biosynthesis and in maturation of the FeM-co deficient MoFe protein (Rangaraj et al., 1999). The first crystallographic structure of Fe protein from *A. vinelandii* demonstrated that a single [4Fe–4S] cysteine ligated cluster is bridged symmetrically between two identical subunits (Fig. 7).

The structure was determined using the native anomalous scattering properties of the Fe atoms of the [4Fe–4S] cluster and a single isomorphous derivative for phasing (Georgiadis et al., 1992). The [4Fe–4S] cluster is an essential component of the electron-transfer network that operates within nitrogenase. During catalysis, MgATP binds to the Fe protein, which then undergoes a gross conformational change that permits it to dock with the MoFe protein (Jang, Seefeldt, & Peters, 2000a). In the nitrogenase complex electrons can be transferred from the reduced [4Fe–4S] cluster of the Fe protein to the MoFe protein (Seefeldt, Hoffman, & Dean, 2012).

Later, higher resolution structures of both the *A. vinelandii* Fe protein (2.2 Å) and *C. pasteurianum* (1.93 Å) were published (Schlessman et al., 1998). *A. vinelandii* Fe protein crystals were grown by the capillary

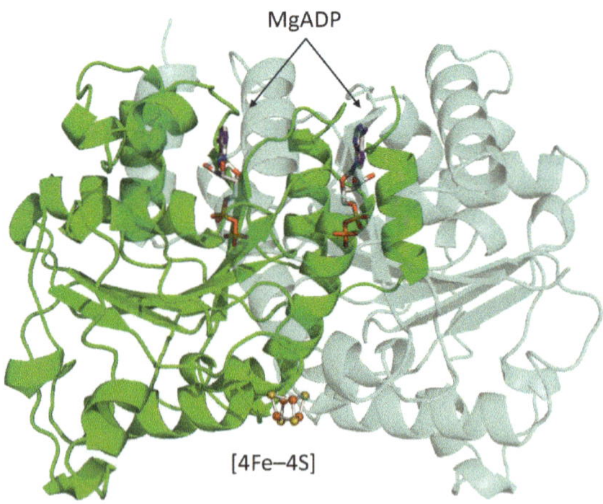

Fig. 7 Structure of nitrogenase Fe protein. Nitrogenase Fe protein composed of two dimers shown in *green* and *cyan*. One monomer is in transparent representation, showing the [4Fe–4S] cluster (presented in *balls* and *sticks*) and bound ADP (*sticks*). *Figure generated from PDB file 1N2C: Schindelin, H., Kisker, C., Schlessman, J. L., Howard, J. B., & Rees, D. C. (1997). Structure of ADP x AIF4(-)-stabilized nitrogenase complex and its implications for signal transduction.* Nature, *387(6631), 370–376. http://dx.doi.org/10.1038/387370a0.*

microbatch method with 30 mL of precipitant solution containing 25%–30% PEG 4000, 140–200 m*M* Na_2MoO_4, 50 m*M* Tris–Cl (pH 8.0), and 15 mL of protein solution containing in 20% glycerol, 450 m*M* NaCl, 2 m*M* $Na_2S_2O_4$, and 50 m*M* Tris–HCl (pH 8.0) in 1.5 mm capillaries. *C. pasteurianum* Fe protein crystals were grown under anaerobic conditions using the microcapillary batch method, with 40 mL of precipitant solution containing 5%–15% glycerol, 18.75%–22.5% PEG 4000, 230–270 m*M* $CaCl_2$, 50 m*M* HEPES (pH 7.5), layered with 10 mL of protein solution containing *C. pasteurianum* Fe protein in 20% glycerol, 2 m*M* $Na_2S_2O_4$, 50 m*M* HEPES, (pH 7.5). The detailed comparison of these structures shows that some regions of the Fe protein have a considerable level of conformational variability in spite of the large degree of structural conservation. In addition, these higher resolution structures enable a more critical analysis of the hydrogen-bonding interactions, which can be modulated by conformational changes in the Fe protein upon nucleotide binding and complex formation with MoFe protein (Table 3).

Structural characterization of the nucleotide-binding sites of the Fe protein shows that they are located at the interface of the two Fe-protein subunits. The structure of the Fe protein in complex with MgADP revealed the details of the protein–ADP interactions as well as the details of Mg^{2+} coordination (Jang et al., 2000a). The overall conformation state of the Fe protein in the presence of MgADP in comparison to the previous structures of the Fe protein in the absence of bound nucleotides is very similar. The most significant structural changes are manifested in the reorientations of amino acid side chains to accommodate MgADP binding and subtle conformational changes within the nucleotide-dependent switch regions (Jang et al., 2000a).

3.2.2 Structural Characterization of Nitrogenase MoFe Protein

Biochemical and biophysical studies of MoFe protein of nitrogenase have showed that the FeMo-co and P-cluster play an important role in substrate reduction and electron/proton transfer (Fig. 8). For the last 20 years significant work has been performed for structure–function characterization of MoFe protein.

The first structural models of the MoFe protein were published in 1992 in the journal *Science* (Kim & Rees, 1992b), and the metal centers and their coordination were presented in an article published the same year in *Nature* (Kim & Rees, 1992a). The crystals of MoFe protein *A. vinelandii* were obtained by the microcapillary batch method (Drenth, Hol, & Wierenga, 1975)

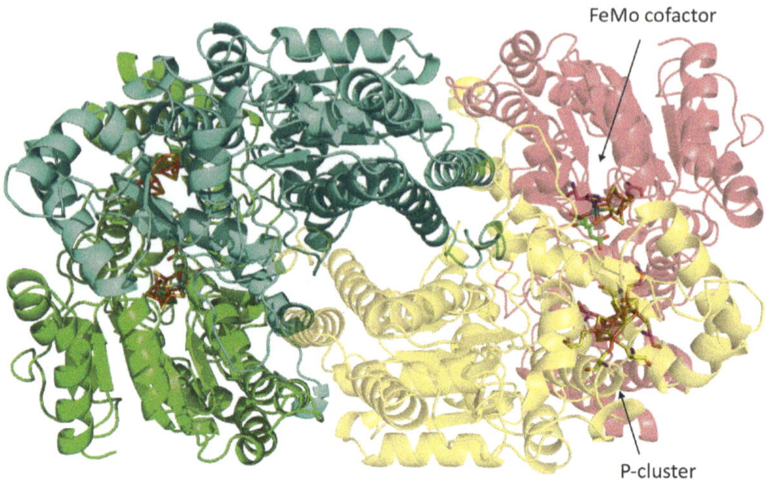

Fig. 8 Structure of nitrogenase MoFe protein and the overall organization of the $\alpha_2\beta_2$ heterodimer of nitrogenase MoFe protein. One heterodimer is transparent, showing the FeMo cofactor and P-cluster (presented in *sticks*). *Figure generated from PDB file 1N2C: Schindelin, H., Kisker, C., Schlessman, J. L., Howard, J. B., & Rees, D. C. (1997). Structure of ADP x AlF4(-)-stabilized nitrogenase complex and its implications for signal transduction.* Nature, *387(6631), 370–376. http://dx.doi.org/10.1038/387370a0.*

from 18% polyethylene glycol (PEG) 4000, 0.14 *M* NaCl, 0.10 *M* Na_2MoO_4, 80 m*M* Tris–HCl, pH 8.0, and ~8 mg/mL protein. Two crystal forms of the *C. pasteurianum* MoFe protein were obtained by crystallization from 15% PEG 4000, 0.21 *M* $MgCl_2$, 80 m*M* Tris–HCl, pH 8.0, and ~8 mg/mL protein. The MoFe-protein crystal structure was determined by the methods of MIR and noncrystallographic symmetry (NCS) averaging within and between crystal forms. Heavy atom derivatives of MoFe protein *A. vinelandii* were prepared by soaking separate crystals in solutions of 5 m*M* ethylmercurithiosalicylate (EMTS), 5 m*M* K_2PtCl_4, and 4 m*M* di-μ-iodobis(ethylene-diamine)-di-platinum (II) nitrate (PIP). Derivatives of $MgCl_2$ and CsCl were obtained by soaking each crystal form in 5 m*M* EMTS (Kim & Rees, 1992a).

3.2.3 Redox Poising of Nitrogenase MoFe Protein: The Role of the P-Cluster in Electron Transfer

The electron transfer chain in nitrogenase complex extends over 30 Å from the [4Fe–4S] of the Fe protein to FeMo cofactor of MoFe protein (Schindelin et al., 1997). The P-cluster functions as an intermediate electron carrier between the external electron donor, the Fe protein, and the FeMo cofactor in the MoFe protein. The P-cluster is coordinated by six cysteinyl residues with the α- and β-subunits each providing three ligands. All P-cluster ligands were identified through amino acid sequence analysis and site-directed mutagenesis studies prior to the structure determination (Brigle, Newton, & Dean, 1985; Dean et al., 1990; Lammers & Haselkorn, 1983; Thony, Kaluza, & Hennecke, 1985).

Results obtained from extensive spectroscopic studies of the MoFe protein from *A. vinelandii*, including EPR, CD, MCD, and Mössbauer, show that the P-cluster could exist in the following oxidized states: the all-ferrous state (P^N), the two-electron oxidized state (P^{OX}), and the one-electron oxidized state (P^{+1}). In order to understand how the structure of MoFe protein changes as a function of redox state, various experiments were undertaken to trap these states (Drennan & Peters, 2003; Mayer et al., 1999; Owens et al., 2016; Peters et al., 1997). Crystals were grown by microcapillary batch diffusion using 30% PEG 8000, 0.2 *M* NaMoO4, and 0.1 *M* Tris (pH 8.5) as precipitating solution. To obtain MoFe protein in the reduced state the crystals were incubated overnight after adding 10 m*M* dithionite to the mother liquor. The resulting solved structure revealed that the P-cluster in the P^N state consists of two [4Fe–4S] cubes covalently integrated together by a shared sulfur atom (Fig. 9). In the P^{OX} state, in which crystals were not treated with dithionite, one of the [4Fe–4S] clusters changes location to the open conformation, where an iron atom forms a new bond to the backbone amide N of α-C88. In contrast, the other iron atom is coordinated to β-S188 (Fig. 9A and C) (Peters et al., 1997). These structural changes in both the α-C88 backbone N and β-S188 hydroxyl group result in stabilization of the P-cluster in an oxidized state and provide evidence for conformational redox gating in nitrogenase. These experiments did not isolate the P^{+1} oxidation state, which was observed only in spectroscopic experiments in native and variant MoFe proteins (Lanzilotta, Christiansen, Dean, & Seefeldt, 1998), and has yet to be observed structurally. Similar redox-associated changes to the coordination of the P-cluster have been observed in the MoFe protein from *Gluconacetobacter diazotrophicus*, which has a Tyr coordination to the P-cluster in contrast to *A. vinlandii's* Phe residue, and an Ala in place of the S188 (Fig. 9B and D). Based on these observations, the O-based (Tyr or Ser) ligand is hypothesized to play an important role as a redox switch for P-cluster of MoFe protein (Owens et al., 2016). This highlights how the structural characterization of protein from different organisms may inform on similar mechanistic underpinnings despite differences in primary amino acid sequence.

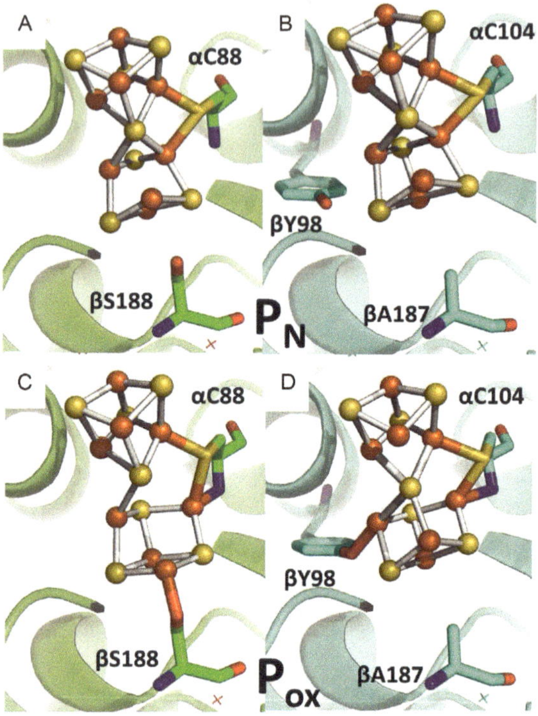

Fig. 9 Structural models of reduced (P_N) and oxidized (P_{OX}) P-clusters of the nitrogenase MoFe protein. Comparison of P-clusters in P_N (A) and P_{OX} (C) from *A. vinelandii* (in *green*) and P_N (B) and P_{OX} (D) from *G. diazotrophicus* (in *cyan*). For clarity the P-clusters are depicted in the same orientation. The αC88 and βS188 (*A. vinelandii* numbering) and αC104 and βY98 (*G. diazotrophicus* numbering) involved in the bonding of P-cluster are shown in *sticks*. P-clusters are shown in *ball* and *sticks*. The MoFe protein from *G. diazotrophicus* has an Ala residue in the position of the Ser ligand to the P-cluster, and βY98 is additionally involved in the bonding of P-cluster. *Figure generated from PDB files for P_N 3MIN and 5KOH; and P_{OX} 2MIN and 5KOJ states for* A. vinelandii *(Peters, J. W., Stowell, M. H., Soltis, S. M., Finnegan, M. G., Johnson, M. K., & Rees, D. C. (1997). Redox-dependent structural changes in the nitrogenase P-cluster.* Biochemistry, *36(6), 1181–1187. http://dx.doi.org/10.1021/bi9626665) and for* G. diazotrophicus *(Owens, C. P., Katz, F. E. H., Carter, C. H., Oswald, V. F., & Tezcan, F. A. (2016). Tyrosine-coordinated P-cluster in G. diazotrophicus nitrogenase: Evidence for the importance of O-based ligands in conformationally gated electron transfer.* Journal of the American Chemical Society, *138(32), 10124–10127. http://dx.doi.org/10.1021/jacs.6b06783), respectively.*

3.2.4 FeMo Cofactor of Nitrogenase MoFe Protein: Insights Into Substrate Binding Site

The FeMo cofactor is located in a cavity between the three domains of the α-subunit (Kim & Rees, 1992a, 1994). The FeMo cofactor is an

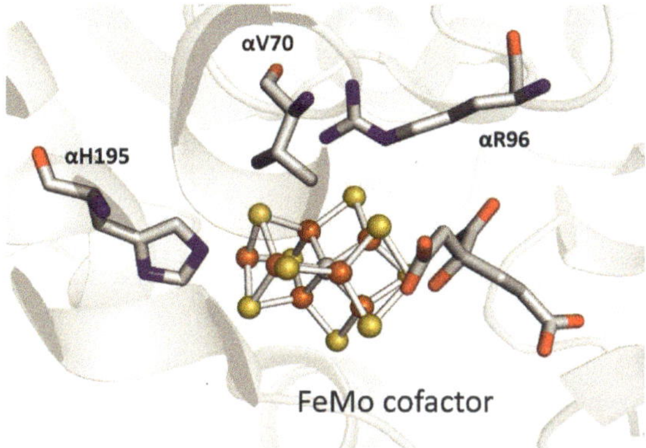

Fig. 10 The structure of FeMo cofactor of MoFe protein from *A. vinelandii*. The αH195, αV70, and αR96 involved in the substrate binding are shown in *sticks*, FeMo cluster is shown in *balls* and *sticks*, R-homocitrate is shown in *sticks*. *Figure generated from PDB file 3U7Q: Spatzal, T., Aksoyoglu, M., Zhang, L., Andrade, S. L., Schleicher, E., Weber, S., (2011). Evidence for interstitial carbon in nitrogenase FeMo cofactor.* Science, 334*(6058), 940. http://dx.doi.org/10.1126/science.1214025.*

unprecedented in nature [7Fe–9S–C–Mo–homocitrate] organometallo cluster. This cofactor can be described as a [Mo–3Fe–3S] subcluster bridged to [4Fe–3S] partial cluster via three Fe bridging sulfides and a carbide moiety (Lancaster et al., 2011; Spatzal et al., 2011) (Fig. 10). The molybdenum atom is covalently coordinated by homocitrate. The coordination and the secondary-sphere environments of FeMo cofactor are very similar among the nitrogenases characterized from different sources (Spatzal et al., 2011; Zhang et al., 2015). A number of amino acid residues such as α-H195, α-V70, and α-R96 (*A. vinelandii* numbering) have been identified through amino acid sequence comparisons and site-directed mutagenesis studies prior to the structure determination (Brigle et al., 1985; Kent, Baines, Gormal, Smith, & Buck, 1990; Scott, Dean, & Newton, 1992) and are proposed to be important for substrate interactions at the FeMo cofactor (Sarma et al., 2010; Scott et al., 1992; Seefeldt et al., 2009).

Significant progress has been made toward characterization of ligand binding to the FeMo cofactor after solving the structure with CO-bound nitrogenase complex. MoFe protein with CO was crystallized by the sitting drop vapor diffusion method at 294 K in an anaerobic chamber containing a 95% Ar and 5% H_2 atmosphere. The reservoir solution contained 21% PEG 4000, 0.75 *M* NaCl, 0.1 *M* imidazole/malate (pH 7.5), 2 m*M* spermine, 1% glycerol, 0.5% 2,2,2-trifluoroethanol, and 2.5 m*M* $Na_2S_2O_4$. All crystallization solutions were saturated with CO and 300 μL (0.12 atm) CO were added to the headspace of each crystallization well before resealing the crystallization tray. Additionally, a seeding strategy was applied to significantly accelerate the crystallization process and improve the overall crystal shape. As a result, the first crystals were obtained 2 h after setting up the crystallization experiment. Crystals of sufficient size for diffraction were harvested after 5 h. Cryo-protection was achieved by transferring crystals into the reservoir solution containing 5% MPD and 5% 2,2,2-trifluoroethanol flushed with CO. Since CO is a noncompetitive inhibitor for all substrates of nitrogenase except protons, and it binds to the partially reduced protein thereby mimicking the binding of substrate to the active site, the structure of nitrogenase with CO bound revealed that this site is close to the α-H195, α-V70, and α-R96 (Spatzal et al., 2014). In order to further confirm that the site of CO binding is the main substrate binding sites, site-directed mutagenesis was performed on the aforementioned the α-H195, α-V70, and α-R96. These variants were tested biochemically for nitrogenase activity as well as structurally characterized, which verified the catalytic incompetence of the variants and showed that substitutions to α-V70 were controlling a substrate access to the FeMo-co active site. The structural comparison of these variants showed that the side chain of α-I70 is blocking access of the substrate to the Fe6 of FeMo cofactor (Sarma et al., 2010).

A crystallographic approach was used to determine the potential substrate access pathways from the protein surface to the FeMo cofactor or the P-cluster (Morrison et al., 2015). The crystals of MoFe protein were grown in 24-well plates using the sitting-drop method at room temperature in an anaerobic chamber with an atmosphere of ~95% argon and ~5% hydrogen. All crystallization solutions were purged with argon prior to use. For *C. pasteurianum* MoFe protein, the reservoir and crystallization solutions consisted of double-distilled water, 23% polyethylene glycol, 0.2 *M* lithium citrate, and 5 m*M* sodium dithionite. After cryo-protection crystals were moved to the pressurization chamber of a homemade gas pressurization device. While lightly venting Xe gas through the device, the pressurization chamber was closed. The crystals were pressurized at 14 atm for 10 − 15 min, after which they were quickly transferred to liquid nitrogen for storage. In the nitrogenase MoFe protein, five potential substrate access pathways from the protein surface to the FeMo cofactor (the active site) or the P-cluster using experimental structures of Xe pressurized into MoFe protein crystals from *A. vinelandii* and *C. pasteurianum* have been identified. In addition, all published structures of the MoFe protein, including those from *Klebsiella pneumoniae*, were analyzed for the presence of nonwater, small molecules bound to the protein interior. Each pathway is based on identification of plausible routes from buried small molecule binding sites to both the protein surface and a metallocluster. Of these five pathways, two have been previously suggested as substrate access pathways. While the small molecule binding sites are not conserved among the three species of MoFe protein, residues lining the pathways are generally conserved, indicating that the proposed pathways may be accessible in all three species. These observations imply that there is unlikely a unique pathway utilized for substrate access from the protein surface to the active site; however, there may be preferred pathways such as those described here (Morrison et al., 2015) (Table 4).

3.3 Nitrogenase MoFe and Fe Protein Complex Structures

The X-ray diffraction analysis of MoFe protein and Fe protein crystals provides important information about their "static" structures. To understand the nitrogen fixation mechanism, it is necessary to obtain knowledge about the "dynamic" of the two proteins during binding and reducing substrates. The ground-work for these studies was laid by a couple of very important biochemical observations. First, the Fe protein and MoFe protein can form a tight-binding complex in the presence of ADP and tetrafluoroaluminate (Duyvis, Wassink, & Haaker, 1996; Renner & Howard, 1996). Second, Fe protein is the only known electron donor to MoFe protein. The first nitrogenase complex structurally characterized was the Fe protein–MoFe protein complex stabilized with bound ADP and tetrafluoroaluminate (Schindelin et al., 1997). The structure revealed the anticipated pairing of the twofold symmetric axis of the Fe protein with the pseudo-twofold symmetric axis formed by the structurally related α- and β-subunits of the MoFe protein (Fig. 6). The resulting complex places the [4Fe–4S] clusters of each Fe protein in close proximity to the P-clusters of the MoFe protein as required for effective electron transfer. The overall structure of the complex is consistent with the results of small angle X-ray scattering studies, which indicated that the overall structure of the complex is maintained in solution (Grossman, Hasnain, Yousafzai, Smith, & Eady, 1997; Grossman et al., 1999).

Later in 2005, and 2015 crystal structures of MoFe protein–Fe protein complexes in different nucleotide states: the nucleotide-free (nf), the MgADP (adp), and the MgAMPPCP (pcp) complexes at resolutions of 2.1 Å, 3.1 Å, and 2.3 Å, respectively, have been determined to identify conformational changes in the nitrogenase complex during ATP turnover (Tezcan et al., 2015, 2005). Crystals of nf-, adp- and pcp-complexes were prepared by the sitting drop vapor diffusion method in an anaerobic chamber at room temperature. Precipitating solutions for the nf-complex contained 16%–18% PEG 10000, and those for pcp- and adp-complexes contained 18%–22% PEG 8000, in addition to 0–50 m*M* NaCl, 100 m*M* Tris buffer (pH 8.5), and 10 mM dithionite. The protein solutions were prepared by mixing appropriate amounts of MoFe protein (~20 mg/mL) and Fe protein stock solutions (~40 mg/mL), both of which were in 100 m*M* Tris (pH 7.75), 200 m*M* NaCl, and 5 m*M* dithionite. Best crystals for nf-, pcp- and adp-complexes were obtained using 1:1.2–1:1.5 MoFe protein:Fe protein ratios (v/v), corresponding to five- to sixfold molar excess of Fe protein over per active site of MoFe protein. The results of the analysis of these complexes revealed that the Fe protein molecules can occupy different interaction sites on the MoFe protein. Three separated docking areas may be identified, all on the surface of the MoFe protein surrounding the twofold axis approximately relating pairs of α- and β-subunits (Tezcan et al., 2015, 2005) (Table 5).

4. CONCLUSIONS

Hydrogenases and nitrogenases are both fascinating enzymes that catalyze energetically intensive reduction of substrates with minimal thermodynamic loss, and therefore represent a rich area for the study of fundamental mechanistic underpinnings. Fortunately, the last several decades have furnished a large body of knowledge of how to poise these oxygen-sensitive enzymes in various redox states, with various substrates and inhibitors, to be structurally solved at ever-increasing resolutions. The basic techniques outlined here may be used as a template from which to base future crystallization studies. In conclusion, improving techniques of both crystal growth and structural analysis are leading to a full mechanistic understanding of nitrogenases and hydrogenases.

ACKNOWLEDGMENT

This work was supported as part of the Biological Electron Transfer and Catalysis (BETCy) Energy Frontier Research Center (EFRC), an EFRC funded by US Department of Energy, Office of Science Grant DE-SC0012518.

REFERENCES

Abou-Hamdan, A., Ceccaldi, P., Lebrette, H., Gutiérrez-Sanz, O., Richaud, P., Cournac, L., et al. (2015). A threonine stabilizes the NiC and NiR catalytic intermediates of [NiFe]-hydrogenase. *Journal of Biological Chemistry*, *290*(13), 8550–8558. http://dx.doi.org/10.1074/jbc.M114.630491.

Adams, M. W. (1987). The mechanisms of H2 activation and CO binding by hydrogenase I and hydrogenase II of Clostridium pasteurianum. *The Journal of Biological Chemistry, 262*(31), 15054–15061.

Adams, M. W. (1990). The structure and mechanism of iron-hydrogenases. *Biochimica et Biophysica Acta, 1020*(2), 115–145.

Adams, M. W., Eccleston, E., & Howard, J. B. (1989). Iron-sulfur clusters of hydrogenase I and hydrogenase II of *Clostridium pasteurianum. Proceedings of the National Academy of Sciences, 86*(13), 4932–4936.

Adamska-Venkatesh, A., Simmons, T. R., Siebel, J. F., Artero, V., Fontecave, M., Reijerse, E., et al. (2015). Artificially maturated [FeFe] hydrogenase from *Chlamydomonas reinhardtii*: A HYSCORE and ENDOR study of a non-natural H-cluster. *Physical Chemistry Chemical Physics, 17*(7), 5421–5430. http://dx.doi.org/10.1039/C4CP05426A.

Allegretti, M., Mills, D. J., McMullan, G., Kühlbrandt, W., & Vonck, J. (2014). Atomic model of the F420-reducing [NiFe] hydrogenase by electron cryo-microscopy using a direct electron detector. *eLife3*, e01963. http://dx.doi.org/10.7554/eLife.01963.

Alonso-Mori, R., Kern, J., Gildea, R. J., Sokaras, D., Weng, T. C., Lassalle-Kaiser, B., et al. (2012). Energy-dispersive X-ray emission spectroscopy using an X-ray free-electron laser in a shot-by-shot mode. *Proceedings of the National Academy of Sciences of the United States of America, 109*(47), 19103–19107. http://dx.doi.org/10.1073/pnas.1211384109.

Artz, J. H., Mulder, D. W., Ratzloff, M. W., Lubner, C. E., Zadvornyy, O. A., LeVan, A. X., et al. (2017). Reduction potentials of [FeFe]-hydrogenase accessory iron–sulfur clusters provide insights into the energetics of proton reduction catalysis. *Journal of the American Chemical Society, 139*(28), 9544–9550. http://dx.doi.org/10.1021/jacs.7b02099.

Baltazar, C. S. A., Marques, M. C., Soares, C. M., DeLacey, A. M., Pereira, I. A. C., & Matias, P. M. (2011). Nickel–iron–selenium hydrogenases—An overview. *European Journal of Inorganic Chemistry, 2011*(7), 948–962. http://dx.doi.org/10.1002/ejic.201001127.

Barty, A., Caleman, C., Aquila, A., Timneanu, N., Lomb, L., White, T. A., et al. (2012). Self-terminating diffraction gates femtosecond X-ray nanocrystallography measurements. *Nature Photonics, 6*, 35–40. http://dx.doi.org/10.1038/nphoton.2011.297.

Benton, P. M. C., & Peters, J. W. (2004). The structures of the nitrogenase proteins and stabilized complexes. In B. E. Smith, R. L. Richards, & N. W. E. (Eds.), *Catalysis for nitrogen fixation: nitrogenases, relevant chemical models, and commercial process* (pp. 77–96). Netherlands: Kluwer Academic Publishers.

Berggren, G., Adamska, A., Lambertz, C., Simmons, T. R., Esselborn, J., Atta, M., et al. (2013). Biomimetic assembly and activation of [FeFe]-hydrogenases. *Nature, 499*(7456), 66–69. http://dx.doi.org/10.1038/nature12239. http://www.nature.com/nature/journal/v499/n7456/abs/nature12239.html#supplementary-information.

Birrell, J. A., Wrede, K., Pawlak, K., Rodriguez-Maciá, P., Rüdiger, O., Reijerse, E. J., et al. (2016). Artificial maturation of the highly active heterodimeric [FeFe] hydrogenase from *Desulfovibrio desulfuricans* ATCC 7757. *Israel Journal of Chemistry, 56*(9–10), 852–863. http://dx.doi.org/10.1002/ijch.201600035.

Boyd, E. S., Hamilton, T. L., Swanson, K. D., Howells, A. E., Baxter, B. K., Meuser, J. E., et al. (2014). [FeFe]-hydrogenase abundance and diversity along a vertical redox gradient in Great Salt Lake, USA. *International Journal of Molecular Sciences, 15*(12), 21947–21966. http://dx.doi.org/10.3390/ijms151221947.

Boyd, E., Schut, G., Adams, M., & Peters, J. (2014). Hydrogen metabolism and the evolution of biological respiration. *Microbe Magazine, 9*(9), 361–367. http://dx.doi.org/10.1128/microbe.9.361.1.

Brecht, M., van Gastel, M., Buhrke, T., Friedrich, B., & Lubitz, W. (2003). Direct detection of a hydrogen ligand in the [NiFe] center of the regulatory H2-sensing hydrogenase from *Ralstonia eutropha* in its reduced state by HYSCORE and ENDOR spectroscopy. *Journal of the American Chemical Society*, *125*(43), 13075–13083. http://dx.doi.org/10.1021/ja036624x.

Brigle, K. E., Newton, W. E., & Dean, D. R. (1985). Complete nucleotide sequence of the *Azotobacter vinelandii* nitrogenase structural gene cluster. *Gene*, *37*(1–3), 37–44.

Broderick, J. B., Byer, A. S., Duschene, K. S., Duffus, B. R., Betz, J. N., Shepard, E. M., et al. (2014). H-Cluster assembly during maturation of the [FeFe]-hydrogenase. *JBIC: Journal of Biological Inorganic Chemistry*, *19*(6), 747–757. http://dx.doi.org/10.1007/s00775-014-1168-8.

Brooke, E. J., Evans, R. M., Islam, S. T. A., Roberts, G. M., Wehlin, S. A. M., Carr, S. B., et al. (2016). Importance of the active site "Canopy" residues in an O2-tolerant [NiFe]-hydrogenase. *Biochemistry*, 132–142. http://dx.doi.org/10.1021/acs.biochem.6b00868.

Burgess, B. K., & Lowe, D. J. (1996). Mechanism of molybdenum nitrogenase. *Chemical Reviews*, *96*(7), 2983–3012.

Burns, R. C., Holsten, R. D., & Hardy, R. W. (1970). Isolation by crystallization of the Mo-Fe protein of *Azotobacter* nitrogenase. *Biochemical and Biophysical Research Communications*, *39*(1), 90–99.

Caserta, G., Adamska-Venkatesh, A., Pecqueur, L., Atta, M., Artero, V., Roy, S., et al. (2016). Chemical assembly of multiple metal cofactors: The heterologously expressed multidomain [FeFe]-hydrogenase from *Megasphaera elsdenii*. *Biochimica et Biophysica Acta (BBA)—Bioenergetics*, *1857*(11), 1734–1740. http://dx.doi.org/10.1016/j.bbabio.2016.07.002.

Chen, J.-S., & Blanchard, D. K. (1978). Isolation and properties of a unidirectional H2-oxidizing hydrogenase from the strictly anaerobic N2-fixing bacterium *Clostridium pasteurianum* W5. *Biochemical and Biophysical Research Communications*, *84*(4), 1144–1150. http://dx.doi.org/10.1016/0006-291X(78)91703-5.

Chen, J.-S., & Mortenson, L. E. (1974). Purification and properties of hydrogenase from *Clostridium pasteurianum* W5. *Biochimica et Biophysica Acta (BBA)—Protein Structure*, *371*(2), 283–298. http://dx.doi.org/10.1016/0005-2795(74)90025-7.

Chiu, H., Peters, J. W., Lanzilotta, W. N., Ryle, M. J., Seefeldt, L. C., Howard, J. B., et al. (2001). MgATP-Bound and nucleotide-free structures of a nitrogenase protein complex between the Leu 127 Delta-Fe-protein and the MoFe-protein. *Biochemistry*, *40*(3), 641–650.

Cohen, J., Kim, K., Posewitz, M., Ghirardi, M. L., Schulten, K., Seibert, M., et al. (2005). Molecular dynamics and experimental investigation of H(2) and O(2) diffusion in [Fe]-hydrogenase. *Biochemical Society Transactions*, *33*(Pt. 1), 80–82. http://dx.doi.org/10.1042/bst0330080.

Cohen, A. E., Soltis, S. M., González, A., Aguila, L., Alonso-Mori, R., Barnes, C. O., et al. (2014). Goniometer-based femtosecond crystallography with X-ray free electron lasers. *Proceedings of the National Academy of Sciences of the United States of America*, *111*(48), 17122–17127. http://dx.doi.org/10.1073/pnas.1418733111.

Danyal, K., Rasmussen, A. J., Keable, S. M., Inglet, B. S., Shaw, S., Zadvornyy, O. A., et al. (2015). Fe protein-independent substrate reduction by nitrogenase MoFe protein variants. *Biochemistry*, *54*(15), 2456–2462. http://dx.doi.org/10.1021/acs.biochem.5b00140.

Dean, D. R., Setterquist, R. A., Brigle, K. E., Scott, D. J., Laird, N. F., & Newton, W. E. (1990). Evidence that conserved residues Cys-62 and Cys-154 within the *Azotobacter vinelandii* nitrogenase MoFe protein alpha-subunit are essential for nitrogenase activity but conserved residues His-83 and Cys-88 are not. *Molecular Microbiology*, *4*(9), 1505–1512.

Dementin, S., Burlat, B., De Lacey, A. L., Pardo, A., Adryanczyk-Perrier, G., Guigliarelli, B., et al. (2004). A glutamate is the essential proton transfer gate during the catalytic cycle of the [NiFe] hydrogenase. *Journal of Biological Chemistry, 279*(11), 10508–10513. http://dx.doi.org/10.1074/jbc.M312716200.

Drennan, C. L., & Peters, J. W. (2003). Surprising cofactors in metalloenzymes. *Current Opinion in Structural Biology, 13*(2), 220–226.

Drenth, J., Hol, W. G., & Wierenga, R. K. (1975). Crystallization and preliminary x-ray investigation of p-hydroxybenzoate hydroxylase from *Pseudomonas fluorescens. The Journal of Biological Chemistry, 250*(13), 5268–5269.

Duyvis, M. G., Wassink, H., & Haaker, H. (1996). Pre-steady-state kinetics of nitrogenase from *Azotobacter vinelandii*. Evidence for an ATP-induced conformational change of the nitrogenase complex as part of the reaction mechanism. *The Journal of Biological Chemistry, 271*(47), 29632–29636.

Dyer, D. H., Rubio, L. M., Thoden, J. B., Holden, H. M., Ludden, P. W., & Rayment, I. (2003). The three-dimensional structure of the core domain of NafY from *Azotobacter vinelandii* determined at 1.8-A resolution. *The Journal of Biological Chemistry, 278*(34), 32150–32156. http://dx.doi.org/10.1074/jbc.M304264200.

Eady, R. R. (1996). Structure – function relationships of alternative nitrogenases. *Chemical Reviews, 96*(7), 3013–3030. http://dx.doi.org/10.1021/cr950057h.

Einsle, O., Tezcan, F. A., Andrade, S. L., Schmid, B., Yoshida, M., Howard, J. B., et al. (2002). Nitrogenase MoFe-protein at 1.16 A resolution: A central ligand in the FeMo-cofactor. *Science, 297*(5587), 1696–1700. http://dx.doi.org/10.1126/science.1073877.

Erbes, D. L., King, D., & Gibbs, M. (1979). Inactivation of hydrogenase in cell-free extracts and whole cells of *Chlamydomonas reinhardi* by oxygen. *Plant Physiology, 63*(6), 1138–1142.

Erisman, J. W., & Sutton, M. A. (2008). Reduced nitrogen in ecology and the environment: Special issue of the ESF-FWF Conference in partnership with LFUI, October 2006. *Environmental Pollution, 154*(3), 357–358. http://dx.doi.org/10.1016/j.envpol.2008.05.001.

Esselborn, J., Muraki, N., Klein, K., Engelbrecht, V., Metzler-Nolte, N., Apfel, U. P., et al. (2016). A structural view of synthetic cofactor integration into [FeFe]-hydrogenases. *Chemical Science*, 7(2), 959–968. http://dx.doi.org/10.1039/C5SC03397G.

Flanagan, L. A., & Parkin, A. (2016). Electrochemical insights into the mechanism of NiFe membrane-bound hydrogenases. *Biochemical Society Transactions, 44*(1), 315–328. http://dx.doi.org/10.1042/bst20150201.

Frielingsdorf, S., Fritsch, J., Schmidt, A., Hammer, M., Löwenstein, J., Siebert, E., et al. (2014). Reversible [4Fe-3S] cluster morphing in an O2-tolerant [NiFe] hydrogenase. *Nature Chemical Biology, 10*(5), 378–385. http://dx.doi.org/10.1038/nchembio.1500. http://www.nature.com/nchembio/journal/v10/n5/abs/nchembio.1500.html#supplementary-information.

Garcin, E., Vernede, X., Hatchikian, E. C., Volbeda, A., Frey, M., & Fontecilla-Camps, J. C. (1999). The crystal structure of a reduced [NiFeSe] hydrogenase provides an image of the activated catalytic center. *Structure*, 7(5), 557–566. http://dx.doi.org/10.1016/S0969-2126(99)80072-0.

Georgiadis, M. M., Komiya, H., Chakrabarti, P., Woo, D., Kornuc, J. J., & Rees, D. C. (1992). Crystallographic structure of the nitrogenase iron protein from *Azotobacter vinelandii. Science, 257*(5077), 1653–1659.

Ghirardi, M. L., Cohen, J., King, P., Schulten, K., Kim, K., & Seibert, M. (2006). [FeFe]-hydrogenases and photobiological hydrogen production. *Proceedings of SPIE, 6340*, 63400X-1-63400X-6.

Greene, B. L., Wu, C.-H., McTernan, P. M., Adams, M. W. W., & Dyer, R. B. (2015). Proton-coupled electron transfer dynamics in the catalytic mechanism of a [NiFe]-hydrogenase. *Journal of the American Chemical Society, 137*(13), 4558–4566. http://dx.doi.org/10.1021/jacs.5b01791.

Grossman, J. G., Hasnain, S. S., Yousafzai, F. K., Smith, B. E., & Eady, R. R. (1997). The first glimpse of a complex of nitrogenase component proteins by solution X-ray scattering: Conformation of the electron transfer transition state complex of *Klebsiella pneumoniae* nitrogenase. *Journal of Molecular Biology*, *266*(4), 642–648. http://dx.doi.org/10.1006/jmbi.1996.0846.

Grossman, J. G., Hasnain, S. S., Yousafzai, F. K., Smith, B. E., Eady, R. R., Schindelin, H., et al. (1999). Comparing crystallographic and solution structures of nitrogenase complexes. *Acta Crystallographica Section D, Biological Crystallography*, *55*(Pt. 4), 727–728.

Hales, B. J. (1990). Alternative nitrogenase. *Advances in Inorganic Biochemistry*, *8*, 165–198.

Hedderich, R. (2004). Energy-converting [NiFe] hydrogenases from archaea and extremophiles: Ancestors of complex I. *Journal of Bioenergetics and Biomembranes*, *36*(1), 65–75. http://dx.doi.org/10.1023/b:jobb.0000019599.43969.33.

Hidalgo, R., Ash, P. A., Healy, A. J., & Vincent, K. A. (2015). Infrared spectroscopy during electrocatalytic turnover reveals the Ni-L active site state during H2 oxidation by a NiFe hydrogenase. *Angewandte Chemie*, *127*(24), 7216–7219. http://dx.doi.org/10.1002/ange.201502338.

Higuchi, Y., Yagi, T., & Yasuoka, N. (1997). Unusual ligand structure in Ni-Fe active center and an additional Mg site in hydrogenase revealed by high resolution X-ray structure analysis. *Structure*, *5*(12), 1671–1680.

Higuchi, Y., Yasuoka, N., Kakudo, M., Katsube, Y., Yagi, T., & Inokuchi, H. (1987). Single crystals of hydrogenase from *Desulfovibrio vulgaris* Miyazaki F. *Journal of Biological Chemistry*, *262*(6), 2823–2825.

Horch, M., Schoknecht, J., Mroginski, M. A., Lenz, O., Hildebrandt, P., & Zebger, I. (2014). Resonance Raman spectroscopy on [NiFe] hydrogenase provides structural insights into catalytic intermediates and reactions. *Journal of the American Chemical Society*, *136*(28), 9870–9873. http://dx.doi.org/10.1021/ja505119q.

Howard, J. B., & Rees, D. C. (1996). Structural basis of biological nitrogen fixation. *Chemical Reviews*, *96*(7), 2965–2982.

Imlay, J. A. (2006). Iron-sulphur clusters and the problem with oxygen. *Molecular Microbiology*, *59*(4), 1073–1082. http://dx.doi.org/10.1111/j.1365-2958.2006.05028.x.

Jang, S. B., Jeong, M. S., Seefeldt, L. C., & Peters, J. W. (2004). Structural and biochemical implications of single amino acid substitutions in the nucleotide-dependent switch regions of the nitrogenase Fe protein from *Azotobacter vinelandii*. *Journal of Biological Inorganic Chemistry*, *9*(8), 1028–1033. http://dx.doi.org/10.1007/s00775-004-0605-5.

Jang, S. B., Seefeldt, L. C., & Peters, J. W. (2000a). Insights into nucleotide signal transduction in nitrogenase: Structure of an iron protein with MgADP bound. *Biochemistry*, *39*(48), 14745–14752.

Jang, S. B., Seefeldt, L. C., & Peters, J. W. (2000b). Modulating the midpoint potential of the [4Fe-4S] cluster of the nitrogenase Fe protein. *Biochemistry*, *39*(4), 641–648.

Jeong, M. S., & Jang, S. B. (2004). Structural basis for the changes in redox potential in the nitrogenase Phe135Trp Fe protein with MgADP Bound. *Molecules and Cells*, *18*(3), 374–382.

Jones, A. K., Lamle, S. E., Pershad, H. R., Vincent, K. A., Albracht, S. P., & Armstrong, F. A. (2003). Enzyme electrokinetics: Electrochemical studies of the anaerobic interconversions between active and inactive states of *Allochromatium vinosum* [NiFe]-hydrogenase. *Journal of the American Chemical Society*, *125*(28), 8505–8514. http://dx.doi.org/10.1021/ja035296y.

Kaiser, J. T., Hu, Y., Wiig, J. A., Rees, D. C., & Ribbe, M. W. (2011). Structure of precursor-bound NifEN: A nitrogenase FeMo cofactor maturase/insertase. *Science*, *331*(6013), 91–94. http://dx.doi.org/10.1126/science.1196954.

Kalms, J., Schmidt, A., Frielingsdorf, S., van der Linden, P., von Stetten, D., Lenz, O., et al. (2016). Krypton derivatization of an O2-tolerant membrane-bound [NiFe] hydrogenase reveals a hydrophobic tunnel network for gas transport. *Angewandte Chemie International Edition*, *55*(18), 5586–5590. http://dx.doi.org/10.1002/anie.201508976.

Kamp, C., Silakov, A., Winkler, M., Reijerse, E. J., Lubitz, W., & Happe, T. (2008). Isolation and first EPR characterization of the [FeFe]-hydrogenases from green algae. *Biochimica et Biophysica Acta (BBA)—Bioenergetics*, *1777*(5), 410–416. http://dx.doi.org/10.1016/j.bbabio.2008.02.002.

Kanai, T., Matsuoka, R., Beppu, H., Nakajima, A., Okada, Y., Atomi, H., et al. (2011). Distinct physiological roles of the three [NiFe]-hydrogenase orthologs in the hyperthermophilic archaeon Thermococcus kodakarensis. *Journal of Bacteriology*, *193*(12), 3109–3116. http://dx.doi.org/10.1128/jb.01072-10.

Kent, H. M., Baines, M., Gormal, C., Smith, B. E., & Buck, M. (1990). Analysis of site-directed mutations in the alpha- and beta-subunits of *Klebsiella pneumoniae* nitrogenase. *Molecular Microbiology*, *4*(9), 1497–1504.

Kim, J., & Rees, D. C. (1992a). Crystallographic structure and functional implications of the nitrogenase molybdenum-iron protein from *Azotobacter vinelandii*. *Nature*, *360*(6404), 553–560.

Kim, J., & Rees, D. C. (1992b). Structural models for the metal centers in the nitrogenase molybdenum-iron protein. *Science*, *257*(5077), 1677–1682.

Kim, J., & Rees, D. C. (1994). Nitrogenase and biological nitrogen fixation. *Biochemistry*, *33*(2), 389–397.

Kitano, M., Inoue, Y., Yamazaki, Y., Hayashi, F., Kanbara, S., Matsuishi, S., et al. (2012). Ammonia synthesis using a stable electride as an electron donor and reversible hydrogen store. *Nature Chemistry*, *4*(11), 934–940. http://dx.doi.org/10.1038/nchem.1476.

Knörzer, P., Silakov, A., Foster, C. E., Armstrong, F. A., Lubitz, W., & Happe, T. (2012). Importance of the protein framework for catalytic activity of [FeFe]-hydrogenases. *Journal of Biological Chemistry*, *287*(2), 1489–1499. http://dx.doi.org/10.1074/jbc.M111.305797.

Lambertz, C., Leidel, N., Havelius, K. G. V., Noth, J., Chernev, P., Winkler, M., et al. (2011). O(2) reactions at the six-iron active site (H-cluster) in [FeFe]-hydrogenase. *The Journal of Biological Chemistry*, *286*(47), 40614–40623. http://dx.doi.org/10.1074/jbc.M111.283648.

Lammers, P. J., & Haselkorn, R. (1983). Sequence of the nifD gene coding for the alpha subunit of dinitrogenase from the cyanobacterium *Anabaena*. *Proceedings of the National Academy of Sciences of the United States of America*, *80*(15), 4723–4727.

Lancaster, K. M., Roemelt, M., Ettenhuber, P., Hu, Y., Ribbe, M. W., Neese, F., et al. (2011). X-ray emission spectroscopy evidences a central carbon in the nitrogenase iron-molybdenum cofactor. *Science*, *334*(6058), 974–977. http://dx.doi.org/10.1126/science.1206445.

Lanzilotta, W. N., Christiansen, J., Dean, D. R., & Seefeldt, L. C. (1998). Evidence for coupled electron and proton transfer in the [8Fe-7S] cluster of nitrogenase. *Biochemistry*, *37*(32), 11376–11384. http://dx.doi.org/10.1021/bi980048d.

Lawson, D. M., & Smith, B. E. (2002). Molybdenum nitrogenases: A crystallographic and mechanistic view. *Metal Ions in Biological Systems*, *39*, 75–119.

Lemon, B. J., & Peters, J. W. (1999). Binding of exogenously added carbon monoxide at the active site of the iron-only hydrogenase (CpI) from *Clostridium pasteurianum*. *Biochemistry*, *38*(40), 12969–12973.

Lenz, O., Bernhard, M., Buhrke, T., Schwartz, E., & Friedrich, B. (2002). The hydrogen-sensing apparatus in *Ralstonia eutropha*. *Journal of Molecular Microbiology and Biotechnology*, *4*(3), 255–262.

Liebgott, P.-P., de Lacey, A. L., Burlat, B., Cournac, L., Richaud, P., Brugna, M., et al. (2011). Original design of an oxygen-tolerant [NiFe] hydrogenase: Major effect of a valine-to-cysteine mutation near the active site. *Journal of the American Chemical Society, 133*(4), 986–997. http://dx.doi.org/10.1021/ja108787s.

Lubitz, W., Ogata, H., Rüdiger, O., & Reijerse, E. (2014). Hydrogenases. *Chemical Reviews, 114*(8), 4081–4148. http://dx.doi.org/10.1021/cr4005814.

Marques, M. C., Coelho, R., Pereira, I. A. C., & Matias, P. M. (2013). Redox state-dependent changes in the crystal structure of [NiFeSe] hydrogenase from *Desulfovibrio vulgaris* Hildenborough. *International Journal of Hydrogen Energy, 38*(21), 8664–8682. http://dx.doi.org/10.1016/j.ijhydene.2013.04.132.

Mayer, S. M., Gormal, C. A., Smith, B. E., & Lawson, D. M. (2002). Crystallographic analysis of the MoFe protein of nitrogenase from a nifV mutant of *Klebsiella pneumoniae* identifies citrate as a ligand to the molybdenum of iron molybdenum cofactor (FeMoco). *The Journal of Biological Chemistry, 277*(38), 35263–35266. http://dx.doi.org/10.1074/jbc.M205888200.

Mayer, S. M., Lawson, D. M., Gormal, C. A., Roe, S. M., & Smith, B. E. (1999). New insights into structure-function relationships in nitrogenase: A 1.6 A resolution X-ray crystallographic study of *Klebsiella pneumoniae* MoFe-protein. *Journal of Molecular Biology, 292*(4), 871–891. http://dx.doi.org/10.1006/jmbi.1999.3107.

McPherson, A., Jr. (1976). The growth and preliminary investigation of protein and nucleic acid crystals for X-ray diffraction analysis. *Methods of Biochemical Analysis, 23*, 249–345.

Melis, A., Seibert, M., & Happe, T. (2004). Genomics of green algal hydrogen research. *Photosynthesis Research, 82*(3), 277–288. http://dx.doi.org/10.1007/s11120-004-2050-2.

Mills, D. J., Vitt, S., Strauss, M., Shima, S., & Vonck, J. (2013). De novo modeling of the F420-reducing [NiFe]-hydrogenase from a methanogenic archaeon by cryo-electron microscopy. *eLife2*, e00218. http://dx.doi.org/10.7554/eLife.00218.

Milne, J. L. S., Borgnia, M. J., Bartesaghi, A., Tran, E. E. H., Earl, L. A., Schauder, D. M., et al. (2013). Cryo-electron microscopy: A primer for the non-microscopist. *The FEBS Journal, 280*(1), 28–45. http://dx.doi.org/10.1111/febs.12078.

Montet, Y., Amara, P., Volbeda, A., Vernede, X., Hatchikian, E. C., Field, M. J., et al. (1997). Gas access to the active site of Ni-Fe hydrogenases probed by X-ray crystallography and molecular dynamics. *Nature Structural & Molecular Biology, 4*(7), 523–526.

Morra, S., Arizzi, M., Valetti, F., & Gilardi, G. (2016). Oxygen stability in the new [FeFe]-hydrogenase from *Clostridium beijerinckii* SM10 (CbA5H). *Biochemistry, 55*(42), 5897–5900. http://dx.doi.org/10.1021/acs.biochem.6b00780.

Morrison, C. N., Hoy, J. A., Zhang, L., Einsle, O., & Rees, D. C. (2015). Substrate pathways in the nitrogenase MoFe protein by experimental identification of small molecule binding sites. *Biochemistry, 54*(11), 2052–2060. http://dx.doi.org/10.1021/bi501313k.

Mulder, D. W., Boyd, E. S., Sarma, R., Lange, R. K., Endrizzi, J. A., Broderick, J. B., et al. (2010). Stepwise [FeFe]-hydrogenase H-cluster assembly revealed in the structure of HydAΔEFG. *Nature, 465*(7295), 248–251. http://www.nature.com/nature/journal/v465/n7295/suppinfo/nature08993_S1.html.

Mulder, D. W., Ortillo, D. O., Gardenghi, D. J., Naumov, A. V., Ruebush, S. S., Szilagyi, R. K., et al. (2009). Activation of HydA(DeltaEFG) requires a preformed [4Fe-4S] cluster. *Biochemistry, 48*(26), 6240–6248. http://dx.doi.org/10.1021/bi9000563.

Mulder, D. W., Shepard, E. M., Meuser, J. E., Joshi, N., King, P. W., Posewitz, M. C., et al. (2011). Insights into FeFe-hydrogenase structure, mechanism, and maturation. *Structure, 19*(8), 1038–1052. http://dx.doi.org/10.1016/j.str.2011.06.008.

Murphy, B. J., Hidalgo, R., Roessler, M. M., Evans, R. M., Ash, P. A., Myers, W. K., et al. (2015). Discovery of dark pH-dependent H+ migration in a [NiFe]-hydrogenase and its mechanistic relevance: mobilizing the hydrido ligand of the Ni-C intermediate.

Journal of the American Chemical Society, *137*(26), 8484–8489. http://dx.doi.org/10.1021/jacs.5b03182.

Neutze, R., Wouts, R., van der Spoel, D., Weckert, E., & Hajdu, J. (2000). Potential for biomolecular imaging with femtosecond X-ray pulses. *Nature*, *406*(6797), 752–757. http://dx.doi.org/10.1038/35021099.

Nicolet, Y., Piras, C., Legrand, P., Hatchikian, C. E., & Fontecilla-Camps, J. C. (1999). Desulfovibrio desulfuricans iron hydrogenase: The structure shows unusual coordination to an active site Fe binuclear center. *Structure*, 7(1), 13–23. http://dx.doi.org/10.1016/S0969-2126(99)80005-7.

Ogata, H., Hirota, S., Nakahara, A., Komori, H., Shibata, N., Kato, T., et al. (2005). Activation process of [NiFe] hydrogenase elucidated by high-resolution X-ray analyses: Conversion of the ready to the unready state. *Structure*, *13*(11), 1635–1642. http://dx.doi.org/10.1016/j.str.2005.07.018.

Ogata, H., Lubitz, W., & Higuchi, Y. (2016). Structure and function of [NiFe] hydrogenases. *Journal of Biochemistry*, *160*(5), 251–258. http://dx.doi.org/10.1093/jb/mvw048.

Ogata, H., Mizoguchi, Y., Mizuno, N., Miki, K., Adachi, S.-I., Yasuoka, N., et al. (2002). Structural studies of the carbon monoxide complex of [NiFe]hydrogenase from *Desulfovibrio vulgaris* Miyazaki F: Suggestion for the initial activation site for dihydrogen. *Journal of the American Chemical Society*, *124*(39), 11628–11635. http://dx.doi.org/10.1021/ja012645k.

Ogata, H., Nishikawa, K., & Lubitz, W. (2015). Hydrogens detected by subatomic resolution protein crystallography in a [NiFe] hydrogenase. *Nature*, *520*(7548), 571–574. http://dx.doi.org/10.1038/nature14110.

Orlova, E. V., & Saibil, H. R. (2011). Structural analysis of macromolecular assemblies by electron microscopy. *Chemical Reviews*, *111*(12), 7710–7748. http://dx.doi.org/10.1021/cr100353t.

Owens, C. P., Katz, F. E. H., Carter, C. H., Oswald, V. F., & Tezcan, F. A. (2016). Tyrosine-coordinated P-cluster in G. diazotrophicus nitrogenase: Evidence for the importance of O-based ligands in conformationally gated electron transfer. *Journal of the American Chemical Society*, *138*(32), 10124–10127. http://dx.doi.org/10.1021/jacs.6b06783.

Pandelia, M.-E., Nitschke, W., Infossi, P., Giudici-Orticoni, M.-T., Bill, E., & Lubitz, W. (2011). Characterization of a unique [FeS] cluster in the electron transfer chain of the oxygen tolerant [NiFe] hydrogenase from *Aquifex aeolicus*. *Proceedings of the National Academy of Sciences of the United States of America*, *108*(15), 6097–6102. http://dx.doi.org/10.1073/pnas.1100610108.

Pandey, A. S., Harris, T. V., Giles, L. J., Peters, J. W., & Szilagyi, R. K. (2008). Dithiomethylether as a ligand in the hydrogenase H-cluster. *Journal of the American Chemical Society*, *130*(13), 4533–4540. http://dx.doi.org/10.1021/ja711187e.

Pardo, A., De Lacey, A. L., Fernandez, V. M., Fan, H. J., Fan, Y., & Hall, M. B. (2006). Density functional study of the catalytic cycle of nickel-iron [NiFe] hydrogenases and the involvement of high-spin nickel(II). *Journal of Biological Inorganic Chemistry*, *11*(3), 286–306. http://dx.doi.org/10.1007/s00775-005-0076-3.

Peters, J. W., Fisher, K., & Dean, D. R. (1995). Nitrogenase structure and function: A biochemical-genetic perspective. *Annual Review of Microbiology*, *49*, 335–366. http://dx.doi.org/10.1146/annurev.mi.49.100195.002003.

Peters, J. W., Lanzilotta, W. N., Lemon, B. J., & Seefeldt, L. C. (1998). X-ray crystal structure of the Fe-only hydrogenase (CpI) from *Clostridium pasteurianum* to 1.8 angstrom resolution. *Science*, *282*(5395), 1853–1858.

Peters, J. W., Schut, G. J., Boyd, E. S., Mulder, D. W., Shepard, E. M., Broderick, J. B., et al. (2015). [FeFe]- and [NiFe]-hydrogenase diversity, mechanism, and maturation.

Biochimica et Biophysica Acta (BBA)—Molecular Cell Research, *1853*(6), 1350–1369. http://dx.doi.org/10.1016/j.bbamcr.2014.11.021.

Peters, J. W., Stowell, M. H., Soltis, S. M., Finnegan, M. G., Johnson, M. K., & Rees, D. C. (1997). Redox-dependent structural changes in the nitrogenase P-cluster. *Biochemistry*, *36*(6), 1181–1187. http://dx.doi.org/10.1021/bi9626665.

Poudel, S., Tokmina-Lukaszewska, M., Colman, D. R., Refai, M., Schut, G. J., King, P. W., et al. (2016). Unification of [FeFe]-hydrogenases into three structural and functional groups. *Biochimica et Biophysica Acta*, *1860*(9), 1910–1921. http://dx.doi.org/10.1016/j.bbagen.2016.05.034.

Rangaraj, P., Ryle, M. J., Lanzilotta, W. N., Goodwin, P. J., Dean, D. R., Shah, V. K., et al. (1999). Inhibition of iron-molybdenum cofactor biosynthesis by L127Delta NifH and evidence for a complex formation between L127Delta NifH and NifNE. *The Journal of Biological Chemistry*, *274*(41), 29413–29419.

Renner, K. A., & Howard, J. B. (1996). Aluminum fluoride inhibition of nitrogenase: Stabilization of a nucleotide.Fe-protein.MoFe-protein complex. *Biochemistry*, *35*(17), 5353–5358. http://dx.doi.org/10.1021/bi960441o.

Sarma, R., Barney, B. M., Keable, S., Dean, D. R., Seefeldt, L. C., & Peters, J. W. (2010). Insights into substrate binding at FeMo-cofactor in nitrogenase from the structure of an alpha-70(Ile) MoFe protein variant. *Journal of Inorganic Biochemistry*, *104*(4), 385–389. http://dx.doi.org/10.1016/j.jinorgbio.2009.11.009.

Schindelin, H., Kisker, C., Schlessman, J. L., Howard, J. B., & Rees, D. C. (1997). Structure of ADP x AIF4(-)-stabilized nitrogenase complex and its implications for signal transduction. *Nature*, *387*(6631), 370–376. http://dx.doi.org/10.1038/387370a0.

Schlessman, J. L., Woo, D., Joshua-Tor, L., Howard, J. B., & Rees, D. C. (1998). Conformational variability in structures of the nitrogenase iron proteins from *Azotobacter vinelandii* and *Clostridium pasteurianum*. *Journal of Molecular Biology*, *280*(4), 669–685. http://dx.doi.org/10.1006/jmbi.1998.1898.

Schmid, B., Ribbe, M. W., Einsle, O., Yoshida, M., Thomas, L. M., Dean, D. R., et al. (2002). Structure of a cofactor-deficient nitrogenase MoFe protein. *Science*, *296*(5566), 352–356. http://dx.doi.org/10.1126/science.1070010.

Schut, G. J., & Adams, M. W. W. (2009). The iron-hydrogenase of *Thermotoga maritima* utilizes ferredoxin and NADH synergistically: A new perspective on anaerobic hydrogen production. *Journal of Bacteriology*, *191*(13), 4451–4457. http://dx.doi.org/10.1128/jb.01582-08.

Schut, G. J., Zadvornyy, O., Wu, C. H., Peters, J. W., Boyd, E. S., & Adams, M. W. (2016). The role of geochemistry and energetics in the evolution of modern respiratory complexes from a proton-reducing ancestor. *Biochimica et Biophysica Acta*, *1857*(7), 958–970. http://dx.doi.org/10.1016/j.bbabio.2016.01.010.

Scott, D. J., Dean, D. R., & Newton, W. E. (1992). Nitrogenase-catalyzed ethane production and CO-sensitive hydrogen evolution from MoFe proteins having amino acid substitutions in an alpha-subunit FeMo cofactor-binding domain. *The Journal of Biological Chemistry*, *267*(28), 20002–20010.

Seefeldt, L. C., Hoffman, B. M., & Dean, D. R. (2009). Mechanism of Mo-dependent nitrogenase. *Annual Review of Biochemistry*, *78*, 701–722. http://dx.doi.org/10.1146/annurev.biochem.78.070907.103812.

Seefeldt, L. C., Hoffman, B. M., & Dean, D. R. (2012). Electron transfer in nitrogenase catalysis. *Current Opinion in Chemical Biology*, *16*(1-2), 19–25. http://dx.doi.org/10.1016/j.cbpa.2012.02.012.

Sen, S., Igarashi, R., Smith, A., Johnson, M. K., Seefeldt, L. C., & Peters, J. W. (2004). A conformational mimic of the MgATP-bound "on state" of the nitrogenase iron protein. *Biochemistry*, *43*(7), 1787–1797. http://dx.doi.org/10.1021/bi0358465.

Sen, S., Krishnakumar, A., McClead, J., Johnson, M. K., Seefeldt, L. C., Szilagyi, R. K., et al. (2006). Insights into the role of nucleotide-dependent conformational change in nitrogenase catalysis: Structural characterization of the nitrogenase Fe protein Leu127 deletion variant with bound MgATP. *Journal of Inorganic Biochemistry, 100*(5-6), 1041–1052. http://dx.doi.org/10.1016/j.jinorgbio.2006.02.016.

Shah, V. K., & Brill, W. J. (1973). Nitrogenase. IV. Simple method of purification to homogeneity of nitrogenase components from *Azotobacter vinelandii*. *Biochimica et Biophysica Acta, 305*(2), 445–454.

Shepard, E. M., Mus, F., Betz, J. N., Byer, A. S., Duffus, B. R., Peters, J. W., et al. (2014). [FeFe]-hydrogenase maturation. *Biochemistry, 53*(25), 4090–4104. http://dx.doi.org/10.1021/bi500210x.

Shi, D., Nannenga, B. L., de la Cruz, M. J., Liu, J., Sawtelle, S., Calero, G., et al. (2016). The collection of MicroED data for macromolecular crystallography. *Nature Protocols, 11*(5), 895–904. http://dx.doi.org/10.1038/nprot.2016.046.

Shima, S., & Thauer, R. K. (2007). A third type of hydrogenase catalyzing H2 activation. *Chemical Record*, 7(1), 37–46. http://dx.doi.org/10.1002/tcr.20111.

Sorlie, M., Christiansen, J., Lemon, B. J., Peters, J. W., Dean, D. R., & Hales, B. J. (2001). Mechanistic features and structure of the nitrogenase alpha-Gln195 MoFe protein. *Biochemistry, 40*(6), 1540–1549.

Spatzal, T., Aksoyoglu, M., Zhang, L., Andrade, S. L., Schleicher, E., Weber, S., et al. (2011). Evidence for interstitial carbon in nitrogenase FeMo cofactor. *Science, 334*(6058), 940. http://dx.doi.org/10.1126/science.1214025.

Spatzal, T., Perez, K. A., Einsle, O., Howard, J. B., & Rees, D. C. (2014). Ligand binding to the FeMo-cofactor: Structures of CO-bound and reactivated nitrogenase. *Science, 345*(6204), 1620–1623. http://dx.doi.org/10.1126/science.1256679.

Spatzal, T., Perez, K. A., Howard, J. B., & Rees, D. C. (2015). Catalysis-dependent selenium incorporation and migration in the nitrogenase active site iron-molybdenum cofactor. *eLife. 4*, e11620. http://dx.doi.org/10.7554/eLife.11620.

Stiebritz, M. T., & Reiher, M. (2012). Hydrogenases and oxygen. *Chemical Science, 3*(6), 1739–1751. http://dx.doi.org/10.1039/C2SC01112C.

Stripp, S. T., Goldet, G., Brandmayr, C., Sanganas, O., Vincent, K. A., Haumann, M., et al. (2009). How oxygen attacks [FeFe] hydrogenases from photosynthetic organisms. *Proceedings of the National Academy of Sciences, 106*(41), 17331–17336. http://dx.doi.org/10.1073/pnas.0905343106.

Strop, P., Takahara, P. M., Chiu, H., Angove, H. C., Burgess, B. K., & Rees, D. C. (2001). Crystal structure of the all-ferrous [4Fe-4S]0 form of the nitrogenase iron protein from *Azotobacter vinelandii*. *Biochemistry, 40*(3), 651–656.

Swanson, K. D., Ratzloff, M. W., Mulder, D. W., Artz, J. H., Ghose, S., Hoffman, A., et al. (2015). [FeFe]-hydrogenase oxygen inactivation is initiated at the H cluster 2Fe subcluster. *Journal of the American Chemical Society, 137*(5), 1809–1816. http://dx.doi.org/10.1021/ja510169s.

Tamagnini, P., Leitão, E., Oliveira, P., Ferreira, D., Pinto, F., Harris, D. J., et al. (2007). Cyanobacterial hydrogenases: Diversity, regulation and applications. *FEMS Microbiology Reviews, 31*(6), 692–720. http://dx.doi.org/10.1111/j.1574-6976.2007.00085.x.

Tersteegen, A., & Hedderich, R. (1999). *Methanobacterium thermoautotrophicum* encodes two multisubunit membrane-bound [NiFe] hydrogenases. *European Journal of Biochemistry, 264*(3), 930–943. http://dx.doi.org/10.1046/j.1432-1327.1999.00692.x.

Tezcan, F. A., Kaiser, J. T., Howard, J. B., & Rees, D. C. (2015). Structural evidence for asymmetrical nucleotide interactions in nitrogenase. *Journal of the American Chemical Society, 137*(1), 146–149. http://dx.doi.org/10.1021/ja511945e.

Tezcan, F. A., Kaiser, J. T., Mustafi, D., Walton, M. Y., Howard, J. B., & Rees, D. C. (2005). Nitrogenase complexes: Multiple docking sites for a nucleotide switch protein. *Science, 309*(5739), 1377–1380. http://dx.doi.org/10.1126/science.1115653.

Thauer, R. K., Kaster, A.-K., Goenrich, M., Schick, M., Hiromoto, T., & Shima, S. (2010). Hydrogenases from methanogenic archaea, nickel, a novel cofactor, and H2 storage.

Annual Review of Biochemistry, 79(1), 507–536. http://dx.doi.org/10.1146/annurev.biochem.030508.152103.

Thony, B., Kaluza, K., & Hennecke, H. (1985). Structural and functional homology between the alpha-subunits and beta-subunits of the nitrogenase MoFe protein as revealed by sequencing the *Rhizobium japonicum* Nifk gene. *Molecular & General Genetics, 198*(3), 441–448. http://dx.doi.org/10.1007/Bf00332937.

van Dijk, C., van Berkel-Arts, A., & Veeger, C. (1983). The effect of re-oxidation on the reduced hydrogenase of *Desulfovibrio vulgaris* strain Hildenborough and its oxygen stability. *FEBS Letters, 156*(2), 340–344. http://dx.doi.org/10.1016/0014-5793(83)80525-0.

Verhagen, M. F., O'Rourke, T., & Adams, M. W. (1999). The hyperthermophilic bacterium, *Thermotoga maritima*, contains an unusually complex iron-hydrogenase: Amino acid sequence analyses versus biochemical characterization. *Biochimica et Biophysica Acta, 1412*(3), 212–229.

Vignais, P. M., & Billoud, B. (2007). Occurrence, classification, and biological function of hydrogenases: An overview. *Chemical Reviews, 107*(10), 4206–4272. http://dx.doi.org/10.1021/cr050196r.

Vignais, P. M., Billoud, B., & Meyer, J. (2001). Classification and phylogeny of hydrogenases. *FEMS Microbiology Reviews, 25*(4), 455–501.

Vitt, S., Ma, K., Warkentin, E., Moll, J., Pierik, A. J., Shima, S., et al. (2014). The F420-reducing [NiFe]-hydrogenase complex from *Methanothermobacter marburgensis*, the first X-ray structure of a group 3 family member. *Journal of Molecular Biology, 426*(15), 2813–2826. http://dx.doi.org/10.1016/j.jmb.2014.05.024.

Volbeda, A., Amara, P., Darnault, C., Mouesca, J.-M., Parkin, A., Roessler, M. M., et al. (2012). X-ray crystallographic and computational studies of the O2-tolerant [NiFe]-hydrogenase 1 from *Escherichia coli*. *Proceedings of the National Academy of Sciences, 109*(14), 5305–5310. http://dx.doi.org/10.1073/pnas.1119806109.

Volbeda, A., Amara, P., Iannello, M., De Lacey, A. L., Cavazza, C., & Fontecilla-Camps, J. C. (2013). Structural foundations for the O2 resistance of *Desulfomicrobium baculatum* [NiFeSe]-hydrogenase. *Chemical Communications, 49*(63), 7061–7063. http://dx.doi.org/10.1039/C3CC43619E.

Volbeda, A., Charon, M.-H., Piras, C., Hatchikian, E. C., Frey, M., & Fontecilla-Camps, J. C. (1995). Crystal structure of the nickel-iron hydrogenase from *Desulfovibrio gigas*. *Nature, 373*(6515), 580–587.

Volbeda, A., Martin, L., Barbier, E., Gutiérrez-Sanz, O., De Lacey, A. L., Liebgott, P.-P., et al. (2015). Crystallographic studies of [NiFe]-hydrogenase mutants: Towards consensus structures for the elusive unready oxidized states. *JBIC: Journal of Biological Inorganic Chemistry, 20*(1), 11–22. http://dx.doi.org/10.1007/s00775-014-1203-9.

Volbeda, A., Martin, L., Cavazza, C., Matho, M., Faber, B. W., Roseboom, W., et al. (2005). Structural differences between the ready and unready oxidized states of [NiFe] hydrogenases. *Journal of Biological Inorganic Chemistry, 10*(3), 239–249. http://dx.doi.org/10.1007/s00775-005-0632-x.

Wang, S., Huang, H., Kahnt, J., & Thauer, R. K. (2013). A reversible electron-bifurcating ferredoxin- and NAD-dependent [FeFe]-hydrogenase (HydABC) in *Moorella thermoacetica*. *Journal of Bacteriology, 195*(6), 1267–1275. http://dx.doi.org/10.1128/jb.02158-12.

Weininger, M. S., & Mortenson, L. E. (1982). Crystallographic properties of the MoFe proteins of nitrogenase from *Clostridium pasteurianum* and *Azotobacter vinelandii*. *Proceedings of the National Academy of Sciences of the United States of America, 79*(2), 378–380.

Yang, K. Y., Haynes, C. A., Spatzal, T., Rees, D. C., & Howard, J. B. (2014). Turnover-dependent inactivation of the nitrogenase MoFe-protein at high pH. *Biochemistry, 53*(2), 333–343. http://dx.doi.org/10.1021/bi4014769.

Zhang, L.-M., Morrison, C. N., Kaiser, J. T., & Rees, D. C. (2015). Nitrogenase MoFe protein from *Clostridium pasteurianum* at 1.08 Å resolution: Comparison with the *Azotobacter vinelandii* MoFe protein. *Acta Crystallographica Section D: Biological Crystallography*, *71*(Pt 2), 274–282. http://dx.doi.org/10.1107/S1399004714025243.

Zhang, X., Sherman, D. M., & Sherman, L. A. (2014). The uptake hydrogenase in the unicellular diazotrophic cyanobacterium *Cyanothece* sp. strain PCC 7822 protects nitrogenase from oxygen toxicity. *Journal of Bacteriology*, *196*(4), 840–849. http://dx.doi.org/10.1128/jb.01248-13.

CHAPTER NINE

Nitrogenase Assembly: Strategies and Procedures

Nathaniel S. Sickerman, Yilin Hu[1], Markus W. Ribbe[1]
University of California, Irvine, Irvine, CA, United States
[1]Corresponding authors: e-mail address: yilinh@uci.edu; mribbe@uci.edu

Contents

Abstract

Nitrogenase is a metalloenzyme system that plays a critical role in biological nitrogen fixation, and the study of how its metallocenters are assembled into functional entities to facilitate the catalytic reduction of dinitrogen to ammonia is an active area of interest. The diazotroph *Azotobacter vinelandii* is especially amenable to culturing and genetic manipulation, and this organism has provided the basis for many insights into the assembly of nitrogenase proteins and their respective metallocofactors. This chapter will cover the basic procedures necessary for growing *A. vinelandii* cultures and subsequent recombinant transformation and protein expression techniques. Furthermore, protocols for nitrogenase protein purification and substrate reduction activity assays are described. These methods provide a solid framework for the assessment of nitrogenase assembly and catalysis.

Methods in Enzymology, Volume 595
ISSN 0076-6879
http://dx.doi.org/10.1016/bs.mie.2017.07.006

1. INTRODUCTION

The subset of prokaryotic organisms that produces the enzyme nitrogenase has been responsible for much of the nitrogen consumed by humans on this planet through its production of a bioavailable nitrogen source (Crews & Peoples, 2004). Nitrogenase is the only enzyme known to convert the stable dinitrogen (N_2) molecule into ammonia (NH_3) and does so through a complex reductive process (Burgess & Lowe, 1996; Hoffman, Lukoyanov, Yang, Dean, & Seefeldt, 2014). That this single enzyme has such an outsize influence on the global nitrogen cycle is testament to its critical importance to life. For years, nitrogenase enzymes from different diazotrophs have been studied to understand how the respective components facilitate the activation and reduction of N_2 (Dean & Jacobson, 1992; Eady, 1996). In particular, *Azotobacter vinelandii* has emerged as a preferred organism for genetic manipulation and protein expression (Brill, 1980). Through the study nitrogenase in *A. vinelandii* and other organisms, a detailed picture of the proteins involved in catalytic substrate reduction has emerged (Burgess & Lowe, 1996; Hoffman et al., 2014). Three different variants of nitrogenase have been discovered so far: the most well-studied type contains molybdenum (Mo'nitrogenase) and is the predominant form expressed by all diazotrophs. The other variants, the vanadium-containing V'nitrogenase and the iron-only Fe'nitrogenase, are also competent toward N_2 reduction, albeit with lower efficiencies, in the order Mo > V > Fe (Eady, 1996). These so-called alternative nitrogenase systems have unique properties and reactivity, but the study Mo'nitrogenase is much more established, and expression of this particular enzyme and its components will be the primary focus of this chapter.

Molybdenum nitrogenase consists of two main protein components. A reductase component called the iron (Fe) protein is composed of two identical 33 kDa subunits each containing a Walker A motif for the binding of magnesium adenosine triphosphate (MgATP) (Georgiadis et al., 1992; Schlessman, Woo, Joshua-Tor, Howard, & Rees, 1998). A single, solvent-exposed Fe_4S_4 cluster binds at the subunit interface of the Fe protein, which is the essential electron transfer partner for the catalytic component, called the MoFe protein (Burgess & Lowe, 1996; Hoffman et al., 2014). The MoFe protein is a 220-kDa $\alpha_2\beta_2$ tetramer that contains two sophisticated iron–sulfur metalloclusters that are critical for catalysis (Ribbe, Hu, Hodgson, & Hedman, 2014). At the $\alpha\beta$-interface of the protein, the subunit-bridging, $[Fe_8S_7]$-core P-cluster controls the flow of reducing equivalents to the active site (Chan, Kim, & Rees, 1993). The active site of Mo'nitrogenase is buried within the α subunit of the MoFe protein and contains a $[MoFe_7S_9C]$(*R*-homocitrate) cofactor called the M-cluster (Einsle et al., 2002; Lancaster et al., 2011); this cofactor receives electrons from the P-cluster and is the site of substrate binding and reduction (Burgess, 1990; Burgess & Lowe, 1996; Hoffman et al., 2014).

Biochemical and structural studies have demonstrated that during catalysis, a transient complex is formed between the reduced Fe and MoFe proteins (Burgess & Lowe, 1996; Schindelin, Kisker, Schlessman, Howard, & Rees, 1997). Moreover, the complexation process is dependent upon the hydrolysis of MgATP by the Fe protein (Bulen & LeComte, 1966). Although the interaction between the two components is dynamic, the full complex can be trapped by the use of a nonhydrolyzable MgATP analog (Schindelin et al., 1997) (Fig. 1). Structural

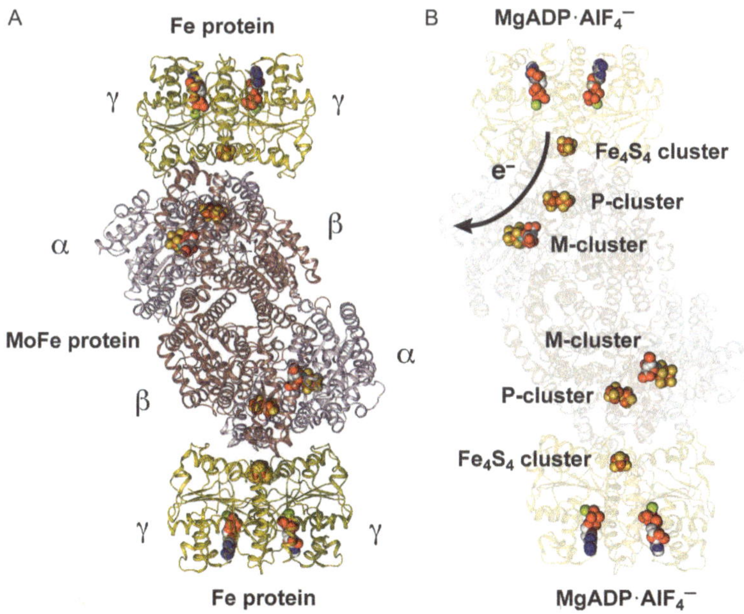

Fig. 1 (A) Representation of a complex between *A. vinelandii* nitrogenase MoFe ($\alpha_2\beta_2$) and Fe (γ_2) protein components. The normally dynamic complex structure is shown trapped with the nonhydrolyzable ATP analog $ADP \cdot AlF_4^-$. (B) Electrons flow as indicated from the Fe_4S_4 cluster of the Fe protein to the MoFe protein P-cluster, then to the M-cluster where catalytic substrate reduction occurs. Atoms are depicted as the following colored spheres: C, *light gray*; N, *blue*; O, *red*; F, *sky blue*; S, *yellow*; Mg, *green*; Al, *medium gray*; Fe, *orange*; Mo, *cyan*. Images made with PYMOL (PDB: 1N2C).

analysis of the Fe/MoFe protein complex shows that the short distance between Fe_4S_4 cluster of the Fe protein and the P-cluster of the MoFe protein should facilitate electron transfer. After electron transfer from its Fe_4S_4 cluster, the Fe protein undergoes a conformational change that leads to the hydrolysis of the two bound MgATP molecules (Schlessman et al., 1998); following MgATP hydrolysis, the Fe protein dissociates from the MoFe protein (Danyal, Dean, Hoffman, & Seefeldt, 2011). The reducing equivalents pass from the reduced P-cluster to the M-cluster, where N_2 binds and is reduced to NH_3 (Burgess & Lowe, 1996; Hoffman et al., 2014). Overall, multiple association and dissociation steps must occur for the reduced Fe protein to deliver enough electrons to effect substrate reduction (Eq. 1).

$$N_2 + 8e^- + 16MgATP + 8H^+ \rightarrow 2NH_3 + 16MgADP + 16P_i \quad (1)$$

Nitrogenase research has focused on understanding the structure–function relationships of the proteins involved in biological nitrogen fixation and their biosynthesis. These proteins include not just the main enzyme components (Fe and MoFe proteins) but also the series of proteins responsible for assembly of the complex P- and M-clusters and many other regulatory, chaperone, and maturation functions (Hu & Ribbe, 2016; Ribbe et al., 2014). Having a reliable method to express and purify these proteins, many of which are sensitive to dioxygen (O_2), has been a continuing challenge in the field (Paustian, Shah, & Roberts, 1990). However, the understanding of nitrogenase protein and metallocofactor assembly has met with considerable success in the past decade through the combined utilization of molecular biology, biochemistry, chemistry, and spectroscopy approaches (Corbett et al., 2004; Hu, Corbett, et al., 2005; Lee, Blank, et al., 2009; Wiig, Hu, & Ribbe, 2011, 2015).

This chapter will first discuss methods to culture and genetically manipulate *A. vinelandii* for the purposes of studying nitrogenase assembly and function (Section 2). Following this, protocols for the growth of large-scale cultures and expression of nitrogenase proteins will be described (Section 3). Next, methods are provided for the isolation and purification of expressed nitrogenase proteins (Section 4). Finally, substrate reduction assays for evaluating protein activity will be outlined (Section 5). In all, these strategies and procedures provide a basis for the study of the many fascinating aspects of nitrogenase assembly and catalysis.

2. MOLECULAR BIOLOGY OF *A. vinelandii*

For Mo'nitrogenase, at least 11 nitrogen fixation (*nif*) genes must work in concert to reduce N_2. The main components of Mo'nitrogenase, Fe protein and MoFe protein, are encoded by *nif*H and *nif*DK, respectively. In addition, a suite of other *nif* genes—*nif*E, *nif*N, *nif*U, *nif*S, *nif*V, *nif*M, *nif*B, and *nif*Q—fulfill essential metallocluster biosynthesis and maturation roles (Brill, 1980; Dean & Jacobson, 1992; Ribbe et al., 2014). Expressing certain nitrogenase proteins in *Escherichia coli* has met success in recent years, but to date, the complex interplay of the required genes has hampered heterologous expression of a full, functional nitrogenase (Fay, Wiig, Lee, & Hu, 2015; Wilcoxen et al., 2016). Accordingly, current molecular biology methods with *E. coli* are not convenient for nitrogenase research, and this drawback has necessitated the use of homologous systems from model organisms such as *Clostridium pasteurianum*, *Klebsiella pneumoniae*, *Azotobacter chroococcum*, and *A. vinelandii* (Dean & Jacobson, 1992; Eady, 1996). The homologous expression of nitrogenase proteins in *A. vinelandii* has been a preferred approach to study the enzyme for several reasons. First, the organism is a strict aerobe and requires no specialized nutrients or conditions for growth (Kennedy & Bishop, 2004). In addition, *A. vinelandii* has an extremely high respiration rate, allowing it to more effectively fix nitrogen due to a high N_2 flux (Curatti, Brown, Ludden, & Rubio, 2005; Kennedy, Rudnick, MacDonald, & Melton, 2005). Despite the increased aeration in vivo, protective proteins and the ability to form an alginate capsule limit the damage of nitrogenase by O_2 (Maier & Moshiri, 2000; Sabra, Zeng, Lünsdorf, & Deckwer, 2000). In fact, the optimal expression and functioning of nitrogenase is carried out in microaerobic conditions (~2% dissolved O_2) (Kelly, Poole, Yates, & Kennedy, 1990; Kennedy et al., 1991).

Finally, in *A. vinelandii*, all three variants of nitrogenase (Mo, V, and Fe) can be expressed, and the genetic tractability of the organism makes it amenable to experimentation through gene deletion and site-directed mutants (Kennedy et al., 1991; Paustian et al., 1990; Ribbe, Hu, Guo, Schmid, & Burgess, 2002). Notably, the use of these types of mutants has contributed to the understanding of metallocofactor biosynthesis and assembly (Ribbe et al., 2014). Making competent *A. vinelandii* cells is a straightforward process that does not involve specialized or expensive reagents, and transformation can be carried out at room temperature (Page & von Tigerstrom, 1979).

In this section, strategies for cell growth, transformation, and recombinant strain selection will be discussed.

2.1 Cultivating Cultures of *A. vinelandii*

Strains of *A. vinelandii* are commercially available from sources such as the American Type Culture Collection (ATCC) and provide a starting point into nitrogenase research. This diazotroph is most commonly grown at an optimal temperature of 30°C in Burk's minimal medium (B-medium) or B-medium that has been supplemented with a fixed nitrogen source (B+-medium) (Curatti et al., 2005; Kennedy et al., 2005; Pena, Campos, & Galindo, 1997). Cell growth can be carried out on solid agar B+-media and in liquid B+-media, and both are necessary for the maintenance, culturing, and transformation of *A. vinelandii* (Dos Santos, 2011). Various preparations of B+-medium have been reported, and slightly different B+-medium formulations can be used. For example, certain growth conditions may be varied if desired, including the use of 10 m*M* urea, which unlike ammonium does not repress nitrogenase expression (Kennedy et al., 2005), or the use of alternative carbon sources such as acetate, glucose, or glycerol in place of sucrose (Esposito & Wilson, 1958; Kennedy et al., 1991; Scrutton & Taylor, 1974). In general, the optimum growth for *A. vinelandii* cultures in liquid B+-medium containing ammonium is 30°C at 300 rpm, with a doubling time of approximately 2 h (Page & von Tigerstrom, 1979). On solid B+-medium grown at 30°C, colonies become visible after about 2 days. Frozen cell stocks are prepared from fresh plates, and when using a frozen cell stock for liquid media inoculation, a sterile pipette tip can be used to transfer a small frozen aliquot.

In the presence of ammonium, *nif* gene expression is repressed (Kennedy et al., 1991); therefore, sufficient amounts of this nitrogen source (10–30 m*M*) should be added to the media to prevent premature expression of nitrogenase proteins. Depending on the experiment, urea can also substitute for ammonium to avoid the repression of nitrogenase genes (Kennedy et al., 1991). Typically, limiting amounts of ammonium are only added to large-scale cultures for the specific intent of nitrogenase expression (Section 3). Under nitrogen-deficient conditions, *nif* genes will become derepressed and nitrogenase expression will occur (Nicholas & Deering, 1976; Shah, Davis, & Brill, 1972): for strains incapable of diazotrophic growth, 2–4 h are optimal to allow for the expression of nitrogenase proteins.

2.1.1 Equipment

- Incubator (New Brunswick, Edison, NJ)
- Incubator–shaker (New Brunswick, Edison, NJ)
- Ultralow temperature (−80°C) freezer (Revco, Asheville, NC)

2.1.2 Buffers and Reagents

- *A. vinelandii* strains from ATCC (Manassas, VA)
- Ultrapure water (dH_2O)
- Burk's minimal medium (B-medium) with ammonium (B+-medium):
  - Prepared from 900 mL of phosphate solution and 100 mL of 10× salts solution (see below) per 1 L
  - Phosphate solution: 0.2 g K_2HPO_4, 0.8 g KH_2PO_4 in 900 mL dH_2O
  - 10× salts solution: 200 g sucrose, 2.0 g $MgSO_4 \cdot 7H_2O$, 0.9 g $CaCl_2 \cdot 2H_2O$, 15 g $NH_4(CH_3CO_2)$ (ammonium acetate), 0.5 m*M* $FeSO_4 \cdot 7H_2O$, 0.1 m*M* $Na_2MoO_4 \cdot 2H_2O$ in 1 L dH_2O
- B+-medium plates: prepared from B+-medium autoclaved with 15 g agar per 1 L of medium
- Storage buffer: sterile-filtered phosphate buffer (0.2 g KH_2PO_4, 0.8 g K_2HPO_4, and 900 mL dH_2O) with 1% v/v dimethylsulfoxide

2.1.3 Procedure

1. Spread cells on a B+-medium plate either directly from frozen cell stocks or from a 2-mL culture incubated 16–20 h at 30°C and 300 rpm.
2. Incubate plate for 2–3 days at 30°C.
3. Use plate for either liquid B+-medium cultures or preparation of frozen cell stocks (Note 1).
4. Store colonized plates at room temperature (Note 2).

2.1.4 Notes

1. Prepare frozen cell stocks by combining a full loop of cells (~8 mm^2) from a B+-medium plate with 1 mL of storage buffer. Disperse the cell material with brief vortexing and store in plastic cryotubes at −80°C.
2. *A. vinelandii* plates are viable for several weeks but are best at <1 week old.

2.2 Transformation of Competent *A. vinelandii* Cells

The first *A. vinelandii* strains were identified in the early 20th century shortly after the discovery of *A. chroococcum* (Beijerinck, 1901; Lipman, 1903, 1908).

However, cultivation and biochemical studies with these strains were limited by the "gummy" nature of the bacterial cultures. One of the earliest *A. vinelandii* strains utilized for genetic manipulation was a "nongummy" phenotype designated OP (Bush & Wilson, 1959). The OP strain was subsequently renamed CA (North Carolina University) and UW (University of Wisconsin), and notable derivative strains include DJ (Deloria Jacobs), MV (mutant *vinelandii*), and YM (Yilin Hu & Markus Ribbe) designations (Kennedy & Bishop, 2004; Lee, Hu, & Ribbe, 2009). Many derivatives of this original strain are still considered as wild-type, even if certain genes have been modified to produce His-tagged and Strep-tagged proteins.

The generation of *A. vinelandii* cells that are competent for DNA incorporation (Fig. 2) is much more straightforward than with *E. coli* and can be accomplished by simply growing the bacterial cultures in Mo- and Fe-deficient media (Page & von Tigerstrom, 1979). Under Fe-starvation conditions, *A. vinelandii* expresses and secretes a fluorescent green siderophore (Page & Grant, 1987). The release of this siderophore has been correlated to the ability of the organism to take up DNA, although the mechanism is unknown. Regardless, the greenish-yellow tint of these *A. vinelandii* cultures can be clearly observed and serves as an indicator of competency.

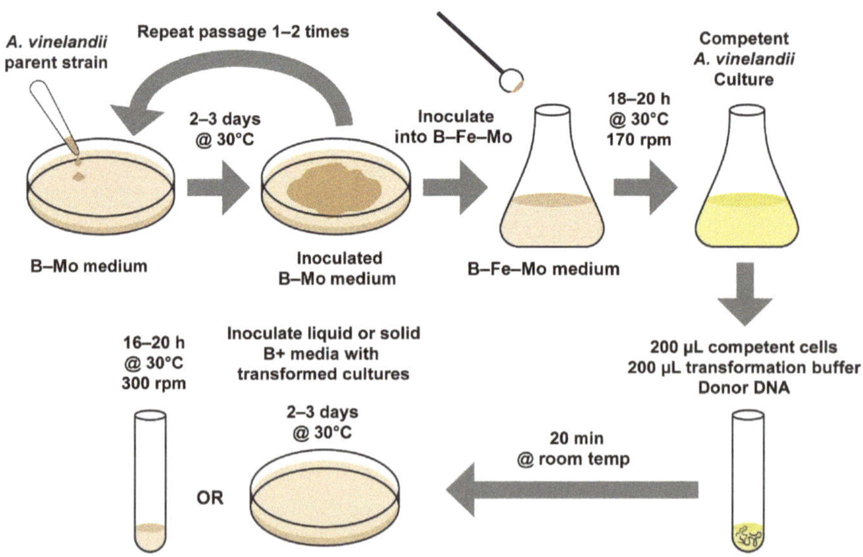

Fig. 2 General schematic for the recombinant DNA transformation of competent *A. vinelandii* cells.

Incubating these green *A. vinelandii* cultures with the appropriate donor DNA should incorporate the genetic modification through double-reciprocal recombination. To improve the chances of recombination, the DNA constructs should contain 150–500 bp of homologous flanking regions on either side of the modified sequence (Dos Santos, 2011; Jacobson, Cash, et al., 1989). The transformation process is complicated by the presence of multiple genome equivalents within late-stage growth cells, which may necessitate larger amounts of DNA to ensure total recombination (Page & von Tigerstrom, 1979).

The type of DNA used in transformation will vary based on the desired genetic modification, phenotype of the parent strain, and the intended selection strategy. Selection approaches are discussed in Section 2.3. Auxotrophic or knockout strains can be generated through the insertion of antibiotic resistance cassettes into the coding sequences of specific genes; transposon mutagenesis can also be utilized for this purpose (Kennedy et al., 1986; Premakumar, Loveless, & Bishop, 1994). These techniques enable direct antibiotic selection or can be used in conjunction with congression methods. Blue and white selection using the *lacZ* gene is possible as well (Section 2.3), but since *A. vinelandii* naturally cannot express β-D-galactosidase, *E. coli lacZ* must be inserted (Joerger, Wolfinger, & Bishop, 1991; Walmsley, Toukdarian, & Kennedy, 1994). This approach can be used to directly insert *lacZ* into *A. vinelandii* to act as a reporter for the expression of nitrogenase genes. Conversely, a parent strain of *A. vinelandii* already containing *lacZ* at the recombination site of interest can be used, whereby transformation will prevent the recombinant strain from expressing β-D-galactosidase (Miller, 1972). Donor DNA can be isolated from *A. vinelandii* strains through conventional methods or can be derived from other organisms, but certain recombinant plasmids used in *E. coli*, including pUC, M13, or pET series, are not compatable in *A. vinelandii* (Kennedy et al., 2005).

2.2.1 Equipment

- Same as Section 2.1
- Sterile polystyrene culture tubes

2.2.2 Buffers and Reagents

- Donor DNA (Note 1)
- Antibiotics solutions (Note 2)

- B–Mo(–Fe) medium: B+-medium lacking $Na_2MoO_4·2H_2O$ (B—Mo) or both $Na_2MoO_4·2H_2O$ and $FeSO_4·7H_2O$ (B–Mo–Fe), in the 10× salts solution
- B-medium plates: prepared from B-, B+-, B–Mo, or B–Mo–Fe media, autoclaved with 15 g agar per 1 L of medium
- Antibiotic-selective B+-medium plates: B+-medium plates with added antibiotics (Note 3)
- Transformation buffer: sterile-filtered 20 m*M* 3-(*N*-morpholino)propanesulfonic acid (MOPS), pH 7.2, 20 m*M* $MgCl_2$

2.2.3 Procedure

1. Inoculate an agar/B–Mo plate with cell stock.
2. Incubate plated cultures for 2–3 days at 30°C.
3. Use cultured cells to inoculate another agar/B–Mo plate.
4. Repeat the passage of cells over Mo-deficient media (steps 2 and 3) at least 1–2 additional times.
5. To a 125-mL flask containing 50 mL B–Mo–Fe media, add approximately a quarter of a loop of cells from the multiple-passage B–Mo plate.
6. Vortex the flask to disperse the cells within the medium.
7. Incubate and shake the culture for 18–20 h at 30°C and 170 rpm.
8. If cultures are competent (expressing fluorescent green siderophore), proceed to step 9. Otherwise, repeat steps 4–7.
9. To a culture tube, combine 200 μL fresh competent cell solution, 200 μL transformation buffer, and 1–10 μg donor DNA (Note 4).
10. Incubate transformation for 20 min at room temperature.
11. Add transformed cells to either plates with appropriate selection media (Section 2.3) or inoculate into B+-media.
12. Incubate plated cultures for 2–3 days at 30°C or B+-media cultures for 16–20 h at 30°C and 300 rpm (Section 2.3).

2.2.4 Notes

1. DNA can be extracted from *A. vinelandii* cultures using standard methods and kits. Different types of genetic material can also be used for transforming competent *A. vinelandii* cells, such as linearized DNA fragments, circular plasmids, or genomic DNA (Page & von Tigerstrom, 1979).
2. Many antibiotics can be used for selection in *A. vinelandii* (Bertsova, Bogachev, & Skulachev, 2001; Fang & Helinski, 1991; Johnson,

Unciuleac, & Dean, 2006; Kelly et al., 1990; Peralta-Gil, Segura, Guzman, Servin-Gonzalez, & Espin, 2002) These include ampicillin (50–100 μg/mL), rifampicin (5–10 μg/mL), and kanamycin (0.5–3 μg/mL).

3. Selective B+-medium plates can be prepared either by addition of antibiotics directly to the plates or by adding the antibiotics to the cooled agar/B+-medium solution immediately before pouring the plates.
4. If congression is being performed, add 1–10 μg of each DNA source.

2.3 *A. vinelandii* Selection

The efficacy of cell transformation is evaluated by growth on selective solid media and subsequent selection of cell colonies. The need for double-reciprocal recombination events to occur within all genome copies means that homologous recombination will only be successful for a small portion of the cells that uptake DNA during transformation (Dos Santos, 2011). The low transformation efficiency also means that colony screening can involve the scoring of thousands of individual populations. Depending on the transformation approach utilized, however, different strategies can be employed to streamline the selection process.

One convenient approach, rescue, takes advantage of N_2 reduction as a screening method, whereby strains with inactive nitrogenases cannot grow on media lacking a fixed nitrogen source. The inactivated DNA region can be replaced by a sequence that could be modified but still restores nitrogenase function. For example, the new sequence could encode for amino acid point mutations or tags, and with the proper flanking regions, can be recombined into the genome (Brigle et al., 1987; Dos Santos, Johnson, Ragle, Unciuleac, & Dean, 2007; Jacobson, Brigle, et al., 1989). The mutant should be capable of N_2 reduction, and its growth on B-media plates is easily assessed. Accordingly, this method is useful for the generation of point mutants and sequence-tagged variants (Brigle et al., 1987; Dos Santos et al., 2007; Jacobson, Brigle, et al., 1989).

The use of antibiotic resistance cassettes in the transformation of *A. vinelandii* enables direct antibiotic selection as a straightforward selection technique. Colony growth, or lack of growth, on selective antibiotic B-medium plates can be quickly and easily evaluated, although care must be taken to use the proper amounts of antibiotic, as *A. vinelandii* requires lower concentrations compared to other organisms (Johnson et al., 2006). Furthermore, the ability of *A. vinelandii* to accumulate multiple copies of its genome means that if essential genes are knocked out by antibiotic resistance cassettes, some copies lacking the antibiotic resistance (and coding for the essential gene) may persist. Multiple passages over the antibiotic-selective media are necessary to apply selective pressure, followed by relieving that pressure using nonselective media (Dos Santos, 2011). Finally, the cells are passed over selective media once more to ensure that antibiotic resistance has not been lost. This technique is an effective way to evaluate the genetic homogeneity and stability of the new strain. A related approach is to replace one antibiotic resistance cassette with another, such that the ability of the cells to grow on selective media will be noticeably altered following transformation (Paustian et al., 1990). Loss of one antibiotic resistance in this way is quickly verified by passage on the appropriate media. Multiple passages over the new and old selective antibiotic media are still necessary to ensure that all the genome copies are homogeneous (Page & von Tigerstrom, 1979).

Rescue and antibiotic selection can also be combined with other methods to improve the screening of transformed colonies. Congression, in which two different types of genetic markers are inserted during transformation, can benefit from this approach (Robinson, Burgess, & Dean, 1986; Wiig et al., 2011). Two separate sources of donor DNA are required for congression, and one set of donor DNA often encodes for antibiotic resistance or other selectable marker. The other donor DNA contains the desired genetic modification, often one that results in a screenable phenotype such as auxotrophy (Hu, Fay, & Ribbe, 2005). After cotransformation and plating the cells on selective antibiotic media as an initial screen, another selective pressure can be used to assess the presence of the second DNA insert. For example, if the substitution knocks out nitrogenase functionality, colonies that survive the antibiotic selection can be tested for lack of growth on solid B-media (Robinson, Dean, & Burgess, 1987).

Additionally, *A. vinelandii* supports inclusion of the *lacZ* gene, which can be used in blue and white screening assays to verify the presence or absence of a gene of interest (Miller, 1972). In these experiments, plates contain X-gal (5-bromo-4-chloro-3-indolyl-β-D-galactopyranoside), which is cleaved by β-D-galactosidase to form the blue insoluble pigment 5-bromo-4-chloro-3-indole. This blue color acts as a qualitative indicator for cell populations that contain the *lacZ* operon. Thus, the presence of blue colonies signals the expression of β-D-galactosidase, and the amount of expression in *A. vinelandii* cultures can be quantified using the Miller assay, which involves spectrophotometrically monitoring the generation of *o*-nitrophenol from *o*-nitrophenylgalactoside (Miller, 1972). However, for certain blue and white screening experiments, the parent strain undergoing recombination already should have the *lacZ* gene incorporated into its genome (Walmsley et al., 1994). Accordingly, replacement of the *lacZ* gene by the desired recombinant sequence would produce white colonies on a B+-medium plate containing X-gal, while untransformed colonies would be blue (Dos Santos et al., 2007; Johnson et al., 2006).

Other routes for selection can include a more traditional genetics approach in which mutations are selected for based on the phenotype. The current speed and decreasing cost of genome sequencing methods then allows for the genotyping of these strains, which can arise either through random mutation or selective pressure (Setubal et al., 2009). For example, the "nongummy" *A. vinelandii* OP phenotype was identified as a distinct mutant that arose from random mutation (Bush & Wilson, 1959), and later analysis of its derivative strains identified a disruption in the *alg*U gene that normally encodes for an alginate regulator (Martinez-Salazar et al., 1996). Exposing *A. vinelandii* cultures to inhibitory conditions or other can also lead to the identification of inhibitor-resistant revertant strains. In one case, a mutant strain was discovered that could grow on nitrogen-deficient plates when subjected to inhibitory amounts (~2%) of C_2H_2 (Christiansen, Cash, Seefeldt, & Dean, 2000). Since C_2H_2 is a strong noncompetative inhibitor of N_2 reduction in *A. vinelandii* (Rivera-Ortiz & Burris, 1975), the genotype was expected to have mutations within its nitrogenase genes. The acetylene-resistant phenotype indeed was found to express a mutant MoFe protein in which C_2H_2 acted only as a competitive inhibitor of N_2 reduction, enabling nitrogen fixation to proceed (Christiansen et al., 2000). Predicting the generation of random mutations and revertant strains a priori is difficult, if not impossible in many cases, but with the proper inventiveness and attention to detail, new aspects of nitrogenase can be investigated. These *A. vinelandii* selection strategies have served the nitrogenase field well and provide a large toolbox with which to produce mutants and screen for new strains.

2.3.1 Equipment

- Same as Section 2.2

2.3.2 Buffers and Reagents

- Same as Section 2.2
- Blue and white screening

- X-gal plates: B+-medium plates with 60 μg/mL of 5-bromo-4-chloro-3-indolyl-β-D-galactopyranoside (X-gal)

2.3.3 Procedure

- Direct antibiotic selection
  1. Culture transformed cells in B+-medium for 16–20 h at 30°C and 300 rpm.
  2. Perform serial dilutions (1:10, 1:100, 1:200, etc.) on antibiotic-treated solid B+-medium plates that correspond to the antibiotic resistance cassette(s) used.
  3. Incubate plates at 30°C for 2–3 days or until suitable colonies have developed.
  4. Isolate individual colonies on the same type of selective B+-medium plates.
  5. Repeat steps 3–4 for at least two passages over the selective media.
  6. Plate colonies on solid, antibiotic-free B+-medium and repeat steps 3–4 but with antibiotic-free medium. Perform at least two passages.
  7. Plate colonies on selective B+-medium.
  8. Incubate plates at 30°C for 2–3 days to evaluate successful growth of the new strain.
- Congression (indirect selection)
  1. Use cells that have been cotransformed with an antibiotic resistance cartridge and genetic modification of interest (Note 1).
  2. Perform direct antibiotic selection (steps 1–5). The desired cotransformed strain will contain both an antibiotic-resistant cassette and the modification of interest.
  3. Score the surviving colonies on B-medium (lacking fixed nitrogen). Select for either surviving or auxotrophic colonies, depending on the desired genetic modification phenotype.
- Blue and white screening
  1. Use transformed cells that have had the *lacZ* gene inserted or replaced, depending on the intended screening approach (Miller, 1972).
  2. Incubate cells at 30°C for 2–3 days on the appropriate media containing X-gal.
  3. Colonies still containing *lacZ* will appear blue, while colonies with inactivated or replaced *lacZ* will appear white (Note 2).

2.3.4 Notes

1. While the first step of the congression selection method for *A. vinelandii* often involves antibiotic resistance selection, evaluation of the second step will depend on the phenotype of the designated genetic modification. For example, if the modification knocks out nitrogen fixation, a lack of growth on B-media plates is an indicator of this auxotrophic phenotype. The converse also applies if the genetic modification restores nitrogen fixation.
2. The blue and white screening method can also be used in conjunction with congression to aid in the selection process.

3. GROWTH OF *A. vinelandii* STRAINS

Once active cultures of *A. vinelandii* have been acquired, larger-scale growth for nitrogenase expression can be performed. Expression of nitrogenase proteins is induced by a deficiency in fixed nitrogen source, most often ammonium (Kennedy et al., 1991). Providing the cells with a minimum amount of ammonium in the media allows the organisms to grow to an optimal density before depleting the nitrogen source and derepressing nitrogenase genes. Based on the *A. vinelandii* strain used, one of two main growth patterns will be observed under normal growth conditions. For strains that express functional nitrogenase, exponential phase growth will proceed until ammonium depletion, and after a brief lag, nitrogen fixation will proceed, leading to another, slower growth phase (Walmsley et al., 1994). For strains that cannot express a functional nitrogenase due to either gene deletion or mutation, after derepression and expression of nitrogenase proteins, no appreciable growth will proceed (Christiansen, Goodwin, Lanzilotta, Seefeldt, & Dean, 1998; Ribbe et al., 2002). Thus, nitrogen-fixing strains are harvested at an appropriate optical density, whereas strains incapable of nitrogen fixation are harvested several hours after the initial exponential phase has ceased to optimize expression of proteins (Shah et al., 1972).

Under certain conditions such as low dissolved O_2 tension, *A. vinelandii* will form large quantities of poly-β-hydroxybutyrate (PHB) polymer (Sabra et al., 2000). The formation of PHB is also prevalent during growth on ammonia but is less common when cells are grown on nitrate or N_2 (Kennedy et al., 2005). Granules of PHB can account for over 70% of the dried cell mass, and cells pastes that have overproduced the polymer will be noticeably whiter in color (Peña et al., 2002). The formation of excessive PHB can also influence the absorbance in optical density measurements, and care must be taken to avoid inclusion of PHB in the lysate, which can complicate protein purification (Section 4).

3.1 Cell Growth

Large-scale growth of *A. vinelandii* cells for nitrogenase protein expression is carried out in a fermentation system with high aeration (Burgess, Jacobs, & Stiefel, 1980). The starter cultures are grown in B+-media, but ammonium is omitted from the fermentor growth media. Consequently, the only fixed nitrogen source is provided from these cultures, and depletion of ammonia will trigger nitrogenase derepression (Kennedy et al., 2005). After an adequate amount of time has passed (2–4h after derepression) or the culture has reached the appropriate optical density, the cell material is collected and frozen for later use.

3.1.1 Equipment

- Incubator/shakers (New Brunswick, Edison, NJ)
- 100–400-L Fermentor (New Brunswick, Edison, NJ)
- UV/vis Spectrophotometer (Agilent, Santa Clara, CA)
- Flow-through centrifuge (Carl Padberg Zentrifugenbau GmbH, DE)
- Sorvall Legend centrifuge (Thermo-Fisher, Waltham, MA)

3.1.2 Buffers and Reagents

- B+-medium (Section 2.1)
- Cell wash buffer: 50 m*M* Tris–HCl, pH 7.4

3.1.3 Procedure

1. Start small-scale cultures from solid B+-medium plates (see Section 2.1) in 5–10 autoclaved 1-L flasks containing B+-media. Grow overnight at 30°C and 300 rpm.
2. Sterilize fermentor with ammonium-deficient B-medium.
3. Inoculate sterilized fermentor with small-scale cultures. Aerate fermentor at 40 ft^3/min.
4. Monitor cell growth by measuring turbidity at 436 nm. Harvest wild-type strains during mid-exponential phase or when OD_{436} reaches approximately 1.0. For strains that cannot fix nitrogen, harvest after exhaustion of ammonia or 2–4 h following end of exponential phase.
5. Harvest cells by centrifugation at 10,000 × *g* for 10 min.

6. Clean harvested cells with cell wash buffer, followed by centrifugation at 10,000 × *g* for 10 min.
7. Store harvested cells at −80°C for future use.

3.2 Crude Extract Preparation

Lysing of *A. vinelandii* cells is an important step in the purification of nitrogenase proteins. After lysing, the nitrogenase proteins will lose the reductive protection conferred by the cellular environment. Thus, beyond this point in the purification process, the solutions of crude extract and proteins will be susceptible to degradation by O_2. Degassing the lysate and storing sealed under an Ar atmosphere is critical to obtaining highly active nitrogenase proteins (Burgess et al., 1980).

3.2.1 Equipment

- Manton Gaulin homogenizer (Boston, MA)

3.2.2 Buffers and Reagents

- Cell wash buffer: 50 m*M* Tris–HCl, pH 7.4

3.2.3 Procedure

1. Resuspend frozen cells (approximately 800 g) in cell wash buffer (1.4 mL buffer per 1 g cells) and allow mixture to thaw while degassing for 1–2 h.
2. Lyse cells in homogenizer at 12,000 lb./in^2 (psi).
3. Degas cell lysate for 1 h, then centrifuge at 14,000 × *g* for 30 min at 4°C.
4. Collect supernatant for immediate use (see Section 4) or store at −20°C overnight in an airtight container (Note 1).

3.2.4 Notes

1. For best results, store under Ar atmosphere.

4. NITROGENASE PROTEIN PURIFICATION

Following the removal of expressed nitrogenase proteins from the cellular matrix (i.e., lysing, see Section 3.2), the individual proteins can be isolated. The crude extracts and isolated protein solutions are sensitive to deactivation by O_2, so care must be taken to perform manipulations in an anaerobic chamber or with a Schlenk line manifold (Burgess et al., 1980). In a typical Schlenk system, a dual manifold setup is used whereby the glass manifolds are connected with three-way ground-glass stopcocks.

In this way, the manifold attachments can be set to and cycled through three different states: under reduced pressure (vacuum), under gas pressure (typically Ar), or static (isolated from the manifold). One manifold connects directly to vacuum and contains a trap (e.g., Virtis solid CO_2 macrotrap) that prevents solvent vapors from reaching the vacuum pump. The other manifold is connected directly to an Ar tank operated at slight overpressure; a mercury release valve (for overpressure), O_2-scrubbing column (solid catalyst to remove O_2 contamination), and mineral oil bubbler (to observe gas flow) are assembled in between the gas source and the manifold. Furthermore, the gas delivery manifold is ballasted by attached 1 L round vessels; the ballast aids in maintaining positive pressure within the manifold while refilling evacuated containers. Schlenk techniques are desirable for protein handling during purification and for activity assays as well (Section 5).

Furthermore, before use, all solutions must be made anaerobic by a combination of degassing and the addition of the reductant sodium dithionite ($Na_2S_2O_4$). To render buffers anaerobic, the buffer solutions are first subjected to vacuum (at least 10^{-4} Torr) with stirring for 1 h. After the degassing step, add the indicated amount of $Na_2S_2O_4$ to the buffers, followed by an additional 10 min of vacuum and then place the flasks under positive pressure Ar. To evaluate if a buffer is anaerobic, add a few drops to a solution of 1 m*M* methyl viologen. Excess reducing equivalents in the buffer (i.e., no O_2 and excess $S_2O_4^{2-}$) will immediately turn the methyl viologen solution from colorless to dark blue (Ensign, 1995); this blue solution will return to colorless after prolonged exposure to air and can be reused to test the other buffers. Additionally, the reductant $S_2O_4^{2-}$ will lose its reductive power over time as it degrades into reactive $H_xS_yO_z$ breakdown products (Mayhew, 1978). The process of $S_2O_4^{2-}$ breakdown is accelerated by more acidic media and is less pronounced at pH 7–8 (Mayhew, 1978), but care should be taken to avoid using solutions more than a day old. If buffers are prepared in advance or intended for storage, $Na_2S_2O_4$ should be omitted initially and added immediately before use.

For nontagged MoFe protein purification, the process is usually performed over the course of 2–3 days (Fig. 3) (Burgess et al., 1980). On the first day, the cell lysate is prepared (Section 3.2) and either used as is or frozen for the next day. Columns can be prepared on the first day as well, which allows the separation of MoFe and Fe proteins (Section 4.1.3, steps 1–7) to proceed early on the second day. Further purification of nontagged MoFe

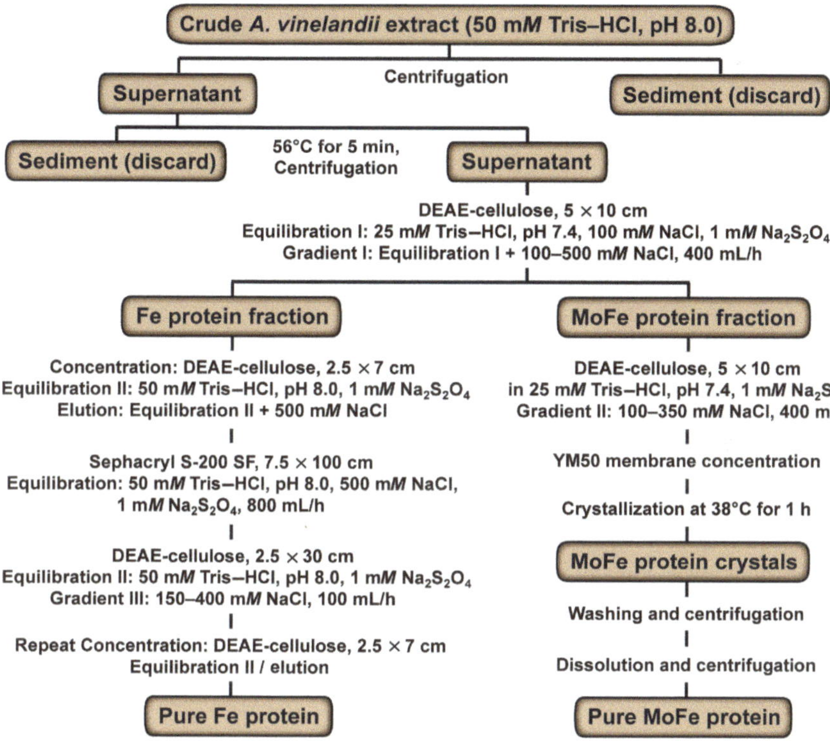

Fig. 3 Flow diagram for the purification of nontagged *A. vinelandii* nitrogenase proteins.

(Section 4.1.3, steps 8–21) can then proceed on the second or third day, and nontagged Fe protein can be isolated separately (Section 4.2).

The purification of nitrogenase proteins has been aided significantly by the development of His-tagged mutants that can be isolated in a single day using immobilized metal affinity column (IMAC) chromatography (Fig. 4; Section 4.3). This rapid method is preferred in many laboratories and can be used in conjunction with anion exchange chromatography to isolated both tagged and nontagged proteins in the same preparation (Christiansen et al., 1998; Lee, Hu, et al., 2009).

The collection of isolated protein solutions involves freezing as pellets (by adding the solution dropwise) and storing under liquid nitrogen. Protein solutions can also be frozen at −80°C, but long-term storage in this way leads to the gradual loss of protein activity and is not recommended. After purified proteins are isolated, they can be subjected to routine biochemical

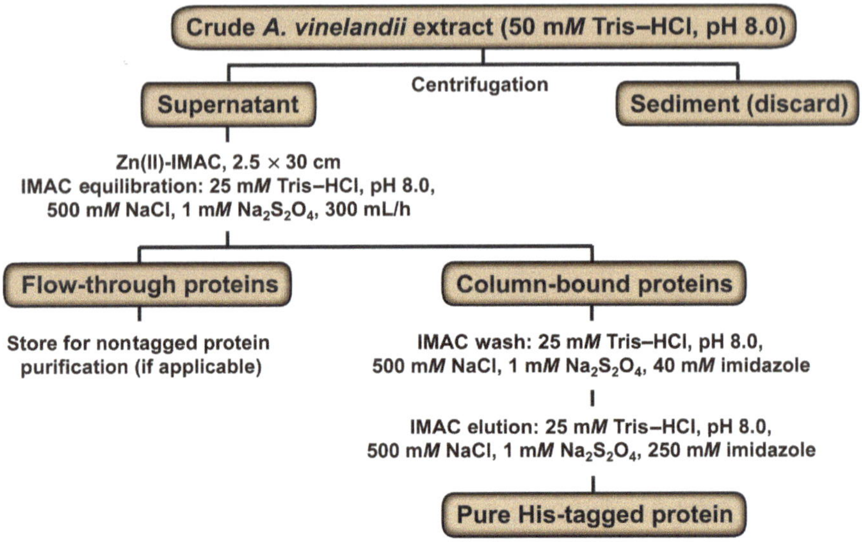

Fig. 4 Flow diagram for the purification of His-tagged *A. vinelandii* nitrogenase proteins.

analyses to evaluate protein concentration (Biuret, Lowry, etc.) and composition (native and denaturing polyacrylamide gel electrophoresis). Additionally, established activity assays are used to determine the catalytic competency of nitrogenase proteins (Section 5) (Newton & Dilworth, 2011).

4.1 Isolation of Nontagged MoFe Protein

The initial processing of cell lysate containing nontagged nitrogenase proteins involves a salt gradient fractionation with weak anion exchange chromatography (WAEC) (Bulen & LeComte, 1966). This step separates the MoFe protein and Fe protein fractions (Section 4.2). Further purification of the MoFe protein involves the removal of additional contaminant proteins and crystallization/precipitation of the MoFe protein (Burgess et al., 1980). The MoFe protein crystals can then be washed and redissolved, and the protein is concentrated and collected. From about 800 g of cells, approximately 600 mg nontagged MoFe protein can be obtained.

4.1.1 Equipment

- Heating/circulating bath
- Sorvall Legend centrifuge (Thermo-Fisher, Waltham, MA)
- Beckman ultracentrifuge (Indianapolis, IN)
- ISCO UA-6 UV–visible spectrophotometer with 405-nm filter (Lincoln, NE)

- Amicon microfiltration concentrators with YM50 membranes (Danver, MA)
- Gilson peristaltic pumps and tubing (Middleton, WI)
- Columns and mesh membranes (GE, Piscataway, NJ)
- Schlenk line manifold
- Erlenmeyer and Schlenk flasks with gas adapters and rubber stoppers

4.1.2 Buffers and Reagents

- Cell lysate (Section 3.2)
- Deoxyribonuclease (DNAse)
- Liquid nitrogen (LN_2)
- Methyl viologen
- DEAE Sepharose fast-flow (GE, Piscataway, NJ) suspended in 50 m*M* Tris–HCl, pH 7.4
- DEAE equilibration buffer I: 25 m*M* Tris–HCl, pH 7.4, 100 m*M* NaCl, 1 m*M* $Na_2S_2O_4$
- Dilution buffer: 25 m*M* Tris–HCl, pH 7.4, 1 m*M* $Na_2S_2O_4$
- Gradient buffer set I (MoFe/Fe protein separation):
    - Buffer IA: 25 m*M* Tris–HCl, pH 7.4, 100 m*M* NaCl, 1 m*M* $Na_2S_2O_4$ in 1.2 L
    - Buffer IB: 25 m*M* Tris–HCl, pH 7.4, 500 m*M* NaCl, 1 m*M* $Na_2S_2O_4$ in 1.2 L
- Gradient buffer set II (MoFe protein purification):
    - Buffer IIA: 25 m*M* Tris–HCl, pH 7.4, 100 m*M* NaCl, 1 m*M* $Na_2S_2O_4$ in 1.2 L
    - Buffer IIB: 25 m*M* Tris–HCl, pH 7.4, 350 m*M* NaCl, 1 m*M* $Na_2S_2O_4$ in 1.2 L
- Crystallization buffers:
    - Buffer A: 25 m*M* Tris–HCl, pH 7.4, 1 m*M* $Na_2S_2O_4$ in 1.0 L
    - Buffer B: 25 m*M* Tris–HCl, pH 7.4, 28 m*M* NaCl, 1 m*M* $Na_2S_2O_4$ in 0.05 L
    - Buffer C: 25 m*M* Tris–HCl, pH 7.4, 250 m*M* NaCl, 1 m*M* $Na_2S_2O_4$ in 0.05 L

4.1.3 Procedure

1. Incubate cell lysate and 10 μg/mL DNAse for 1 h at room temperature with constant stirring (Note 1).
2. Heat cell lysate to 56°C for 5 min with constant stirring. Undesired proteins will precipitate from the cell lysate at this point.

3. Centrifuge cell lysate at 70,000 × *g* for 45 min at 4°C.
4. Collect supernatant and load onto anaerobic, preequilibrated DEAE column. Best results come from a 5 × 10-cm column preequilibrated with anaerobic DEAE equilibration buffer I (Note 2).
5. After cell lysate is completely loaded, wash column with two to three column volumes of DEAE equilibration buffer I.
6. Elute proteins from the column with a linear NaCl gradient using gradient buffer set I as follows: Pump buffer IB into buffer IA at 200 mL/h; simultaneously, load buffer IA onto the column at 400 mL/h. Monitor protein elution at 405 nm: MoFe protein and Fe protein elute from the column at approximately 120 and 220 m*M* NaCl, respectively. Collect the MoFe protein and Fe protein fractions separately.
7. Freeze the Fe protein fraction as pellets in LN_2 for future use (Section 4.2). The MoFe protein fraction can be frozen if needed or used immediately by following the steps below.
8. Dilute MoFe protein fraction 1:1 with dilution buffer and load onto new 5 × 10-cm anaerobic DEAE column (see step 4).
9. Elute proteins from the column with a linear NaCl gradient using gradient buffer set II as follows: Pump buffer IIB into buffer IIA at 200 mL/h; simultaneously, load buffer IIA onto the column at 400 mL/h. Monitor protein elution at 405 nm: MoFe protein should elute from the column at approximately 120 m*M* NaCl.
10. Anaerobically transfer partially pure MoFe protein to Amicon concentrators that are lined with YM50 membranes and prewashed with crystallization buffer A.
11. Stir concentrator contents gently at approximately 24 psi under constant argon flow to reach a protein solution of 5–10 mL.
12. Dilute concentrated MoFe protein solution with crystallization buffer A to reach a solution concentration of 28 m*M* NaCl (dilution factor of 4.4).
13. Incubate crystallization mixture at 38°C for 1 h to allow for the formation of MoFe protein crystals.
14. Transfer the MoFe protein crystal solution into centrifuge bottles with the centrifuge rotor preequilibrated to 38°C, then centrifuge at 20,200 × *g* for 15 min at 38°C.
15. Discard the supernatant and wash the sediment by gently stirring the mixture in prewarmed 38°C crystallization buffer B.
16. Centrifuge the MoFe protein crystal slurry again at 20,200 × *g* for 15 min at 38°C.

17. Discard the supernatant and dissolve the pellet in approximately 8 mL of prechilled 4°C crystallization buffer C.
18. Equilibrate solution on ice for approximately 10 min, allowing crystals to fully dissolve.
19. Centrifuge the chilled protein solution at 25,000 × *g* for 20 min at 4°C.
20. Transfer the supernatant, now containing pure MoFe protein, into an anaerobic Schlenk flask.
21. Freeze the pure MoFe protein as pellets in LN_2.

4.1.4 Notes

1. While not necessary, addition of DNAse decreases the lysate viscosity, thus avoiding pressure issues and improving the performance of column manipulations.
2. For protein concentration, the resins Q-sepharose and DEAE sepharose can be used interchangeably.

4.2 Isolation of Nontagged Fe Protein

The purification of nontagged Fe protein can occur immediately after the purification of nontagged MoFe protein (Section 4.1), or multiple batches of frozen, dilute nontagged Fe protein can be accumulated and combined for efficiency. The overall process involves concentration of the dilute eluate followed by fractionation on a size-exclusion column to separate adventitious MoFe protein and additional flavodoxin and ferrodoxin proteins (Burgess et al., 1980). Subsequent WAEC, concentration, and collection steps yield pure Fe protein in high concentrations. From about 800 g of cells, approximately 500 mg nontagged Fe protein can be obtained.

4.2.1 Equipment

- Same as Section 4.1 with the exception of the YM50 membranes

4.2.2 Buffers and Reagents

- Fe protein fraction (Section 4.1)
- Sephacryl S-200 superfine (GE, Piscataway, NJ) suspended in 50 m*M* Tris–HCl, pH 7.4, 500 m*M* NaCl, 1 m*M* $Na_2S_2O_4$
- Sephacryl equilibration buffer: 50 m*M* Tris–HCl, pH 7.4, 500 m*M* NaCl, 1 m*M* $Na_2S_2O_4$
- DEAE equilibration buffer II: 50 m*M* Tris–HCl, pH 7.4, 1 m*M* $Na_2S_2O_4$

- Gradient buffer set III (Fe protein purification):
  - Buffer IIIA: 50 m*M* Tris–HCl, pH 7.4, 150 m*M* NaCl, 1 m*M* $Na_2S_2O_4$ in 0.35 L
  - Buffer IIIB: 50 m*M* Tris–HCl, pH 7.4, 400 mM NaCl, 1 m*M* $Na_2S_2O_4$ in 0.35 L
- Elution buffer: 50 m*M* Tris–HCl, pH 7.4, 500 m*M* NaCl, 1 m*M* $Na_2S_2O_4$

4.2.3 Procedure

1. Dilute thawed Fe protein fraction (Section 4.1) twofold with DEAE equilibration buffer II and load onto preequilibrated 2.5 × 7-cm DEAE column.
2. Wash column with two to three column volumes of DEAE equilibration buffer II.
3. Elute with elution buffer in the reverse direction of loading to concentrate the fraction to approximately 30 mL (Note 1).
4. Load concentrated Fe protein fraction onto 7.5 × 100-cm Sephacryl S-200 SF size-exclusion column preequilibrated with Sephacryl equilibration buffer. Operate the pump at 800 mL/h overnight and monitor elution of Fe protein at 405 nm.
5. Dilute eluted Fe protein fourfold with DEAE equilibration buffer II and load onto preequilibrated 2.5 × 30 cm DEAE column.
6. Wash column with one column volume of DEAE equilibration buffer II.
7. Elute Fe protein from the column with a linear NaCl gradient using gradient buffer set III as follows: Pump buffer IIIB into buffer IIIA at 50 mL/h; simultaneously, load buffer IIIA onto the column at 100 mL/h. Monitor Fe protein elution at 405 nm.
8. Concentrate the purified Fe protein by repeating steps 1–3.
9. Freeze the pure Fe protein as pellets in LN_2.

4.2.4 Notes

1. Reverse elution is an effective method to obtain higher eluted protein concentrations. However, if the column membrane becomes dislodged during the loading process, reverse elution is not recommended, as it would lead to the leaking of resin from the column.

4.3 Isolation of His-Tagged Nitrogenase Proteins

The overall process for purification of His-tagged nitrogenase proteins is much easier and time-effective than with nontagged proteins and requires the passage of cell lysate over a properly equilibrated IMAC column (Christiansen et al., 1998). After a short wash step, proteins can be eluted from IMAC column with a high-concentration imidazole solution. The protein yields vary depending on the type of tagged protein and the extent of expression, but obtaining roughly 600 mg of His-tagged, wild-type MoFe protein from 400 g of cells is not uncommon.

4.3.1 Equipment

- Same as Section 4.1 with the exception of the heating bath, centrifuge, and YM50 membranes

4.3.2 Buffers and Reagents

- Cell lysate (Section 3.2)
- dH_2O
- DNAse
- Phenylmethanesulfonyl fluoride (PMSF, Note 1)
- IMAC Sepharose 6 FF (GE, Piscataway, NJ) suspended in 25 m*M* Tris–HCl, pH 7.9
- Zn(II) solution: 25 m*M* Tris–HCl, pH 8.0, 100 m*M* $ZnSO_4{\cdot}6H_2O$
- Equilibration buffer: 25 m*M* Tris–HCl, pH 8.0, 500 m*M* NaCl, 1 m*M* $Na_2S_2O_4$
- Wash buffer: 25 m*M* Tris–HCl, pH 8.0, 500 m*M* NaCl, 40 m*M* imidazole, 1 m*M* $Na_2S_2O_4$
- Elution buffer: 25 m*M* Tris–HCl, pH 8.0, 500 m*M* NaCl, 250 m*M* imidazole, 1 m*M* $Na_2S_2O_4$

4.3.3 Procedure

1. Prepare cell lysate containing His-tagged protein according to Section 3.2. To accommodate IMAC resin-binding capacity, use only 400 g of cells.
2. Incubate cell lysate with 500 m*M* NaCl, 10 μg/mL DNAse, and 0.2 m*M* PMSF for 45 min at room temperature with constant stirring.
3. Meanwhile, prepare Zn(II)-IMAC column as follows: Pack 2.5 × 30 cm Zn(II)-IMAC resin and wash briefly with dH_2O; pass three column volumes of Zn(II) solution to charge resin; pass three column volumes of dH_2O to remove excess Zn(II); and equilibrate column with five column volumes of equilibration buffer.
4. Load cell lysate onto Zn(II)-charged IMAC resin preequilibrated with equilibration buffer. Operate the pump at 300 mL/h and collect column

flow-through in airtight container for purification of nontagged nitrogenase proteins (Sections 4.1 and 4.2), if applicable (Note 2).

5. Wash loaded column with 5–10 column volumes of wash buffer to remove nonspecifically bound proteins.
6. Elute protein from column with elution buffer, monitoring elution at 405 nm.
7. Freeze the pure His-tagged protein as pellets in LN_2.

4.3.4 Notes

1. Protease inhibitors other than PMSF can be used, but care should be taken to avoid metal-chelating reagents (e.g., bestatin and ETDA).
2. Depending on the nitrogenase proteins being expressed, it may be possible to obtain both tagged and nontagged proteins from a single cell batch. If His-tagged, wild-type MoFe protein is used, for example, then the column flow-through can be collected and used to purify nontagged Fe protein.

5. NITROGENASE PROTEIN ACTIVITY ASSAYS

Assessing the quality of purified nitrogenase proteins can accomplished by a variety of established substrate reduction assays. The six general criteria needed to evaluate these steady-state reactions are as follows: (1) reactions must be carried out under strictly anaerobic conditions; (2) a suitable reductant must be used; (3) inclusion of a MgATP regeneration system is necessary; (4) the buffer system used must be compatible with the enzyme and not interfere with product quantification; (5) the appropriate substrate or inhibitor must be included in the assay; and (6) the reactions require timely initiation and termination with the appropriate reagents (Burgess et al., 1980; Burris, 1972).

The exclusion of O_2 from the enzyme assays requires an inert and anaerobic gas. Since N_2 is not considered an inert gas for nitrogenase, the use of argon of sufficient quality (low O_2 content) fulfills this role (Daesch & Mortenson, 1968). Rubber-sealed assay vials are preferred for their ease of creating an airtight seal while providing a means to render the vials anaerobic, to add reagents, and for headspace gas sampling (Burris, 1972). The vials can be flushed with the appropriate gas, or a Schlenk line or similar apparatus can be used to cycle between vacuum and gas to effectively evacuate and refill the vials.

The most common reductant used in nitrogenase chemistry is $Na_2S_2O_4$. The electron source is used both to reduce the Fe protein and to remove adventitious O_2 from the solution (Mayhew, 1978). Optional electron sources for more specialized purposes include titanium(III) citrate, europium(II) diethylenetriaminepentaacetate, or reduced flavodoxin (Duyvis, Wassink, & Haaker, 1998; Lee, Hu, & Ribbe, 2012; Seefeldt & Ensign, 1994; Vincent et al., 2003). The reductant is typically added to the sealed assay vials after they have been degassed.

The requirement of the reduced Fe protein to hydrolyze two molecules of MgATP per electron transferred necessitates a constant source of MgATP in reactivity assays (Burgess & Lowe, 1996). Phosphorylation of the hydrolyzed product magnesium adenosine diphosphate (MgADP) to regenerate MgATP is often accomplished by a creatine phosphokinase/creatine phosphate system (Burgess et al., 1980). With creatine phosphate as a sacrificial phosphate source, levels of MgATP can be kept at concentrations suitable for steady-state analysis, and the inhibitory effects of MgADP accumulation can be avoided (Thorneley, Bergstroem, Eady, & Lowe, 1989). Ideal buffers for nitrogenase assays must maintain the solution at the appropriate pH range for optimal enzyme activity (normally $\sim$7–8) and not interfere with the product quantification methods (Pham & Burgess, 1993). This consideration depends on the compatibility between the functional groups of the buffer and the particular detection method to be used.

A variety of substrates and inhibitors have been evaluated with nitrogenase, and these compounds can be added to the assay vials either as a gas or a solution (Burris, 1972; Newton & Dilworth, 2011). Gaseous substrates and inhibitors can be delivered to the assay vials as the last step of the degassing process, after the headspace has been rendered inert. In the presence of no added substrate (under Ar atmosphere), catalysis with nitrogenase will produce H_2 (see Section 5.2), which incidentally is an unavoidable side product of other substrate reduction reactions such as N_2 and C_2H_2 (Burgess & Lowe, 1996; Simpson & Burris, 1984). It should also be noted that some substrates will act as inhibitors at higher concentrations: anions such as azide and cyanide represent examples of situational inhibitors, and care should be taken to ensure that their concentration is low enough for proper steady-state reactivity (Guth & Burris, 1983; Hwang, Chen, & Burris, 1973; Rivera-Ortiz & Burris, 1975; Rubinson, Burgess, Corbin, & Dilworth, 1985).

Finally, the timing of substrate reduction activity assays must be carefully controlled by the initiation and termination of these reactions. The reactions are often initiated by the injection of one or more essential components, such as MoFe protein, Fe protein, MgATP, or $MgCl_2$ (Newton & Dilworth, 2011). After allowing the reaction to proceed for up to 10 min, quenching occurs by the addition of either acid or a metal chelator: acids such as H_2SO_4 and trichloroacetic acid (TCA) are effective at immediate termination of the reaction, but chelators such as disodium ethylenediaminetetraacetate (Na_2EDTA) are preferred for assays in which product or substrate quantification may be affected by acid-catalyzed hydrolysis (Dilworth, Eldridge, & Eady, 1992; Dilworth & Fisher, 1998). For example, creatine phosphate is susceptible to hydrolysis at low pH, and if this compound is to be quantified, the reactions should be quenched with Na_2EDTA or another chelator instead of acid.

The protocols for the activity assays and quantification methods described below are not intended to be exhaustive; rather, these are general guidelines that may vary depending on the purpose of the experiment (Fig. 5). Other electron sources may be used, and the order of reagent addition and reaction times may be adjusted as needed. Also, if timecourse experiments are performed, the headspace of assay vials can be sampled without quenching the reactions described in Section 5.1. Methods for the quantification of products from benchmarking assays with C_2H_2, N_2, and H^+ will be discussed in Section 5.2. The activities of knockouts and other modified mutants are especially important toward the understanding of nitrogenase, and these assays can be used in conjunction with other maturation methods to evaluate metallocluster biosynthesis and assembly (Kaiser, Hu, Wiig, Rees, & Ribbe, 2011; Lee, Blank, et al., 2009; Yoshizawa et al., 2009).

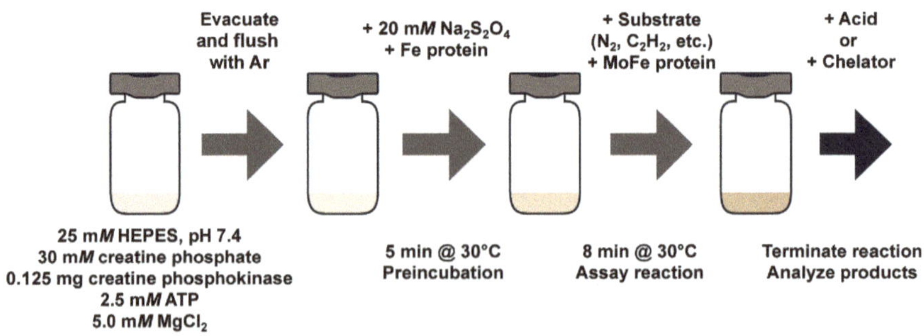

Fig. 5 General schematic for substrate reduction activity assays with nitrogenase proteins.

5.1 Substrate Reduction Assays

Since the discovery of acetylene (C_2H_2) as an alternative substrate for nitrogenase, the small molecule has been employed as a useful benchmark for enzymatic activity (Dilworth, 1966; Rivera-Ortiz & Burris, 1975; Schöllhorn & Burris, 1967). Acetylene is isoelectronic to N_2 and undergoes reduction to ethylene (C_2H_4) in the presence of wild-type Mo'nitrogenase (Eq. 2) (Hardy, Holsten, Jackson, & Burns, 1968). In addition, V- and Fe'nitrogenase can fully reduce C_2H_2 to ethane (C_2H_6; Eq. 3), and similar results have been observed with certain genetically modified MoFe variants (Dilworth, Eady, Robson, & Miller, 1987; Scott, Dean, & Newton, 1992).

$$HC{\equiv}CH + 2H^+ + 2e^- \rightarrow H_2C{=}CH_2 \quad (2)$$

$$HC{\equiv}CH + 4H^+ + 4e^- \rightarrow H_3C{-}CH_3 \quad (3)$$

The production of H_2, as previously stated, is a necessary and concomitant by-product of C_2H_2 and N_2 reduction (Burgess & Lowe, 1996; Jensen & Burris, 1985; Simpson & Burris, 1984). Under an Ar atmosphere, however, protons are the sole recipients of the nitrogenase electron flux, and only H_2 is produced (Burgess, Wherland, Newton, & Stiefel, 1981). Measuring the amount of H_2 produced as both a by-product and singular product is an important gauge of nitrogenase activity (Newton & Dilworth, 2011).

The reduction of the canonical substrate N_2 to produce NH_3 and H_2 is carried out under pure N_2 (Burris, 1972; Jensen & Burris, 1985). The overall approach for all the substrate activity assays is roughly similar, but the intended analyte will determine the reagent used to terminate the reaction. One notable difference for assays with N_2 is that if NH_3 is to be measured, the vials will need to be opened prior to work-up and analysis. This is in contrast to assays under 10% C_2H_2 or Ar, in which the sealed vials are necessary for headspace measurements (Burris, 1972; Newton & Dilworth, 2011).

5.1.1 Equipment

- Schlenk line manifold
- Water bath shaker
- 9.5-mL assay vials with rubber serum stoppers and metal caps
- Gastight syringes (Hamilton, Reno, NV)

5.1.2 Buffers and Reagents

- High purity Ar gas
- Ultra-high purity N_2 gas
- High purity C_2H_2 gas (100% or 10%, Ar balance)
- *N*-(2-hydroxy-ethyl)piperazine-*N*′-2-ethanesulfonic acid (HEPES)
- ATP
- Creatine phosphate
- Creatine phosphokinase
- $MgCl_2$
- $Na_2S_2O_4$
- Assay termination reagent: 500 m*M* Na_2EDTA, pH 7.5, 2.5 *M* H_2SO_4, or 10% TCA

5.1.3 Procedure

1. Prepare assay vials to contain the following reagent amounts per 1 mL (Note 1): 25 m*M* HEPES, pH 7.4, 2.5 m*M* ATP, 30 m*M* creatine phosphate, 0.125 mg creatine phosphokinase, 5.0 m*M* $MgCl_2$.
2. Evacuate and flush assay vials with the appropriate gas to render the contents anaerobic (Note 2).
3. Inject 20 μmol $Na_2S_2O_4$ into each vial, followed by Fe protein.
4. Preincubate assay vials for 5 min at 30°C (Note 3).
5. Initiate reaction with MoFe protein and let assays react for 8 min at 30°C.
6. Terminate reaction with 250 μL of the appropriate assay termination reagent (total solution volume will be 1.25 mL).
7. Save sealed assay vials for either headspace or solution analysis.

5.1.4 Notes

1. The volume of components added to the assay vial solution should be calculated such that the final volume will only reach 1 mL after the reaction is initiated with MoFe protein.
2. The final gas added to the vials depends on the intended substrate: for acetylene reduction, 10% C_2H_2 in Ar (0.1 atm C_2H_2 and 0.9 atm Ar); for N_2 reduction (NH_3 production), 100% (1 atm) pure N_2; for H^+ reduction (H_2 production), 100% (1 atm) Ar.
3. The preincubation process effectively "loads" the Fe protein with MgATP to ensure steady-state kinetics when the MoFe protein is added.

5.2 Reduction Assay Product Quantification

There are many methods and techniques available to isolate and evaluate the products from nitrogenase catalysis, and many are described in great detail elsewhere (Burris, 1972; Newton & Dilworth, 2011). Cursory descriptions of the more conventional approaches for the detection of C_2H_4, NH_3, and H_2 can be found below.

Typically, a gas chromatograph (GC) equipped with appropriate column and a flame ionization detector (FID) is suitable for the evaluation of C_2H_4, C_2H_6, and other hydrocarbon products (Lee, Hu, & Ribbe, 2010). The GC columns mentioned in this section have good separation properties for the analytes of interest (Ashby, Dilworth, & Thorneley, 1987). Similarly, a GC equipped with a thermal conductivity detector (TCD) can be used to analyze H_2 production (Burris, 1972). For analysis, an aliquot of headspace is removed from the assay vials and injected directly into a GC that is equipped with the proper column and carrier gas. These experiments require premixed gas standards for quantification, but calibration and analysis are straightforward.

One reason for the use of C_2H_2 as a benchmark for nitrogenase reactivity is the difficulty in measuring NH_3 from N_2 reduction (Beecher & Whitten, 1970). First and foremost, unlike the direct quantification of C_2H_4 and H_2, the determination of NH_3 involves indirect methods, adding more potential sources of error (Beecher & Whitten, 1970; Maryan & Vorley, 1979; O'Donovan, 1971). Additionally, many compounds external to the activity assay are known to interfere with the quantification of NH_3. For example, items such as filter paper and tobacco smoke-covered clothing can affect NH_3 quantification, as can protein breakdown products, dithionite, certain buffers, and creatine, depending on the analytical method used (Beecher & Whitten, 1970). Methods have been devised to isolate NH_3 prior to quantification such as desalting and microdistillation, but aspects of these approaches also can serve as sources of error (Burris, 1972). If the detection of NH_3 from N_2 reduction is required, analysis invariably involves a solution-based colorimetric approach (Beecher & Whitten, 1970; Burris, 1972).

One method for NH_3 analysis involves the use of a K_2HgI_4 solution (Nessler's reagent) that produces a yellow–orange complex in the presence of NH_3 (Crosby, 1968). The resulting complex can be spectrophotometrically monitored at 490 nm, but the method has limitations in terms of the amount of NH_3 that can be detected. Higher concentrations of NH_3 will produce a brown precipitate that can confound the analysis; the use of colloidal stabilizers is somewhat useful to this effect, but the amount of NH_3 that can be detected is still limiting (Newton & Dilworth, 2011).

Another approach for NH_3 determination is the *o*-phthalaldehyde (OPT) method. This technique utilizes the reactivity of OPT with NH_3

in the presence of a thiol such as 2-mercaptoethanol to produce a yellow species (Corbin, 1984). Fluorimetry is used to measure the amount of product formed with a rapid liquid chromatography method that uses a short C_{18} column (Corbin, 1984). However, the OPT method has some drawbacks, including a sensitivity of the reagents to light and the formation of multiple fluorescent products associated with the reaction of OPT and ammonia, which can complicate quantification (Newton & Dilworth, 2011). Furthermore, access to a dedicated HPLC and fluorimeter for this purpose may be limiting for some laboratories.

A more common analytical technique for the measurement of NH_3 is the indophenol method. This method is very sensitive and involves the reaction of basified phenol with sodium nitroprusside and sodium hypochlorite (bleach) to produce a reagent that forms blue-colored indophenol in the presence of ammonia (Chaykin, 1969). The product can be measured at 625–630 nm, and freshly preparing the reagent solution is preferred due to its short shelf-life, even when stored cold (Maryan & Vorley, 1979). Indophenol-testing solutions will darken over time due to oxidation, causing high background absorbance. Additionally, while the color of the indophenol solution is more stable than the Nessler and OPT methods, the test is subject to certain interferences such as N-containing buffers and dithionite (O'Donovan, 1971). The latter compound is a necessary component in nitrogenase activity assays, but the interference of this reductant can be circumvented by aerating the assay vials after the reaction is completed (Dilworth et al., 1992; Dilworth & Fisher, 1998).

5.2.1 Equipment

- Gastight syringes (Hamilton, Reno, NV)
- 1-cm quartz cuvettes
- C_2H_4 and C_2H_6 quantification
  - Gas chromatograph with flame ionization detector (GC-FID) and N_2 or He as carrier gas (Thermo-Fisher, Waltham, MA)
  - GC column: Porapak N or activated alumina (Restek, Bellefonte, PA)
- NH_3 quantification
  - UV–visible spectrophotometer (Agilent, Santa Clara, CA)
  - OPT method:
    - HPLC with Fluorimeter (Waters, Milford, MA)
    - HPLC column: C_{18} guard column (Bio-Rad, Hercules, CA)

- H_2 quantification
  - Gas chromatograph with thermal conductivity detector (GC-TCD) and Ar as carrier gas (Thermo-Fisher, Waltham, MA)
  - GC column: 5 or 13 Å molecular sieves (Supelco, Bellefonte, PA)

5.2.2 Buffers and Reagents

- Quenched activity assay vials, still sealed (Section 5.1)
- C_2H_4 and C_2H_6 quantification
  - C_2H_4 and C_2H_6 gas calibration mixtures, 1000 ppm in He (Scott Specialty Gases, Plumsteadville, PA)
- NH_3 quantification
  - Nessler's reagent method
    - K_2HgI_4 with 0.1% gum ghatti or similar colloid (Nessler's reagent)
  - OPT method:
    - OPT solution: 40 m*M* OPT in 5 mL 95% ethanol, 7 m*M* 2-mercaptoethanol, all in 45 mL 200 m*M* potassium phosphate buffer, pH 7.3 (Note 1)
    - OPT mobile phase: acetonitrile, 10 m*M* potassium phosphate buffer, pH 7.3
  - Indophenol method:
    - 5% (w/v) phenol, 2.5% (w/v) NaOH (phenolate solution)
    - 0.02% (w/v) $Na_2[Fe(CN)_4(NO)_2]$ (sodium nitroprusside)
    - 25 mL 5% NaClO (sodium hypochlorite) in 225 mL dH_2O
- H_2 quantification
  - H_2 gas calibration mixture, 1% in N_2 (Scott Specialty Gases, Plumsteadville, PA)

5.2.3 Procedure

- NH_3 quantification
  - Nessler's reagent method
    1. Treat sample with Nessler's reagent to a final Hg concentration of 2 m*M*.
    2. Incubate reaction for 1 min at room temperature.
    3. Measure sample absorbance at 490 nm.
  - OPT method
    1. Add 25 μL of sample to 1 mL of OPT solution.
    2. Incubate reaction in the dark for 60 min at room temperature.
    3. Inject 25 μL of incubated reaction into HPLC with 1.7 mL/min flow rate.

4. Monitor fluorescence based on excitation and emission wavelengths of 410 and 472 nm, respectively.

- Indophenol method
  1. Treat sample with 300 μL phenolate solution.
  2. Immediately add 450 μL each of sodium nitroprusside and hypochlorite solutions.
  3. Incubate reaction for 40 min at room temperature (Note 2).
  4. Measure sample absorbance at 630 nm. Use ammonium sulfate or ammonium chloride as a standard.

5.2.4 Notes

1. The OPT solution must be stored cold and in the dark, preferably under Ar. When stored in this way, freshly prepared OPT reagent is stable for at least 1 week.
2. Some interfering reagents for the indophenol method include dithionite, creatine, and N-based buffers such as tris(hydroxymethyl) aminomethane (Tris).

6. SUMMARY AND CONCLUSION

Strains and variants of *A. vinelandii* have been invaluable in the research of nitrogenase and its related proteins. Growth of *A. vinelandii* cultures does not require any special conditions or reagents, enabling large-scale growth with minimal media components (Kennedy et al., 2005). Moreover, modulating the amount of fixed nitrogen that is added is an effective means of controlling and inducing nitrogenase expression. Many molecular biology and genetic strategies are suitable for the manipulation of *A. vinelandii*, including the use of antibiotic resistance cassettes, transposon mutagenesis, congression, and blue and white screening via *lacZ*. That *A. vinelandii* can produce multiple copies of its genome complicates the transformation and selection process, but judicious use of selective media can ensure the genetic homogeneity of transformed strains (Dos Santos, 2011). The screening of transformed strains is also aided by the ability of *A. vinelandii* colonies that express a functional nitrogenase to survive on media lacking nitrogen source. Thus, wild-type and certain knock-out strains can be easily differentiated on nitrogen-deficient media.

The metalloproteins of nitrogenase are highly sensitive to O_2, so protein purification protocols are more complicated than normal benchtop isolation procedures (Burgess et al., 1980). The proper usage of Schlenk techniques and the convenience of the reductant dithionite are essential to the isolation of highly pure and active nitrogenase proteins. The purification of both non-tagged and His-tagged nitrogenase proteins is possible: the isolation of non-tagged proteins requires WAEC and size-exclusion chromatographic steps and takes several days, and the His-tagged purifications can be performed in a single day using IMAC chromatography. Both methods work well to obtain large amounts of catalytically active protein, and flow-through from IMAC experiments can yield additional nontagged proteins after adequate processing.

The evaluation of nitrogenase protein quality and functionality is carried out by a series of substrate reduction activity assays (Burris, 1972). These assays require precise conditions in order to properly compare the results to benchmark assays with the wild-type nitrogenase proteins. While one would expect the detection of NH_3 to be the primary mode of determining nitrogenase activity, this approach is hampered by interferences and indirect detection techniques that rely on colorimetric assays (Beecher & Whitten, 1970). Ammonia detection can still be accomplished, but the reduction of C_2H_2 to C_2H_4 (and C_2H_6, in some cases) can be detected directly and reliably using GC-FID methods (Burris, 1972). The production of H_2 from protons is an unavoidable process during nitrogenase catalysis, and the gas can be easily measured with GC-TCD instrumentation. The detection of many other substrates and products is also possible in nitrogenase assays, depending on the experiment at hand.

Overall, these strategies and procedures with *A. vinelandii* Mo-nitrogenase represent a small portion of the ever-expanding field of biological nitrogen fixation. From an enzymatic standpoint, though, *A. vinelandii* has provided, and continues to provide, a vast array of molecular and mechanistic insights into the activation of N_2 and its related biosynthetic and reductive capabilities (Hoffman et al., 2014; Ribbe et al., 2014).

ACKNOWLEDGMENTS

Work in our laboratory is supported by National Institutes of Health Grant GM-67626 (M.W.R. and Y.H.).

REFERENCES

Ashby, G. A., Dilworth, M. J., & Thorneley, R. N. F. (1987). Klebsiella pneumoniae nitrogenase. Inhibition of hydrogen evolution by ethylene and the reduction of ethylene to ethane. *Biochemical Journal*, *247*(3), 547–554. http://dx.doi.org/10.1042/bj2470547.

Beecher, G. R., & Whitten, B. K. (1970). Ammonia determination: Reagent modification and interfering compounds. *Analytical Biochemistry*, *36*(1), 243–246. http://dx.doi.org/10.1016/0003-2697(70)90356-8.
Beijerinck, M. W. (1901). Ueber oligonitrophile mikroben. *Zentralblatt für Bakteriologie, Parasitenkunde, Infektionskrankheiten und Hygiene*, *7*, 561–582.
Bertsova, Y. V., Bogachev, A. V., & Skulachev, V. P. (2001). Noncoupled NADH: Ubiquinone oxidoreductase of Azotobacter vinelandii is required for diazotrophic growth at high oxygen concentrations. *Journal of Bacteriology*, *183*(23), 6869–6874. http://dx.doi.org/10.1128/JB.183.23.6869-6874.2001.
Brigle, K. E., Setterquist, R. A., Dean, D. R., Cantwell, J. S., Weiss, M. C., & Newton, W. E. (1987). Site-directed mutagenesis of the nitrogenase MoFe protein of Azotobacter vinelandii. *Proceedings of the National Academy of Sciences of the United States of America*, *84*(20), 7066–7069.
Brill, W. J. (1980). Biochemical genetics of nitrogen fixation. *Microbiological Reviews*, *44*(3), 449–467.
Bulen, W. A., & LeComte, J. R. (1966). The nitrogenase system from Azotobacter: Two-enzyme requirement for nitrogen reduction, ATP-dependent hydrogen evolution, and ATP hydrolysis. *Proceedings of the National Academy of Sciences of the United States of America*, *56*(3), 979–986. http://dx.doi.org/10.1073/pnas.56.3.979.
Burgess, B. K. (1990). The iron-molybdenum cofactor of nitrogenase. *Chemical Reviews*, *90*(8), 1377–1406. http://dx.doi.org/10.1021/cr00106a002.
Burgess, B. K., Jacobs, D. B., & Stiefel, E. I. (1980). Large-scale purification of high activity Azotobacter vinelandii nitrogenase. *Biochimica et Biophysica Acta—Enzymology*, *614*(1), 196–209. http://dx.doi.org/10.1016/0005-2744(80)90180-1.
Burgess, B. K., & Lowe, D. J. (1996). Mechanism of molybdenum nitrogenase. *Chemical Reviews*, *96*(7), 2983–3011. http://dx.doi.org/10.1021/CR950055X.
Burgess, B. K., Wherland, S., Newton, W. E., & Stiefel, E. I. (1981). Nitrogenase reactivity: Insight into the nitrogen-fixing process through hydrogen-inhibition and HD-forming reactions. *Biochemistry*, *20*(18), 5140–5146. http://dx.doi.org/10.1021/bi00521a007.
Burris, R. H. (1972). [37] Nitrogen fixation—Assay methods and techniques. *Methods in Enzymology*, *24*, 415–431. http://dx.doi.org/10.1016/0076-6879(72)24088-5.
Bush, J. A., & Wilson, P. W. (1959). A non-gummy chromogenic strain of Azotobacter vinelandii. *Nature*, *184*(4683), 381.
Chan, M. K., Kim, J., & Rees, D. C. (1993). The nitrogenase iron-molybdenum-cofactor and P-cluster pair: 2.2 Å resolution structures. *Science*, *260*(5109), 792–794. http://dx.doi.org/10.1126/science.8484118.
Chaykin, S. (1969). Assay of nicotinamide deamidase. Determination of ammonia by the indophenol reaction. *Analytical Biochemistry*, *31*, 375–382. http://dx.doi.org/10.1016/0003-2697(69)90278-4.
Christiansen, J., Cash, V. L., Seefeldt, L. C., & Dean, D. R. (2000). Isolation and characterization of an acetylene-resistant nitrogenase. *The Journal of Biological Chemistry*, *275*(15), 11459–11464. http://dx.doi.org/10.1074/jbc.275.15.11459.
Christiansen, J., Goodwin, P. J., Lanzilotta, W. N., Seefeldt, L. C., & Dean, D. R. (1998). Catalytic and biophysical properties of a nitrogenase Apo-MoFe protein produced by a nifB-deletion mutant of Azotobacter vinelandii. *Biochemistry*, *37*(36), 12611–12623. http://dx.doi.org/10.1021/BI981165B.
Corbett, M. C., Hu, Y., Naderi, F., Ribbe, M. W., Hedman, B., & Hodgson, K. O. (2004). Comparison of iron-molybdenum cofactor-deficient nitrogenase MoFe proteins by X-ray absorption spectroscopy: Implications for P-cluster biosynthesis. *The Journal of Biological Chemistry*, *279*(27), 28276–28282. http://dx.doi.org/10.1074/jbc.M403156200.
Corbin, J. L. (1984). Liquid chromatographic-fluorescence determination of ammonia from nitrogenase reactions: A 2-min assay. *Applied and Environment Microbiology*, *47*(5), 1027–1030.

Crews, T. E., & Peoples, M. B. (2004). Legume versus fertilizer sources of nitrogen: Ecological tradeoffs and human needs. *Agriculture, Ecosystems & Environment, 102*(3), 279–297. http://dx.doi.org/10.1016/j.agee.2003.09.018.

Crosby, N. T. (1968). Determination of ammonia by the Nessler method in waters containing hydrazine. *Analyst, 93*(1107), 406–408. http://dx.doi.org/10.1039/AN9689300406.

Curatti, L., Brown, C. S., Ludden, P. W., & Rubio, L. M. (2005). Genes required for rapid expression of nitrogenase activity in Azotobacter vinelandii. *Proceedings of the National Academy of Sciences of the United States of America, 102*(18), 6291–6296. http://dx.doi.org/10.1073/pnas.0501216102.

Daesch, G., & Mortenson, L. E. (1968). Sucrose catabolism in Clostridium pasteurianum and its relation to N2 fixation. *Journal of Bacteriology, 96*(2), 346–351.

Danyal, K., Dean, D. R., Hoffman, B. M., & Seefeldt, L. C. (2011). Electron transfer within nitrogenase: Evidence for a deficit-spending mechanism. *Biochemistry, 50*(43), 9255–9263. http://dx.doi.org/10.1021/bi201003a.

Dean, D. R., & Jacobson, M. R. (1992). Biochemical genetics of nitrogenase. In G. Stacey, R. H. Burris, & H. J. Evan (Eds.), *Biological nitrogen fixation* (pp. 763–834). New York: Chapman and Hall.

Dilworth, M. J. (1966). Acetylene reduction by nitrogen-fixing preparations from Clostridium pasteurianum. *Biochimica et Biophysica Acta, 127*(2), 285–294. http://dx.doi.org/10.1016/0304-4165(66)90383-7.

Dilworth, M. J., Eady, R. R., Robson, R. L., & Miller, R. W. (1987). Ethane formation from acetylene as a potential test for vanadium nitrogenase in vivo. *Nature, 327*(6118), 167–168. http://dx.doi.org/10.1038/327167a0.

Dilworth, M. J., Eldridge, M. E., & Eady, R. R. (1992). Correction for creatine interference with the direct indophenol measurement of ammonia in steady-state nitrogenase assays. *Analytical Biochemistry, 207*(1), 6–10. http://dx.doi.org/10.1016/0003-2697(92)90491-O.

Dilworth, M. J., & Fisher, K. (1998). Elimination of creatine interference with the indophenol measurement of NH3 produced during nitrogenase assays. *Analytical Biochemistry, 256*(2), 242–244. http://dx.doi.org/10.1006/abio.1997.2517.

Dos Santos, P. C. (2011). Molecular biology and genetic engineering in nitrogen fixation. In M. W. Ribbe (Ed.), *Methods in molecular biology: Vol. 766* (pp. 81–92). New York, NY: Springer.

Dos Santos, P. C., Johnson, D. C., Ragle, B. E., Unciuleac, M.-C., & Dean, D. R. (2007). Controlled expression of nif and isc iron-sulfur protein maturation components reveals target specificity and limited functional replacement between the two systems. *Journal of Bacteriology, 189*(7), 2854–2862. http://dx.doi.org/10.1128/JB.01734-06.

Duyvis, M. G., Wassink, H., & Haaker, H. (1998). Nitrogenase of Azotobacter vinelandii: Kinetic analysis of the Fe protein redox cycle. *Biochemistry, 37*(50), 17345–17354. http://dx.doi.org/10.1021/BI981509Y.

Eady, R. R. (1996). Structure-function relationships of alternative nitrogenases. *Chemical Reviews, 96*(7), 3013–3030. http://dx.doi.org/10.1021/CR950057H.

Einsle, O., Tezcan, F. A., Andrade, S. L. A., Schmid, B., Yoshida, M., Howard, J. B., et al. (2002). Nitrogenase MoFe-protein at 1.16 Å resolution: A central ligand in the FeMo-cofactor. *Science, 297*, 1696–1700. http://dx.doi.org/10.1126/science.1073877.

Ensign, S. A. (1995). Reactivity of carbon monoxide dehydrogenase from Rhodospirillum rubrum with carbon dioxide, carbonyl sulfide, and carbon disulfide. *Biochemistry, 34*(16), 5372–5381. http://dx.doi.org/10.1021/bi00016a008.

Esposito, R. G., & Wilson, P. W. (1958). Acetate as a calcium-sparing factor in nitrogen fixation by Azotobacter vinelandii. *Proceedings of the National Academy of Sciences of the United States of America, 44*, 472–476. http://dx.doi.org/10.1073/pnas.44.5.472.

Fang, F. C., & Helinski, D. R. (1991). Broad-host-range properties of plasmid RK2: Importance of overlapping genes encoding the plasmid replication initiation protein TrfA.

Journal of Bacteriology, *173*(18), 5861–5868. http://dx.doi.org/10.1128/jb.173.18.5861-5868.1991.

Fay, A. W., Wiig, J. A., Lee, C. C., & Hu, Y. (2015). Identification and characterization of functional homologs of nitrogenase cofactor biosynthesis protein NifB from methanogens. *Proceedings of the National Academy of Sciences of the United States of America*, *112*(48), 14829–14833.

Georgiadis, M. M., Komiya, H., Chakrabarti, P., Woo, D., Kornuc, J. J., & Rees, D. C. (1992). Crystallographic structure of the nitrogenase iron protein from Azotobacter vinelandii. *Science*, *257*(5077), 1653–1659. http://dx.doi.org/10.1126/science.1529353.

Guth, J. H., & Burris, R. H. (1983). Inhibition of nitrogenase-catalyzed ammonia formation by hydrogen. *Biochemistry*, *22*(22), 5111–5122. http://dx.doi.org/10.1021/bi00291a010.

Hardy, R. W. F., Holsten, R. D., Jackson, E. K., & Burns, R. C. (1968). The acetylene-ethylene assay for nitrogen fixation: Laboratory and field evaluation. *Plant Physiology*, *43*(8), 1185–1207. http://dx.doi.org/10.1104/pp.43.8.1185.

Hoffman, B. M., Lukoyanov, D., Yang, Z.-Y., Dean, D. R., & Seefeldt, L. C. (2014). Mechanism of nitrogen fixation by nitrogenase: The next stage. *Chemical Reviews*, *114*(8), 4041–4062. http://dx.doi.org/10.1021/cr400641x.

Hu, Y., Corbett, M. C., Fay, A. W., Webber, J. A., Hedman, B., Hodgson, K. O., et al. (2005). Nitrogenase reactivity with P-cluster variants. *Proceedings of the National Academy of Sciences of the United States of America*, *102*(39), 13825–13830. http://dx.doi.org/10.1073/pnas.0506967102.

Hu, Y., Fay, A. W., & Ribbe, M. W. (2005). Identification of a nitrogenase FeMo cofactor precursor on NifEN complex. *Proceedings of the National Academy of Sciences of the United States of America*, *102*(9), 3236–3241. http://dx.doi.org/10.1073/pnas.0409201102.

Hu, Y., & Ribbe, M. W. (2016). Biosynthesis of the metalloclusters of nitrogenases. *Annual Review of Biochemistry*, *85*(1), 455–483. http://dx.doi.org/10.1146/annurev-biochem-060614-034108.

Hwang, J. C., Chen, C. H., & Burris, R. H. (1973). Inhibition of nitrogenase-catalyzed reductions. *Biochimica et Biophysica Acta—Bioenergetics*, *292*(1), 256–270. http://dx.doi.org/10.1016/0005-2728(73)90270-3.

Jacobson, M. R., Brigle, K. E., Bennett, L. T., Setterquist, R. A., Wilson, M. S., Cash, V. L., et al. (1989). Physical and genetic map of the major nif gene cluster from Azotobacter vinelandii. *Journal of Bacteriology*, *171*(2), 1017–1027. http://dx.doi.org/10.1128/jb.171.2.1017-1027.1989.

Jacobson, M. R., Cash, V. L., Weiss, M. C., Laird, N. F., Newton, W. E., & Dean, D. R. (1989). Biochemical and genetic analysis of the nifUSVWZM cluster from Azotobacter vinelandii. *Molecular and General Genetics*, *219*(1), 49–57. http://dx.doi.org/10.1007/BF00261156.

Jensen, B. B., & Burris, R. H. (1985). Effect of high pN2 and high pD2 on ammonia production, hydrogen evolution, and hydrogen deuteride formation by nitrogenases. *Biochemistry*, *24*(5), 1141–1147. http://dx.doi.org/10.1021/bi00326a012.

Joerger, R. D., Wolfinger, E. D., & Bishop, P. E. (1991). The gene encoding dinitrogenase reductase 2 is required for expression of the second alternative nitrogenase from Azotobacter vinelandii. *Journal of Bacteriology*, *173*(14), 4440–4446. http://dx.doi.org/10.1128/jb.173.14.4440-4446.1991.

Johnson, D. C., Unciuleac, M.-C., & Dean, D. R. (2006). Controlled expression and functional analysis of iron-sulfur cluster biosynthetic components within Azotobacter vinelandii. *Journal of Bacteriology*, *188*(21), 7551–7561. http://dx.doi.org/10.1128/JB.00596-06.

Kaiser, J. T., Hu, Y., Wiig, J. A., Rees, D. C., & Ribbe, M. W. (2011). Structure of precursor-bound NifEN: A nitrogenase FeMo cofactor maturase/insertase. *Science*, *331*(6013), 91–94. http://dx.doi.org/10.1126/science.1196954.

Kelly, M. J. S., Poole, R. K., Yates, M. G., & Kennedy, C. (1990). Cloning and mutagenesis of genes encoding the cytochrome bd terminal oxidase complex in Azotobacter vinelandii: Mutants deficient in the cytochrome d complex are unable to fix nitrogen in air. *Journal of Bacteriology, 172*(10), 6010–6019. http://dx.doi.org/10.1128/jb.172.10.6010-6019.1990.

Kennedy, C., Bali, A., Blanco, G., Contreras, A., Drummond, M., Merrick, M., et al. (1991). Regulation of expression of genes for three nitrogenases in Azotobacter vinelandii. In M. Polsinelli, R. Materassi, & M. Vincenzini (Eds.), *Developments in plant and soil sciences: Vol. 48*. (pp. 13–23). Dordrecht: Springer Netherlands.

Kennedy, C., & Bishop, P. (2004). Genetics of nitrogen fixation and related aspects of metabolism in species of Azotobacter: History and current status. In W. E. Newton (Ed.), *Nitrogen fixation: Origins, applications, and research progress: Vol. 2* (pp. 27–52). Dordrecht: Kluwer Academic Publishers.

Kennedy, C., Gamal, R., Humphrey, R., Ramos, J., Brigle, K., & Dean, D. (1986). The nifH, nifM and nifN genes of Azotobacter vinelandii: Characterisation by Tn5 mutagenesis and isolation from pLAFR1 gene banks. *Molecular and General Genetics MGG, 205*(2), 318–325. http://dx.doi.org/10.1007/BF00430445.

Kennedy, C., Rudnick, P., MacDonald, M. L., & Melton, T. (2005). Genus III. Azotobacter Beijerinck 1901, 567al. In D. J. Brenner, R. K. Noel, J. T. Stanley, & G. M. Garrity (Eds.), *Bergey's manual of systematic bacteriology—The proteobacteria* (pp. 384–402). New York, NY: Springer.

Lancaster, K. M., Roemelt, M., Ettenhuber, P., Hu, Y., Ribbe, M. W., Neese, F., et al. (2011). X-ray emission spectroscopy evidences a central carbon in the nitrogenase iron-molybdenum cofactor. *Science, 334*(6058), 974–977. http://dx.doi.org/10.1126/science.1206445.

Lee, C. C., Blank, M. A., Fay, A. W., Yoshizawa, J. M., Hu, Y., Hodgson, K. O., et al. (2009). Stepwise formation of P-cluster in nitrogenase MoFe protein. *Proceedings of the National Academy of Sciences of the United States of America, 106*(44), 18474–18478. http://dx.doi.org/10.1073/pnas.0909149106.

Lee, C. C., Hu, Y., & Ribbe, M. W. (2009). Unique features of the nitrogenase VFe protein from Azotobacter vinelandii. *Proceedings of the National Academy of Sciences of the United States of America, 106*(23), 9209–9214. http://dx.doi.org/10.1073/pnas.0904408106.S9209/9201-S9209/9203.

Lee, C. C., Hu, Y., & Ribbe, M. W. (2010). Vanadium nitrogenase reduces CO. *Science, 329*(5992), 642. http://dx.doi.org/10.1126/science.1191455.

Lee, C. C., Hu, Y., & Ribbe, M. W. (2012). ATP-independent substrate reduction by nitrogenase P-cluster variant. *Proceedings of the National Academy of Sciences of the United States of America, 109*(18), 6922–6926. http://dx.doi.org/10.1073/pnas.1202429109.

Lipman, J. G. (1903). Experiments on the transformation and fixation of nitrogen by bacteria. *Annual Report of the New Jersey State Agricultural Experiment Station, 24*, 217–285.

Lipman, J. G. (1908). Azotobacter studies. *Annual Report of the New Jersey State Agricultural Experiment Station*, 137–143.

Maier, R. J., & Moshiri, F. (2000). Role of the Azotobacter vinelandii nitrogenase-protective shethna protein in preventing oxygen-mediated cell death. *Journal of Bacteriology, 182*(13), 3854–3857.

Martinez-Salazar, J. M., Moreno, S., Najera, R., Boucher, J. C., Espin, G., Soberon-Chavez, G., et al. (1996). Characterization of the genes coding for the putative sigma factor AlgU and its regulators MucA, MucB, MucC, and MucD in Azotobacter vinelandii and evaluation of their roles in alginate biosynthesis. *Journal of Bacteriology, 178*(7), 1800–1808. http://dx.doi.org/10.1128/jb.178.7.1800-1808.1996.

Maryan, P. S., & Vorley, W. T. (1979). An improved spectrophotometric method for the determination of ammonia with particular relevance to in vitro nitrogenase activity. *Laboratory Practice, 28*(3), 251–252.

Mayhew, S. G. (1978). The redox potential of dithionite and SO − 2 from equilibrium reactions with flavodoxins, methyl viologen and hydrogen plus hydrogenase. *European Journal of Biochemistry*, *85*(2), 535–547. http://dx.doi.org/10.1111/j.1432-1033.1978.tb12269.x.

Miller, J. H. (1972). *Experiments in molecular genetics*. Cold Spring Harbor, New York: Cold Spring Harbor Laboratories.

Newton, W. E., & Dilworth, M. J. (2011). Assays of nitrogenase reaction products. In M. W. Ribbe (Ed.), *Methods in molecular biology: Vol. 766* (pp. 105–127). New York, N.Y: Springer.

Nicholas, D. J. D., & Deering, J. V. (1976). Repression, derepression and activation of nitrogenase in Azotobacter vinelandii. *Australian Journal of Biological Sciences*, *29*(1), 147–161.

O'Donovan, D. J. (1971). Inhibition of the indophenol reaction in the spectrophotometric determination of ammonia. *Clinica Chimica Acta*, *32*(1), 59–61. http://dx.doi.org/10.1016/0009-8981(71)90463-3.

Page, W. J., & Grant, G. A. (1987). Effect of mineral iron on the development of transformation competence in Azotobacter vinelandii. *FEMS Microbiology Letters*, *41*(3), 257–261. http://dx.doi.org/10.1111/j.1574-6968.1987.tb02207.x.

Page, W. J., & von Tigerstrom, M. (1979). Optimal conditions for transformation of Azotobacter vinelandii. *Journal of Bacteriology*, *139*(3), 1058–1061.

Paustian, T. D., Shah, V. K., & Roberts, G. P. (1990). Apodinitrogenase: Purification, association with a 20-kilodalton protein, and activation by the iron-molybdenum cofactor in the absence of dinitrogenase reductase. *Biochemistry*, *29*(14), 3515–3522. http://dx.doi.org/10.1021/bi00466a014.

Pena, C., Campos, N., & Galindo, E. (1997). Changes in alginate molecular mass distributions, broth viscosity and morphology of Azotobacter vinelandii cultured in shake flasks. *Applied Microbiology and Biotechnology*, *48*(4), 510–515.

Peña, C., Miranda, L., Segura, D., Núñez, C., Espín, G., & Galindo, E. (2002). Alginate production by Azotobacter vinelandii mutants altered in poly-β-hydroxybutyrate and alginate biosynthesis. *Journal of Industrial Microbiology and Biotechnology*, *29*(5), 209–213. http://dx.doi.org/10.1038/sj.jim.7000310.

Peralta-Gil, M., Segura, D., Guzman, J., Servin-Gonzalez, L., & Espin, G. (2002). Expression of the Azotobacter vinelandii poly-β-hydroxybutyrate biosynthetic phbBAC operon is driven by two overlapping promoters and is dependent on the transcriptional activator PhbR. *Journal of Bacteriology*, *184*(20), 5672–5677. http://dx.doi.org/10.1128/JB.184.20.5672-5677.2002.

Pham, D. N., & Burgess, B. K. (1993). Nitrogenase reactivity: Effects of pH on substrate reduction and carbon monoxide inhibition. *Biochemistry*, *32*(49), 13725–13731. http://dx.doi.org/10.1021/bi00212a043.

Premakumar, R., Loveless, T. M., & Bishop, P. E. (1994). Effect of amino acid substitutions in a potential metal-binding site of AnfA on expression from the anfH promoter in Azotobacter vinelandii. *Journal of Bacteriology*, *176*(19), 6139–6142. http://dx.doi.org/10.1128/jb.176.19.6139-6142.1994.

Ribbe, M. W., Hu, Y., Guo, M., Schmid, B., & Burgess, B. K. (2002). The FeMoco-deficient MoFe protein produced by a nifH deletion strain of Azotobacter vinelandii shows unusual P-cluster features. *The Journal of Biological Chemistry*, *277*(26), 23469–23476. http://dx.doi.org/10.1074/jbc.M202061200.

Ribbe, M. W., Hu, Y., Hodgson, K. O., & Hedman, B. (2014). Biosynthesis of nitrogenase metalloclusters. *Chemical Reviews*, *114*(8), 4063–4080. http://dx.doi.org/10.1021/cr400463x.

Rivera-Ortiz, J. M., & Burris, R. H. (1975). Interactions among substrates and inhibitors of nitrogenase. *Journal of Bacteriology*, *123*(2), 537–545.

Robinson, A. C., Burgess, B. K., & Dean, D. R. (1986). Activity, reconstitution, and accumulation of nitrogenase components in Azotobacter vinelandii mutant strains containing defined deletions within the nitrogenase structural gene cluster. *Journal of Bacteriology, 166*(1), 180–186.

Robinson, A. C., Dean, D. R., & Burgess, B. K. (1987). Iron-molybdenum cofactor biosynthesis in Azotobacter vinelandii requires the iron protein of nitrogenase. *The Journal of Biological Chemistry, 262*(29), 14327–14332.

Rubinson, J. F., Burgess, B. K., Corbin, J. L., & Dilworth, M. J. (1985). Nitrogenase reactivity: Azide reduction. *Biochemistry, 24*(2), 273–283. http://dx.doi.org/10.1021/bi00323a006.

Sabra, W., Zeng, A. P., Lünsdorf, H., & Deckwer, W. D. (2000). Effect of oxygen on formation and structure of *Azotobacter vinelandii* alginate and its role in protecting nitrogenase. *Applied and Environmental Microbiology, 66*(9), 4037–4044.

Schindelin, H., Kisker, C., Schlessman, J. L., Howard, J. B., & Rees, D. C. (1997). Structure of ADP·AlF4-stabilized nitrogenase complex and its implications for signal transduction. *Nature, 387*(6631), 370–376. http://dx.doi.org/10.1038/387370a0.

Schlessman, J. L., Woo, D., Joshua-Tor, L., Howard, J. B., & Rees, D. C. (1998). Conformational variability in structures of the nitrogenase iron proteins from Azotobacter vinelandii and Clostridium pasteurianum. *Journal of Molecular Biology, 280*(4), 669–685. http://dx.doi.org/10.1006/jmbi.1998.1898.

Schöllhorn, R., & Burris, R. H. (1967). Acetylene as a competitive inhibitor of N-2 fixation. *Proceedings of the National Academy of Sciences of the United States of America, 58*(1), 213–216.

Scott, D. J., Dean, D. R., & Newton, W. E. (1992). Nitrogenase-catalyzed ethane production and carbon monoxide-sensitive hydrogen evolution from molybdenum-iron proteins having amino acid substitutions in an α-subunit iron-molybdenum cofactor-binding domain. *The Journal of Biological Chemistry, 267*(28), 20002–20010.

Scrutton, M. C., & Taylor, B. L. (1974). Isolation and characterization of pyruvate carboxylase from Azotobacter vinelandii OP. *Archives of Biochemistry and Biophysics, 164*(2), 641–654. http://dx.doi.org/10.1016/0003-9861(74)90076-9.

Seefeldt, L. C., & Ensign, S. A. (1994). A continuous, spectrophotometric activity assay for nitrogenase using the reductant titanium(III) citrate. *Analytical Biochemistry, 221*(2), 379–386. http://dx.doi.org/10.1006/abio.1994.1429.

Setubal, J. C., dos Santos, P., Goldman, B. S., Ertesvåg, H., Espin, G., Rubio, L. M., et al. (2009). Genome sequence of Azotobacter vinelandii, an obligate aerobe specialized to support diverse anaerobic metabolic processes. *Journal of Bacteriology, 191*(14), 4534–4545. http://dx.doi.org/10.1128/jb.00504-09.

Shah, V. K., Davis, L. C., & Brill, W. J. (1972). Nitrogenase. I. Repression and derepression of the iron-molybdenum and iron proteins of nitrogenase in Azotobacter vinelandii. *Biochimica et Biophysica Acta—Bioenergetics, 256*(2), 498–511. http://dx.doi.org/10.1016/0005-2728(72)90078-3.

Simpson, F. B., & Burris, R. H. (1984). A nitrogen pressure of 50 atmospheres does not prevent evolution of hydrogen by nitrogenase. *Science, 224*(4653), 1095.

Thorneley, R. N. F., Bergstroem, N. H. J., Eady, R. R., & Lowe, D. J. (1989). Vanadium nitrogenase of Azotobacter chroococcum. Magnesium ATP-dependent electron transfer within the protein complex. *Biochemical Journal, 257*(3), 789–794. http://dx.doi.org/10.1042/bj2570789.

Vincent, K. A., Tilley, G. J., Quammie, N. C., Streeter, I., Burgess, B. K., Cheesman, M. R., et al. (2003). Instantaneous, stoichiometric generation of powerfully reducing states of protein active sites using Eu(ii) and polyaminocarboxylate ligands. *Chemical Communications*, 2590–2591. http://dx.doi.org/10.1039/B308188E.

Walmsley, J., Toukdarian, A., & Kennedy, C. (1994). The role of regulatory genes nifA, vnfA, anfA, nfrX, ntrC, and rpoN in expression of genes encoding the three nitrogenases of Azotobacter vinelandii. *Archives of Microbiology*, *162*(6), 422–429. http://dx.doi.org/10.1007/BF00282107.

Wiig, J. A., Hu, Y., & Ribbe, M. W. (2011). NifEN-B complex of Azotobacter vinelandii is fully functional in nitrogenase FeMo cofactor assembly. *Proceedings of the National Academy of Sciences of the United States of America*, *108*(21), 8623–8627. http://dx.doi.org/10.1073/pnas.1102773108.S8623/8621-S8623/8622.

Wiig, J. A., Hu, Y., & Ribbe, M. W. (2015). Refining the pathway of carbide insertion into the nitrogenase M-cluster. *Nature Communications*, *6*, 8034. http://dx.doi.org/10.1038/ncomms9034.

Wilcoxen, J., Arragain, S., Scandurra, A. A., Jimenez-Vicente, E., Echavarri-Erasun, C., Pollmann, S., et al. (2016). Electron paramagnetic resonance characterization of three iron-sulfur clusters present in the nitrogenase cofactor maturase NifB from Methanocaldococcus infernus. *Journal of the American Chemical Society*, *138*, 7468–7471. http://dx.doi.org/10.1021/jacs.6b03329.

Yoshizawa, J. M., Blank, M. A., Fay, A. W., Lee, C. C., Wiig, J. A., Hu, Y., et al. (2009). Optimization of FeMoco maturation on NifEN. *Journal of the American Chemical Society*, *131*(26), 9321–9325. http://dx.doi.org/10.1021/ja9035225.

CHAPTER TEN

TsrM as a Model for Purifying and Characterizing Cobalamin-Dependent Radical *S*-Adenosylmethionine Methylases

Anthony J. Blaszczyk*, Roy X. Wang*, Squire J. Booker*,†,1
*The Pennsylvania State University, University Park, PA, United States
†Howard Hughes Medical Institute, The Pennsylvania State University, University Park, PA, United States
[1]Corresponding author: e-mail address: squire@psu.edu

Contents

Methods in Enzymology, Volume 595
ISSN 0076-6879
http://dx.doi.org/10.1016/bs.mie.2017.07.007

Abstract

Cobalamin-dependent radical *S*-adenosylmethionine (SAM) methylases play vital roles in the de novo biosynthesis of many antibiotics, cofactors, and other important natural products, yet remain an understudied subclass of radical SAM enzymes. In addition to a [4Fe–4S] cluster that is ligated by three cysteine residues, these enzymes also contain an N-terminal cobalamin-binding domain. In vitro studies of these enzymes have been severely limited because many are insoluble or sparingly soluble upon their overproduction in *Escherichia coli*. This solubility issue has led a number of groups either to purify the protein from inclusion bodies or to purify soluble protein that often lacks proper cofactor incorporation. Herein, we use TsrM as a model to describe methods that we have used to generate soluble protein that is purified in an active form with both cobalamin and [4Fe–4S] cluster cofactors bound. Additionally, we highlight the methods that we developed to characterize the enzyme following purification.

1. INTRODUCTION

In 2001, a bioinformatics study identified the radical *S*-adenosylmethionine (SAM) superfamily of enzymes, which, at the time, contained 600 members catalyzing more than 30 different reactions (Sofia, Chen, Hetzler, Reyes-Spindola, & Miller, 2001). To date, the radical SAM (RS) superfamily is composed of almost 114,000 individual protein sequences, catalyzing more than 85 distinct reactions (Akiva et al., 2014). All RS enzymes harbor a [4Fe–4S] cluster coordinated by three conserved cysteines that are most commonly found in a CxxxCxxC motif (Sofia et al., 2001). In almost all cases, the reduced form of the cluster, $[4Fe–4S]^+$, is used to catalyze the reductive cleavage of SAM to methionine and a 5′-deoxyadenosine 5′-radical (5′-dA$^•$) (Frey & Booker, 2001; Frey, Hegeman, & Ruzicka, 2008), which initiates catalysis by abstracting a hydrogen atom from a target substrate. RS enzymes have been predicted and/or shown to catalyze a vast range of biologically important reactions, which include sulfur insertion and thioether bond formation, methylation and methythiolation of unactivated carbon and/or phosphinate phosphorous centers, oxidative decarboxylation of unactivated carbon centers and other C–C bond fragmentations, C–C bond formation, C–P bond fragmentation, cyclopropanation, epimerization, hydroxylation, metallocofactor assembly and organometallic bond formation, oxidation, mutase reactions, and complex rearrangements. These reactions are involved in numerous critical biological pathways, such as DNA/RNA biosynthesis and repair, protection against viral invasion, the biosynthesis of a large number of enzyme cofactors and clinically relevant natural products, nucleic acid and protein maturation, and antibiotic resistance to antibiotic biosynthesis, among many others (Booker & Grove, 2010; Broderick, Duffus, Duschene, & Shepard, 2014; Landgraf, McCarthy, & Booker, 2016).

Methylation of unactivated carbon and phosphinate phosphorous centers is uniquely carried out by the subfamily of RS enzymes known as RS methylases. In most instances, the added methyl group is derived from SAM; however, the mechanisms by which the methyl group is added are distinctly different from that of traditional SAM-dependent methyltransferases. RS methylases have been organized into five classes to date: A, B, C, D, and E (Bauerle, Schwalm, & Booker, 2015; Hu & Ribbe, 2016; Zhang, van der Donk, & Liu, 2012). Class A RS methylases are comprised of RlmN and Cfr, which methylate the C2 or C8 amidine carbons of specific adenosine nucleobases in ribosomal RNA (rRNA) or transfer RNA (tRNA) (Kowalak, Bruenger, & McCloskey, 1995; Toh, Xiong, Bae, & Mankin, 2008). Class B methylases require a cobalamin cofactor in addition to the [4Fe–4S] cluster shared by all RS enzymes. These enzymes appear to be the most versatile of the RS methylases, because they methylate sp^2- and sp^3-hybridized carbon atoms as well as phosphinate phosphorus atoms, while Class A, C, and D RS methylases only methylate sp^2-hybridized carbon atoms (Zhang et al., 2012). Class C RS methylases are predicted to use two simultaneously bound SAM molecules to methylate sp^2-hybridized carbon centers during the biosynthesis of several complex natural products. Class D methylases also methylate sp^2-hybridized carbon centers but have been suggested to use methylenetetrahydrofolate as the methyl source (Allen, Xu, & White, 2014). The recently annotated Class E RS methylases, by contrast, are involved in the maturation of the M cluster of nitrogenase and appear to methylate a sulfide ion as part of a metallocluster (Hu & Ribbe, 2016).

Class B RS methylases catalyze key steps in the biosynthesis of a number of biomolecules, including protein cofactors, antibiotics, herbicides, and other natural products, such as bacteriochlorophyll (Chew & Bryant, 2007), pactamycin (Kudo, Kasama, Hirayama, & Eguchi, 2007), mitomycin C (Mao, Varoglu, & Sherman, 1999), moenomycin A (Ostash, Saghatelian, & Walker, 2007), fosfomycin (Hidaka, Goda, et al., 1995; Kuzuyama, Hidaka, Kamigiri, Imai, & Seto, 1992; Woodyer et al., 2006), thienamycin (Nunez, Mendez, Brana, Blanco, & Salas, 2003), gentamicin (Kim et al., 2008), clorobiocin (Anderle et al., 2007), fortimicin (Dairi,

Ohta, Hashimoto, & Hasegawa, 1992; Kuzuyama, Seki, Dairi, Hidaka, & Seto, 1995), thiostrepton (Kelly, Pan, & Li, 2009), chondrochloren (Rachid, Scharfe, Blocker, Weissman, & Muller, 2009), polytheonamide (Parent et al., 2016), cystobactamids (Baumann et al., 2014; Wang, Schnell, Baumann, Muller, & Begley, 2017), watasemycin (Inahashi et al., 2017), and bialaphos (Hidaka, Hidaka, Kuzuyama, & Seto, 1995; Kamigiri, Hidaka, Imai, Murakami, & Seto, 1992). However, mechanistic investigations of these enzymes have been limited, due in large part to their inherent insolubility upon overproduction in *Escherichia coli*. Seven Class B RS methylases have been studied in vitro; however, their mechanistic characterization has been scant. Fom3, GenK, and PhpK were solubilized from inclusion bodies by treatment with urea and then refolded and reconstituted with their iron–sulfur (Fe–S) cluster and cobalamin cofactors. However, this strategy resulted in enzymes with poor cofactor incorporation and low activity (Allen & Wang, 2014; Kim et al., 2013; Werner et al., 2011). Three other Class B methylases, PoyC, ThnK, and TsrM, were amenable to purification from soluble crude lysate in an active form that contained both cofactors (Blaszczyk et al., 2016; Marous et al., 2015; Parent et al., 2016), but only when overproduced in a specialized medium containing ethanolamine, which has been suggested to increase cobalamin uptake in *E. coli* (Bandarian & Matthews, 2004). Lastly, another Class B RS methylase, CysS, has been purified from soluble crude lysate and reconstituted with cobalamin to yield an active form of the enzyme that contains the Fe–S cluster as well as cobalamin (Wang et al., 2017).

In this chapter, we describe strategies that we have developed for overproducing, purifying, and characterizing RS methylases, with a particular focus on TsrM (Blaszczyk et al., 2016), which methylates C2 of tryptophan (Trp) as part of the biosynthesis of the quinaldic acid moiety of the thiopeptide natural product thiostrepton (Li & Kelly, 2010; Pierre et al., 2012). A previous chapter in Methods in Enzymology described our strategies for overproducing and characterizing Fe–S-containing enzymes, with a particular focus on those in the RS superfamily (Lanz et al., 2012). In this work, we combine these two strategies but focus on characterization of the cobalamin cofactor.

2. OVERPRODUCTION AND PURIFICATION OF TsrM

Our initial attempts to purify TsrM focused on using the same overproduction and purification strategies previously shown to be successful for other RS enzymes (Lanz et al., 2012); however, these methods resulted in poor quantities of protein that was typically less than 10% pure. Almost all of the protein was produced in inclusion bodies. The *tsrM* gene was eventually cloned into a pSUMO vector, wherein TsrM would be overproduced as a C-terminal fusion with SUMO, a small protein-based tag (~12 kDa) used to modify proteins posttranslationally during various cellular processes. SUMO has been shown to increase the solubility of proteins to which it is fused and can be excised from the fusion construct by a very specific protease, Ulp1, which does not leave residual amino acids on the target protein (Marblestone et al., 2006). The N-terminal hexahistidine (His_6) tag on the SUMO protein allows for facile purification by immobilized metal affinity chromatography (IMAC). Upon cleavage of the fusion construct by Ulp1, the affinity resin can be used to capture SUMO and other proteins that bind nonspecifically to it. However, only after overproduction of the SUMO-TsrM fusion protein in the RosettaBlue (DE3) pLysS strain of *E. coli* was TsrM able to be purified to homogeneity in reasonable yields (10 mg per 16 L of original culture). Below is a detailed description of the methods used to overproduce TsrM and purify it to homogeneity.

2.1 Gene Cloning and Expression Strategy

The *tsrM* gene from *Streptomyces laurentii* was cloned into the pSUMO vector (LifeSensors) to generate a construct that encodes a SUMO-TsrM fusion protein with an N-terminal His_6 tag (Blaszczyk et al., 2016). The pSUMO vector is constructed from a pET24 vector. The pET cloning system, coupled with the strong T7 promoter, allows for facile cloning and robust overproduction of proteins (Panavas, Sanders, & Butt, 2009). The pSUMO vector also contains a kanamycin (Kan) resistance marker for in vivo selection, which, in addition, allows for coexpression of genes contained on plasmid pDB1282, because this plasmid confers ampicillin (Amp) resistance. pDB1282 harbors the *isc* operon from *Azotobacter vinelandii*, which encodes proteins involved in Fe–S cluster assembly (Cicchillo et al., 2004; Zheng, Cash, Flint, & Dean, 1998). The RosettaBlue (DE3) pLysS strain of *E. coli* was cotransformed with pDB1282 and pSUMO-TsrM, giving rise to a strain that confers resistance to Kan, Amp, and chloramphenicol (Cam), which is encoded on the pLysS plasmid. The RosettaBlue (DE3) pLysS strain is similar to the commonly used BL21 (DE3) strain of *E. coli* but encodes additional tRNAs that are rarely used in *E. coli*. In addition, pLysS expresses low levels of T7 lysozyme, which inhibits T7 RNA polymerase, thereby reducing its basal expression levels (Dubendorf & Studier, 1991).

Overproduction of TsrM is performed in M9-ethanolamine medium supplemented with cobalamin. This medium is similar to M9-minimal media, but lacks ammonium chloride and glucose, which serve as nitrogen and carbon sources, respectively. It has been shown that culturing *E. coli* under conditions in which ethanolamine serves as the sole source of nitrogen and carbon leads to upregulation of the adenosylcobalamin (AdoCbl)-dependent ethanolamine ammonium lyase (Blackwell & Turner, 1978), which catalyzes the conversion of ethanolamine to acetaldehyde and ammonia (Babior, Moss, Orme-Johnson, & Beinert, 1974). More importantly, when M9-ethanolamine media is supplemented with cobalamin, the intracellular cobalamin concentration is augmented (Bandarian & Matthews, 2004). This increase in cobalamin concentration can be leveraged to overproduce TsrM with a bound cobalamin cofactor.

2.2 Protocol for Expression of the *tsrM* Gene

1. Introduce pDB1282 and pSUMO-TsrM by cotransformation of *E. coli* RosettaBlue (DE3) pLysS using standard procedures with at least 1 μg of each plasmid. Select transformants on Luria–Bertani (LB) plates containing Kan, Amp, and Cam at final concentrations of 50, 100, and 35 μg/mL, respectively.
2. Use a single colony to inoculate 200 mL of LB medium containing Kan, Amp, and Cam at 50, 100, and 35 μg/mL, respectively. Incubate at 37°C and shake at 250 rpm for ~24 h. In addition, place four 6 L flasks, each containing 4 L of M9-ethanolamine medium (Table 1) in an incubator-shaker to allow for equilibration to 37°C.
3. Add the following to each flask just prior to inoculation: 50 μg/mL Kan, 100 μg/mL Amp, 35 μg/mL Cam, 54 mL of 10 × cobalamin supplement (Table 1), and 100 mL of 1 *M* ethanolamine (Table 1).
4. Inoculate each 6 L flask with ~2 mL of the 200 mL starter culture grown to an OD_{600} of 2.0.
5. At an OD_{600} of 0.3 (~15 h), add L-arabinose to a final concentration of 0.1% (w/v), and $FeCl_3$ and L-cysteine to final concentrations of 25 and 150 μ*M*, respectively.
6. At an OD_{600} of 0.6, cool flasks to ~20°C in an ice bath and adjust the temperature in the incubator to 20°C. Once the medium is equilibrated

Table 1 Composition of M9-Ethanolamine Minimal Medium

4 L of 5× M9 Salts[a]	**g**
Na_2HPO_4 (anhydrous)	136
KH_2PO_4	60
NaCl	10
1 L of 10,000× Micronutrient Mixture	**mg/L**
$(NH_4)_2Mo_7O_2$	37.1
$CoCl_2 \cdot 6H_2O$	71.4
H_3BO_3	247
$CuSO_4 \cdot 5H_2O$	25
$MnCl_2 \cdot 4H_2O$	198
$ZnSO_4 \cdot 7H_2O$	28.8
500 mL of 10× Cobalamin Supplement[b]	**Quantity**
10,000 × Micronutrient mixture	4 mL
$MgSO_4 \cdot 7H_2O$	9.2 g
$CaCl_2 \cdot 2H_2O$	0.6 g
Thiamine	37 mg
Hydroxocobalamin	126 mg
1 *M* Ethanolamine[b]	**Volume**
12 *M* Stock HCl	78 mL
16 *M* Stock ethanolamine	60 mL
Water	862 mL
pH to 7.4 with HCl or ethanolamine	
Volumes for 4 L of M9-Ethanolamine Minimal Medium	
5 × M9 salts	
5 × Amino Acid Mixture (Table 2)	800 mL
10 × Cobalamin supplement	54 mL
1 *M* Ethanolamine	100 mL
ddH_2O[c]	2246 mL

[a]After salts dissolve, adjust the pH of the medium to 7.4 with HCl and KOH; adjust volume to 4 L with water.
[b]Solutions should be sterilized by filtration and added just before inoculation.
[c]Medium should be autoclaved after addition to ddH_2O in a 6-L Erlenmeyer flask.

Table 2 Amino Acid Mixture[a]

4 L of 5× Amino Acid	**Quantity (g)**
Alanine	1.426
Arginine monohydrochloride	0.9656
Asparagine	1.065
Aspartic acid	1.057
Cysteine monohydrochloride monohydrate	0.2422
Glutamic acid	2.224
Glutamine	1.755
Glycine	1.201
Histidine monohydrochloride monohydrate	0.839
Isoleucine	1.050
Leucine	2.099
Lysine monohydrochloride	1.462
Methionine	0.5970
Phenylalanine	1.322
Proline	0.9210
Serine	21.02
Threonine	0.9530
Tryptophan	0.4090
Tyrosine	0.7250
Valine	1.407

[a]Amino acids should be added directly to M9-Ethanolamine Minimal Medium solution prior to autoclaving the medium.

(about 20 min), return the flasks to the shaker and induce expression of the *tsrM* gene by adding isopropyl-β-D-thiogalactopyranoside (IPTG) to a final concentration of 800 μ*M*. Add $FeCl_3$ and L-cysteine again to final concentrations of 25 and 150 μ*M*, respectively.

7. Allow cultures to incubate for ~16 h at 20°C while shaking at 180 rpm.
8. Harvest the bacteria by centrifugation and freeze the resulting cell paste (~1.5 g/L) in liquid N_2 and store in a liquid N_2 dewar.

2.3 Purification of TsrM

In addition to enhancing the solubility of TsrM during overproduction, the SUMO-TsrM fusion construct allows for an additional purification step following cleavage with Ulp1 (Fig. 1). After initial purification by IMAC, the protein is judged to be ~30% pure by SDS-PAGE analysis. Native TsrM is then generated upon cleavage with Ulp1 and can be collected in the flow-through after reapplying the cleavage mixture to the IMAC matrix (Marblestone et al., 2006). The SUMO tag and proteins that bind to the matrix nonspecifically are recaptured during this step. An SDS-PAGE analysis of the TsrM purification process is displayed in Fig. 2.

One of the greatest challenges of isolating RS enzymes is their instability upon exposure to oxygen. To address this issue, purification of TsrM and

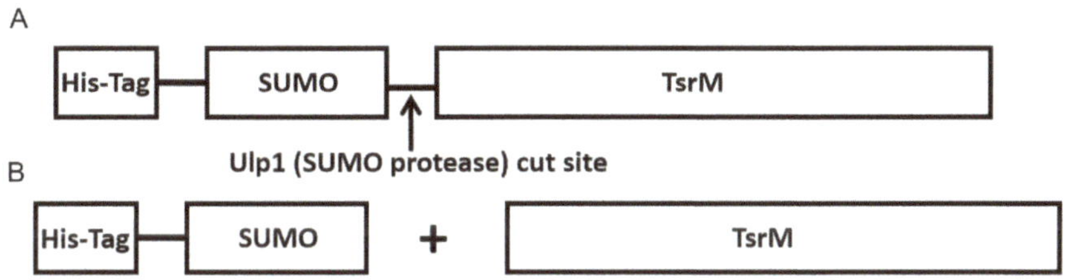

Fig. 1 SUMO-TsrM fusion construct (A) before and (B) after cleavage.

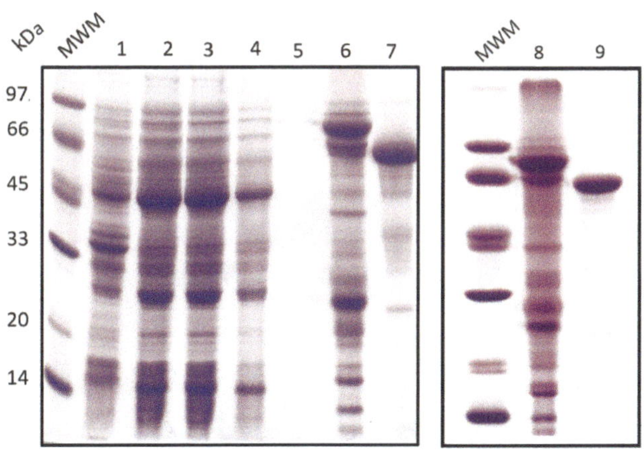

Fig. 2 SDS-PAGE gel of TsrM purification. The lanes are as follows: (1) pellet, (2) crude lysate, (3) flow-through, (4) wash 1, (5) wash 2, (6) eluate, (7) postcleavage flow-through, (8) eluate, and (9) post-S200.

all subsequent manipulations of the protein are done in a Coy anaerobic chamber to exclude oxygen, as has been previously described (Lanz et al., 2012). Additionally, the purification process is completed without interruption while keeping the protein chilled during each step. The following steps detail the protocol that is used to isolate homogeneous TsrM in its native form.

1. In an anaerobic chamber, prepare 500 mL each of Lysis Buffer (50 m*M* HEPES, pH 7.5, 300 m*M* KCl, 5% glycerol, 10 m*M* β-mercaptoethanol (BME), 10 m*M* imidazole), Wash Buffer (50 m*M* HEPES, pH 7.5, 300 m*M* KCl, 5% glycerol, 10 m*M* BME, 20 m*M* imidazole), Elution Buffer (50 m*M* HEPES, pH 7.5, 300 m*M* KCl, 15% glycerol, 10 m*M* BME, 300 m*M* imidazole), Cleavage Buffer (50 m*M* HEPES, pH 7.5, 300 m*M* KCl, 15% glycerol, 10 m*M* BME), and Gel-Filtration Buffer (50 m*M* HEPES, pH 7.5, 300 m*M* KCl, 15% glycerol, 5 m*M* DTT) with water that is oxygen free. Except for centrifugation, all subsequent steps are performed in an anaerobic chamber.
2. In the anaerobic chamber, resuspend 50 g of frozen cell paste in 150 mL Lysis Buffer containing lysozyme (1 mg/mL), and DNase I (0.1 mg/mL) in a 250-mL metal sonication cup. Hydroxocobalamin (OHCbl) at 0.2 mg/mL can also be included if necessary. However, in TsrM, cobalamin incorporation takes place during expression; inclusion of OHCbl in the Lysis Buffer has no additional effect.
3. Once the cell paste is completely resuspended (~30 min), insert the metal sonication cup into a crystallizing dish packed with ice, ensuring not to introduce any ice (which contains oxygen) into the sonication cup.
4. Sonicate for 45 s at 65% output using a Fisher 1 550 sonic dismembrator. Repeat for at least six cycles with a 7-min interval between each cycle to allow for temperature reequilibration (~8°C).
5. Pour the lysate into centrifuge tubes. Cap the tubes tightly and seal the edges with vinyl tape before removing from the anaerobic chamber.
6. Centrifuge the lysate at 45,000 × *g* for 1 h at 4°C. During centrifugation, equilibrate a column of Ni-NTA resin with the chilled Lysis Buffer.
7. After centrifugation, reintroduce the centrifuge tubes into the anaerobic chamber and carefully decant the supernatant into a clean 250-mL bottle. Remove a 100-μL sample of the supernatant and a sample of the pellet for SDS-PAGE analysis.

8. Load the supernatant onto the column. Proceed to wash with 150 mL of prechilled Wash Buffer. Take 100 μL samples of the flow-through both from the loaded lysate and from the two washes for SDS-PAGE analysis.
9. Elute the protein from the column with prechilled Elution Buffer. Once the colored band begins to elute, collect fractions until all of the protein (brown/black band) elutes from the column. Concentrate the protein by centrifugal ultrafiltration at $3500 \times g$ at 4°C in an Amicon centrifugal filter device (10 kDa molecular weight cutoff) to a final volume of ≤2.5 mL. While the protein is concentrating, equilibrate a PD-10 column with prechilled Cleavage Buffer.
10. Collect a 20-μL sample of the concentrated protein for SDS-PAGE analysis. Exchange the protein into prechilled Cleavage Buffer using the preequilibrated PD-10 column. Treat the flow-through with ~400 μg of the commercially available Ulp1 protease. Keep the cleavage reaction overnight on ice to generate native TsrM. Note that a chemical reconstitution step can take place simultaneously at this time by adding $FeCl_3 \cdot 6H_2O$, $Na_2S \cdot 9H_2O$, and additional BME to the cleavage reaction, as described in detail later.
11. After allowing the cleavage reaction to proceed for ~15 h, equilibrate a Ni-NTA column with 100 mL of prechilled Cleavage Buffer.
12. Load the prechilled cleavage reaction onto the Ni-NTA column and elute TsrM with additional Cleavage Buffer. Collect the red/brown band of native TsrM (may be very faint in color). The SUMO protein and Ulp1 protease, both of which contain His_6 tags, should remain bound to the Ni-NTA resin.
13. Concentrate TsrM by centrifugal ultrafiltration at $3500 \times g$ at 4°C in an Amicon centrifugal filter device (10 kDa molecular weight cutoff) to a final volume of ≤2.5 mL. While the protein is concentrating, equilibrate the PD-10 column with Gel-Filtration Buffer. Elute the nonspecifically bound proteins on the Ni-NTA resin with Elution Buffer.
14. Collect 20 μL of native TsrM for SDS-PAGE analysis. Exchange the protein into Gel-Filtration Buffer using the equilibrated PD-10 column.
15. Concentrate the protein by centrifugal ultrafiltration with the Amicon filter devices if desired. Aliquot and flash-freeze the protein in liquid N_2 before storing in a liquid N_2 dewar.

2.4 Chemical Reconstitution of As-Isolated TsrM

Although the overproduction and purification methods described earlier result in TsrM that binds cobalamin stoichiometrically and contains 3.4 and 3.1 mol of iron and sulfide, respectively, reconstitution can enhance Fe–S cluster incorporation. To avoid an additional overnight incubation step, Ulp1 can be added to the chemical reconstitution reaction, such that cleavage and reconstitution take place simultaneously, given that both steps require long incubation periods. Although DTT has been used in the reconstitution of other RS enzymes, this reconstitution method requires an additional IMAC step following cleavage of the SUMO-TsrM construct. During this step, the reductant BME is used, because DTT can reduce the Ni^{2+} ions in the Ni-NTA resin when present at concentrations greater than 1 m*M* (Lanz et al., 2012).

1. Introduce the following into the anaerobic chamber: 27 mg of $FeCl_3 \cdot 6H_2O$ (270.3 g/mol), 24 mg of $Na_2S \cdot 9H_2O$ (240.2 g/mol), 17.5 μL of stock BME (14.3 *M*), and a 15-mL conical tube.
2. Combine the following into the conical tube: 6.03 mL of Cleavage Buffer, 1 mL of 1 *M* HEPES buffer, pH 7.5, and the 17.5 μL of BME (14.3 *M*).
3. Dissolve the $FeCl_3 \cdot 6H_2O$ and $Na_2S \cdot 9H_2O$ in 1 mL of water. Add 50 μL of the Na_2S solution followed by 50 μL of the $FeCl_3$ solution to the conical tube.
4. Place the mixture on ice for ~15 min before adjusting the pH, if necessary, to ~7.0.
5. Add 2.85 mL of the cleavage reaction, which includes SUMO-TsrM and Ulp1, and incubate on ice overnight for ~12 h.
6. Centrifuge the reconstituted protein at 10,000 × *g* to remove any insoluble aggregates.
7. Apply the supernatant onto the Ni-NTA resin and collect the flow-through, which should contain native TsrM.
8. Concentrate the flow-through to ≤2.5 mL by centrifugal ultrafiltration as described earlier.
9. Exchange the concentrated protein into Gel-Filtration Buffer using a PD-10 column.
10. Load the protein onto an S200 column equilibrated in Gel-Filtration Buffer and connected to an ÅKTA fast protein liquid chromatography system (housed in an anaerobic chamber) for further purification by size-exclusion chromatography.

3. DETERMINATION OF COFACTOR STOICHIOMETRY

3.1 Strategy for Determining the Stoichiometry and Forms of Iron–Sulfur Clusters

Determining the correct number of Fe–S clusters per polypeptide is critical for accurate characterization of RS enzymes, which is typically performed in concert with Mössbauer and electron paramagnetic resonance (EPR) spectroscopies (Pandelia, Lanz, Booker, & Krebs, 2015). We follow protocols that have been previously described for determining the amount of iron and sulfide per protein in other RS enzymes (Lanz et al., 2012). A brief description of these methods follows:

1. Following purification, the concentration of enzyme is determined by the Bradford assay using bovine serum albumin, fraction V, (BSA) as a standard (Bradford, 1976).
2. Amino acid analysis is conducted to determine a correction factor for TsrM when BSA is used as a standard for Bradford analysis.
3. Quantification of the amount of iron and acid-labile sulfide per protein is done using established procedures (Beinert, 1978, 1983).

Although determining the amount of iron and sulfide per protein is useful, further characterization is required to determine the types of Fe–S clusters bound to an enzyme, which is done using EPR and ^{57}Fe-Mössbauer spectroscopies. Preparation of samples for EPR and Mössbauer spectroscopy is carried out using methods that we have previously described (Lanz et al., 2012).

3.2 Strategy for Determining Stoichiometry of Bound Cobalamin

The ability to determine the stoichiometry of cobalamin bound to Class B RS methylases is an important prerequisite for any in vitro mechanistic studies. These enzymes are predicted to bind a methylcobalamin (MeCbl) cofactor, whose methyl moiety is transferred to an sp^2- or sp^3-hybridized carbon center or phosphinate phosphorous atom (Zhang et al., 2012). In studies of TsrM purified without cobalamin, it has been shown that activity is dependent upon addition of MeCbl to the reaction (Pierre et al., 2012). Although MeCbl is an essential cofactor for Class B RS methylases, information regarding the stoichiometry of bound cobalamin per polypeptide has only been reported for the enzymes PoyC and TsrM (Blaszczyk et al., 2016; Parent et al., 2016). The protocol below describes our strategy for determining the stoichiometry of bound cobalamin per TsrM polypeptide.

1. Determine the concentration of purified enzyme by Bradford analysis.
2. Dilute TsrM in water to ~11 μ*M* in a final volume of 900 mL.
3. To the sample, add 100 μL of 1 *M* KCN to bring the final concentrations of enzyme and KCN to be ~10 μ*M* and 100 m*M*, respectively. This treatment allows for the formation of the dicyanocobalamin adduct, which can be easily detected and quantified because of its unique UV–vis spectrum.
4. To release the dicyanocobalamin from the enzyme, boil the sample for 5 min.
5. Centrifuge the sample at 14,000 × *g* for 5 min to separate the dicyanocobalamin from precipitated protein.
6. Using UV–vis spectroscopy, determine the concentration of dicyanocobalamin, which has an extinction coefficient of 30,800 M^{-1}/cm at 367 nm (Ljungdahl, LeGall, & Lee, 1973).
7. Determine the amount of cobalamin per protein by dividing the concentration of dicyanocobalamin by the concentration of protein.
   * The absorbance spectrum of TsrM treated with potassium cyanide is displayed in Fig. 3.

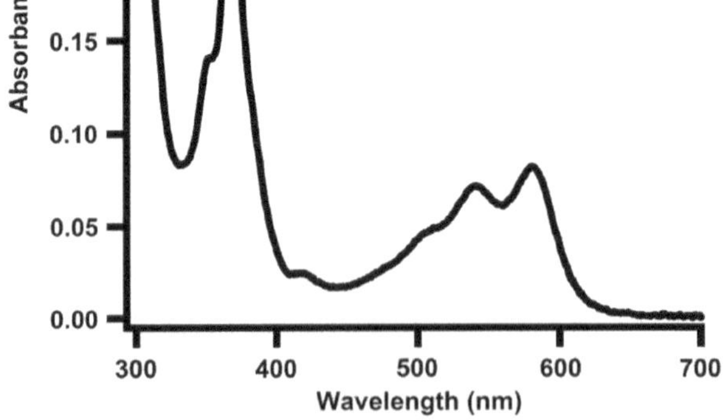

Fig. 3 UV–vis spectrum of as-isolated TsrM (9.5 μ*M*) treated with potassium cyanide.

4. CHARACTERIZATION OF COBALAMIN FORMS

Cobalamin exists in three different oxidation states—cob(I)alamin, cob(II)alamin, or cob(III)alamin—and some Class B RS methylases have been proposed to access each of these states during a reaction cycle (Zhang et al., 2012). In Class B RS methylases, cob(I)alamin is widely accepted to act as a supernucleophile, attacking the methyl group of SAM via an S_N2 displacement mechanism to generate *S*-adenosylhomocysteine (SAH) and methylcob(III)alamin. The next step of the reaction, which involves the transfer of a methyl group from MeCbl to substrate, remains elusive for many of the Class B RS methylases. In the first proposed mechanism, a substrate radical, presumably generated upon abstraction of a hydrogen atom by a 5′-dA$^{\bullet}$, attacks the methyl moiety of MeCbl, with subsequent homolysis of the Co–C bond to yield cob(II)alamin. The cob(II)alamin is then reduced to reform cob(I)alamin, which initiates another round of catalysis by attacking the methyl moiety of SAM. Class B RS methylases also methylate sp^2-hybridized carbon centers in pyrrole-containing substrates, albeit via a mechanism that appears to be distinctly different from that employed for enzymes that methylate sp^3-hybridized carbon centers. The model system for this reaction is TsrM, which methylates C2 of Trp to form 2-methyltryptophan (2-MeTrp). This enzyme appears to be distinct from all other characterized RS enzymes, because it does not catalyze a reductive cleavage of SAM to give a 5′-dA$^{\bullet}$. The reaction appears to involve two polar S_N2 displacements. In the first step, a cob(I)alamin species attacks the methyl moiety of SAM to generate MeCbl and SAH. In the second step, the *pi* electrons at C2 of Trp attack the methyl moiety of MeCbl to yield the methylated product upon loss of the C2 proton. This second step appears to be facilitated by proton abstraction from the indole nitrogen of Trp (Blaszczyk, Wang, Silakov, Ho, & Booker, 2017). The ability to monitor different forms of cobalamin throughout the reaction is critical in elucidating the mechanisms employed by these enzymes. Furthermore, when characterizing cobalamin during a reaction, it is necessary to remove excess cobalamin from reactions to ensure that any changes observed reflect enzyme-bound forms of the cofactor.

4.1 Strategies for Characterizing Cobalamin

Cob(I)alamin has a d^8 electron configuration at its Co^{+1} metal center, and when monitored by UV–vis spectroscopy, displays a sharp absorbance peak at 390 nm (Banerjee, Frasca, Ballou, & Matthews, 1990; Banerjee & Ragsdale, 2003). Cob(I)alamin is extremely prone to oxidation, so UV–vis analysis must be performed in an anaerobic cuvette or in an anaerobic chamber to retain the reduced form of the cofactor. Because the redox potential for the cob(II)alamin/cob(I)alamin couple is −610 mV, a low-potential reductant is generally required to generate cob(I)alamin (Banerjee, Harder, Ragsdale, & Matthews, 1990; Lexa & Saveant, 1983). However, mild reductants such as BME and DTT have also been shown to reduce cob(II)alamin in the presence of OHCbl (Jarrett, Goulding, Fluhr, Huang, & Matthews, 1997).

Unlike cob(I)alamin and cob(III)alamin, cob(II)alamin has a non-integer spin state ($S=½$) and can be observed by EPR spectroscopy. Cob(II)alamin typically contains a five-coordinate cobalt center that has one occupied and one unoccupied axial ligand. Usually, either a conserved histidine (His) or the dimethylbenzimidazole moiety of cobalamin binds as the lower-axial ligand to generate a His-on or base-on form of cob(II)alamin, respectively. His-on and base-on cob(II)alamin have a distinguishable EPR spectrum that exhibits superhyperfine coupling from the ^{14}N ($I=1$) that is coordinated to the cobalt ion. However, as-isolated TsrM does not show this superhyperfine coupling but exhibits a unique EPR spectrum that is consistent with an atom in the lower-axial position. Based on these observations, as well as similarities with two cobalamin-dependent dehalogenases, which have been shown by X-ray crystallography to have water molecules bound to their lower-axial positions (Bommer et al., 2014; Payne et al., 2015), an oxygen atom ($I=0$) from a water molecule was proposed to be coordinated in the lower-axial position (Blaszczyk et al., 2016). The other common form of cob(II)alamin is referred to as base-off, which typically has a water molecule bound to the upper-axial ligand. This form of the cofactor also has a distinct EPR spectrum that can easily be distinguished.

Cob(III)alamin is most commonly found as MeCbl, AdoCbl, or OHCbl in Nature. All three of these forms can be distinguished by UV–vis spectroscopy; however, their UV–vis spectra can be perturbed from that exhibited in free solution when the cobalamin form is bound to an enzyme. Free MeCbl, AdoCbl, and OHCbl display absorption features at 520 nm ($\varepsilon_{520}=8600\,M^{-1}$/cm), 262 nm ($\varepsilon_{262}=35{,}100\,M^{-1}$/cm), and 351 nm ($\varepsilon_{351}=26{,}500\,M^{-1}$/cm), respectively, in aqueous solutions (Dawson, Elliott, Elliott, & Jones, 1987). As is the case in TsrM, MeCbl possessing neither a His nor a dimethylbenzimidazole moiety in its lower-axial position loses its absorption feature at 520 nm and displays a new peak at 450 nm (Menon & Ragsdale, 1998). Fig. 4 displays a UV–vis spectrum of TsrM with a reduced cobalamin cofactor, as evident by the absorption peak at 390 nm, which corresponds to cob(I)alamin. The addition of SAM to the sample results in a UV–vis spectrum with a new peak at ~450 nm, which corresponds to the formation of MeCbl. Liquid chromatography with detection by tandem mass spectrometry (LC-MS/MS) is an alternative method that allows for characterization of MeCbl, AdoCbl, and OHCbl. By using standards with known concentrations of all three forms of cob(III)alamin, LC-MS/MS can be used for their accurate quantification.

4.2 Protocol for Preparation of Samples for EPR Analysis of Cob(II)alamin

1. Freeze a small container of isopentane in liquid N_2 until it becomes a slush.
2. Dilute the protein sample to a final concentration of ~200 μ*M* in 200 μL. In addition to an as-isolated sample, also make samples with 1 m*M*

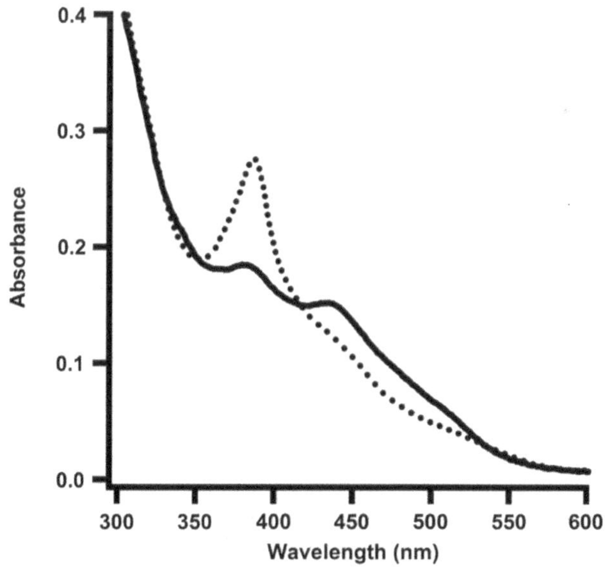

Fig. 4 UV–vis spectrum of TsrM treated with dithionite. Reduced form of enzyme (*dotted line*) shows a predominant peak at 390 nm, corresponding to cob(I)alamin. Upon addition of SAM (*solid line*), the cob(I)alamin peak decreases and a new peak arises at 450 nm, which corresponds to the formation of methylcobalamin in its base-off/His-off form.

substrate and/or SAM to monitor spectral perturbations that reflect changes in cobalamin states.

3. Remove sample, place in an EPR tube, and immediately freeze by immersing in isopentane that is cooled with liquid N_2.
4. Record an EPR spectrum of the sample at ~70 K.
5. To quantify the amount of cob(II)alamin present, make a 1-m*M* copper (II) sulfate standard and record an EPR spectrum under identical conditions. Calculate the double integral and determine the concentration of cob(II)alamin in each sample using protocols previously described (Aasa & Vänngård, 1975).
   * EPR spectra of TsrM are displayed in Fig. 5.

4.3 Sample Preparation Protocol for Analysis of MeCbl, AdoCbl, and OHCbl in Reactions

1. Because of the susceptibility of MeCbl and AdoCbl to photolysis, reactions should be performed both in the dark and in an anaerobic chamber. To protect the cobalamin from light, all samples should be prepared in amber tubes.
2. To release enzyme-bound cobalamin from the protein, quench samples in H_2SO_4 at a final concentration of 150 m*M*.
3. Centrifuge samples for 15 min at 14,000 × *g* to remove precipitated protein.
4. In the dark, remove the soluble fraction and pipette at least 40 μL into amber MS vials.
5. After preparing samples, LC-MS/MS analysis should be performed immediately to avoid breakdown of cob(III)alamin forms.

4.4 Protocol for Quantification of Cobalamin by HPLC Coupled to Electrospray-Ionization Mass Spectrometry

1. Prepare standards of MeCbl, AdoCbl, and OHCbl in the dark in an anaerobic chamber. Note that the concentration of each form of cobalamin can be determined using its known extinction coefficient or by treating with potassium cyanide and boiling to form the dicyanocobalamin adduct, which is not light or oxygen sensitive.
2. In the dark, create a standard curve containing 0.031–64 μ*M* MeCbl, AdoCbl, and OHCbl and 100 μ*M* Tyr (internal standard). Generate a standard curve by making a serial dilution starting from a sample containing 64 μ*M* of each form of cobalamin that is diluted in twofold increments, as previously described (Lanz et al., 2012).

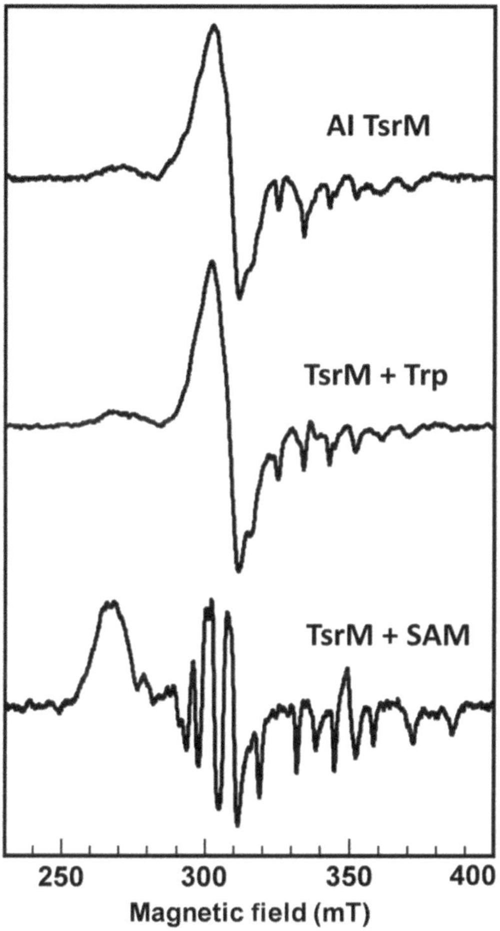

Fig. 5 EPR spectra of as-isolated TsrM, TsrM in the presence of Trp, and TsrM in the presence of SAM.

3. Centrifuge samples for 15 min at 14,000 × *g* and, in the dark, aliquot 60 μL into amber MS vials.
4. Analyze the samples on an Agilent Technologies Zorbax Extend-C18 RRHD column (2.1 mm × 50 mm, 1.8 μm particle size) equilibrated in 95% solvent A (0.1% formic acid, pH 2.6) and 5% solvent B (acetonitrile) at a flow rate of 0.35 mL/min. Apply a gradient of 5%–20% solvent B from 0.4 to 0.8 min, and then a gradient of 20%–41% solvent B from 0.8 to 1.6 min before returning to 5% solvent B from 1.6 to 1.9 min. Allow the column to reequilibrate for 0.5 min under initial conditions before subsequent sample injections. MeCbl, AdoCbl, and OHCbl elute at approximately 1.70, 1.53, and 1.17 min, respectively.

5. Use MS/MS to detect MeCbl, AdoCbl, and OHCbl by electrospray ionization in positive mode (ESI^+). MeCbl and AdoCbl are detected in the +2 charge state with *m/z* values of 673.0 and 790.6, respectively, with an optimal fragmentor voltage of 104 V for the former and 108 V for the latter. OHCbl is detected in its +2 charge state after the loss of a water molecule using a 120-V fragmentor voltage. The transitions and collision energies used for MeCbl, AdoCbl, and OHCbl are listed in Table 3. Note that the optimal voltages and collision energies may vary depending on the instrument used.
   * Elution profile of cobalamin standards analyzed by MS/MS is displayed in Fig. 6.

Table 3 Parameters for MS/MS Analysis of Cobalamin

	Transition (*m/z*)	Collision Energy
Hydroxocobalamin	664.9→635.8	11
	664.9→147.0	41
Methylcobalamin	673.0→665.6	17
	673.0→147.0	57
Adenosylcobalamin	790.6→665.6	17
	790.6→147.0	75

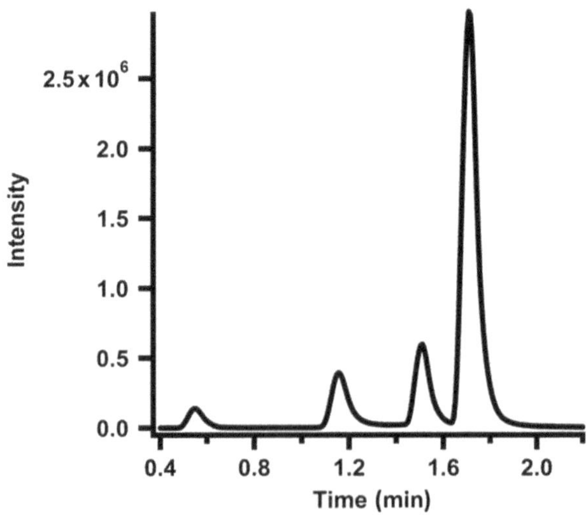

Fig. 6 Elution profile of cobalamin standard curve monitored by MS/MS. Retention times for Tyr, OHCbl, AdoCbl, and MeCbl are 0.48, 1.17, 1.53, and 1.70 min, respectively.

5. QUANTITATIVE ANALYSIS OF TURNOVER

Some Class B RS methylases generate 5′-dA, SAH, and their respective products stoichiometrically. These products—as well as respective substrates—can be quantified by LC-MS/MS as described later for a TsrM-catalyzed reaction, wherein Trp, MeTrp, SAH, SAM, and 5′-dA are monitored.

5.1 Sample Preparation Protocol for Analysis of Reaction Products by HPLC Coupled to Electrospray-Ionization Mass Spectrometry

1. Assays should contain final concentrations of the following in a total volume of 150 μL: 50 m*M* HEPES, pH 7.5, 100 m*M* KCl, 5% glycerol, 1 m*M* SAM, 1 m*M* substrate, 0.1–1 μ*M* TsrM, 100 μ*M* Tyr (internal standard), and reductant. The *E. coli* flavodoxin/flavodoxin reductase reducing system—which contains 25 μ*M* flavodoxin, 10 μ*M* flavodoxin reductase, and 1 m*M* NADPH, is typically used as a source of electrons—but can be substituted with other chemical reductants (e.g., dithionite). DTT also can be used as a reductant for the TsrM-catalyzed reaction; however, corresponding reaction rates are about twofold lower.
2. Before initiating the reaction by adding substrate, incubate the enzyme with reductant and SAM for 15 min at ambient temperature to allow for methylation of the cobalamin cofactor.
3. Initiate the reaction by adding substrate, then remove 20 μL aliquots at specified time points, and quench the reactions with an equal volume of 100 m*M* H_2SO_4.
4. Centrifuge the samples for 15 min at 14,000 × *g* before removing supernatants and placing them into MS vials.

5.2 Protocol for Quantification of Reaction Products by HPLC Coupled to Electrospray-Ionization Mass Spectrometry

1. Prepare standards of SAM, SAH, 5′-dA, Trp, and MeTrp using extinction coefficients for each molecule. SAM, SAH, and 5′-dA display absorption features at 260 nm with extinction coefficients of 15,400, 16,000, and 15,400 M^{-1}/cm, respectively, while Trp and MeTrp display absorption features at 278 nm with an extinction coefficient of 5560 M^{-1}/cm (Dawson et al., 1987). Using these standards, generate a

Table 4 Parameters for LC-MS/MS Analysis of Reaction Products

	Retention Time (min)	Parent Ion	Fragmentor Energy (V)	Product Ion 1	Collision Energy 1	Product Ion 2	Collision Energy 2
MeTrp	6.6	202.0	130	160.1	10	132.0	21
Trp	6.2	188	130	146.1	10	118	21
SAM	0.8	399.1	90	250.1	14		
SAH	1.4	385.4	136.0	100	24	134.0	24
5′-dA	2.8	252.1	136.0	90	13	119.0	50
Tyr (IS)	1.7	182.0	165.0	60	3	123.0	15

standard curve containing 0.5–250 μ*M* of each molecule and 100 μ*M* Tyr (internal standard).

2. Analyze samples on an Agilent Technologies Zorbax Extend-C18 RRHD column (4.6 mm × 50 mm, 1.8 μm particle size) equilibrated in 90.5% solvent A (0.1% formic acid, pH 2.6) and 9.5% solvent B (methanol). Apply a gradient of 9.5%–9.8% solvent B from 0.2 to 3.3 min, and then a gradient of 9.8%–30% solvent B from 3.3 to 4.1 min. Next, increase solvent B to 60% from 4.1 to 4.7 min and hold constant from 4.7 to 5.1 min before returning to 9.5% solvent B from 5.1 to 5.5 min. Allow the column to reequilibrate for 3.2 min under initial conditions before subsequent sample injections.
3. Detect products using MS/MS with electrospray ionization in positive mode. All products are detected in their +1 charge state. Table 4 displays the *m/z* values (parent and daughter), fragmentor voltages, collision energies, and retention times of all molecules that are monitored in a TsrM reaction.
4. Quantify product formation. Shown in Fig. 7 is the time-dependent formation of MeTrp, SAH, and 5′-dA in TsrM-catalyzed reaction.

6. CONCLUSIONS

Class B RS methylases catalyze the methylation of sp^2- and sp^3-hybridized carbon centers and phosphinate phosphorous centers in myriad biomolecules that play critical roles in human health and disease. Their characterization and mechanistic interrogation can lead to new insight into how cobalamin functions within enzyme-catalyzed reactions and also facilitate

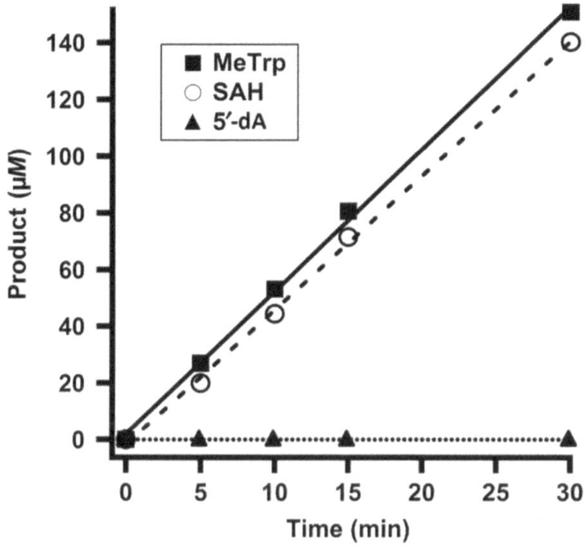

Fig. 7 Time-dependent formation of MeTrp (*solid line*), SAH (*dashed line*), and 5′-dA (*dotted line*) monitored by LC-MS/MS in a reaction with TsrM (1 µ*M*), SAM (1 m*M*), Trp (1 m*M*), flavodoxin (25 µ*M*), flavodoxin reductase (10 µ*M*), and NADPH (1 m*M*).

the development of biomolecules that have better pharmacological properties. Class B RS methylases, however, have been somewhat resistant to overproduction and purification in a homogenous and soluble form, which has hampered their study. The successful purification of some of these enzymes can be enhanced when they are produced as fusion proteins with SUMO and under conditions that facilitate cobalamin entry into the cell.

ACKNOWLEDGMENTS

This work was supported by the National Institutes of Health (GM-103268, GM-101957, GM-122595, and AI-111419) and Penn State Funds for Undergraduate Research (R.X.W.). S.J.B. is an investigator of the Howard Hughes Medical Institute.

REFERENCES

Aasa, R., & Vänngård, T. (1975). EPR signal intensity and powder shapes: A reexamination. *Journal of Magnetic Resonance (1969)*, *19*(3), 308–315. http://dx.doi.org/10.1016/0022-2364(75)90045-1.

Akiva, E., Brown, S., Almonacid, D. E., Barber, A. E., Custer, A. F., Hicks, M. A., et al. (2014). The structure–function linkage database. *Nucleic Acids Research*, *42*(Database issue), D521–D530. http://dx.doi.org/10.1093/nar/gkt1130.

Allen, K. D., & Wang, S. C. (2014). Initial characterization of Fom3 from Streptomyces wedmorensis: The methyltransferase in fosfomycin biosynthesis. *Archives of Biochemistry and Biophysics*, *543*, 67–73. http://dx.doi.org/10.1016/j.abb.2013.12.004.

Allen, K. D., Xu, H., & White, R. H. (2014). Identification of a unique radical S-adenosylmethionine methylase likely involved in methanopterin biosynthesis in Methanocaldococcus jannaschii. *Journal of Bacteriology*, *196*(18), 3315–3323. http://dx.doi.org/10.1128/jb.01903-14.

Anderle, C., Alt, S., Gulder, T., Bringmann, G., Kammerer, B., Gust, B., et al. (2007). Biosynthesis of clorobiocin: Investigation of the transfer and methylation of the pyrrolyl-2-carboxyl moiety. *Archives of Microbiology*, *187*(3), 227–237. http://dx.doi.org/10.1007/s00203-006-0190-9.

Babior, B. M., Moss, T. H., Orme-Johnson, W. H., & Beinert, H. (1974). The mechanism of action of ethanolamine ammonia-lyase, a B-12-dependent enzyme. The participation of paramagnetic species in the catalytic deamination of 2-aminopropanol. *The Journal of Biological Chemistry*, *249*(14), 4537–4544.

Bandarian, V., & Matthews, R. G. (2004). Measurement of energetics of conformational change in cobalamin-dependent methionine synthase. *Methods in Enzymology*, *380*, 152–169. http://dx.doi.org/10.1016/s0076-6879(04)80007-7.

Banerjee, R. V., Frasca, V., Ballou, D. P., & Matthews, R. G. (1990). Participation of Cob(I)alamin in the reaction catalyzed by methionine synthase from Escherichia coli: A steady-state and rapid reaction kinetic analysis. *Biochemistry*, *29*(50), 11101–11109. http://dx.doi.org/10.1021/bi00502a013.

Banerjee, R. V., Harder, S. R., Ragsdale, S. W., & Matthews, R. G. (1990). Mechanism of reductive activation of cobalamin-dependent methionine synthase: An electron paramagnetic resonance spectroelectrochemical study. *Biochemistry*, *29*(5), 1129–1135. http://dx.doi.org/10.1021/bi00457a005.

Banerjee, R., & Ragsdale, S. W. (2003). The many faces of vitamin B12: Catalysis by cobalamin-dependent enzymes. *Annual Review of Biochemistry*, *72*, 209–247. http://dx.doi.org/10.1146/annurev.biochem.72.121801.161828.

Bauerle, M. R., Schwalm, E. L., & Booker, S. J. (2015). Mechanistic diversity of radical S-adenosylmethionine (SAM)-dependent methylation. *The Journal of Biological Chemistry*, *290*(7), 3995–4002. http://dx.doi.org/10.1074/jbc.R114.607044.

Baumann, S., Herrmann, J., Raju, R., Steinmetz, H., Mohr, K. I., Huttel, S., et al. (2014). Cystobactamids: Myxobacterial topoisomerase inhibitors exhibiting potent antibacterial activity. *Angewandte Chemie (International Ed. in English)*, *53*(52), 14605–14609. http://dx.doi.org/10.1002/anie.201409964.

Beinert, H. (1978). Micro methods for the quantitative determination of iron and copper in biological material. *Methods in Enzymology*, *54*, 435–445.

Beinert, H. (1983). Semi-micro methods for analysis of labile sulfide and of labile sulfide plus sulfane sulfur in unusually stable iron-sulfur proteins. *Analytical Biochemistry*, *131*(2), 373–378.

Blackwell, C. M., & Turner, J. M. (1978). Microbial metabolism of amino alcohols. Formation of coenzyme B12-dependent ethanolamine ammonia-lyase and its concerted induction in Escherichia coli. *Biochemical Journal*, *176*(3), 751–757.

Blaszczyk, A. J., Silakov, A., Zhang, B., Maiocco, S. J., Lanz, N. D., Kelly, W. L., et al. (2016). Spectroscopic and electrochemical characterization of the iron-sulfur and cobalamin cofactors of TsrM, an unusual radical S-adenosylmethionine methylase. *Journal of the American Chemical Society*, *138*(10), 3416–3426. http://dx.doi.org/10.1021/jacs.5b12592.

Blaszczyk, A. J., Wang, B., Silakov, A., Ho, J. V., & Booker, S. J. (2017). The importance of N1 of tryptophan in the mechanism of TsrM, a cobalamin-dependent radical S-adenosylmethionine methylase. *The Journal of Biological Chemistry*. Accepted. http://www.jbc.org/content/early/2017/07/26/jbc.M117.778548.abstract.

Bommer, M., Kunze, C., Fesseler, J., Schubert, T., Diekert, G., & Dobbek, H. (2014). Structural basis for organohalide respiration. *Science*, *346*(6208), 455–458. http://dx.doi.org/10.1126/science.1258118.

Booker, S. J., & Grove, T. L. (2010). Mechanistic and functional versatility of radical SAM enzymes. *F1000 Biology Reports*, *2*, 52. http://dx.doi.org/10.3410/B2-52.

Bradford, M. M. (1976). A rapid and sensitive method for the quantitation of microgram quantities of protein utilizing the principle of protein-dye binding. *Analytical Biochemistry*, *72*, 248–254.

Broderick, J. B., Duffus, B. R., Duschene, K. S., & Shepard, E. M. (2014). Radical S-adenosylmethionine enzymes. *Chemical Reviews*, *114*(8), 4229–4317. http://dx.doi.org/10.1021/cr4004709.

Chew, A. G., & Bryant, D. A. (2007). Chlorophyll biosynthesis in bacteria: The origins of structural and functional diversity. *Annual Review of Microbiology*, *61*, 113–129. http://dx.doi.org/10.1146/annurev.micro.61.080706.093242.

Cicchillo, R. M., Iwig, D. F., Jones, A. D., Nesbitt, N. M., Baleanu-Gogonea, C., Souder, M. G., et al. (2004). Lipoyl synthase requires two equivalents of S-adenosyl-L-methionine to synthesize one equivalent of lipoic acid. *Biochemistry*, *43*(21), 6378–6386. http://dx.doi.org/10.1021/bi049528x.

Dairi, T., Ohta, T., Hashimoto, E., & Hasegawa, M. (1992). Self cloning in Micromonospora olivasterospora of fms genes for fortimicin A (astromicin) biosynthesis. *Molecular & General Genetics*, *232*(2), 262–270.

Dawson, R. M., Elliott, D. C., Elliott, W. H., & Jones, K. M. (1987). Data for biochemical research (third edition), Oxford Science Publications, OUP, Oxford, 1986. ISBN 0-19-855299-8. *Biochemical Education*, *15*(2), 97. http://dx.doi.org/10.1016/0307-4412(87)90110-5.

Dubendorf, J. W., & Studier, F. W. (1991). Controlling basal expression in an inducible T7 expression system by blocking the target T7 promoter with lac repressor. *Journal of Molecular Biology*, *219*(1), 45–59. http://dx.doi.org/10.1016/0022-2836(91)90856-2.

Frey, P. A., & Booker, S. J. (2001). Radical mechanisms of S-adenosylmethionine-dependent enzymes. *Advances in Protein Chemistry*, *58*, 1–45.

Frey, P. A., Hegeman, A. D., & Ruzicka, F. J. (2008). The radical SAM superfamily. *Critical Reviews in Biochemistry and Molecular Biology*, *43*(1), 63–88. http://dx.doi.org/10.1080/10409230701829169.

Hidaka, T., Goda, M., Kuzuyama, T., Takei, N., Hidaka, M., & Seto, H. (1995). Cloning and nucleotide sequence of fosfomycin biosynthetic genes of Streptomyces wedmorensis. *Molecular & General Genetics*, *249*(3), 274–280.

Hidaka, T., Hidaka, M., Kuzuyama, T., & Seto, H. (1995). Sequence of a P-methyltransferase-encoding gene isolated from a bialaphos-producing Streptomyces hygroscopicus. *Gene*, *158*(1), 149–150.

Hu, Y., & Ribbe, M. W. (2016). Maturation of nitrogenase cofactor—The role of a class E radical SAM methyltransferase NifB. *Current Opinion in Chemical Biology*, *31*, 188–194. http://dx.doi.org/10.1016/j.cbpa.2016.02.016.

Inahashi, Y., Zhou, S., Bibb, M. J., Song, L., Al-Bassam, M. M., Bibb, M. J., et al. (2017). Watasemycin biosynthesis in Streptomyces venezuelae: Thiazoline C-methylation by a type B radical-SAM methylase homologue. *Chemical Science*, *8*(4), 2823–2831. http://dx.doi.org/10.1039/C6SC03533G.

Jarrett, J. T., Goulding, C. W., Fluhr, K., Huang, S., & Matthews, R. G. (1997). Purification and assay of cobalamin-dependent methionine synthase from Escherichia coli. *Methods in Enzymology*, *281*, 196–213.

Kamigiri, K., Hidaka, T., Imai, S., Murakami, T., & Seto, H. (1992). Studies on the biosynthesis of bialaphos (SF-1293) 12. C-P bond formation mechanism of bialaphos: Discovery of a P-methylation enzyme. *Journal of Antibiotics (Tokyo)*, *45*(5), 781–787.

Kelly, W. L., Pan, L., & Li, C. (2009). Thiostrepton biosynthesis: Prototype for a new family of bacteriocins. *Journal of the American Chemical Society*, *131*(12), 4327–4334. http://dx.doi.org/10.1021/ja807890a.

Kim, H. J., McCarty, R. M., Ogasawara, Y., Liu, Y.-N., Mansoorabadi, S. O., LeVieux, J., et al. (2013). GenK-catalyzed C-6′ methylation in the biosynthesis of gentamicin: Isolation and characterization of a cobalamin-dependent radical SAM enzyme. *Journal of the American Chemical Society*, *135*(22), 8093–8096. http://dx.doi.org/10.1021/ja312641f.

Kim, J. Y., Suh, J. W., Kang, S. H., Phan, T. H., Park, S. H., & Kwon, H. J. (2008). Gene inactivation study of gntE reveals its role in the first step of pseudotrisaccharide modifications in gentamicin biosynthesis. *Biochemical and Biophysical Research Communications*, *372*(4), 730–734. http://dx.doi.org/10.1016/j.bbrc.2008.05.133.

Kowalak, J. A., Bruenger, E., & McCloskey, J. A. (1995). Posttranscriptional modification of the central loop of domain V in Escherichia coli 23 S ribosomal RNA. *The Journal of Biological Chemistry*, *270*(30), 17758–17764.

Kudo, F., Kasama, Y., Hirayama, T., & Eguchi, T. (2007). Cloning of the pactamycin biosynthetic gene cluster and characterization of a crucial glycosyltransferase prior to a unique cyclopentane ring formation. *Journal of Antibiotics (Tokyo)*, *60*(8), 492–503. http://dx.doi.org/10.1038/ja.2007.63.

Kuzuyama, T., Hidaka, T., Kamigiri, K., Imai, S., & Seto, H. (1992). Studies on the biosynthesis of fosfomycin. 4. The biosynthetic origin of the methyl group of fosfomycin. *Journal of Antibiotics (Tokyo)*, *45*(11), 1812–1814.

Kuzuyama, T., Seki, T., Dairi, T., Hidaka, T., & Seto, H. (1995). Nucleotide sequence of fortimicin KL1 methyltransferase gene isolated from Micromonospora olivasterospora, and comparison of its deduced amino acid sequence with those of methyltransferases involved in the biosynthesis of bialaphos and fosfomycin. *Journal of Antibiotics (Tokyo)*, *48*(10), 1191–1193.

Landgraf, B. J., McCarthy, E. L., & Booker, S. J. (2016). Radical S-adenosylmethionine enzymes in human health and disease. *Annual Review of Biochemistry*, *85*, 485–514. http://dx.doi.org/10.1146/annurev-biochem-060713-035504.

Lanz, N. D., Grove, T. L., Gogonea, C. B., Lee, K. H., Krebs, C., & Booker, S. J. (2012). RlmN and AtsB as models for the overproduction and characterization of radical SAM proteins. *Methods in Enzymology*, *516*, 125–152. http://dx.doi.org/10.1016/b978-0-12-394291-3.00030-7.

Lexa, D., & Saveant, J. M. (1983). The electrochemistry of vitamin B12. *Accounts of Chemical Research*, *16*(7), 235–243. http://dx.doi.org/10.1021/ar00091a001.

Li, C., & Kelly, W. L. (2010). Recent advances in thiopeptide antibiotic biosynthesis. *Natural Product Reports*, *27*(2), 153–164. http://dx.doi.org/10.1039/b922434c.

Ljungdahl, L. G., LeGall, J., & Lee, J. P. (1973). Isolation of a protein containing tightly bound 5-methoxybenzimidazolylcobamide (factor 3 m) from Clostridium thermoaceticum. *Biochemistry*, *12*(9), 1802–1808.

Mao, Y., Varoglu, M., & Sherman, D. H. (1999). Molecular characterization and analysis of the biosynthetic gene cluster for the antitumor antibiotic mitomycin C from Streptomyces lavendulae NRRL 2564. *Chemistry & Biology*, *6*(4), 251–263. http://dx.doi.org/10.1016/s1074-5521(99)80040-4.

Marblestone, J. G., Edavettal, S. C., Lim, Y., Lim, P., Zuo, X., & Butt, T. R. (2006). Comparison of SUMO fusion technology with traditional gene fusion systems: Enhanced expression and solubility with SUMO. *Protein Science: A Publication of the Protein Society*, *15*(1), 182–189. http://dx.doi.org/10.1110/ps.051812706.

Marous, D. R., Lloyd, E. P., Buller, A. R., Moshos, K. A., Grove, T. L., Blaszczyk, A. J., et al. (2015). Consecutive radical S-adenosylmethionine methylations form the ethyl side chain in thienamycin biosynthesis. *Proceedings of the National Academy of Sciences of the United States of America*, *112*(33), 10354–10358. http://dx.doi.org/10.1073/pnas.1508615112.

Menon, S., & Ragsdale, S. W. (1998). Role of the [4Fe-4S] cluster in reductive activation of the cobalt center of the corrinoid iron-sulfur protein from Clostridium thermoaceticum during acetate biosynthesis. *Biochemistry*, *37*(16), 5689–5698. http://dx.doi.org/10.1021/bi9727996.

Nunez, L. E., Mendez, C., Brana, A. F., Blanco, G., & Salas, J. A. (2003). The biosynthetic gene cluster for the beta-lactam carbapenem thienamycin in Streptomyces cattleya. *Chemistry & Biology*, *10*(4), 301–311.

Ostash, B., Saghatelian, A., & Walker, S. (2007). A streamlined metabolic pathway for the biosynthesis of moenomycin A. *Chemistry & Biology*, *14*(3), 257–267. http://dx.doi.org/10.1016/j.chembiol.2007.01.008.

Panavas, T., Sanders, C., & Butt, T. R. (2009). SUMO fusion technology for enhanced protein production in prokaryotic and eukaryotic expression systems. In H. D. Ulrich (Ed.), *SUMO protocols* (pp. 303–317). Totowa, NJ: Humana Press.

Pandelia, M. E., Lanz, N. D., Booker, S. J., & Krebs, C. (2015). Mossbauer spectroscopy of Fe/S proteins. *Biochimica et Biophysica Acta*, *1853*(6), 1395–1405. http://dx.doi.org/10.1016/j.bbamcr.2014.12.005.

Parent, A., Guillot, A., Benjdia, A., Chartier, G., Leprince, J., & Berteau, O. (2016). The B12-radical SAM enzyme PoyC catalyzes valine Cbeta-methylation during polytheonamide biosynthesis. *Journal of the American Chemical Society*, *138*(48), 15515–15518. http://dx.doi.org/10.1021/jacs.6b06697.

Payne, K. A., Quezada, C. P., Fisher, K., Dunstan, M. S., Collins, F. A., Sjuts, H., et al. (2015). Reductive dehalogenase structure suggests a mechanism for B12-dependent dehalogenation. *Nature*, *517*(7535), 513–516. http://dx.doi.org/10.1038/nature13901.

Pierre, S., Guillot, A., Benjdia, A., Sandstrom, C., Langella, P., & Berteau, O. (2012). Thiostrepton tryptophan methyltransferase expands the chemistry of radical SAM enzymes. *Nature Chemical Biology*, *8*(12), 957–959. http://dx.doi.org/10.1038/nchembio.1091.

Rachid, S., Scharfe, M., Blocker, H., Weissman, K. J., & Muller, R. (2009). Unusual chemistry in the biosynthesis of the antibiotic chondrochlorens. *Chemistry & Biology*, *16*(1), 70–81. http://dx.doi.org/10.1016/j.chembiol.2008.11.005.

Sofia, H. J., Chen, G., Hetzler, B. G., Reyes-Spindola, J. F., & Miller, N. E. (2001). Radical SAM, a novel protein superfamily linking unresolved steps in familiar biosynthetic pathways with radical mechanisms: Functional characterization using new analysis and information visualization methods. *Nucleic Acids Research*, *29*(5), 1097–1106.

Toh, S.-M., Xiong, L., Bae, T., & Mankin, A. S. (2008). The methyltransferase YfgB/RlmN is responsible for modification of adenosine 2503 in 23S rRNA. *RNA*, *14*(1), 98–106. http://dx.doi.org/10.1261/rna.814408.

Wang, Y., Schnell, B., Baumann, S., Muller, R., & Begley, T. P. (2017). Biosynthesis of branched alkoxy groups: Iterative methyl group alkylation by a cobalamin-dependent radical SAM enzyme. *Journal of the American Chemical Society*, *139*(5), 1742–1745. http://dx.doi.org/10.1021/jacs.6b10901.

Werner, W. J., Allen, K. D., Hu, K., Helms, G. L., Chen, B. S., & Wang, S. C. (2011). In vitro phosphinate methylation by PhpK from Kitasatospora phosalacinea. *Biochemistry*, *50*(42), 8986–8988. http://dx.doi.org/10.1021/bi201220r.

Woodyer, R. D., Shao, Z., Thomas, P. M., Kelleher, N. L., Blodgett, J. A., Metcalf, W. W., et al. (2006). Heterologous production of fosfomycin and identification of the minimal biosynthetic gene cluster. *Chemistry & Biology*, *13*(11), 1171–1182. http://dx.doi.org/10.1016/j.chembiol.2006.09.007.

Zhang, Q., van der Donk, W. A., & Liu, W. (2012). Radical-mediated enzymatic methylation: A tale of two SAMS. *Accounts of Chemical Research*, *45*(4), 555–564. http://dx.doi.org/10.1021/ar200202c.

Zheng, L., Cash, V. L., Flint, D. H., & Dean, D. R. (1998). Assembly of iron-sulfur clusters. Identification of an iscSUA-hscBA-fdx gene cluster from Azotobacter vinelandii. *The Journal of Biological Chemistry*, *273*(21), 13264–13272.

CHAPTER ELEVEN

Mechanism-Based Strategies for Structural Characterization of Radical SAM Reaction Intermediates

Katherine M. Davis*,†, Amie K. Boal†,1
*Princeton University, Princeton, NJ, United States
†The Pennsylvania State University, University Park, PA, United States
[1]Corresponding author: e-mail address: akb20@psu.edu

Contents

Abstract

X-ray crystallographic characterization of enzymes at different stages in their reaction cycles can provide unique insight into the reaction pathway, the number and type of intermediates formed, and their structural context. The known mechanistic diversity in the radical *S*-adenosylmethionine (SAM) superfamily of enzymes makes it an appealing target for such studies as more than 100,000 sequences have been identified to date with wide-ranging reactivities that share a pattern of complex radical-mediated chemistry. Here, we review selected examples of radical SAM enzyme crystal structures representative of reactant, product, and intermediate state complexes with a particular emphasis on the strategies employed to capture these states. Broader application of structural characterization techniques to analyze mechanism and substrate specificity is certain to play an important role as more members of this family become tractable for biochemical study.

Methods in Enzymology, Volume 595
ISSN 0076-6879
http://dx.doi.org/10.1016/bs.mie.2017.07.008

1. INTRODUCTION

Enzymes utilize carefully choreographed dynamics to catalyze the diverse and often challenging reactions that are necessary for life. These controlled motions link structure to function and are key components in our understanding of enzyme reactivity. Although the advent of macromolecular X-ray crystallography during the past century has provided unprecedented advancements in the field of structural biology, the technical challenges associated with trapping short-lived intermediate states have limited our mechanistic interpretation. For example, in enzymes that use radical-based chemistry, electron transfer intermediates are often transient and associated with subtle conformational changes that are difficult to visualize. While nature ordinarily employs molecular oxygen to generate reactive radical species, anaerobic systems rely on alternative mechanisms for the formation of highly oxidizing intermediates. One of the most abundant of these mechanisms is the homolytic cleavage of *S*-adenosylmethionine (SAM) by the radical SAM superfamily.

Ubiquitous across all domains of life, radical SAM enzymes catalyze a multitude of unusual biosynthetic transformations using a reduced [4Fe–4S] cluster chelated by a canonical three-cysteine motif (CX_3CX_2C). Donation of an electron from the Fe/S cluster to SAM, coordinated as a bidentate ligand to the cluster, results in the generation of methionine (Met) and a 5′-deoxyadenosyl radical (5′-dA•). The subsequent abstraction of a substrate hydrogen atom by 5′-dA• initiates the desired modification reaction to yield both product and 5′-deoxyadenosine (5′-dA) (Booker & Grove, 2010; Broderick, Duffus, Duschene, & Shepard, 2014; Dowling, Vey, Croft, & Drennan, 2012). Following classification of the superfamily in 2001 (Sofia, Chen, Hetzler, Reyes-Spindola, & Miller, 2001), significant effort was made to establish this basic mechanistic scheme across a diverse subset of radical SAM enzymes. However, we have barely scratched the surface with respect to understanding how these fascinating enzymes elaborate and modify this pathway to modulate their individual reactivities.

Various experimental approaches have been applied toward this question, with the ultimate goal of determining a complete step-by-step sequence of elementary reactions. The use of biochemical techniques such as enzymatic assays and site-directed mutagenesis has helped to elucidate radical SAM enzyme kinetics and identify key active site residues, while spectroscopic methods have allowed for the investigation of hypothesized transient intermediates (Broderick et al., 2014). For example, the formation of a cross-linked reaction intermediate, in which an auxiliary Fe/S cluster of the radical SAM enzyme LipA is covalently bound to its *n*-octanoyl substrate, was recently suggested in a pivotal study utilizing Mössbauer and electron paramagnetic resonance spectroscopies (Lanz et al., 2014). Although these techniques are extremely informative, they only probe the local environment surrounding a particular radical species or transition metal and technical limitations prevent the observation of certain proposed intermediates (Pandelia, Lanz, Booker, & Krebs, 2015). Structural methods, such as X-ray crystallography, not only overcome these hurdles but also allow for visualization of the active site in its entirety, making them largely complementary to spectroscopic approaches in defining the reaction mechanism for a given metalloenzyme.

In this chapter, we present three key strategies for trapping the catalytic states of radical SAM enzymes *in crystallo* using relevant literature as a guide. We first discuss methods to obtain snapshots of pre- and postturnover conformations via the addition of substrates, products, or their respective analogs. We then address the challenges associated with obtaining intermediate state structures, including cocrystallization and soaking techniques that may allow for the formation of serendipitous intermediate state conformations, culminating in strategies to stabilize short-lived intermediates. In particular, we delve into the specifics of rationally designing a crystallization experiment to capture proposed catalytic states using a priori knowledge of the reaction mechanism.

2. OBTAINING PRE- AND POSTTURNOVER SNAPSHOTS

As with all in vitro techniques, the act of crystallizing an enzyme inherently removes it from its physiological context. Thus, mechanistic interpretation of a protein structure requires careful consideration of the observed interactions and their relevance to the active complex. The presence of substrates/products bound to the enzyme lends confidence to the mechanistic relevance of a given conformation and provides a pre- or postturnover framework in which to place the structural model. Furthermore, this approach can even facilitate initial crystallization trials due to the often stabilizing effect of ligand binding on disordered or flexible regions of the protein (Hassell et al., 2007). Substrate-bound structures have resulted in a number of key breakthroughs for the field of radical SAM enzymology. Selected examples are detailed below.

2.1 The Structure of a Substrate-Bound MoaA Complex

The radical SAM enzyme MoaA catalyzes the first step in molybdenum cofactor (Moco) biosynthesis by abstracting a hydrogen atom from the 3′-C of guanosine 5′-triphosphate (GTP) to generate 3′,8-cyclic-7,8-dihydro-GTP (Hover, Tonthat, Schumacher, & Yokoyama, 2015; Mehta et al., 2013). Subsequent modification by the synthase MoaC results in the formation of the critical pyranopterin ring required for the production of cyclic pyranopterin (cPMP). As with most radical SAM enzymes, concentric rings of six sequential α-helices and β-strands form the catalytic core of MoaA, in which rearrangement of the guanosine triphosphate substrate occurs proximal to the active site [4Fe–4S] cluster (Hänzelmann & Schindelin, 2004). MoaA was the third radical SAM enzyme to be structurally characterized and an initial X-ray structure, solved in the absence of GTP, helped to establish this partial triosephosphate isomerase (TIM) barrel as the common fold in radical SAM enzymes (Table 1, PDB accession codes 1TV7, 1TV8; Broderick et al., 2014). Auxiliary domains appended at the N- or C-terminus often augment the canonical fold, facilitating diversification of substrate scope and reactivity. Indeed, MoaA, like several other members of the superfamily, contains a cysteine-rich C-terminal extension (Grell, Goldman, & Drennan, 2015; Menéndez, Igloi, Henninger, & Brandsch, 1995). Early biochemical work on the human ortholog MOCS1A further supported the existence of a unique second [4Fe–4S] cluster within the auxiliary domain (Hänzelmann et al., 2004). Although determination of the structure confirmed this hypothesis, new questions quickly arose regarding the function of the additional Fe/S cofactor. A possible role in substrate binding was proposed based on the presence of an open coordination site similar to that of the canonical cluster; however, the SAM-bound structure proved inconclusive (Hänzelmann et al., 2004). Subsequent cocrystallization with the substrate GTP not only resolved this fundamental mechanistic question (Fig. 1), but also identified key components of the active site that contribute to the enzyme's high substrate specificity (Table 1, PDB accession code 2FB3).

This compelling follow-up structure by Hänzelmann and Schindelin illustrates GTP binding to the unique iron of the auxiliary cluster via the N1 nitrogen in the pyrimidine ring and the exocyclic amino group of

Table 1 Summary of Crystal Structures Representative of Radical SAM Reactant, Product, and Intermediate State Complexes

Enzyme	No. Fe/S Clusters	Organism	PDBID	Res. (Å)	Cluster/s[a]	SAM	SAH	Met	5′-dA	Ions[b]	Substrate	Product	Analogs[c]	Intermediate
MoaA	2 × [4Fe–4S]	*Staphylococcus aureus*	1TV7	2.80	✓✓									
			1TV8[d]	2.20	✓✓	✓								
			2FB2[e]	2.25	✓✓	✓								
			2FB3	2.35	✓✓			✓	✓		GTP			
ThiC	1 × [4Fe–4S]	*Caulobacter crescentus*	3EPM	2.79						Zn(II)			HMP	
			3EPN	2.11									IMR	
			3EPO	2.10								HMP-P		
		Arabidopsis thaliana	4N7Q	1.60						Co(II)				
		Caulobacter crescentus	4S2A	2.93	✓									
		Arabidopsis thaliana	4S25	1.45	✓		✓			Zn(II), Cl^-			IMR	
			4S26	1.85	✓		✓			Zn(II), Cl^-			IMR	
			4S27	1.27	✓			✓	✓	Fe(II), Cl^-		AIR		
			4S28	1.25	✓		✓			Fe(II), Cl^-		AIR		
			4S29	1.38						Fe(II)			IMR	

Continued

Table 1 Summary of Crystal Structures Representative of Radical SAM Reactant, Product, and Intermediate State Complexes—cont'd

Enzyme	No. Fe/S Clusters	Organism	PDBID	Res. (Å)	Cluster/s	SAM	SAH	Met	5′-dA	Ions	Substrate	Product	Analogs	Intermediate
QueE	1 × [4Fe–4S]	*Burkholderia multivorans*	4NJG	2.60	✓	✓							6CP	
			4NJH	1.90	✓	✓					CPH_4			
			4NJI	2.20	✓	✓				Mg^{2+}	CPH_4			
			4NJJ	2.70	✓	✓				Mn(II)	CPH_4			
			4NJK	1.91	✓	✓				Mg^{2+}		CDG		
		Bacillus subtilis	5TGS	2.55	✓			✓						
			5TH5	2.41	✓			✓						6CP-dA ester
SuiB	3 × [4Fe–4S]	*Streptococcus suis*	5V1Q	2.50	✓✓✓									
			5V1S	2.49	✓✓✓	✓								
			5V1T	2.10	✓✓✓	✓		✓			SuiA			
HydE	1 × [4Fe–4S], 1 × [2Fe–2S]	*Thermotoga maritima*	3CIW	1.35	✓		✓			Cl^-				
			3CIX	1.70	✓✓		✓							
			3IIX	1.25	✓			✓	✓	Cl^-				
			3IIZ	1.62	✓✓	✓								
			5FEP	1.45	✓	✓					(2*R*,4*R*)-MeTDA			

			5FES	1.27	✓	✓					(2*R*-4*R*)-MeSeTDA			
			5FEW[f]	1.17	✓			✓		Cl^-		SAC		
			5FEX	1.32	✓			✓	✓	Cl^-			Se-Cys	
			5FEZ[g]	1.35	✓			✓	✓		(2*R*-4*R*)-MeSeTDA			
			5FF0 [g]	1.49	✓			✓		Cl^-		SAC		
			5FF2	1.47	✓		✓						(2*R*,4*R*)-TDA	
			5FF3	1.18	✓✓		✓			Cl^-			4*R*–TCA	
			5FF4[g]	1.35	✓		✓						(2*R*,4*R*)-TMeTDA	
LipA	2 × [4Fe–4S]	*Thermosynechococcus elongatus*	4U0O[h]	1.60	✓✓									5′-MTA
			4U0P	1.62	✓✓		✓							
		Mycobacterium tuberculosis	5EXI[h]	2.28	✓✓									
			5EXJ[h]	1.64	✓✓									
			5EXK[i]	1.86	✓✓			✓	✓					6-Thiooctanoyl[j]

Continued

Table 1 Summary of Crystal Structures Representative of Radical SAM Reactant, Product, and Intermediate State Complexes—cont'd

Enzyme	No. Fe/S Clusters	Organism	PDBID	Res. (Å)	Cluster/s	SAM	SAH	Met	5′-dA	Ions	Substrate	Product	Analogs	Intermediate
RlmN	1 × [4Fe–4S]	*Escherichia coli*	3RFA	2.05	✓	✓								
			3RF9	2.20	✓									
			4PL1 [k]	2.58	✓	✓								
			4PL2[k]	2.20	✓									
			5HR6 [k,l]	2.88	✓			✓	✓					$tRNA^{Glu}$ cross-link
			5HR7 [k,m]	2.40	✓			✓	✓					$tRNA^{Glu}$ cross-link
BioB	1 × [4Fe–4S], 1 × [2Fe–2S]	*Escherichia coli*	1R30	3.40	✓	✓					DTB			

[a]Checkmarks correspond to each cluster observed.
[b]Key active site ions discussed.
[c]Analogs of substrates or products.
[d]DTT bound to Aux cluster.
[e]Arg17/266/268/Ala triple mutant.
[f]Some occupancy for SAH and (2*R*,4*R*)-MeTDA.
[g]Fe/Se clusters.
[h]DTT bound to radical SAM cluster.
[i]1 × [4Fe–4S], 1 × [3Fe–4S].
[j]Peptide cross-linked to Aux cluster.
[k]C118A mutant.
[l]In vivo.
[m]In vitro.

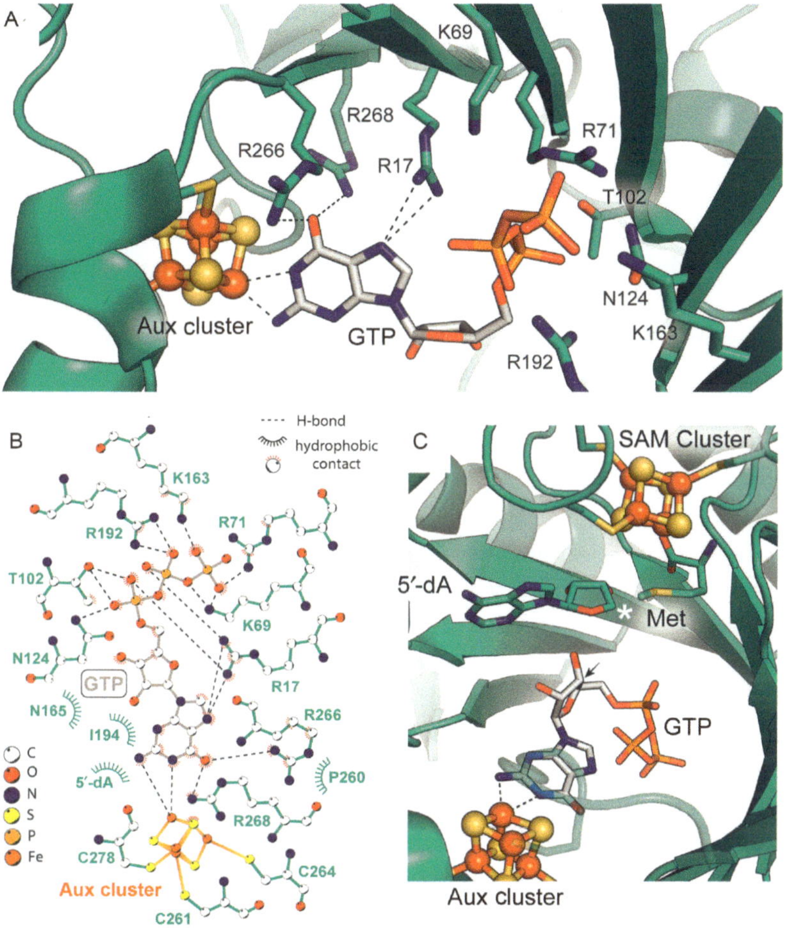

Fig. 1 Substrate binding in MoaA (PDB accession code 2FB3). (A) A zoomed-in view of the active site depicts the bidentate binding mode of GTP to the auxiliary cluster. Important active site residues are shown in *stick representation*. (B) A 2-D interaction map illustrates the extensive hydrogen bonding network with the triphosphate moiety. Hydrogen bonds are shown for distances of 3.35 Å or less. (C) A zoomed-out view of the barrel interior shows the arrangement of cleaved products Met and 5′-dA in relation to GTP. Although the critical 5′-C (identified with an *asterisk*) was not modeled due to disorder, it would clearly be positioned proximal to the C3′ position of the substrate, in preparation for H-atom abstraction.

guanine (Hänzelmann & Schindelin, 2006). Numerous hydrogen bonds with the triphosphate moiety explain the low affinity for GDP and GMP, while the observed bidentate binding mode to the auxiliary cluster reveals the structural basis of purine specificity (Fig. 1A and B). Migration of the exocyclic amino group from C2 to C6, as in ATP for example, would eliminate a coordination interaction and weaken hydrogen bonding to R266 and R268. Likewise, MoaA's high affinity for the unnatural nucleotide XTP (xanthosine triphosphate) can be rationalized as replacing one ligating group (C2–NH_2) with another equally acceptable option (C2=O), while maintaining the second-sphere H-bonding interactions. The nucleotide coordination mode in both cases resembles that of SAM to the canonical cluster, involving a bidentate ligand with coordination via O/N atoms. At the time, direct substrate binding to an auxiliary [4Fe–4S] cluster was unprecedented and remains relatively rare among structurally and biochemically characterized radical SAM enzymes, with only five examples reported to date (Mulliez, Duarte, Arragain, Fontecave, & Atta, 2017).

The structure of MoaA in complex with GTP was determined using a combination of cocrystallization and soaking techniques. To ensure high occupancy of substrate within the active site, GTP was included at high concentration (tens of millimolar) in both the crystallization condition and the cryoprotection soak. While the cosubstrate SAM was also present in the cryosoak, the concurrent addition of sodium dithionite resulted in the reduction of the radical SAM cluster and allowed for the initiation of turnover *in crystallo* to yield Met and 5′-dA (Fig. 1C; Table 1). Intriguingly, the resultant structure contains GTP rather than product, perhaps due to a disparity between the concentrations of the substrates (Hänzelmann & Schindelin, 2006). Additional examples in which fortuitous intermediate states were trapped by triggering catalysis *in crystallo* are presented in Section 3.

To obtain a true preturnover conformation, the presence of all relevant substrates is required in the absence of a catalytic trigger. For radical SAM enzymes, this is often achieved by working with predominantly oxidized clusters, unable to reductively cleave SAM. Examples of this type of preturnover structure include that of the glycyl radical activating enzyme of pyruvate formate-lyase (PFL-AE) (Vey et al., 2008), lysine 2,3-aminomutase (LAM) (Lepore, Ruzicka, Frey, & Ringe, 2005), biotin synthase (BioB) (Berkovitch, Nicolet, Wan, Jarrett, & Drennan, 2004), the DOIA dehydrogenase BtrN (Goldman, Grove, Booker, & Drennan, 2013), and the anaerobic sulfatase maturating enzyme anSMEcpe (Goldman, Grove, Sites, et al., 2013). Although it is preferable to utilize natural substrates, substrate analogs are also useful for trapping the reactant state without concern for unintended enzyme turnover. To obtain the preturnover structure of LAM referenced earlier, the enzyme was actually cocrystallized in the presence of *S*-adenosylhomocysteine (SAH); however, SeSAM was observed bound to the cluster instead. Such a phenomenon is not wholly unexpected as SAM is often endogenously synthesized and copurifies with the enzymes (Betz et al., 2015; Yan & Fujimori, 2011). In the case of LAM, SeSAM was likely incorporated during recombinant expression and was not displaced during cocrystallization with the substrate analog, SAH (Lepore et al., 2005). Nevertheless, there are many instances in which SAH and other substrate/product analogs have proven invaluable (Dowling et al., 2014; Fenwick et al., 2015; Harmer et al., 2014; Nicolet, Zeppieri, Amara, & Fontecilla-Camps, 2014; Rohac et al., 2016).

2.2 The Structures of Product-Bound ThiC and QueE Complexes

Substrate, product, and analog-bound structures are especially effective in the study of noncanonical radical SAM enzymes, for which the substrate-binding site and location of active site components can be ambiguous. Such structures of ThiC and QueE not only allowed for the identification of their respective active sites, but also further assisted in the elucidation of unprecedented mechanistic features (Table 1; Chatterjee et al., 2008; Dowling et al., 2014; Fenwick et al., 2015). ThiC catalyzes the complex rearrangement of 4-aminoimidazole ribonucleotide (AIR) to 4-amino-5-hydroxymethyl-2-methylpyrimidine phosphate (HMP-P) in the biosynthetic pathway of thiamin (Jurgenson, Begley, & Ealick, 2009). Unlike archetypal radical SAM enzymes in which the [4Fe–4S] cluster is chelated by an N-terminal CX_3CX_2C motif, ThiC contains a modified cluster-binding loop, CX_2CX_4C near its C-terminus. Initial attempts to visualize the holoenzyme failed due to disorder in this cluster-binding domain; however, electron density for HMP-P was detected within the barrel (β_8/α_8) far from the C-terminus. Intriguingly, the position of bound product proximal to the C-terminus of a different protomer suggested a domain-swapped dimer as the functional form (Fig. 2), an observation that was subsequently corroborated by alignments to structurally homologous adenosylcobalamin enzymes (Chatterjee et al., 2008).

More recently, the cluster-binding domain was stabilized via the addition of the substrate analog SAH and the SAM cleavage products, Met and 5′-dA, into the crystallization conditions of ThiC from *Arabidopsis thaliana* and *Caulobacter crescentus* (Table 1, PDB accession codes 4S25, 4S26, 4S27, 4S28; Fenwick et al., 2015). These results verify the hypotheses derived from the HMP-P product-bound model (Table 1, PDB accession code 3EPO; Chatterjee et al., 2008), while further differentiating ThiC from

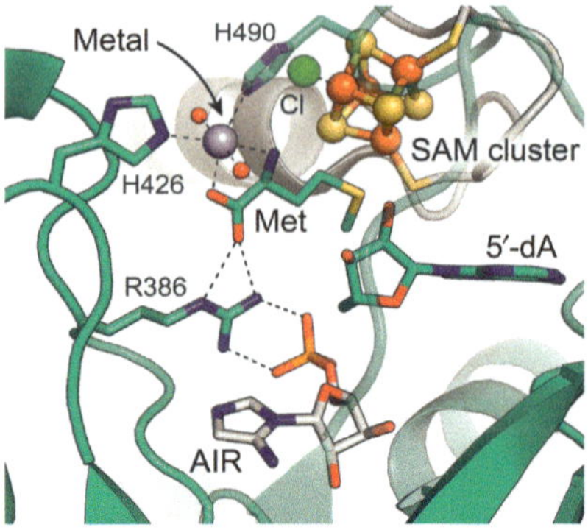

Fig. 2 The noncanonical active site of ThiC (PDB accession code 4S27). Positioned in the substrate-binding site via a domain-swapping mechanism, the Fe/S cluster is ligated to another protomer, depicted in *gray*. Rather than being ligated to the unique Fe of the [4Fe–4S] cluster, however, the amino and carboxyl groups of Met coordinate a mononuclear transition metal ion in ThiC. The open coordination site of the cluster is instead occupied by a chloride ion.

other members of the superfamily. Surprisingly, the amino- and carboxyl groups of SAH coordinate a separate mononuclear transition metal center rather than the unique Fe of the canonical cluster, which instead binds a chloride ion (Fenwick et al., 2015). Depending on the metal ions available during crystallization, this site appears to reconstitute with Co(II), Zn(II), or Fe(II) in different ThiC structures (Table 1). The relevant ion for in vivo activity is unknown. To validate this singular binding configuration, SAM cleavage products were also cocrystallized with the enzyme. Good agreement between the SAH and Met/5′-dA structures, in addition to strict conservation of an arginine observed to form a salt bridge with the carboxylate (Fig. 2), strongly supports this conformation as physiologically relevant rather than an artifact of crystallization or use of the SAM analog, SAH.

Equally compelling is the series of QueE structures characterizing both pre- and postturnover reaction intermediates/mimics (Table 1; Fig. 3; Dowling et al., 2014). A key component in 7-deazapurine biosynthesis, QueE catalyzes the transformation of 6-carboxy-5,6,7,8-tetrahydropterin (CPH_4) to 7-carboxy-7-deazaguanine (CDG), essential for the production of the modified nucleoside queuosine, as well as numerous natural products (McCarty, Somogyi, Lin, Jacobsen, & Bandarian, 2009). Like ThiC (Chatterjee et al., 2008), QueE from *Bacillus multivorans* forms a physiological homodimer (Dowling et al., 2014); however, its core fold is remarkably abbreviated from the partial TIM barrel motif (β_6/α_6) commonly found in radical SAMs (Broderick et al., 2014). The catalytic core contains two short loops and a 3_{10}-helix in the place of three characteristic α-helices to yield a final β_6/α_3 fold (Dowling et al., 2014). In the absence of substrates and products, identification of the substrate-binding site within this relatively open configuration would have been difficult.

Instead cocrystallization experiments with CPH_4 and the substrate analog 6-carboxypterin (6CP) not only resolved this critical question, but also elucidated the enzyme's puzzling dependence on Mg^{2+} (Table 1, PDB accession codes 4NJG, 4NJH, 4NJI, 4NJJ, 4NJK). In the 6CP-bound structure, hydrogen bonding interactions between components of the pyrimidine ring and a C-terminal extension facilitated significant occlusion of the otherwise solvent exposed active site. Complete closure was only achieved in the presence of the natural substrates/products CPH_4 and CDG, both of which chelate a mononuclear ion believed to be the biochemically identified catalytic Mg^{2+} (Fig. 3; Dowling et al., 2014; McCarty, Krebs, & Bandarian, 2013). This result, in conjunction with the product-bound structure, allowed for the proposal of a revised mechanism for QueE in which Mg^{2+} plays an active role in catalysis, possibly facilitating H-atom abstraction via optimal substrate positioning relative to the SAM cofactor and/or functioning as a Lewis acid. Substitution of other metals (e.g., Na^+) in place of Mg^{2+} altered key interactions with the substrate and lowered catalytic activity, thereby solidifying the ion's structural role. While most members of the superfamily consume SAM stoichiometrically, QueE (McCarty et al., 2013) and others (Cheek & Broderick, 2002; Chen, Walsby, Hoffman, & Frey, 2003; Ruzicka & Frey, 2007) are able to regenerate SAM during the course of turnover. The structural basis for these distinct mechanisms of SAM utilization are not well understood. To account for the catalytic use of SAM by QueE, Dowling et al. modeled a proposed reaction intermediate, constrained by the crystallographically observed position of substrate CPH_4 (Dowling et al., 2014). The CPH_4 exocyclic nitrogen radical of this putative intermediate was located near the 5′-C of SAM, a juxtaposition that is consistent with regeneration of SAM. Even in the absence of an experimentally determined transition state structure, the static pre- and postturnover mimics of QueE proved invaluable in deciphering several important features of the reaction mechanism.

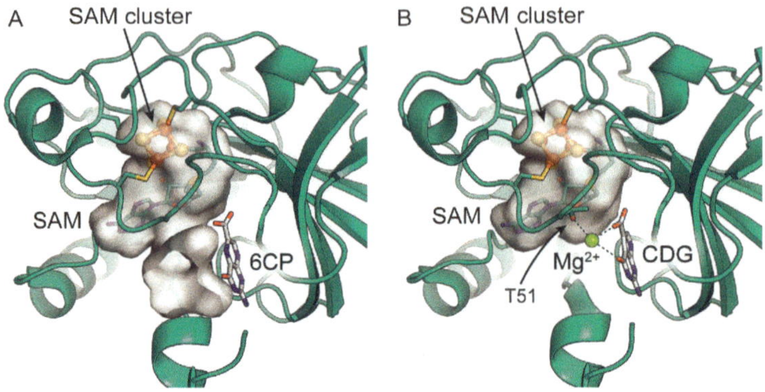

Fig. 3 Closure of the active site in QueE. Surface representations of the active site cavity with (A) substrate analog, 6CP (PDB accession code 4NJG), vs (B) product, CDG (PDB accession code 4NJK), bound highlight the potential role of Mg^{2+} to occlude the active site during catalysis.

3. SERENDIPITOUS STRUCTURES OF REACTION INTERMEDIATES

Although great insight into enzyme function can be inferred from the crystallization of steady-state systems, structures of reaction intermediates can help confidently bridge the mechanistic gap between pre- and post-turnover conformations. The short-lived nature of most catalytic intermediates precludes their explicit crystallization. However, a number of methods exist that allow for, and even encourage, the adventitious crystallization of typically transient conformations. Cocrystallization of a reduced radical SAM reactant complex, e.g., via the addition of a strong reductant such as sodium dithionite, supports active turnover. While this typically yields a postcleavage structure (Hänzelmann & Schindelin, 2006), premature termination of the reaction due to disproportionate concentrations of reaction components (e.g., substrate, SAM, or reductant) occasionally results in the crystallization of an unexpected intermediate state (Davis et al., 2017). Selected examples are reviewed later; however, this method is extremely unpredictable, and in many cases, difficult to design rationally.

By contrast, initiation of turnover *in crystallo* eliminates the uncertainty in crystal composition and establishes a means by which to control intermediate state selection. The relative inefficiency of radical SAM enzymes, with the exclusion of LAM and glutamate 2,3-aminomutase (EAM) (Broderick et al., 2014), enables manual cryotrapping of intermediate states. Crystals of the reactant complex are typically soaked in a cryoprotectant solution containing the catalytic trigger, e.g., sodium dithionite, before being flash-frozen after a period of time, Δt. The selected time interval dictates the point at which the catalytic cycle is arrested, and in conjunction with biochemical assays, can be used to predict the identity of the trapped intermediate. As this procedure is not an exact science, a series of time intervals is often collected for comparison with the preturnover complex. Section 3.2 discusses the logistics behind such an experiment in greater detail using an extensive structural study of HydE by Rohac et al. as a guide.

3.1 Cocrystallization Yields Intermediate State Structures of MoaA, QueE, SuiB

There are relatively few examples of structurally characterized radical SAM enzymes for which cocrystallization has yielded a serendipitous intermediate state structure. In the case of MoaA, described in Section 2.1, the addition of sodium dithionite clearly induced SAM cleavage. However, the resultant structure contained the GTP substrate instead of an intermediate or product (Fig. 1C; Table 1; Hänzelmann & Schindelin, 2006). Such a conformation cannot be classified as a reactant or product state, but rather a preparatory intermediate, post- one turnover and pre- the next. Similar behavior was observed in the RiPP (ribosomally synthesized and posttranslationally modified peptide) biosynthetic enzyme SuiB, a close ortholog of StrB, known to install a Lys-to-Trp cross-link between two unactivated carbons of its precursor peptide substrate SuiA (Davis et al., 2017; Schramma, Bushin, & Seyedsayamdost, 2015; Schramma & Seyedsayamdost, 2017). As reduction of the enzyme during reconstitution prevented isolation of the reactant complex, and no suitable condition was found with peptide substrate only, cocrystallization of a reaction mixture was attempted (Davis et al., 2017). The final structure was expected to minimally contain the SAM cleavage products, and indeed, electron density for Met was consistent with tridentate chelation to the unique Fe of the radical SAM cluster. Curiously, density contiguous with that of the putative 5′-dA product extended past the nominal position of the 5′-C. The authors conclude that a new molecule of SAM replaced the 5′-deoxyadenosine product in preparation for subsequent turnovers, while the cluster had yet to release Met (Fig. 4B; Table 1, PDC accession code 5V1T). This unusual arrangement of products and substrates had not been observed previously and may shed light on the order of operations required to regenerate the active site. Indeed, it suggests that cosubstrate dissociation and rebinding may be initiated by the loss of 5′-dA, while the

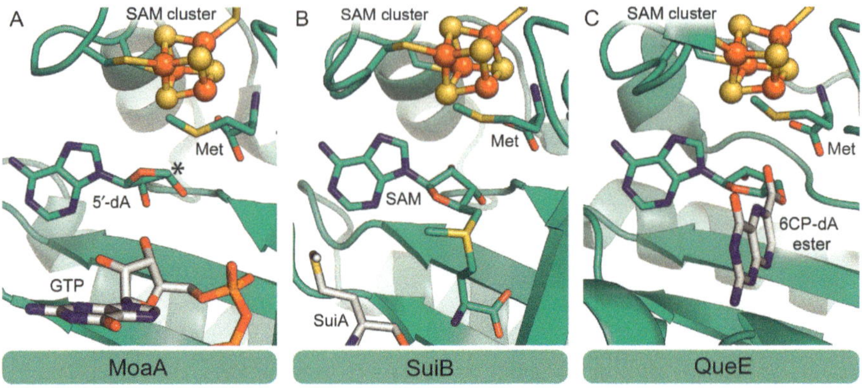

Fig. 4 SAM cleavage observed *in crystallo*. (A) Homolytic cleavage of SAM occurred during crystallization of MoaA (PDB accession code 2FB3) due to the addition of the strong reductant sodium dithionite. Note that the unmodeled 5′-C is again marked with an *asterisk*. (B) Prereduction of SuiB (PDB accession code 5V1T) prohibited crystallization of a steady-state reactant complex and likely resulted in a single turnover event followed by exchange of 5′-dA for a new molecule of intact SAM during cocrystallization with the peptide substrate. (C) Cocrystallization of QueE (PDB accession code 5TH5) with the substrate analog 6CP resulted in the formation of a 6CP-dA ester via a heterolytic, rather than reductive, mechanism. In all panels the modified substrate is shown in *white* for clarity.

subsequent displacement of Met coordinated to the [4Fe–4S] cluster could be precipitated by SAM rebinding within the Ado-binding pocket.

Perhaps the most unexpected modification produced by cocrystallization of a reactant complex occurred in the absence of reductant with a nonnative substrate. Recent structures of QueE from *Bacillus subtilis* cocrystallized with the substrate analog 6CP suggested the formation of a covalent adduct between 6CP and 5′-dA (Table 1, PDB accession code 5TH5; Bruender et al., 2017). While the impetus for formation of the 6CP-dA ester remains unclear, its crystallographic detection was confirmed through biochemical analysis and led to the discovery of yet another product, 6-deoxyadenosylpterin, generated upon addition of SAM and reductant. As in the RNA methylases RlmN and Cfr (Yan et al., 2011), it appears QueE is capable of using SAM to execute both radical and heterolytic transformations (Bruender et al., 2017). This dual reactivity is rare among radical SAM enzymes and is made even more unusual by its action on an alternative substrate. Building on a recent observation of nonreductive chemistry as a primary function in the adenosylcobalamin-dependent radical SAM enzyme TsrM (Blaszczyk et al., 2016), Bruender et al. suggest that similar chemistries may exist in other members of the superfamily as well.

3.2 Initiating Turnover in Crystals of HydE

Should a given lattice support enzyme turnover, initiating the reaction *in crystallo* is perhaps the ideal means to monitor the structural dynamics of enzyme catalysis. This field has exploded in recent years with the development of next-generation X-ray sources, such as free electron lasers, that work on the principle of "detection-before-destruction" to collect real-time, room temperature diffraction data during active turnover (Chapman et al., 2011; Levantino, Yorke, Monteiro, Cammarata, & Pearson, 2015; Neutze, Wouts, van der Spoel, Weckert, & Hajdu, 2000). While such facilities are necessary for the structural characterization of short-lived intermediate states, the relative sluggishness of radical SAM enzymes enables the use of manual cryotrapping to supply time resolution in conventional crystallography settings (Broderick et al., 2014). Rohac et al. applied this underutilized technique to the [FeFe]-hydrogenase maturase HydE, putatively involved in the synthesis of an azadithiolate-bridging molecule required for the assembly of the di-iron center (Nicolet, Martin, Tron, & Fontecilla-Camps, 2010). While the natural substrate of HydE has not yet been determined, exploratory studies aimed at identifying candidates among common metabolites unexpectedly discovered reactivity with certain 1,3-thiazolidine species. These molecules do not facilitate activation of [FeFe]-hydrogenase; however, their stable ligation within the active site makes them attractive alternatives in the quest to understand C–S bond formation by HydE (Rohac et al., 2016).

In particular, assays with (2*R*,4*R*)-2-methyl-1,3-thiazolidine-2,4-dicarboxylic acid ((2*R*,4*R*)-MeTDA) resulted in the formation of *S*-adenosyl-L-cysteine (SAC) via a condensation reaction between the 5′-dA moiety of SAM and the cysteinyl component of (2*R*,4*R*)-MeTDA. High-resolution crystals of the HydE reactant complex were soaked with sodium dithionite and flash-frozen over a series of time points to monitor this reaction (Table 1, PDB accession codes 5F**). Crystals of HydE grown in the presence of Met and SAC were also measured as a control for the final product. To detect subtle differences between the structures, data for each time delay were compared to the preturnover conformation via isomorphous difference Fourier methods, i.e. $F_{obs}(\Delta t_x) - F_{obs}(t=0)$ (Rohac et al., 2016). Double difference electron density maps are more accurate than traditional difference maps as they eliminate model bias that would otherwise be introduced through multiple phase calculations (Wang, Askerka, Brudvig, & Batista, 2017). Using this approach, the authors determined that changes to the enzyme were confined

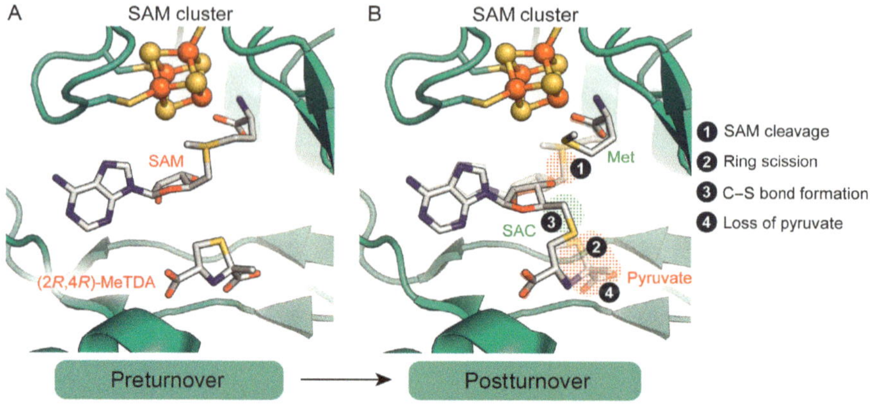

Fig. 5 Structures inform mechanism in HydE. A comparison of the active site of HydE (A) pre- (PDB accession code 5FEP) and (B) postturnover (PDB accession code 5FEW) highlights the four primary changes to the active site identified by collecting a series of structures throughout turnover.

to the active site and consisted of SAM cleavage, 5′-C–S bond formation with the substrate, C–S bond scission in the thiazolidine ring, and loss of the pyruvate moiety, respectively (Fig. 5). Furthermore, each intermediate conformation could be described as some combination of reactant and product states. Together with computational studies, these results suggest that HydE reacts directly with the substrate sulfur (Rohac et al., 2016), differentiating it from other C–S bond forming enzymes in the superfamily, which first generate a carbon-centered radical (Anton et al., 2008; Berkovitch et al., 2004; Miller et al., 2000; Pierrel, Douki, Fontecave, & Atta, 2004). This unique mechanism further distinguishes HydE as only the second characterized radical SAM enzyme in which abstraction of a hydrogen atom by the 5′-dA radical is not the first step in the catalytic cycle (Mahanta, Fedoseyenko, Dairi, & Begley, 2013; Rohac et al., 2016).

4. MECHANISM-DRIVEN DESIGN TO TRAP TRANSIENT REACTION INTERMEDIATES

While the probability of acquiring intermediate state structures for enzymes of unknown mechanism can be improved via the aforementioned techniques, success is ultimately unpredictable. The problem becomes infinitely more tractable, however, given a basic mechanistic framework. Even for radical SAM enzymes that have not been biochemically characterized in detail, a putative mechanistic proposal enables the development of a more focused approach to trap hypothesized intermediate states. Stoichiometric considerations, point mutations, and meticulously selected substrate analogs are all tools in the arsenal of a structural biologist that can be applied to stall catalysis at an appropriate point. These methods attempt to target a particular catalytic state intentionally rather than "fishing" for intermediates that accumulate by accident.

Although potentially applicable to a wide range of systems, certain enzymes may be more or less amenable to this mechanism-driven design. The presence of cross-linked and/or stable reaction intermediates, for example, are encouraging features that facilitate isolation of the relevant species for crystallization. In fact, the prevalence of enzyme–substrate cross-linked species within the radical SAM superfamily provided the basis for recent pioneering studies of LipA (McLaughlin et al., 2016) and RlmN (Schwalm, Grove, Booker, & Boal, 2016) in which key reaction intermediates were successfully trapped for structural analysis. These pivotal studies serve as a proof of concept for the application of mechanistic knowledge to isolate particular intermediates from the catalytic cycles of radical SAM enzymes. The innovative methods used in these illuminating studies will be discussed next.

4.1 The Crystal Structure of a Cross-Linked Intermediate in LipA

Lipoyl synthase, LipA, provides the final modifications in the biosynthesis of the lipoyl cofactor (N^6-(6*R*-6,8-dimercapto-octanoyl)-lysyl), an essential component of numerous oxidative reactions critical for metabolism. In particular, LipA catalyzes the insertion of two sulfurs at the C6 and C8 positions of octanoic acid bound to a lysine ε-amino group within a carrier protein (Douglas, Kriek, Bryant, & Roach, 2006; Miller et al., 2000). Although the origin of the sulfurs installed by LipA and the related radical SAM enzyme BioB has been studied extensively (Cicchillo & Booker, 2005; Ugulava, Sacanell, & Jarrett, 2001), the resultant theory that these enzymes scavenge sulfurs from their respective auxiliary clusters was originally controversial. The identification of a unique auxiliary [4Fe–4S] cluster-binding motif in LipA, characterized by ligation of a nearby serine, bolstered the unconventional hypothesis (Cicchillo et al., 2004; Harmer et al., 2014). However, the absence of bound substrate limited mechanistic interpretation of the crystal structure from *Thermosynechococcus elongatus* (Table 1, PDB accession codes 4U0O, 4U0P; Harmer et al., 2014). It is well known that to generate the final lipoyl cofactor, the reaction requires two equivalents of SAM, corresponding to the sequential additions of sulfur along the fatty acid chain (Cicchillo et al., 2004). Recent biochemical work by Lanz et al. provided the first concrete evidence of a cannibalistic mechanism through the discovery of a cross-link formed between octanoyllysine and the auxiliary cluster of LipA, following the first SAM cleavage event (Lanz et al., 2014). The generation of this covalently bonded species was subsequently confirmed in multiple orthologous enzymes (Lanz et al., 2014, 2016).

To further investigate this extraordinary reaction, crystallographic studies of the catalytic intermediate from *Mycobacterium tuberculosis* quickly followed, facilitated by the incredible stability of the cross-link (Table 1, PDB accession codes 5EXI, 5EXJ, 5EXK; McLaughlin et al., 2016). Given the complications inherent to crystallizing protein complexes, it is often preferable to synthesize a truncated peptidyl mimic of the natural protein substrate. In this case, an 8-mer substrate fragment was incubated with LipA, excess reductant, and one equivalent of SAM to generate the cross-linked species (Fig. 6). The kinetics of C–S bond formation in the reaction mixture were measured via LC-MS (liquid chromatography-mass spectrometry) and applied toward the design of an appropriate crystallography protocol. Although attempts were also made to visualize the reactant complex via cocrystallization with SAM and octanoyl substrate (e.g., the native octanoyl-H protein or a shortened peptide mimic), all trials resulted in substrate-free structures (McLaughlin et al., 2016).

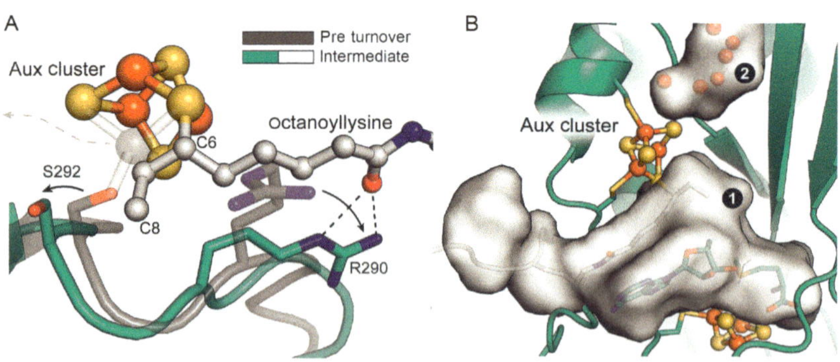

Fig. 6 Cannibalization of the auxiliary cluster in LipA—PDB accession codes 5EXJ (pre-turnover) and 5EXK (intermediate). (A) A zoomed-in view of the active site depicts key changes that facilitate cross-link formation. (B) A surface representation of the primary channels providing access to the active site allows for visualization of the substrate-binding pocket ① as well as the putative exit channel ② for the Fe ion lost during cannibalization of the cluster.

In agreement with the substrate docking simulations computed previously (Harmer et al., 2014), density for the covalently bonded peptide was observed within a deep channel and precisely positioned between the two Fe/S clusters (Fig. 6B). A comparison of the intermediate complex and substrate-free structures revealed that this configuration requires significant contraction of the active site, accompanied by rearrangement of a key arginine side chain. The loosely coordinating serine residue is no longer ligated to the auxiliary cluster of the cross-linked species and, consistent with Mössbauer spectra reported previously (Lanz et al., 2014), the corresponding Fe ion is absent (Fig. 6A). Based on the visualization of the cross-link, the authors conclude that the retention of this Fe would impede the subsequent sulfur insertion at C8 due to steric clash with substrate C7. Furthermore, the structure reveals a solvent-filled channel on the back side of the substrate-binding site in which the ion could exit the active site (McLaughlin et al., 2016). These combined observations greatly enhance our understanding of the LipA mechanism and provide a springboard for future experiments to characterize the second half of this fascinating reaction.

4.2 The Crystal Structure of a Cross-Linked Intermediate in RlmN

While it was possible to exploit the SAM stoichiometry of lipoyl cofactor biosynthesis to suspend catalysis, such an approach cannot be universally extended to other radical SAM systems. Instead, other engineered deviations from the wild-type reaction are often required to stably trap intermediates for structural characterization. In the case of RlmN, this modification came in the form of a rational point mutation (Schwalm et al., 2016). RlmN utilizes SAM as both a methyl donor and a radical source to catalyze the methylation of the sp^2-hybridized C2-carbon of an adenosine nucleotide in both ribosomal and transfer RNAs (Yan et al., 2010). This rare dual-specificity (Benítez-Páez, Villarroya, & Armengod, 2012) combined with strong homology to the radical SAM methylase Cfr, known to confer antibiotic resistance, makes elucidation of the RlmN mechanism particularly compelling (Kehrenberg, Schwarz, Jacobsen, Hansen, & Vester, 2005; Long, Poehlsgaard, Kehrenberg, Schwarz, & Vester, 2006). Unlike the majority of SAM-dependent methyltransferases, which act on nucleophilic targets (Struck, Thompson, Wong, & Micklefield, 2012), these enzymes first methylate a Cys side chain at position 355 (mC355) and then generate a protein-bound methylene radical at this site to attack the relatively inert carbons of adenine. The resultant transient protein/RNA cross-link is then cleaved, a process that is facilitated by a strictly conserved active site cysteine residue (Grove et al., 2011). Biochemical and spectroscopic studies determined that the covalent RNA–protein adduct persists indefinitely upon mutation of the conserved Cys (C118 in RlmN) to Ala or Ser (Grove et al., 2011; McCusker et al., 2012; Silakov et al., 2014).

Although previous attempts to visualize the reactant complex of RlmN were stymied by the flexibility of the RNA substrate (Table 1; Boal et al., 2011), discovery of a covalent C118A–RlmN/RNA complex (McCusker et al., 2012; Schwalm et al., 2016; Silakov et al., 2014) provided a promising means by which to stabilize the nucleic acid component. Heterologous overexpression of C118A–RlmN yielded significant quantities of an in vivo cross-linked complex that was separated from unreacted protein via anion exchange chromatography and confirmed as an adduct with increased molecular weight by SDS-PAGE analysis. Crystallization of this species unexpectedly yielded density consistent with tRNA^{Glu}, rather than the anticipated rRNA substrate (Fig. 7; Table 1, PDB accession codes 5HR6,

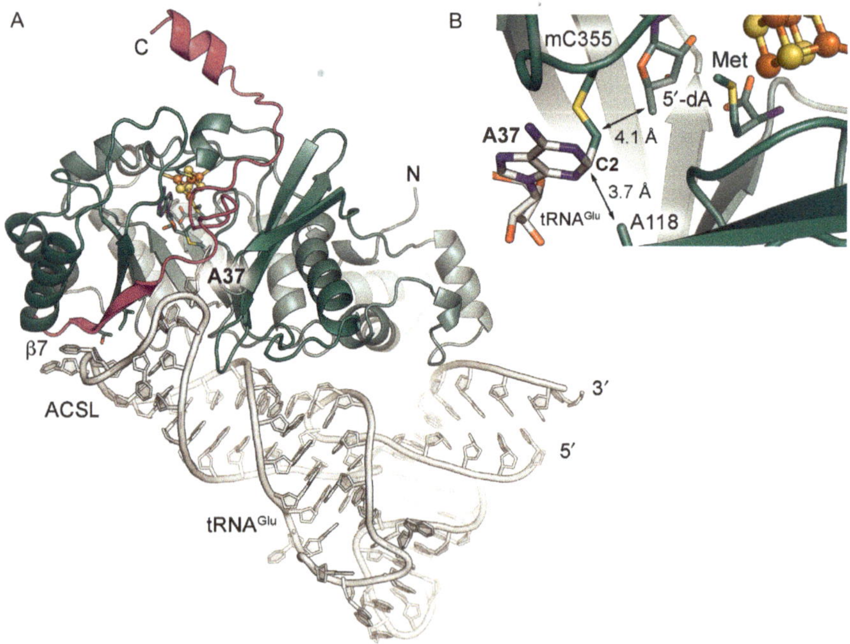

Fig. 7 A stable cross-link enables capture of a protein–RNA complex in RlmN—PDB accession code 5HR7 (in vitro). (A) An overall view of the C118A RlmN•tRNA^{Glu} complex showing the unusual base-flipping mechanism mediated by the β7 motif. (B) A zoomed-in view of the active site showing the enzyme–substrate cross-link and its proximity to the site of 5′-dA• formation and the side chain of residue 118, active site components integral to cross-link formation and resolution, respectively.

5HR7). Improvements in crystallographic resolution were achieved by generating the intermediate state with $tRNA^{Glu}$ in vitro. Free protein, which eluted in distinct fractions from the in vitro cross-linked species, was incubated with approximately one equivalent of $tRNA^{Glu}$, excess SAM, as well as flavodoxin and its associated reductase (Schwalm et al., 2016). The in vivo reducing system was selected to limit unproductive SAM cleavage known to occur in RlmN and other radical SAM enzymes in the presence of strong reductants like sodium dithionite (Yan et al., 2010). Excess NADPH was subsequently used to trigger the reaction. As with the complex generated in vivo, anion exchange chromatography was used to purify the in vitro cross-linked intermediate (Schwalm et al., 2016).

Crystallization of the in vitro intermediate confirmed the presence of the predicted covalent adduct via the appearance of positive difference density between C2 of adenosine 37 and the methylated cysteine, mC355 (Fig. 7B). Consistent with biochemical results that unambiguously defined C355 as the site of initial methylation, substrate-free models placed this conserved cysteine proximal to the methyl group of SAM (Boal et al., 2011; Silakov et al., 2014). In these structures, the positioning mC355 for $-CH_3$ activation via radical generation at the 5′C of SAM remained unclear. The structure of the intermediate complex appears to resolve this issue, as an extensive hydrogen bonding network forms between the protein and RNA substrate. The network induces rearrangement of the flexible loop containing mC355 (Schwalm et al., 2016), relocating the methyl group within 5 Å of the site of 5′-dA• formation. Loop remodeling also facilitates adenine recognition and RNA substrate positioning, both of which are necessary to achieve selective modification at C2. Although the C118A mutation does not allow for visualization of the associated thiol side chain, previously proposed to be critical in cross-link resolution (Grove et al., 2011), the variant structure suggests that C118 would be perfectly poised to accept a proton from the newly methylated C2 atom in the wild-type enzyme, establishing its role as a base in the reaction pathway (Fig. 7B; Silakov et al., 2014).

More generally, the structures of the C118A–RlmN/$tRNA^{Glu}$ complex provide key insights into the unique mode of substrate recognition employed by the enzyme and represent the first view of a radical SAM enzyme in complex with another large macromolecule. RlmN significantly refolds the anticodon stem loop (ACSL) of $tRNA^{Glu}$ using a key secondary structure that is ancillary to the α_6/β_6 radical SAM core. A seventh beta strand (β7), containing several small β-branched amino acids, crosses the front of the open barrel structure and provides a rigid scaffold against which the RNA substrate unfolds to facilitate the remarkable base-flip required to place A37 in the active site (Fig. 7A; Schwalm et al., 2016). The new conformation of the ACSL results in outward splay of four additional bases in the ACSL, some of which also undergo significant modification by other tRNA modifying enzymes. This refolding strategy may explain RlmN's dual specificity for tRNA and rRNA substrates. The A2503 site methylated in the large subunit of the ribosome also resides in a sequence that is dense with posttranscriptionally modified bases and further targeted by hydroxylases, methylases, and pseudouridine synthases (Green & Noller, 1996). The refolding capacity of RlmN may therefore serve an additional function, with the newly installed methyl group stabilizing a new conformation that allows for easy access to neighboring bases by other posttranslational modification machinery. In RlmN, crystallization of an otherwise transient species via mutagenesis not only cemented crucial aspects of the enzymatic mechanism, but it also provided new understanding of the in vivo substrate scope of RlmN and the strategies employed to address the complex molecular recognition problems faced by radical SAM enzymes that target large structured macromolecular substrates. The mechanism-guided mutagenesis approach employed here could be applied widely to other radical SAM enzymes that have been characterized in a similar level of detail and may be particularly useful in structure solution of superfamily members that operate on nucleic acids or other proteins.

5. CONCLUSIONS

As our understanding of the radical SAM superfamily grows, so too does the list of attractive targets for intermediate state crystallization. Numerous potential candidates can already be found among biochemically characterized systems. The success of mechanism-driven intermediate state crystallization in LipA and RlmN makes their homologs particularly attractive for further study. Biotin synthase, for example, utilizes two equivalents of SAM to catalyze the conversion of dethiobiotin (DTB) to biotin—an essential vitamin important for metabolism and cell growth (Marquet, Tse Sum Bui, & Florentin, 2001). Like LipA, BioB has been proposed to cannibalize its auxiliary [2Fe–2S] cluster to procure the sulfur atom subsequently installed at unactivated carbons on DTB (Farrar, Siu, Howell, & Jarrett, 2010; Tse Sum Bui et al., 1998). The mechanism by which this occurs most probably involves the formation of a covalent cross-link between DTB and the auxiliary [2Fe–2S] cluster (Cosper et al., 2004; Fugate & Jarrett, 2012). Reminiscent of the intermediate observed in LipA, this species could conceivably be trapped via the substoichiometric addition of SAM. This system is particularly appealing as crystallization conditions for BioB are known (Table 1; Berkovitch et al., 2004).

In conclusion, the case studies and protocols described herein provide a broad foundation for the crystallization of radical SAM intermediates. Appropriate selection of substrate analogs, or the careful mutation of a single amino acid, may be the difference between a good structure and a great one. While the ideal approach is largely enzyme dependent, crystallization of reactant and product complexes ensures the accurate mechanistic interpretation of structural models. Trapping nominally short-lived catalytic intermediate states of enzymes is still in its infancy, but the impact of its application to the radical SAM superfamily will likely be immense.

ACKNOWLEDGMENTS

Research in the author's laboratory on radical SAM enzymes has been supported by NIH Grants GM100011 and GM119707. K.M.D. is grateful for support from the Arnold O. Beckman Postdoctoral Fellowship Program.

REFERENCES

Anton, B. P., Saleh, L., Benner, J. S., Raleigh, E. A., Kasif, S., & Roberts, R. J. (2008). RimO, a MiaB-like enzyme, methylthiolates the universally conserved Asp88 residue of ribosomal protein S12 in *Escherichia coli*. *Proceedings of the National Academy of Sciences of the United States of America*, *105*, 1826–1831.

Benítez-Páez, A., Villarroya, M., & Armengod, M.-E. (2012). The *Escherichia coli* RlmN methyltransferase is a dual-specificity enzyme that modifies both rRNA and tRNA and controls translational accuracy. *RNA*, *18*, 1783–1795.

Berkovitch, F., Nicolet, Y., Wan, J. T., Jarrett, J. T., & Drennan, C. L. (2004). Crystal structure of biotin synthase, an *S*-adenosylmethionine-dependent radical enzyme. *Science*, *303*, 76–79.

Betz, J. N., Boswell, N. W., Fugate, C. J., Holliday, G. L., Akiva, E., Scott, A. G., et al. (2015). [FeFe]-hydrogenase maturation: Insights into the role HydE plays in dithiomethylamine biosynthesis. *Biochemistry*, *54*, 1807–1818.

Blaszczyk, A. J., Silakov, A., Zhang, B., Maiocco, S. J., Lanz, N. D., Kelly, W. L., et al. (2016). Spectroscopic and electrochemical characterization of the iron–sulfur and cobalamin cofactors of TsrM, an unusual radical *S*-adenosylmethionine methylase. *Journal of the American Chemical Society*, *138*, 3416–3426.

Boal, A. K., Grove, T. L., McLaughlin, M. I., Yennawar, N. H., Booker, S. J., & Rosenzweig, A. C. (2011). Structural basis for methyl transfer by a radical SAM enzyme. *Science*, *332*, 1089–1092.

Booker, S. J., & Grove, T. L. (2010). Mechanistic and functional versatility of radical SAM enzymes. *F1000 Biology Reports*, *2*, 52.

Broderick, J. B., Duffus, B. R., Duschene, K. S., & Shepard, E. M. (2014). Radical *S*-adenosylmethionine enzymes. *Chemical Reviews*, *114*, 4229–4317.

Bruender, N. A., Grell, T. A. J., Dowling, D. P., McCarty, R. M., Drennan, C. L., & Bandarian, V. (2017). 7-Carboxy-7-deazaguanine synthase: A radical *S*-adenosyl-L-methionine enzyme with polar tendencies. *Journal of the American Chemical Society*, *139*, 1912–1920.

Chapman, H. N., Fromme, P., Barty, A., White, T. A., Kirian, R. A., Aquila, A., et al. (2011). Femtosecond x-ray protein nanocrystallography. *Nature*, *470*, 73–77.

Chatterjee, A., Li, Y., Zhang, Y., Grove, T. L., Lee, M., Krebs, C., et al. (2008). Reconstitution of ThiC in thiamine pyrimidine biosynthesis expands the radical SAM superfamily. *Nature Chemical Biology*, *4*, 758–765.

Cheek, J., & Broderick, J. B. (2002). Direct H atom abstraction from spore photoproduct C-6 initiates DNA repair in the reaction catalyzed by spore photoproduct lyase: Evidence for a reversibly generated adenosyl radical intermediate. *Journal of the American Chemical Society*, *124*, 2860–2861.

Chen, D., Walsby, C., Hoffman, B. M., & Frey, P. A. (2003). Coordination and mechanism of reversible cleavage of *S*-adenosylmethionine by the [4Fe-4S] center in lysine 2,3-aminomutase. *Journal of the American Chemical Society*, *125*, 11788–11789.

Cicchillo, R. M., & Booker, S. J. (2005). Mechanistic investigations of lipoic acid biosynthesis in *Escherichia coli*: Both sulfur atoms in lipoic acid are contributed by the same lipoyl synthase polypeptide. *Journal of the American Chemical Society*, *127*, 2860–2861.

Cicchillo, R. M., Iwig, D. F., Jones, A. D., Nesbitt, N. M., Baleanu-Gogonea, C., Souder, M. G., et al. (2004). Lipoyl synthase requires two equivalents of *S*-adenosyl-L-methionine to synthesize one equivalent of lipoic acid. *Biochemistry*, *43*, 6378–6386.

Cosper, M. M., Jameson, G. N. L., Hernández, H. L., Krebs, C., Huynh, B. H., & Johnson, M. K. (2004). Characterization of the cofactor composition of *Escherichia coli* biotin synthase. *Biochemistry*, *43*, 2007–2021.

Davis, K. M., Schramma, K. R., Hansen, W. A., Bacik, J. P., Khare, S. D., Seyedsayamdost, M. R., et al. (2017). Structures of the peptide-modifying radical SAM enzyme SuiB elucidate the basis of substrate recognition. *Proceedings of the National Academy of Sciences of the United States of America*. in revision.

Douglas, P., Kriek, M., Bryant, P., & Roach, P. L. (2006). Lipoyl synthase inserts sulfur atoms into an octanoyl substrate in a stepwise manner. *Angewandte Chemie International Edition*, *45*, 5197–5199.

Dowling, D. P., Bruender, N. A., Young, A. P., McCarty, R. M., Bandarian, V., & Drennan, C. L. (2014). Radical SAM enzyme QueE defines a new minimal core fold and metal-dependent mechanism. *Nature Chemical Biology*, *10*, 106–112.

Dowling, D. P., Vey, J. L., Croft, A. K., & Drennan, C. L. (2012). Structural diversity in the AdoMet radical enzyme superfamily. *Biochimica et Biophysica Acta*, *1824*, 1178–1195.

Farrar, C. E., Siu, K. K. W., Howell, P. L., & Jarrett, J. T. (2010). Biotin synthase exhibits burst kinetics and multiple turnovers in the absence of inhibition by products and product-related biomolecules. *Biochemistry*, *49*, 9985–9996.

Fenwick, M. K., Mehta, A. P., Zhang, Y., Abdelwahed, S. H., Begley, T. P., & Ealick, S. E. (2015). Non-canonical active site architecture of the radical SAM thiamin pyrimidine synthase. *Nature Communications*, *6*, 6480.

Fugate, C. J., & Jarrett, J. T. (2012). Biotin synthase: Insights into radical-mediated carbon–sulfur bond formation. *Biochimica et Biophysica Acta*, *1824*, 1213–1222.

Goldman, P. J., Grove, T. L., Booker, S. J., & Drennan, C. L. (2013). X-ray analysis of butirosin biosynthetic enzyme BtrN redefines structural motifs for AdoMet radical chemistry. *Proceedings of the National Academy of Sciences of the United States of America*, *110*, 15949–15954.

Goldman, P. J., Grove, T. L., Sites, L. A., McLaughlin, M. I., Booker, S. J., & Drennan, C. L. (2013). X-ray structure of an AdoMet radical activase reveals an anaerobic solution for formylglycine posttranslational modification. *Proceedings of the National Academy of Sciences of the United States of America*, *110*, 8519–8524.

Green, R., & Noller, H. F. (1996). In vitro complementation analysis localizes 23S rRNA posttranscriptional modifications that are required for *Escherichia coli* 50S ribosomal subunit assembly and function. *RNA*, *2*, 1011–1021.

Grell, T. A. J., Goldman, P. J., & Drennan, C. L. (2015). SPASM and twitch domains in *S*-adenosylmethionine (SAM) radical enzymes. *The Journal of Biological Chemistry*, *290*, 3964–3971.

Grove, T. L., Benner, J. S., Radle, M. I., Ahlum, J. H., Landgraf, B. J., Krebs, C., et al. (2011). A radically different mechanism for *S*-adenosylmethionine-dependent methyltransferases. *Science*, *332*, 604–607.

Hänzelmann, P., Hernández, H. L., Menzel, C., García-Serres, R., Huynh, B. H., Johnson, M. K., et al. (2004). Characterization of MOCS1A, an oxygen-sensitive iron-sulfur protein involved in human molybdenum cofactor biosynthesis. *The Journal of Biological Chemistry*, *279*, 34721–34732.

Hänzelmann, P., & Schindelin, H. (2004). Crystal structure of the *S*-adenosylmethionine-dependent enzyme MoaA and its implications for molybdenum cofactor deficiency in humans. *Proceedings of the National Academy of Sciences of the United States of America*, *101*, 12870–12875.

Hänzelmann, P., & Schindelin, H. (2006). Binding of 5′-GTP to the C-terminal FeS cluster of the radical *S*-adenosylmethionine enzyme MoaA provides insights into its mechanism. *Proceedings of the National Academy of Sciences of the United States of America*, *103*, 6829–6834.

Harmer, J. E., Hiscox, M. J., Dinis, P. C., Fox, S. J., Iliopoulos, A., Hussey, J. E., et al. (2014). Structures of lipoyl synthase reveal a compact active site for controlling sequential sulfur insertion reactions. *The Biochemical Journal*, *464*, 123–133.

Hassell, A. M., An, G., Bledsoe, R. K., Bynum, J. M., Carter, H. L., Deng, S.-J. J., et al. (2007). Crystallization of protein–ligand complexes. *Acta Crystallographica Section D: Biological Crystallography*, *63*, 72–79.

Hover, B. M., Tonthat, N. K., Schumacher, M. A., & Yokoyama, K. (2015). Mechanism of pyranopterin ring formation in molybdenum cofactor biosynthesis. *Proceedings of the National Academy of Sciences of the United States of America*, *112*, 6347–6352.

Jurgenson, C. T., Begley, T. P., & Ealick, S. E. (2009). The structural and biochemical foundations of thiamin biosynthesis. *Annual Review of Biochemistry*, *78*, 569–603.

Kehrenberg, C., Schwarz, S., Jacobsen, L., Hansen, L. H., & Vester, B. (2005). A new mechanism for chloramphenicol, florfenicol and clindamycin resistance: Methylation of 23S ribosomal RNA at a2503. *Molecular Microbiology*, *57*, 1064–1073.

Lanz, N. D., Lee, K.-H., Horstmann, A. K., Pandelia, M.-E., Cicchillo, R. M., Krebs, C., et al. (2016). Characterization of lipoyl synthase from *Mycobacterium tuberculosis*. *Biochemistry*, *55*, 1372–1383.

Lanz, N. D., Pandelia, M.-E., Kakar, E. S., Lee, K.-H., Krebs, C., & Booker, S. J. (2014). Evidence for a catalytically and kinetically competent enzyme–substrate cross-linked intermediate in catalysis by lipoyl synthase. *Biochemistry*, *53*, 4557–4572.

Lepore, B. W., Ruzicka, F. J., Frey, P. A., & Ringe, D. (2005). The x-ray crystal structure of lysine-2,3-aminomutase from *Clostridium subterminale*. *Proceedings of the National Academy of Sciences of the United States of America*, *102*, 13819–13824.

Levantino, M., Yorke, B. A., Monteiro, D. C. F., Cammarata, M., & Pearson, A. R. (2015). Using synchrotrons and XFELs for time-resolved x-ray crystallography and solution scattering experiments on biomolecules. *Current Opinion in Structural Biology*, *35*, 41–48.

Long, K. S., Poehlsgaard, J., Kehrenberg, C., Schwarz, S., & Vester, B. (2006). The Cfr rRNA methyltransferase confers resistance to phenicols, lincosamides, oxazolidinones, pleuromutilins, and streptogramin A antibiotics. *Antimicrobial Agents and Chemotherapy*, *50*, 2500–2505.

Mahanta, N., Fedoseyenko, D., Dairi, T., & Begley, T. P. (2013). Menaquinone biosynthesis: Formation of aminofutalosine requires a unique radical SAM enzyme. *Journal of the American Chemical Society*, *135*, 15318–15321.

Marquet, A., Tse Sum Bui, B., & Florentin, D. (2001). Biosynthesis of biotin and lipoic acid. *Vitamins and Hormones*, *61*, 51–101.
McCarty, R. M., Krebs, C., & Bandarian, V. (2013). Spectroscopic, steady-state kinetic, and mechanistic characterization of the radical SAM enzyme QueE, which catalyzes a complex cyclization reaction in the biosynthesis of 7-deazapurines. *Biochemistry*, *52*, 188–198.
McCarty, R. M., Somogyi, Á., Lin, G., Jacobsen, N. E., & Bandarian, V. (2009). The deazapurine biosynthetic pathway revealed: *In vitro* enzymatic synthesis of PreQ0 from guanosine 5′-triphosphate in four steps. *Biochemistry*, *48*, 3847–3852.
McCusker, K. P., Medzihradszky, K. F., Shiver, A. L., Nichols, R. J., Yan, F., Maltby, D. A., et al. (2012). Covalent intermediate in the catalytic mechanism of the radical *S*-adenosyl-L-methionine methyl synthase RlmN trapped by mutagenesis. *Journal of the American Chemical Society*, *134*, 18074–18081.
McLaughlin, M. I., Lanz, N. D., Goldman, P. J., Lee, K.-H., Booker, S. J., & Drennan, C. L. (2016). Crystallographic snapshots of sulfur insertion by lipoyl synthase. *Proceedings of the National Academy of Sciences of the United States of America*, *113*, 9446–9450.
Mehta, A. P., Hanes, J. W., Abdelwahed, S. H., Hilmey, D. G., Hänzelmann, P., & Begley, T. P. (2013). Catalysis of a new ribose carbon-insertion reaction by the molybdenum cofactor biosynthetic enzyme MoaA. *Biochemistry*, *52*, 1134–1136.
Menéndez, C., Igloi, G., Henninger, H., & Brandsch, R. (1995). A pAO1-encoded molybdopterin cofactor gene (MoaA) of *Arthrobacter nicotinovorans*: Characterization and site-directed mutagenesis of the encoded protein. *Archives of Microbiology*, *164*, 142–151.
Miller, J. R., Busby, R. W., Jordan, S. W., Cheek, J., Henshaw, T. F., Ashley, G. W., et al. (2000). *Escherichia coli* LipA is a lipoyl synthase: *In vitro* biosynthesis of lipoylated pyruvate dehydrogenase complex from octanoyl-acyl carrier protein. *Biochemistry*, *39*, 15166–15178.
Mulliez, E., Duarte, V., Arragain, S., Fontecave, M., & Atta, M. (2017). On the role of additional [4Fe-4S] clusters with a free coordination site in radical-SAM enzymes. *Frontiers in Chemistry*, *5*, 17.
Neutze, R., Wouts, R., van der Spoel, D., Weckert, E., & Hajdu, J. (2000). Potential for biomolecular imaging with femtosecond x-ray pulses. *Nature*, *406*, 752–757.
Nicolet, Y., Martin, L., Tron, C., & Fontecilla-Camps, J. C. (2010). A glycyl free radical as the precursor in the synthesis of carbon monoxide and cyanide by the [FeFe]-hydrogenase maturase HydG. *FEBS Letters*, *584*, 4197–4202.
Nicolet, Y., Zeppieri, L., Amara, P., & Fontecilla-Camps, J. C. (2014). Crystal structure of tryptophan lyase (NosL): Evidence for radical formation at the amino group of tryptophan. *Angewandte Chemie International Edition*, *53*, 11840–11844.
Pandelia, M.-E., Lanz, N. D., Booker, S. J., & Krebs, C. (2015). Mössbauer spectroscopy of Fe/S proteins. *Biochimica et Biophysica Acta*, *1853*, 1395–1405.
Pierrel, F., Douki, T., Fontecave, M., & Atta, M. (2004). MiaB protein is a bifunctional radical-*S*-adenosylmethionine enzyme involved in thiolation and methylation of tRNA. *The Journal of Biological Chemistry*, *279*, 47555–47563.
Rohac, R., Amara, P., Benjdia, A., Martin, L., Ruffié, P., Favier, A., et al. (2016). Carbon–sulfur bond-forming reaction catalysed by the radical SAM enzyme HydE. *Nature Chemistry*, *8*, 491–500.
Ruzicka, F. J., & Frey, P. A. (2007). Glutamate 2,3-aminomutase: A new member of the radical SAM superfamily of enzymes. *Biochimica et Biophysica Acta*, *1774*, 286–296.
Schramma, K. R., Bushin, L. B., & Seyedsayamdost, M. R. (2015). Structure and biosynthesis of a macrocyclic peptide containing an unprecedented lysine-to-tryptophan crosslink. *Nature Chemistry*, *7*, 431–437.
Schramma, K. R., & Seyedsayamdost, M. R. (2017). Lysine-tryptophan-crosslinked peptides produced by radical SAM enzymes in pathogenic *streptococci*. *ACS Chemical Biology*, *12*, 922–927.

Schwalm, E. L., Grove, T. L., Booker, S. J., & Boal, A. K. (2016). Crystallographic capture of a radical *S*-adenosylmethionine enzyme in the act of modifying tRNA. *Science, 352*, 309–312.

Silakov, A., Grove, T. L., Radle, M. I., Bauerle, M. R., Green, M. T., Rosenzweig, A. C., et al. (2014). Characterization of a cross-linked protein–nucleic acid substrate radical in the reaction catalyzed by RlmN. *Journal of the American Chemical Society, 136*, 8221–8228.

Sofia, H. J., Chen, G., Hetzler, B. G., Reyes-Spindola, J. F., & Miller, N. E. (2001). Radical SAM, a novel protein superfamily linking unresolved steps in familiar biosynthetic pathways with radical mechanisms: Functional characterization using new analysis and information visualization methods. *Nucleic Acids Research, 29*, 1097–1106.

Struck, A.-W., Thompson, M. L., Wong, L. S., & Micklefield, J. (2012). *S*-adenosylmethionine-dependent methyltransferases: Highly versatile enzymes in biocatalysis, biosynthesis and other biotechnological applications. *Chembiochem, 13*, 2642–2655.

Tse Sum Bui, B., Florentin, D., Fournier, F., Ploux, O., Méjean, A., & Marquet, A. (1998). Biotin synthase mechanism: On the origin of sulphur. *FEBS Letters, 440*, 226–230.

Ugulava, N. B., Sacanell, C. J., & Jarrett, J. T. (2001). Spectroscopic changes during a single turnover of biotin synthase: Destruction of a [2Fe-2S] cluster accompanies sulfur insertion. *Biochemistry, 40*, 8352–8358.

Vey, J. L., Yang, J., Li, M., Broderick, W. E., Broderick, J. B., & Drennan, C. L. (2008). Structural basis for glycyl radical formation by pyruvate formate-lyase activating enzyme. *Proceedings of the National Academy of Sciences of the United States of America, 105*, 16137–16141.

Wang, J., Askerka, M., Brudvig, G. W., & Batista, V. S. (2017). Insights into photosystem II from isomorphous difference fourier maps of femtosecond x-ray diffraction data and quantum mechanics/molecular mechanics structural models. *ACS Energy Letters, 2*, 397–407.

Yan, F., & Fujimori, D. G. (2011). RNA methylation by radical SAM enzymes RlmN and Cfr proceeds via methylene transfer and hydride shift. *Proceedings of the National Academy of Sciences of the United States of America, 108*, 3930–3934.

Yan, F., LaMarre, J. M., Röhrich, R., Wiesner, J., Jomaa, H., Mankin, A. S., et al. (2010). RlmN and Cfr are radical SAM enzymes involved in methylation of ribosomal RNA. *Journal of the American Chemical Society, 132*, 3953–3964.

CHAPTER TWELVE

A Polymerase With Potential: The Fe–S Cluster in Human DNA Primase

Marilyn E. Holt, Lauren E. Salay, Walter J. Chazin[1]
Center for Structural Biology, Vanderbilt University, Nashville, TN, United States
[1]Corresponding author: e-mail address: walter.chazin@vanderbilt.edu

Contents

Abstract

Replication of DNA in eukaryotes is primarily executed by the combined action of processive DNA polymerases δ and ε. These enzymes cannot initiate synthesis of new DNA without the presence of a primer on the template ssDNA. The primers on both the leading and lagging strands are generated by DNA polymerase α–primase (pol–prim). DNA primase is a DNA-dependent RNA polymerase that synthesizes the first ~10 nucleotides and then transfers the substrate to polymerase α to complete primer synthesis. The mechanisms governing the coordination and handoff between primase and polymerase α are largely unknown. Isolated DNA primase contains a $[4Fe–4S]^{2+}$ cluster that has been shown to serve as a redox switch modulating DNA binding affinity. This discovery suggests a mechanism for modulating the priming activity of primase and handoff to polymerase α. In this chapter, we briefly discuss the current state of knowledge of primase structure and function, including the role of its iron–sulfur cluster.

Methods in Enzymology, Volume 595
ISSN 0076-6879
http://dx.doi.org/10.1016/bs.mie.2017.07.002

This is followed by providing the methods for expressing, purifying, and biophysically/structurally characterizing primase and its iron–sulfur cluster-containing domain, p58C.

ABBREVIATIONS

APS ammonium persulfate
CD circular dichroism
CV column volume
EMSA electrophoretic mobility gel shift assay
FPLC fast protein liquid chromatography
HSQC heteronuclear single quantum coherence
K_D equilibrium dissociation constant
LB Luria-Bertani
MWCO molecular weight cut-off
Ni-NTA nickel-nitrilotriacetic acid
NMR nuclear magnetic resonance
nts nucleotides
pol–prim DNA polymerase α–primase
pol α DNA polymerase α
pol δ DNA polymerase δ
pol ε DNA polymerase ε
rcf relative centrifugal force
SDS-PAGE sodium dodecyl sulfate polyacrylamide gel electrophoresis
ssDNA single-stranded DNA
TAE Tris–acetate–EDTA
TB Terrific Broth
TBE Tris–borate–EDTA buffer
TEMED *N,N,N′,N′*-tetramethylethylenediamine

1. INTRODUCTION

DNA replication is a multistep biochemical process that is necessary for the propagation of genetic information and maintenance of life. In all living, nonviral organisms, the majority of DNA is copied by processive DNA polymerases δ and ε (pols δ and ε). However, these polymerases are incapable of synthesizing DNA strands de novo and require an approximately 30 oligonucleotide "primer" bound to the template strand prior to initiating synthesis (Burgers, 2009). Primases—the sole class of polymerase capable of synthesizing primers on single-stranded DNA (ssDNA)—fill this critical functional gap by initiating polymerization of the complementary strand on the template. In eukaryotes, the primase is a DNA-dependent RNA polymerase associated with DNA polymerase α (pol α). DNA primase generates the first 7–12 nucleotides (nts) of the primer before handing off the primed template to pol α, which extends the initial RNA primer by ~20 nts before transferring the primed substrate to pols δ and ε (Baranovskiy & Tahirov, 2017; Muzi-Falconi, Giannattasio, Foiani, & Plevani, 2003; Pellegrini, 2012). In vitro, primase can effectively prime ssDNA substrates both within the context of the pol–prim heterotetramer and by itself (Sheaff & Kuchta, 1993).

Although the roles of primase and pols α, δ, and ε in DNA replication are relatively well defined (Yeeles, Janska, Early, & Diffley, 2017), the mechanisms by which these proteins are coordinated in the replicative machinery are still unclear. Interestingly, there is evidence for iron–sulfur clusters in each of these polymerases. Despite years of extensive biochemical characterization, the iron–sulfur clusters in the human and yeast primases were initially identified only in 2007, when their existence was demonstrated with UV–Vis spectroscopy, electron paramagnetic resonance spectroscopy, and mass spectrometry (Klinge, Hirst, Maman, Krude, & Pellegrini, 2007; Weiner et al., 2007). The crystal structure of cluster-loaded human primase has since been characterized (Agarkar, Babayeva, Pavlov, & Tahirov, 2011; Baranovskiy et al., 2015; Sauguet, Klinge, Perera, Maman, & Pellegrini, 2010; Vaithiyalingam, Warren, Eichman, & Chazin, 2010). Iron–sulfur clusters in yeast pols α, δ, and ε were identified in 2010 in yeast studies and confirmed with UV–Vis spectroscopy (Netz et al., 2011). The presence of the cluster in pol ε was since validated by a separate group using extended X-ray fine absorption structure (Jain et al., 2014). Intriguingly, crystal structures of human pol α do not contain an iron–sulfur cluster (Baranovskiy, Babayeva, et al., 2016; Suwa et al., 2015); however, this may simply be due to cluster sensitivity during purification under aerobic conditions. This is consistent with the history of iron–sulfur cluster-containing enzymes, as most require anaerobic conditions and careful handling during purification (Rouault, 2015).

The role of iron–sulfur clusters at the replication fork remains obscure. It has been suggested that they may behave as structural cofactors or as sensors for oxidative stress (Fuss, Tsai, Ishida, & Tainer, 2015; Liu & Huang, 2015). However, since iron–sulfur cluster generation and insertion requires a significant commitment of energy and resources from the cell (Ye & Rouault, 2010), it is possible that they have a more significant mechanistic role in replication. A novel hypothesis for iron–sulfur cluster function has recently been proposed on the basis of electrochemical and biophysical data showing that the cluster in human DNA primase acts as a redox-dependent switch for substrate binding (O'Brien et al., 2017). This suggests a novel mechanism in which charge transfer through the substrate helix enables communication and handoff between primase and pol α. While consistent with a number of previous studies (Baranovskiy, Babayeva, et al., 2016; Baranovskiy, Zhang, et al., 2016; Kuchta & Stengel, 2010; Zerbe & Kuchta, 2002; Zhang, Baranovskiy, Tahirov, & Pavlov, 2014), considerable future effort will be required to test and refine this provocative hypothesis.

Eukaryotic primases are composed of a regulatory subunit and a catalytic subunit. In humans, these subunits are referred to as p58 and p48, respectively. Primase and pol α form a heterotetramer, linked via interactions between p58 and the p180 catalytic subunit of pol α (Fig. 1). While closely associated with pol α in vivo, primase can be independently expressed and purified, retaining its priming activity. Limited proteolysis and X-ray

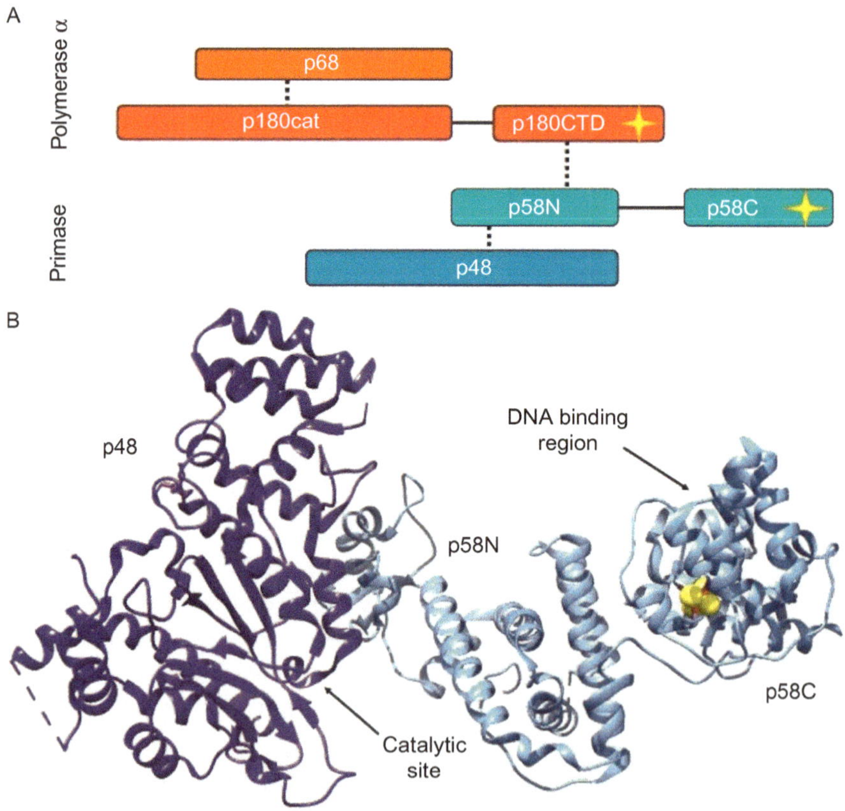

Fig. 1 (A) Domain map of the DNA polymerase α–primase. *Dotted lines* show where domains interact. *Yellow stars* are shown for iron–sulfur clusters. (B) The crystal structure of full-length primase (PDBID 4RR2). The iron–sulfur cluster in p58C is represented with *yellow spheres*.

crystallography of full-length primase revealed two stable fragments: one containing p48 and the amino terminal domain of p58 (p48/58N), and the other the iron–sulfur cluster-containing domain, p58C (Fig. 1). Biochemical and structural studies have shown that p48 contains the catalytic apparatus, while p58C contributes to template binding and is required for "counting" the 7–12 nts primer length (Arezi, Kirk, Copeland, & Kuchta, 1999; Kuchta, Reid, & Chang, 1990; Sheaff & Kuchta, 1993; Zerbe & Kuchta, 2002). Furthermore, it has been shown that the iron–sulfur cluster in p58C is required for replication initiation and cell cycle progression (Liu & Huang, 2015). Here, we describe our strategies for generating full-length human DNA primase and the p58C domain with the iron–sulfur cluster intact. This is followed by the application of a range of methods to characterize the biochemical and structural properties of primase and the isolated p58C domain. These include X-ray crystallography to determine high-resolution structures, electrophoretic mobility gel shift assay (EMSA), and fluorescence anisotropy approaches to measure affinities for template and primed template substrates, nuclear magnetic resonance (NMR) to characterize the binding of NTP cofactors, and circular dichroism (CD), NMR, and X-ray crystallography to validate that no significant structural effects arise from site-directed mutations.

2. EXPRESSION AND PURIFICATION

For bacterial expression of primase and the isolated p58C [4Fe–4S] cluster domain, we use a lac operon-based system with a plasmid containing N-terminal 6xHis tags (Copeland, 1997). Following standard protocols, *Escherichia coli* cells are transformed with this plasmid, grown to a high optical density, and compelled to generate the protein of interest by the addition of isopropyl β-D-1-thiogalactopyranoside (IPTG), which causes constitutive transcription of the downstream gene. The cells are then harvested and lysed, and the protein is purified from this lysate.

Often, iron–sulfur clusters are lost during the purification step; therefore, purification must be performed carefully. Some factors that can be optimized to promote [4Fe–4S] stability are pH, the access to atmospheric oxygen, and the timing of the purification (Rouault, 2015). In fact, relative to most [4Fe–4S] cluster-containing proteins, primase is rather stable because its cluster is buried deep within the protein matrix. Hence, primase represents a favorable case; both full-length primase and the isolated p58C domain do not require purification under anaerobic conditions and maintain their cluster-loaded state in the absence of oxidizing reagents. Below, we describe optimized expression and purification protocols for the isolated p58C domain and full-length human DNA primase.

2.1 p58C

2.1.1 Equipment

- Autoclave
- 0.45 and 0.2 μm syringe filters
- Plastic syringes
- Appropriate flasks and Petri dishes as needed
- Shaking incubator
- Homogenizer
- Sonication system
- Centrifuges
- SDS-PAGE equipment
- 10 kDa MWCO concentrators (Amicon)
- 10 mL Ni-NTA column (Thermo Scientific)
- 5 mL HiTrap HP Heparin column (GE Healthcare)
- 120 mL Superdex S75 column (GE Healthcare)
- FPLC system

2.1.2 Reagents and Solutions

- pBG100 plasmid containing p58C (residues 272–464) with a H3C-cleavable N-terminal 6xHis tag as described in Vaithiyalingam et al. (2010)
- BL21 DE3 competent cells
- LB/agar plates with 35 μg/mL kanamycin
- Terrific Broth (TB) growth media (autoclaved immediately after mixing)
- 50 mg/mL 0.2 μm filtered kanamycin stock
- Lysozyme
- DNase I
- Protease inhibitor tablets (Roche)
- H3C protease
- Reagents for purification buffers (see Table 1)

2.1.3 Protocol

Expression

1. Transform p58C into BL21 (DE3) chemically competent cells using a standard transformation protocol.

Table 1 Purification Buffers for p58C

p58C Lysis Buffer	50 m*M* NaH_2PO_4 (pH 7.8), 1 *M* NaCl, 10 m*M* Imidazole
Nickel Buffer A	50 m*M* NaH_2PO_4 (pH 7.8), 1 *M* NaCl, 20 m*M* Imidazole
Nickel Buffer B	50 m*M* NaH_2PO_4 (pH 7.8), 1 *M* NaCl, 250 m*M* Imidazole
SE Buffer	20 m*M* MES (pH 6.5), 50 m*M* NaCl, 2 m*M* DTT

2. Using 50 μL of transformed solution for plating, grow ~16 h at 37°C on a LB/agar plate supplemented with kanamycin.
3. Pick a single colony from the freshly transformed plate and inoculate a 50–200 mL TB culture to a final concentration of 50 μg/mL kanamycin.
4. Grow this culture ~16 h at 37°C.
5. In the morning, inoculate each culture flask with 10 mL of the overnight culture per 1 L of TB and grow the culture at 37°C.
6. While shaking the culture, monitor cell growth. Once the OD_{600} is approximately 0.5 absorbance units, move the flasks into a 21°C incubator with shaking and wait 20–30 min for the cells to acclimate to this temperature.
7. When the cells reach an OD_{600} of 1–1.5, induce protein overexpression with IPTG to a final concentration of 0.5 m*M*.
8. Allow the culture to grow ~16 h.
9. Harvest the cells by centrifugation at 12,000 rcf for 20 min and freeze at −80°C.

Purification

1. Prepare the buffers as described in Table 1.
2. Thaw frozen cell pellet to room temperature and resuspend in 2 mL lysis buffer per gram of pellet.
3. Add 1 protease inhibitor tablet, 2 mg lysozyme per gram of pellet, and 10 μg DNase I per mL of solution.
4. Begin lysis with a Dounce homogenizer. 5–6 strokes are typically sufficient, or until it looks homogeneous.
5. Sonicate on ice at 50% amplitude for 5 s on, 5 s rest for 10 min.
6. Spin down at 50,000 rcf for 20 min at 4°C.
7. Filter the supernatant with a 0.45 μm filter.
8. Equilibrate Ni-NTA column with 2 column volumes (CVs) of Nickel Buffer A and load supernatant onto the Ni-NTA column. For a 3 L growth, we typically use 10 mL of Ni-NTA resin.

9. Wash the column with 3–4 CVs of Nickel Buffer A.
10. Elute with 3–4 CVs of Nickel Buffer B. Collect 5 mL fractions.
11. Perform SDS-PAGE analysis on fractions and pool p58C-containing fractions.
12. Dialyze into Nickel Buffer A for 12–16 h at 4°C in the presence of H3C protease to remove the 6xHis tag.
13. Load dialyzed and His tag-cleaved protein onto Ni-NTA column that has been preequilibrated with Nickel Buffer A.
14. Wash the column with 3–4 CVs of Nickel Buffer A and elute remainder with 3–4 CVs of Nickel Buffer B.
15. Perform SDS-PAGE analysis of all samples and pool p58C-containing fractions. In the Ni-NTA repass step, p58C typically elutes in the flow-through and wash fractions.
16. Concentrate sample to 1–2 mL using a 10,000 MWCO concentrator and equilibrate the Superdex S75 size exclusion column with SE Buffer on an FPLC system.
17. Load concentrated sample onto the S75 column.
18. Program the FPLC system to flow at 1.0 mL/min and collect 2 mL fractions.
19. Analyze the fractions by SDS-PAGE.
20. Pool the most pure fractions, flash-freeze, and store at −80°C.

2.1.4 Notes

1. During the purification steps, if the protein has expressed appropriately, p58C's rich brown color can be used to observe progress on the columns. To ensure protein is not lost during protein elution, always run buffer over the Ni-NTA column until the eluate is colorless.
2. p58C can be induced at 25°C for a higher yield; however, at this induction temperature, a small (~10 kDa) contaminant is often observed that we have only been able to remove with a heparin column. In short, the protocol we use for this extra purification step is as follows:
   - Dialyze samples into wash buffer containing 20 m*M* HEPES at pH 7.2, 200 m*M* NaCl, and 2 m*M* DTT and load the dialyzed sample onto preequilibrated heparin column.
   - Load dialyzed sample onto the preequilibrated column at 2 mL/min and wash with 4 CVs of wash buffer.
   - Run a 150 min gradient of 0%–40% elution buffer (20 m*M* HEPES at pH 7.2, 1 *M* NaCl, and 2 m*M* DTT). p58C typically elutes at 10%–20% elution buffer.

2.2 Primase

2.2.1 Equipment

- Autoclave
- 0.45 and 0.2 μm syringe filters
- Plastic syringes
- Appropriate flasks and Petri dishes as needed
- Shaking incubator
- Homogenizer
- Sonication system
- Centrifuges
- SDS-PAGE equipment
- 30 kDa MWCO concentrators (15 mL volume)
- 10 mL Ni-NTA gravity column
- 5 mL HiTrap Heparin HP column (GE Healthcare)
- 120 mL Sephadex S200 column (GE Healthcare)
- FPLC system

2.2.2 Reagents and Solutions

- pBG100 plasmid containing p48 with a H3C-cleavable N-terminal 6xHis tag as described in Weiner et al. (2007) and a pET-Duet plasmid containing p58 cloned into site two using the *Nde*I and *Xho*I cleavage sites as described in O'Brien et al. (2017)
- BL21 DE3 (RIL) competent cells
- LB/agar plates with 35 μg/mL kanamycin and 35 μg/mL ampicillin
- 2XYT growth media (20 g tryptone, 10 g yeast extract, and 10 g NaCl per liter of media) supplemented with 0.4% glycerol, autoclaved immediately after mixing
- 50 mg/mL kanamycin sterifiltered prior to use
- 100 mg/mL ampicillin sterifiltered prior to use
- Lysozyme
- DNase I
- Protease inhibitor tablets (Roche)
- Reagents required for purification buffers (see Table 2)

2.2.3 Protocol

Expression

1. Cotransform p48 and p58 constructs into BL21 RIL (DE3) chemically competent cells using a standard transformation protocol.

Table 2 Purification Buffers for Primase

Primase Lysis Buffer	50 m*M* Tris (pH 8.0), 500 m*M* NaCl, 20 m*M* Imidazole, 0.1% IGEPAL CA-630, 5% glycerol
Nickel Buffer A	20 m*M* Tris (pH 7.2), 500 m*M* NaCl, 30 m*M* Imidazole
Nickel Buffer B	20 m*M* Tris (pH 7.2), 500 m*M* NaCl, 300 m*M* Imidazole
Dialysis Buffer	20 m*M* Tris (pH 7.2), 300 m*M* NaCl, 20 m*M* Imidazole, 5 m*M* BME
Dilution Buffer	20 m*M* HEPES (pH 7.2), 5% glycerol
Heparin Buffer A	20 m*M* HEPES (pH 7.2), 50 m*M* NaCl, 3% glycerol
Heparin Buffer B	20 m*M* HEPES (pH 7.2), 1 *M* NaCl, 3% glycerol
SE Buffer	20 m*M* HEPES (pH 7.2), 150 m*M* NaCl, 5% glycerol (or 100 m*M* ammonium sulfate), 2 m*M* DTT

2. Plate 50 μL of transformed solution on a LB/agar plate supplemented with kanamycin and ampicillin and grow ~16 h at 37°C.
3. Pick a single colony from the freshly transformed plate and inoculate a 50–200 mL 2XYT culture containing a final concentration of 50 μg/mL kanamycin and 100 μg/mL ampicillin.
4. Grow this culture ~16 h at 37°C.
5. In the morning, inoculate each 1 L 2XYT culture flask with 10 mL of the overnight culture and grow the culture at 37°C.
6. While shaking the culture, monitor cell growth. Once the OD_{600} is approximately 0.5 absorbance units, move the flasks into an 18°C incubator with shaking and wait 20–30 min for the cells to acclimate to this temperature.
7. When the cells reach an OD_{600} of 1–1.5 absorbance units, induce with IPTG to a final concentration of 0.5 m*M*.
8. Allow the culture to grow for 16 h.
9. Harvest the cells by centrifugation at 12,000 rcf for 20 min and freeze at −80°C.

Purification

1. Prepare the buffers as described in Table 2.
2. Thaw frozen cell pellet to room temperature and resuspend in Primase Lysis Buffer.

3. Add 1 protease inhibitor tablet, 2 mg lysozyme per gram of pellet, and 10 μg DNase I per mL of solution.
4. Begin lysis with a Dounce homogenizer. 5–6 strokes are typically sufficient, or until it looks homogeneous.
5. Sonicate on ice at 50% amplitude for 5 s on, 5 s rest for 10 min.
6. Centrifuge lysate at 50,000 rcf for 20 min at 4°C.
7. Filter the resulting supernatant with a 0.45 μm syringe filter before slowly loading it into a Ni-NTA column preequilibrated with Nickel Buffer A. For a 6 L prep, we typically use 10 mL of Ni-NTA resin.
8. Wash with 3–4 CVs of Nickel Buffer A and then elute the protein with 3–4 CVs of Nickel Buffer B; collect eluate as 2–3 mL fractions.
9. Perform SDS-PAGE for all fractions, including flow-through, wash, and eluate.
10. Pool primase-containing fractions, add H3C protease to cleave the His tag on p48, and dialyze overnight at 4°C into 4 L of prechilled Dialysis Buffer.
11. If the solution contains light precipitation or particles, filter the solution with a 0.2 μm filter.
12. Repass over Ni-NTA column, using the same buffers as above. Primase should elute in the flow-through and wash steps.
13. Perform SDS-PAGE for all fractions including flow-through, wash, and eluate.
14. Pool primase-containing fractions and then slowly dilute the protein with Dilution Buffer (while keeping the sample in an ice bucket) to reduce the salt concentration to 150 m*M*.
15. Load onto heparin column equilibrated in Heparin Buffer A.
16. After washing with 5–10 CVs of 10% Heparin Buffer B, run a 30 CV gradient of 20%–70% Heparin Buffer B. Primase typically elutes at 40%–50% Heparin Buffer B.
17. Perform SDS-PAGE for elution fractions and pool primase-containing fractions.
18. Concentrate pooled fractions to 2 mL using a 30K MWCO centrifugal concentrator. Filter the solution before loading into S200 size exclusion column preequilibrated with SE Buffer.
19. Perform an isocratic elution, collecting 2 mL fractions.
20. Analyze all fractions from elution peaks by SDS-PAGE.
21. Pool the most pure fractions, concentrate to desired concentration, and store at −80°C in SE Buffer.

2.2.4 Notes

1. If poor expression is obtained with the first colony picked from the transformation plate in Step 3 of expression, screen more colonies to identify one that expresses better.
2. Although primase expresses well in either 2XYT or TB media, supplementing with 8 mL 50% glycerol per liter of media boosts expression.
3. Both primase and p58C are usually iron-loaded when grown in a standard rich growth medium. However, supplementing with 0.1 mg/mL of ferric citrate and ammonium iron (III) citrate provides more consistent expression levels.
4. Do not allow cells to grow for longer than 17 h after induction with IPTG.
5. All purification steps should be performed at 4°C or on ice. Buffers should be chilled before use.
6. In our hands, primase tends to be most stable in buffers at pH 7.0–7.5, containing 2% glycerol, and with salt (either NaCl or KCl) concentrations of 100 m*M* or higher. As high ionic strength is detrimental to DNA binding activity, we typically store primase in a higher ionic strength buffer (such as the SE Buffer described in Table 2), and buffer exchange the protein into a lower ionic strength buffer immediately before use in functional assays.
7. A small amount of precipitation is usually observed in Step 14, but less than the amount observed when dialyzing directly into a low-salt buffer. The solution can either be syringe-filtered or centrifuged to remove the precipitate.

3. DNA BINDING ASSAYS

A variety of methods have been developed to measure binding affinity (Pollard, 2010), which is usually reported in terms of an equilibrium dissociation constant (K_D). The two most common approaches for assaying the binding of DNA to proteins are the EMSA and changes in the fluorescence anisotropy of DNA labeled with a fluorescent probe. In the following, we describe the application of these two approaches for determining the DNA substrate affinity of human DNA primase and the isolated p58C domain.

3.1 EMSA

EMSA is a standard method of estimating the binding affinity of a protein for a nucleic acid (Garner & Revzin, 1981; Hellman & Fried, 2007). In this assay, gel electrophoresis is used to separate the bound and unbound populations of a nucleic acid substrate that has been incubated with the protein of interest. The percentage of bound substrate is plotted vs protein concentration, forming a binding curve from which the K_D can be determined. This technique is typically best for complexes with K_Ds in the submicromolar range, as tight, distinct bands for both bound and unbound substrates will be observed.

3.1.1 Reagents and Solutions

All reagents should be "molecular biology grade" or better.

- Stock solution of protein in appropriate DNA binding buffer (see below)
- Fluorescently labeled target nucleic acid
- TEMED (*N,N,N′,N′*-tetramethylethylenediamine), stored at 4°C
- APS (ammonium persulfate), stored at 4°C
- 40% stock solution of 37.5:1 bis:acrylamide (Bio-Rad), stored at 4°C
- 10× Tris–acetate–EDTA (TAE) electrophoresis buffer: 400 m*M* Tris and 25 m*M* EDTA brought to pH 7.2 with acetic acid
- 10× Dye + glycerol stock solution: 50% v/v glycerol, 0.001% w/v bromophenol blue, and 0.001% w/v xylene cyanol FF

3.1.2 Equipment

- Electrophoresis power supply; commercial power supplies sold for SDS-PAGE or DNA sequencing applications are suitable
- Standard electrophoresis apparatus
- Plastic syringe
- Opaque 1.5 mL microcentrifuge tubes
- Plastic wrap; relatively low "cling" wraps, such as Saran Wrap®, are preferred.
- Typhoon Imager (GE Healthcare)
- Data analysis software for graphing and fitting the data (e.g., GraphPad Prism)

3.1.3 Procedure

1. Set up gel apparatus and ensure that it is clean.
2. Prepare the polymerization mixture (Table 3). Add polymerizing agents last, as the gel will polymerize quickly thereafter.

Table 3 Components of 5% 37.5:1 Bis:Acrylamide, 1 × TAE Gel

40% 37.5:1 Bis:acrylamide	5 mL
10 × TAE Buffer (pH 7.2)	4 mL
H_2O	30.7 mL
TEMED	30 μL
APS	300 μL

3. Pour the polymerizing mixture into gel casting assembly, avoiding bubble formation. Insert the well-forming comb immediately and allow the gel to polymerize.
4. Remove comb and bottom spacer (if used) from gel and mount the gel in the vertical electrophoresis apparatus.
5. Fill buffer reservoirs with electrophoresis running buffer, which should match the buffer used in the gel.
6. Conduct preelectrophoresis at the running voltage.
7. Prepare sample mixtures of protein and DNA in opaque microcentrifuge tubes and equilibrate for 15–20 min at 4°C. Low temperature is used to minimize any adverse effects on protein stability/solubility. This can be completed while the gel is undergoing preelectrophoresis.
8. At the end of the equilibration period, rinse the wells of the gel with the plastic syringe and load the samples.
9. Load tracking dye in flanking, unused wells as a marker, and perform electrophoresis. Reduce the voltage if the gel becomes warm during electrophoresis.
10. At the end of electrophoresis, remove the gel and plate assembly from the electrophoresis device.
11. Gently separate the plates, leaving the gel adherent to one of the plates. Wrap the gel and plate in a single layer of plastic wrap. The wrapped surface of the gel should be free of air bubbles and wrinkles, as these will interfere with imaging.
12. Image gel with Typhoon Imager.
13. Plot ratio of bound DNA or total DNA vs protein concentration and fit the data to the binding equation to determine the K_D. We use the equation $A = ((A_{max}{}^*x)/(K_D + x))$, in which A is the signal, x is the protein concentration, and K_D is the dissociation constant.

3.1.4 Notes

1. Buffer optimization is very important. A buffer that has worked well for primase is 20 m*M* HEPES at pH 7.2, 75 m*M* NaCl, 5% glycerol, 1 m*M* TCEP, 0.1% IGEPAL CA-630 detergent (Sigma), 0.1 mg/mL BSA, and 2 μ*M* ssDNA competitive inhibitor (inhibitor sequence: 5′-AAA AAA AAA AAA AAA AAA AAA AAG ATA AAA AAA-3′ as in Baranovskiy, Zhang, et al., 2016). For p58C, 20 m*M* MES at pH 6.5, 50 m*M* NaCl, 5% glycerol, 2 m*M* TCEP, 0.1% IGEPAL CA-630, 0.1 mg/mL BSA, and 2 μ*M* ssDNA competitive inhibitor.
2. The inclusion of 5% glycerol in EMSA buffers is to facilitate sample loading. One can also mix the sample with the dye + glycerol stock solution, but we have found that the dye generates an artifact when imaging FAM-labeled substrates.
3. Nucleic acids can be labeled with radioactive phosphate or fluorescent tags. Both Cy3- and FAM-labeled nucleic acid substrates have been successfully used with primase and p58C.
4. Precast native gels can also be used.
5. Electrophoresis gel and running buffers can be varied. Best results are obtained when the pH of the gel and running buffer matches that of the buffer used for the binding reaction.
6. Primase is more stable in buffers with pH $\leq$8.0, so the use of TBE is not recommended.
7. In the experiment shown in Fig. 1 (reaction compositions given in Table 5), 15 μL was loaded. However, smaller sample volumes can be used. A practical lower limit is imposed by the need to cover the bottom surface of the well with a uniform layer of sample solution; this can be as small as 5 μL.
8. To minimize dissociation of the complex, the electrophoresis interval should be as short as possible for resolution of the species of interest. Primase typically requires at least 2.5 h to completely leave the well, whereas p58C can be successfully electrophoresed in 2 h or less.
9. Care must be taken to avoid "well-shifts," in which the bound substrate does not leave the well (see 250 and 500 n*M* points in Fig. 2), as these points cannot be included in quantification. This phenomenon can occur for multiple reasons, including protein aggregation and gel conditions. If increasing run time and decreasing polyacrylamide percentage do not resolve the issue, the addition of a small amount of detergent (such as 0.1% IGEPAL CA-630) may help.

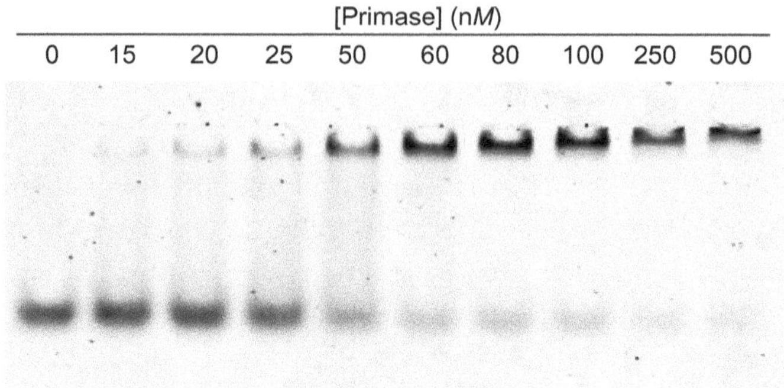

Fig. 2 Representative autoradiogram of EMSA of human DNA primase with RNA-primed DNA substrate. RNA sequence: 5′-triphosphate-GGC UCG GA-3′. DNA sequence: 5′-Cy3-AAA CAC CGA GCC AAC ATA-3′.

3.2 Fluorescence Anisotropy

Fluorescence anisotropy provides a viable, and theoretically more quantitative, alternative to EMSA for characterizing DNA binding affinities (Rossi & Taylor, 2011). Anisotropy is induced in a fluorescent probe attached to molecule, upon interaction with a binding partner (Fig. 3). The degree of anisotropy is proportional to the extent of binding at equilibrium. We measure the fluorescence anisotropy at a fixed quantity of fluorescently labeled DNA over a range of protein concentrations. A standard binding equation is then fit to the data, from which the K_D can be extracted. This method is particularly suitable for complexes or variants with K_D values in the micromolar range, which are difficult to measure by the EMSA approach (Fig. 3).

3.2.1 Reagents and Solutions

All reagents should be "molecular biology grade" or better

- Stock solution of protein in appropriate DNA binding buffer
- Fluorescently labeled target nucleic acid in the same buffer

3.2.2 Equipment

- 396-well plates appropriate for fluorescence measurements (Cornell)
- Opaque 1.5 mL microcentrifuge tubes

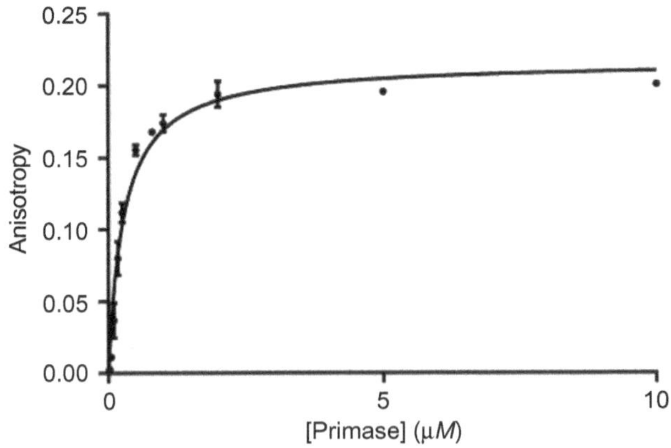

Fig. 3 Binding titration from fluorescence anisotropy assay of human DNA primase with a fluorescent ssDNA template. Sequence: 5′-TTT TTT TTT TTT TTT TTT TTT TTT T-3′.

- Fluorescence plate imager with an appropriate light filter and the capacity to perform fluorescence anisotropy (GE Healthcare)
- Data analysis software for graphing and fitting the data (e.g., GraphPad Prism)

3.2.3 Procedure

1. Prepare sample mixtures of protein and DNA in opaque microcentrifuge tubes and equilibrate for 15–20 min at 4°C. This can be completed as one would for EMSAs, except that the total reaction volume should be 80 μL (see Table 4).
2. Pipette 25 μL of each titration point into three wells on the 396-well plate.
3. Image using plate imager.
4. Plot anisotropy vs protein concentration and fit these data with an appropriate binding equation as described in Section 3.1.3.

3.2.4 Notes

1. Buffer optimization and temperature control are very important for obtaining consistent reproducible data. A buffer that has worked well for primase is 20 m*M* HEPES at pH 7.2, 75 m*M* NaCl, 5% glycerol, and 2 m*M* DTT. For p58C, 20 m*M* MES at pH 6.5, 75 m*M* NaCl, and 2 m*M* DTT. We typically run primase fluorescence anisotropy assays at 21°C, but use low temperature during equilibration to minimize any adverse effects on protein stability/solubility.

Table 4 Calculations for DNA Binding Assay

Final [Primase] (n*M*)	Volume of Protein (μL)	Volume of Substrate (μL)	Volume of Buffer (μL)
0	0.0	0.4	19.6
15	0.4	0.4	19.2
25	0.7	0.4	18.9
50	1.3	0.4	18.3
75	2.0	0.4	17.6
100	2.7	0.4	16.9
250	6.7	0.4	12.9
500	13.3	0.4	6.3
Total	27.1	3.2	129.7

Reaction volume: 20 μL; stock [primase]: 750 n*M*; stock [substrate]: 500 n*M*; final [substrate]: 10 n*M*.

2. We use FITC-labeled DNA oligos, with excitation and emission wavelengths of 485 and 520 nm, respectively. The fluorophore used for this assay can be optimized based on the sensitivity of the plate reader.
3. The concentration of fluorescent DNA should be approximately 10-fold lower than the expected K_D to ensure that measurements are made in the equilibrium binding regime.
4. Temperature is another factor that can be optimized. If using short dsDNA oligos, low temperature may be critical for ensuring that the DNA remains annealed. However, anisotropy can vary as a function of temperature; it is necessary to remain consistent with this variable between experiments.

4. STRUCTURAL CHARACTERIZATION

Structural studies are instrumental in understanding iron–sulfur cluster type and how it is coordinated to the protein matrix. Historically, iron–sulfur clusters have been difficult to identify and structurally characterize because of their intrinsic lability and sensitivity; a combination of methods is often used to prove the existence of the cluster, but X-ray crystallography remains the key to detailed knowledge of cluster (Fuss et al., 2015).

This section describes the application of CD, NMR, and X-ray crystallography to characterize the biochemical and structural properties of the isolated p58C domain. These procedures are applicable for study of the wild-type proteins and mutants, as well as to yeast p58C. Procedures for structural studies of full-length primase are described elsewhere (Baranovskiy et al., 2015; Thompson, Ehlinger, & Chazin, 2017).

4.1 Circular Dichroism

CD spectroscopy is a quick and relatively straightforward method to analyze secondary structure content (e.g., helices, beta sheets, random coil) and tertiary structure related to packing of aromatic side chains in a hydrophobic core (Greenfield, 2006). Secondary structures and aromatic residues result in characteristic contributions to the CD spectrum in the range of 190–260 and 250–330 nm, respectively (Fig. 4). Interestingly, iron–sulfur clusters also display unique spectra in the visible region (600–350 nm); this can be used to help identify the iron–sulfur cluster type (Stephens et al., 1978). Here, we focus on the commonly used near- and far-UV CD, while CD in the visible region will not be discussed. The key approach involves comparing near- and far-UV CD spectra of different proteins to determine if they are structurally similar.

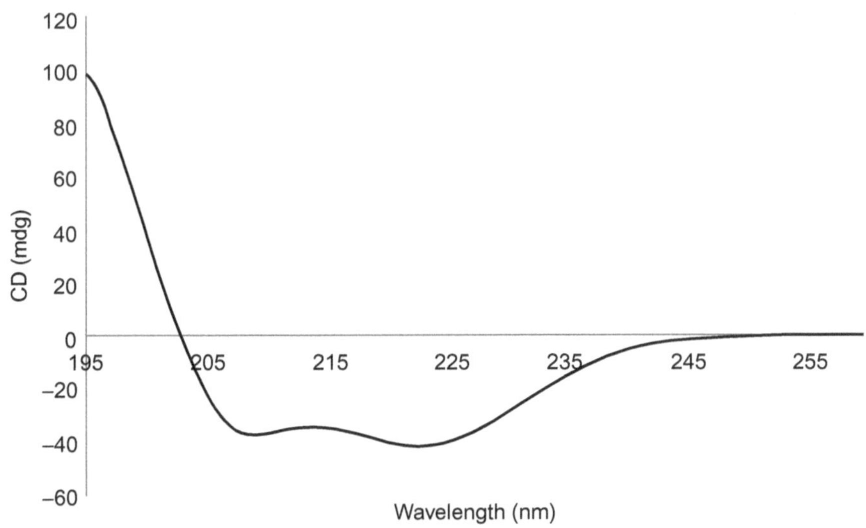

Fig. 4 Circular dichroism spectrum of p58C acquired at room temperature in 10 m*M* KH_2PO_4 (pH 7.2).

4.1.1 Buffers and Reagents

- Stock of p58C in appropriate buffer

4.1.2 Equipment

- CD spectropolarimeter
- Quartz cuvette
- Syringe
- Syringe filter (0.2 μm)

4.1.3 Procedure for Far-UV CD

1. Dilute p58C in buffer to a final concentration of 0.5–1 mg/mL and filter.
2. Measure the spectrum from 260 to 190 nm in increments of 0.5 nm at a rate of 50 nm/min with a response time of 4 s and a bandwidth of 1 nm. Collect buffer blank prior to collecting protein spectrum. We get best results with an accumulation of 3 spectra, which are averaged and smoothed.

4.1.4 Notes

1. For p58C, we use 10 m*M* KH_2PO_4 (pH 7.2) as our buffer system.
2. If the high tension voltage rises above 600 V during the course of the scan, the data are unusable. If this occurs, the problem can be solved by either changing buffers or reducing the protein concentration.

4.1.5 Procedure for Near-UV CD

1. Dilute p58C in buffer to a final concentration of 1–2 mg/mL and filter.
2. Measure the spectrum from 330 to 250 nm in increments of 0.5 nm at a rate of 20 nm/min with a response time of 4 s and a bandwidth of 1 nm. Collect buffer blank prior to collecting protein spectrum. We get best results with an accumulation of 5 spectra, which are averaged.

4.1.6 Notes

1. For p58C, we use 10 m*M* KH_2PO_4 (pH 7.2) as our buffer system.
2. The signal from this region is often very weak. A large cuvette (1 cm pathlength) and higher concentrations of protein can compensate for this. However, if the high tension voltage rises above 600 V during the course of the scan, the data are unusable. If this occurs, the protein concentration is too high or the buffer is unsuitable.
3. Small variations in the chromophores' local chemical environment are clearly reflected in the spectra. This is good for identifying destabilizing

mutations, but necessitates nearly perfect buffer matching between samples.

4.2 Nuclear Magnetic Resonance Spectroscopy

NMR is a powerful tool for characterizing the structure and dynamics of proteins (Cavanagh, Fairbrother, Palmer, Rance, & Skelton, 2007). NMR in solution can be used for a range of applications, spanning from basic analytical biochemistry to determination of structural dynamics at atomic resolution. One very basic application of NMR is to obtain a rapid overview of the tertiary structure (or lack thereof) of a protein. The most common approach involves acquiring heteronuclear two-dimensional correlation experiments (typically $^{15}N–^{1}H$ HSQC) on ^{15}N-enriched samples. Such spectra provide a direct reflection of the extent of folding of a polypeptide chain and a ready means to determine whether a mutation causes perturbation of the tertiary structure.

Because NMR signals are highly sensitive to any changes in their electronic environment, NMR is widely used to monitor binding events (Takeuchi & Wagner, 2006). These experiments involve titration of a ligand into a protein solution and acquisition of the NMR spectrum at each point. The changes in the NMR signals (chemical shift perturbations, CSPs) can be used to map binding sites. CSPs can also be plotted vs the substrate concentration to generate a binding curve, from which a K_D can be extracted. The NMR-based approach is particularly valuable for characterizing weak binding in the μM to mM range.

NMR with iron–sulfur cluster-containing proteins can be challenging because the paramagnetism of the cluster causes rapid relaxation, reflected as severe line broadening that reduces the ability to observe signals, particularly for residues near the paramagnetic site. Despite this, the partial NMR spectra can be used for basic biochemical analyses (Fig. 5), such as validating the structural integrity of p58C mutants and demonstrating binding of ligands to p58C (Vaithiyalingam et al., 2010; Weiner et al., 2007). Methods for detailed NMR analysis of proteins with paramagnetic centers have been developed (Bertini, Luchinat, Parigi, & Pierattelli, 2005), but these have yet to be applied to primase.

4.2.1 Reagents and Solutions

- $^{2}H_2O$ (Sigma Aldrich)
- Solution of ^{15}N-enriched p58C in buffer suitable for NMR

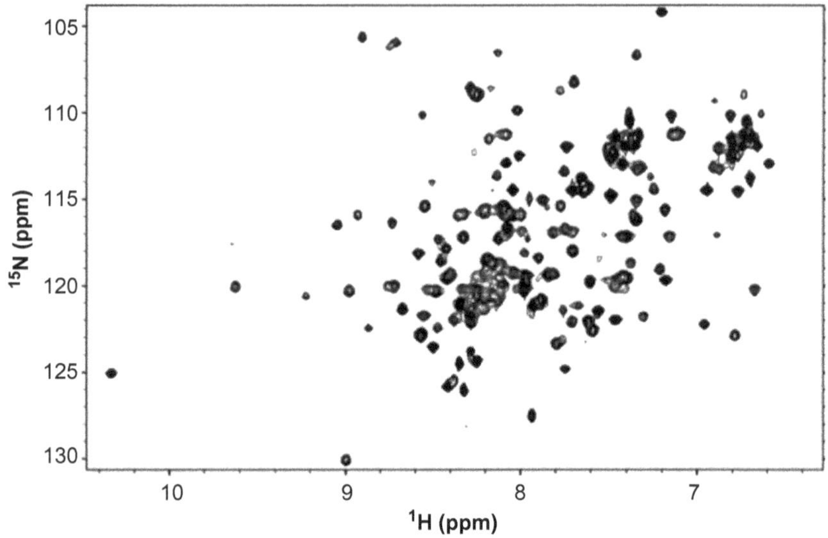

Fig. 5 2D ^{15}N–^{1}H HSQC NMR spectrum of 75 μ*M* ^{15}N-enriched p58C in 20 m*M* MES at pH 6.5, 50 m*M* NaCl, and 2 m*M* DTT, acquired at 25°C on an 800 MHz Bruker AVANCE II spectrometer. Despite the paramagnetic center, there are a significant number of resolved peaks that serve as valuable probes for monitoring the binding of ligands (e.g., NTPs, metals, substrates) and any differences in tertiary structure induced by mutations.

4.2.2 Equipment

- NMR tube
- NMR Pasteur pipette with rubber bulb
- NMR spectrometer (field $\geq$500 MHz)
- Appropriate software for data collection and analysis

4.2.3 Procedure

1. Add $^{2}H_2O$ to ~100 μm protein in NMR buffer to a final concentration of approximately 5% and pipette protein solution into clean NMR tube.
2. After ensuring that the airflow is on, place tube in the spinner and carefully lower into position in the spectrometer.
3. Lock on the ^{2}H signal from the $^{2}H_2O$ and then tune and shim the magnet.
4. Calibrate 180° pulse and receiver gain.
5. Acquire 1D ^{1}H spectrum.
6. Acquire 2D ^{15}N–^{1}H HSQC spectrum.

4.2.4 Notes

1. A good buffer for heteronuclear NMR of p58C is 20 m*M* MES at pH 6.5, 50 m*M* NaCl, and 2 m*M* DTT. Low-salt buffers are preferred for NMR spectroscopy, as increased ionic strength can result in signal loss.
2. There should be sufficient sample volume that the sample covers the height of the probe. For a 5 mm tube in Bruker spectrometers, this will be approximately 550 μL; for a 3 mm tube, 170 μL is sufficient.
3. 2D $^{15}N–^{1}H$ HSQC spectra for a 100 μ*M* p58C sample acquired on an AVANCE II 600 MHz spectrometer at 25°C required 64 scans with 128 t_1 increments.

4.3 X-ray Crystallography

X-ray crystallography is the premier method for determining three-dimensional structures of proteins at atomic resolution. It has been used to determine structures of DNA primase and the isolated p58C domain (Agarkar et al., 2011; Baranovskiy, Babayeva, et al., 2016; Baranovskiy et al., 2015; Sauguet et al., 2010; Vaithiyalingam et al., 2010). These structures have provided insight into the location of the [4Fe–4S] domain in p58C, the binding of primed template to p58C, and the overall organization of the full-length protein in the absence of cofactors and substrates.

Crystallization. We use hanging-drop vapor diffusion to crystallize human p58C. The probability of obtaining crystals is highest when using a matrix of varying lithium sulfate and PEG 3350. Optimal well solutions for wild-type p58C are 100 m*M* Tris–HCl (pH 8.5), 100–400 m*M* Li_2SO_4, and 16%–26% PEG 3350. p58C variants usually require further optimization.

4.3.1 Equipment

- 24-well hanging-drop crystallization plates with sealant (Hampton)
- Siliconized glass slides (Hampton)
- Tweezers
- UV–Vis spectrophotometer
- 10 kDa MWCO concentrator
- Benchtop centrifuge
- 16°C incubator
- Liquid nitrogen dewar
- Cryo-loop

4.3.2 Buffers and Reagents

- All reagents should be filtered with a 0.2 µm filter
- 2–4.5 mg purified p58C (60 µL of 30–75 mg/mL)
- Protein crystallography buffer: 20 m*M* MES (pH 6.5), 50 m*M* NaCl
- Nuclease-free water
- 1 *M* Tris–HCl (pH 8.5)
- 2 *M* Li_2SO_4
- 50% Polyethylene Glycol 3350 (PEG 3350)
- Glycerol
- Liquid nitrogen

4.3.3 Protocol

1. Prepare the crystallization plate by removing a small slice of vacuum grease from the well. This ensures that well is properly sealed when placing the glass slide on top of the vacuum grease.
2. Create the matrix described above by adding reagents as indicated in Table 5.
3. Mix gently and set aside.
4. Next, prepare the protein for crystallization. Use a 10 kDa MWCO concentrator and a benchtop centrifuge to concentrate p58C to 35–75 mg/mL (1.5–3.0 m*M*). The total volume of protein needed is 60 µL.
5. Combine 2 µL of the protein solution and 2 µL of well solution on a siliconized glass slide. Immediately use the tweezers to place the slide, drop-side down, on the well. Press on the edges of the glass slide and rotate it, making sure to fill the gap in the vacuum grease created in Step 1. This ensures an airtight seal. Repeat this for each of the 24 wells.
6. Place in a 16°C incubator. Crystals will grow in 5–7 days.
7. After crystal growth, crystals can be scooped with cryo-loops, making sure to obtain only a single crystal per loop. Crystals can then be soaked in 20% glycerol (as a cryoprotectant), flash frozen in liquid nitrogen, and stored in a liquid nitrogen dewar (Fig. 6).

4.3.4 Notes

1. Drops should be set with p58C in a precrystallization buffer: 20 m*M* MES (pH 6.5) and 50 m*M* NaCl. Dialysis before storage is preferred. If buffer exchange is performed in a concentrator, concentrate until the volume has decreased by 80% and dilute to original volume with precrystallization buffer. Repeat at least 4 times.

Table 5 Crystallization Conditions Matrix

	1	2	3	4	5	6
A	100 m*M* Tris 8.5, 16% PEG 3350, 100 m*M* Li_2SO_4	100 m*M* Tris 8.5, 18% PEG 3350, 100 m*M* Li_2SO_4	100 m*M* Tris 8.5, 20% PEG 3350, 100 m*M* Li_2SO_4	100 m*M* Tris 8.5, 22% PEG 3350, 100 m*M* Li_2SO_4	100 m*M* Tris 8.5, 24% PEG 3350, 100 m*M* Li_2SO_4	100 m*M* Tris 8.5, 26% PEG 3350, 100 m*M* Li_2SO_4
B	100 m*M* Tris 8.5, 16% PEG 3350, 200 m*M* Li_2SO_4	100 m*M* Tris 8.5, 18% PEG 3350, 200 m*M* Li_2SO_4	100 m*M* Tris 8.5, 20% PEG 3350, 200 m*M* Li_2SO_4	100 m*M* Tris 8.5, 22% PEG 3350, 200 m*M* Li_2SO_4	100 m*M* Tris 8.5, 24% PEG 3350, 200 m*M* Li_2SO_4	100 m*M* Tris 8.5, 26% PEG 3350, 200 m*M* Li_2SO_4
C	100 m*M* Tris 8.5, 16% PEG 3350, 300 m*M* Li_2SO_4	100 m*M* Tris 8.5, 18% PEG 3350, 300 m*M* Li_2SO_4	100 m*M* Tris 8.5, 20% PEG 3350, 300 m*M* Li_2SO_4	100 m*M* Tris 8.5, 22% PEG 3350, 300 m*M* Li_2SO_4	100 m*M* Tris 8.5, 24% PEG 3350, 300 m*M* Li_2SO_4	100 m*M* Tris 8.5, 26% PEG 3350, 300 m*M* Li_2SO_4
D	100 m*M* Tris 8.5, 16% PEG 3350, 400 m*M* Li_2SO_4	100 m*M* Tris 8.5, 18% PEG 3350, 400 m*M* Li_2SO_4	100 m*M* Tris 8.5, 20% PEG 3350, 400 m*M* Li_2SO_4	100 m*M* Tris 8.5, 22% PEG 3350, 400 m*M* Li_2SO_4	100 m*M* Tris 8.5, 24% PEG 3350, 400 m*M* Li_2SO_4	100 m*M* Tris 8.5, 26% PEG 3350, 400 m*M* Li_2SO_4

2. Concentration is measured from absorbance of the protein at 280 nm with a UV–Vis spectrophotometer. Because of the high concentration, it is necessary to dilute a sample of protein at least 100 × into buffer, so that absorbance is less than 1, to accurately quantify the concentration. The extinction coefficient of WT p58C used to estimate concentration is 16,000 $M^{-1}\,cm^{-1}$ or 0.7 $(mg/mL)^{-1}\,cm^{-1}$. Alternatively, concentration of the cluster-loaded protein can be quantified by measuring the absorbance at 410 nm with the extinction coefficient 17,000 $M^{-1}\,cm^{-1}$.
3. The best crystals are obtained when the protein is concentrated to 50–75 mg/mL; crystallization from lower concentrations (25–30 mg/mL) is possible but takes more time. The concentration step induces dimerization via a disulfide bond between the Cys 449 of two p58C molecules in the

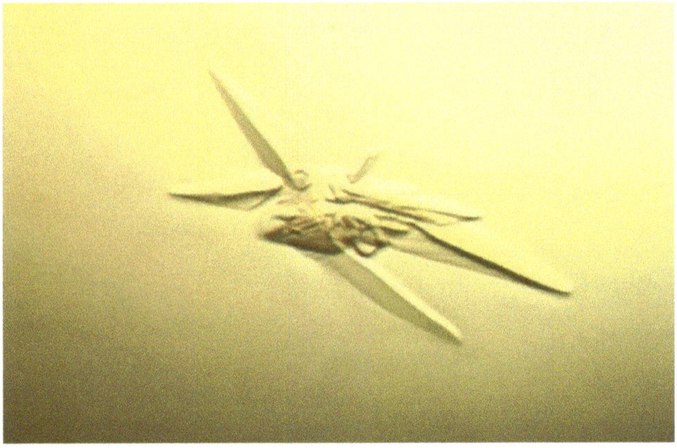

Fig. 6 Crystal of p58C obtained in the conditions described in Section 4.3. Note the *brown color* that arises from the iron–sulfur cluster.

asymmetric unit. Once the protein assumes this dimeric form, the protein can be diluted into the protein crystallography buffer as necessary.

4. If p58C mutants are recalcitrant to forming crystals, it is possible to use crushed wild-type crystals to streak-seed the drops. Lowering the temperature to 4°C is an alternative approach.

4.4 Data Collection and Structure Determination

4.4.1 Equipment

- X-ray generator
- Cryo-stat
- Goniometer
- Detector
- Crystallography data processing software

4.4.2 Reagents

- Cryo-protected crystals of p58C

4.4.3 Procedure

1. Mount the crystals on the goniometer, making sure to use liquid nitrogen and the cryo-stat to ensure that the crystal is always cold.
2. Center the crystal so that it remains in the beam when rotated 360°.
3. Optimize the detector position empirically by taking snapshots of the crystal and moving the detector to obtain optimal reflection dispersion while maintaining the high-resolution reflections.

4. Once the crystal and detector position are optimized, collect data using standard procedures.
5. Index the data. These crystals typically form in the *C*2 space group.
6. Scale the data. Take care to reject frames with a χ^2 value >10 and to adjust the error model so that <0.5% of reflections are rejected.
7. Use molecular replacement to obtain phase information.
8. Create a model using the processed data and the phase information.
9. Build and refine the model in an iterative way, taking care not to overbuild the model.

4.4.4 Notes

1. Nearly all of our data have been collected at a synchrotron beamline, although it is possible to use in-house X-ray generators if the crystal is large and diffracts well.
2. Typical parameters for data collection at the synchrotron beamline are: 1 s exposure, 200° of data, and a delta of 1°. Detector distance will vary based on the diffraction and the data collection system. The crystals form in the *C*2 space group; it is therefore necessary to collect at least 180° of data.
3. Two programs have been used for data processing: HKL2000 (Otwinowski & Minor, 1997) and XDS (Kabsch, 2010).
4. Since p58C contains a [4Fe–4S] cluster, it is possible to use anomalous scattering at the iron or sulfur edges to phase the data.
5. For best results, we use the PDB model 3L9Q to phase the data using molecular replacement.
6. Software used to create and refine models is available in the program suites PHENIX (Adams et al., 2010) and CCP4 (Winn et al., 2011). We use Coot (Emsley, Lohkamp, Scott, & Cowtan, 2010) to build and validate models.
7. An overbuilt model is evident from a $R_{free} - R_{work}$ of >0.05.

5. CONCLUDING REMARKS

The identification of iron–sulfur clusters in key enzymes at the replication fork has opened the door for speculation on their function(s) (Arnold, Grodick, & Barton, 2016; Fuss et al., 2015; Jain et al., 2014; Liu & Huang, 2015; O'Brien et al., 2017). To advance understanding of their roles in replication, it is important to further characterize the properties of these proteins in vitro and in vivo, taking care to recognize the redox sensitivity of the clusters. This report summarizes the first steps to characterize in detail the in vitro properties of DNA primase. Ultimately, the generation and characterization of pol–prim and other iron–sulfur cluster-containing polymerases will clarify the role(s) of iron–sulfur clusters at the replication fork. Since primase plays a critical role in initiation of DNA synthesis, the stability of its iron–sulfur cluster makes it a good test system to explore the interplay between iron–sulfur electrochemistry and the biological function of replicative polymerases.

REFERENCES

Adams, P. D., Afonine, P. V., Bunkoczi, G., Chen, V. B., Davis, I. W., Echols, N., et al. (2010). PHENIX: A comprehensive Python-based system for macromolecular structure solution. *Acta Crystallographica. Section D, Biological Crystallography*, *66*(Pt. 2), 213–221. http://dx.doi.org/10.1107/S0907444909052925.

Agarkar, V. B., Babayeva, N. D., Pavlov, Y. I., & Tahirov, T. H. (2011). Crystal structure of the C-terminal domain of human DNA primase large subunit: Implications for the mechanism of the primase-polymerase alpha switch. *Cell Cycle*, *10*(6), 926–931. http://dx.doi.org/10.4161/cc.10.6.15010.

Arezi, B., Kirk, B. W., Copeland, W. C., & Kuchta, R. D. (1999). Interactions of DNA with human DNA primase monitored with photoactivatable cross-linking agents: Implications for the role of the p58 subunit. *Biochemistry*, *38*(39), 12899–12907.

Arnold, A. R., Grodick, M. A., & Barton, J. K. (2016). DNA charge transport: From chemical principles to the cell. *Cell Chemical Biology*, *23*(1), 183–197. http://dx.doi.org/10.1016/j.chembiol.2015.11.010.

Baranovskiy, A. G., Babayeva, N. D., Zhang, Y., Gu, J., Suwa, Y., Pavlov, Y. I., et al. (2016). Mechanism of concerted RNA-DNA primer synthesis by the human primosome. *The Journal of Biological Chemistry*, *291*(19), 10006–10020. http://dx.doi.org/10.1074/jbc.M116.717405.

Baranovskiy, A. G., & Tahirov, T. H. (2017). Elaborated action of the human primosome. *Genes (Basel)*, *8*(2), 62. http://dx.doi.org/10.3390/genes8020062.

Baranovskiy, A. G., Zhang, Y., Suwa, Y., Babayeva, N. D., Gu, J., Pavlov, Y. I., et al. (2015). Crystal structure of the human primase. *The Journal of Biological Chemistry*, *290*(9), 5635–5646. http://dx.doi.org/10.1074/jbc.M114.624742.

Baranovskiy, A. G., Zhang, Y., Suwa, Y., Gu, J., Babayeva, N. D., Pavlov, Y. I., et al. (2016). Insight into the human DNA primase interaction with template-primer. *The Journal of Biological Chemistry*, *291*(9), 4793–4802. http://dx.doi.org/10.1074/jbc.M115.704064.

Bertini, I., Luchinat, C., Parigi, G., & Pierattelli, R. (2005). NMR spectroscopy of paramagnetic metalloproteins. *Chembiochem*, *6*(9), 1536–1549. http://dx.doi.org/10.1002/cbic.200500124.

Burgers, P. M. (2009). Polymerase dynamics at the eukaryotic DNA replication fork. *The Journal of Biological Chemistry*, *284*(7), 4041–4045. http://dx.doi.org/10.1074/jbc.R800062200.

Cavanagh, J., Fairbrother, W. J., Palmer, A. G., III, Rance, M., & Skelton, N. J. (2007). *Protein NMR spectroscopy: Principles and practice* (2nd ed.). New York: Academic Press.

Copeland, W. C. (1997). Expression, purification, and characterization of the two human primase subunits and truncated complexes from Escherichia coli. *Protein Expression and Purification*, *9*(1), 1–9. http://dx.doi.org/10.1006/prep.1996.0665.

Emsley, P., Lohkamp, B., Scott, W. G., & Cowtan, K. (2010). Features and development of Coot. *Acta Crystallographica. Section D, Biological Crystallography, 66*(Pt. 4), 486–501. http://dx.doi.org/10.1107/S0907444910007493.

Fuss, J. O., Tsai, C. L., Ishida, J. P., & Tainer, J. A. (2015). Emerging critical roles of Fe-S clusters in DNA replication and repair. *Biochimica et Biophysica Acta, 1853*(6), 1253–1271. http://dx.doi.org/10.1016/j.bbamcr.2015.01.018.

Garner, M. M., & Revzin, A. (1981). A gel electrophoresis method for quantifying the binding of proteins to specific DNA regions: Application to components of the Escherichia coli lactose operon regulatory system. *Nucleic Acids Research, 9*(13), 3047–3060.

Greenfield, N. J. (2006). Using circular dichroism spectra to estimate protein secondary structure. *Nature Protocols, 1*(6), 2876–2890. http://dx.doi.org/10.1038/nprot.2006.202.

Hellman, L. M., & Fried, M. G. (2007). Electrophoretic mobility shift assay (EMSA) for detecting protein-nucleic acid interactions. *Nature Protocols, 2*(8), 1849–1861. http://dx.doi.org/10.1038/nprot.2007.249.

Jain, R., Vanamee, E. S., Dzikovski, B. G., Buku, A., Johnson, R. E., Prakash, L., et al. (2014). An iron-sulfur cluster in the polymerase domain of yeast DNA polymerase epsilon. *Journal of Molecular Biology, 426*(2), 301–308. http://dx.doi.org/10.1016/j.jmb.2013.10.015.

Kabsch, W. (2010). Integration, scaling, space-group assignment and post-refinement. *Acta Crystallographica. Section D, Biological Crystallography, 66*(Pt. 2), 133–144. http://dx.doi.org/10.1107/S0907444909047374.

Klinge, S., Hirst, J., Maman, J. D., Krude, T., & Pellegrini, L. (2007). An iron-sulfur domain of the eukaryotic primase is essential for RNA primer synthesis. *Nature Structural & Molecular Biology, 14*(9), 875–877. http://dx.doi.org/10.1038/nsmb1288.

Kuchta, R. D., Reid, B., & Chang, L. M. (1990). DNA primase. Processivity and the primase to polymerase alpha activity switch. *The Journal of Biological Chemistry, 265*(27), 16158–16165.

Kuchta, R. D., & Stengel, G. (2010). Mechanism and evolution of DNA primases. *Biochimica et Biophysica Acta, 1804*(5), 1180–1189. http://dx.doi.org/10.1016/j.bbapap.2009.06.011.

Liu, L., & Huang, M. (2015). Essential role of the iron-sulfur cluster binding domain of the primase regulatory subunit Pri2 in DNA replication initiation. *Protein & Cell, 6*(3), 194–210. http://dx.doi.org/10.1007/s13238-015-0134-8.

Muzi-Falconi, M., Giannattasio, M., Foiani, M., & Plevani, P. (2003). The DNA polymerase alpha-primase complex: Multiple functions and interactions. *TheScientificWorldJournal, 3*, 21–33. http://dx.doi.org/10.1100/tsw.2003.05.

Netz, D. J., Stith, C. M., Stumpfig, M., Kopf, G., Vogel, D., Genau, H. M., et al. (2011). Eukaryotic DNA polymerases require an iron-sulfur cluster for the formation of active complexes. *Nature Chemical Biology, 8*(1), 125–132. http://dx.doi.org/10.1038/nchembio.721.

O'Brien, E., Holt, M. E., Thompson, M. K., Salay, L. E., Ehlinger, A. C., Chazin, W. J., et al. (2017). The [4Fe4S] cluster of human DNA primase functions as a redox switch using DNA charge transport. *Science, 355*(6327), eaag1789. http://dx.doi.org/10.1126/science.aag1789.

Otwinowski, Z., & Minor, W. (1997). Processing of X-ray diffraction data collected in oscillation mode. *Methods in Enzymology, 276*, 307–326.

Pellegrini, L. (2012). The Pol alpha-primase complex. *Sub-Cellular Biochemistry, 62*, 157–169. http://dx.doi.org/10.1007/978-94-007-4572-8_9.

Pollard, T. D. (2010). A guide to simple and informative binding assays. *Molecular Biology of the Cell, 21*(23), 4061–4067. http://dx.doi.org/10.1091/mbc.E10-08-0683.

Rossi, A. M., & Taylor, C. W. (2011). Analysis of protein-ligand interactions by fluorescence polarization. *Nature Protocols, 6*(3), 365–387. http://dx.doi.org/10.1038/nprot.2011.305.

Rouault, T. A. (2015). Iron-sulfur proteins hiding in plain sight. *Nature Chemical Biology*, *11*(7), 442–445. http://dx.doi.org/10.1038/nchembio.1843.

Sauguet, L., Klinge, S., Perera, R. L., Maman, J. D., & Pellegrini, L. (2010). Shared active site architecture between the large subunit of eukaryotic primase and DNA photolyase. *PLoS One*, *5*(4), e10083. http://dx.doi.org/10.1371/journal.pone.0010083.

Sheaff, R. J., & Kuchta, R. D. (1993). Mechanism of calf thymus DNA primase: Slow initiation, rapid polymerization, and intelligent termination. *Biochemistry*, *32*(12), 3027–3037.

Stephens, P. J., Thomson, A. J., Dunn, J. B., Keiderling, T. A., Rawlings, J., Rao, K. K., et al. (1978). Circular dichroism and magnetic circular dichroism of iron-sulfur proteins. *Biochemistry*, *17*(22), 4770–4778.

Suwa, Y., Gu, J., Baranovskiy, A. G., Babayeva, N. D., Pavlov, Y. I., & Tahirov, T. H. (2015). Crystal structure of the human Pol alpha B subunit in complex with the C-terminal domain of the catalytic subunit. *The Journal of Biological Chemistry*, *290*(23), 14328–14337. http://dx.doi.org/10.1074/jbc.M115.649954.

Takeuchi, K., & Wagner, G. (2006). NMR studies of protein interactions. *Current Opinion in Structural Biology*, *16*(1), 109–117. http://dx.doi.org/10.1016/j.sbi.2006.01.006.

Thompson, M. K., Ehlinger, A. C., & Chazin, W. J. (2017). Chapter three: Analysis of functional dynamics of modular multidomain proteins by SAXS and NMR. In B. F. Eichman (Ed.), *Methods in enzymology: Vol. 592* (pp. 49–76): Amsterdam: Elsevier.

Vaithiyalingam, S., Warren, E. M., Eichman, B. F., & Chazin, W. J. (2010). Insights into eukaryotic DNA priming from the structure and functional interactions of the 4Fe-4S cluster domain of human DNA primase. *Proceedings of the National Academy of Sciences of the United States of America*, *107*(31), 13684–13689. http://dx.doi.org/10.1073/pnas.1002009107.

Weiner, B. E., Huang, H., Dattilo, B. M., Nilges, M. J., Fanning, E., & Chazin, W. J. (2007). An iron-sulfur cluster in the C-terminal domain of the p58 subunit of human DNA primase. *The Journal of Biological Chemistry*, *282*(46), 33444–33451. http://dx.doi.org/10.1074/jbc.M705826200.

Winn, M. D., Ballard, C. C., Cowtan, K. D., Dodson, E. J., Emsley, P., Evans, P. R., et al. (2011). Overview of the CCP4 suite and current developments. *Acta Crystallographica. Section D, Biological Crystallography*, *67*(Pt. 4), 235–242. http://dx.doi.org/10.1107/S0907444910045749.

Ye, H., & Rouault, T. A. (2010). Human iron-sulfur cluster assembly, cellular iron homeostasis, and disease. *Biochemistry*, *49*(24), 4945–4956. http://dx.doi.org/10.1021/bi1004798.

Yeeles, J. T., Janska, A., Early, A., & Diffley, J. F. (2017). How the eukaryotic replisome achieves rapid and efficient DNA replication. *Molecular Cell*, *65*(1), 105–116. http://dx.doi.org/10.1016/j.molcel.2016.11.017.

Zerbe, L. K., & Kuchta, R. D. (2002). The p58 subunit of human DNA primase is important for primer initiation, elongation, and counting. *Biochemistry*, *41*(15), 4891–4900.

Zhang, Y., Baranovskiy, A. G., Tahirov, T. H., & Pavlov, Y. I. (2014). The C-terminal domain of the DNA polymerase catalytic subunit regulates the primase and polymerase activities of the human DNA polymerase alpha-primase complex. *The Journal of Biological Chemistry*, *289*(32), 22021–22034. http://dx.doi.org/10.1074/jbc.M114.570333.

AUTHOR INDEX

Note: Page numbers followed by "*f*" indicate figures, "*t*" indicate tables, and "*s*" indicate schemes.

D

E

G

I

L

M

N

O

S

T

U

V

SUBJECT INDEX

Note: Page numbers followed by "*f*" indicate figures, and "*t*" indicate tables.

I

FSC
www.fsc.org
MIX
Papier | Fördert gute Waldnutzung
FSC® C083411

Kolibri 360 GmbH
Ferdinand-Jühlke-Straße 7
99095 Erfurt, Deutschland
produktsicherheit@kolibri360.de